The Respiratory Burst and Its Physiological Significance

Edited by
Anthony J. Sbarra
St. Margaret's Hospital for Women
and Tufts University School of Medicine
Boston, Massachusetts

and
Robert R. Strauss
Clinical Pathology Facility, Inc.
Pittsburgh, Pennsylvania

Plenum Press • New York and London

Library of Congress Cataloging in Publication Data

The Respiratory burst and its physiological significance / edited by Anthony J. Sbarra and Robert R. Strauss.
p. cm.
Includes bibliographies and index.
ISBN 0-306-42883-0
1. Phagocytes—Metabolism. 2. Cell respiration. 3. Active oxygen—Physiological effect. 4. Phagocytosis. I. Sbarra, Anthony J. II. Strauss, Robert R.
[DNLM: 1. Oxygen Consumption. 2. Phagocytes—physiology. WF 110 R434]
QR185.8.P45R47 1988
616.07'9—dc19
DNLM/DLC 88-22476
for Library of Congress CIP

Printed in the United States of America

Contributors

PAOLO BELLAVITE • *Institute of General Pathology, University of Verona, 37134 Verona, Italy*

GIORGIO BERTON • *Institute of General Pathology, University of Verona, 37134 Verona, Italy*

BEN G. J. M. BOLSCHER • *Central Laboratory of the Netherlands Red Cross Blood Transfusion Service and Laboratory of Experimental and Clinical Immunology, University of Amsterdam, 1006 AK Amsterdam, The Netherlands*

NIELS BORREGAARD • *Department of Medicine and Hematology C, Gentofte Hospital, University of Copenhagen, DK-2900 Hellerup, Denmark*

BRADLEY E. BRITIGAN • *Department of Medicine, University of North Carolina, Chapel Hill, North Carolina 27514*

CURTIS L. CETRULO • *Department of Maternal Fetal Medicine, St. Margaret's Hospital for Women, Boston, Massachusetts 02125 and Department of Obstetrics and Gynecology, Tufts University School of Medicine, Boston, Massachusetts 02125*

ANJAN CHAUDHURY • *Department of Maternal Fetal Medicine, St. Margaret's Hospital for Women, Boston, Massachusetts 02125 and Department of Obstetrics and Gynecology, Tufts University School of Medicine, Boston, Massachusetts 02125*

MYRON S. COHEN • *Departments of Medicine and Microbiology and Immunology, Division of Infectious Diseases, University of North Carolina, Chapel Hill, North Carolina 27514*

ALDO DOBRINA • *Institute of General Pathology, University of Trieste, 34127 Trieste, Italy*

STEFANO DUSI • *Institute of General Pathology, University of Verona, 37134 Verona, Italy*

GABRIELLA FÓRIS • *First Department of Medicine, University Medical School, 4012 Debrecen, Hungary*

TAMÁS FÜLÖP, Jr. • *First Department of Medicine, University Medical School, 4012 Debrecen, Hungary*

LEO I. GORDON • *Department of Medicine, Section of Hematology/Oncology and the Cancer Center, Northwestern University Medical School, Chicago, Illinois 60611*

KENT J. JOHNSON • *Department of Pathology, University of Michigan Medical School, Ann Arbor, Michigan 48109-0650*

JOSEPH L. KENNEDY, JR. • *Department of Pediatics, St. Margaret's Hospital for Women, Boston, Massachusetts 02125*

MARK S. KEMPNER • *Department of Medicine, Tufts–New England Medical Center, Boston, Massachusetts 02111*

ÉVA M. KOVÁCS • *First Department of Medicine, University Medical School, 4012 Debrecen, Hungary*

ANDRÁS LEÖVEY • *First Department of Medicine, University Medical School, 4012 Debrecen, Hungary*

CHARLES J. LOCKWOOD • *Department of Maternal Fetal Medicine, St. Margaret's Hospital for Women, Boston, Massachusetts 02125 and Department of Obstetrics and Gynecology, Tufts University School of Medicine, Boston, Massachusetts 02125*

FARID LOUIS • *Department of Pathology, St. Margaret's Hospital for Women, Boston, Massachusetts 02125*

RENÉ LUTTER • *Central Laboratory of the Netherlands Red Cross Blood Transfusion Service and Laboratory of Experimental and Clinical Immunology, University of Amsterdam, 1006 AK Amsterdam, The Netherlands*

RENZO MENEGAZZI • *Institute of General Pathology, University of Trieste, 34127 Trieste, Italy*

JÓZSEF T. NAGY • *First Department of Medicine, University Medical School, 4012 Debrecen, Hungary*

VICTOR A. NAJJAR • *Department of Molecular Biology and Microbiology, Tufts University School of Medicine, Boston, Massachusetts 02111*

PETER J. O'BRIEN • *Faculty of Pharmacy, University of Toronto, Toronto, Ontario M5S 1A1, Canada*

GEORGE PARAGH • *First Department of Medicine, University Medical School, 4012 Debrecen, Hungary*

PIERLUIGI PATRIARCA • *Institute of General Pathology, University of Trieste, 34127 Trieste, Italy*

MARK T. PETERS • *Department of Maternal Fetal Medicine, St. Margaret's Hospital for Women, Boston, Massachusetts 02125 and Department of Obstetrics and Gynecology, Tufts University School of Medicine, Boston, Massachusetts 02125*

DIRK ROOS • *Central Laboratory of the Netherlands Red Cross Blood Transfusion Service and Laboratory of Experimental and Clinical Immunology, University of Amsterdam, 1006 AK Amsterdam, The Netherlands*

ARTHUR L. SAGONE, Jr. • *Department of Medicine and Pharmacology, Division of Hematology and Oncology, Ohio State University, College of Medicine, Columbus, Ohio 43210*

ANTHONY J. SBARRA • *Department of Medical Research and Laboratories, St. Margaret's Hospital for Women, Boston, Massachusetts 02125 and Department of Obstetrics and Gynecology, Tufts University School of Medicine, Boston, Massachusetts 02125*

RUDOLF E. SCHOPF • *Department of Dermatology, Johannes Gutenberg University, D-6500 Mainz, West Germany*

BENNETT M. SHAPIRO • *Department of Biochemistry, University of Washington, Seattle, Washington 98195*

CHRIS J. SHAKR • *Department of Maternal Fetal Medicine, St. Margaret's Hospital for Women, Boston, Massachusetts 02125 and Department of Obstetrics and Gynecology, Tufts University School of Medicine, Boston, Massachusetts 02125*

EMIL SKAMENE • *Division of Clinical Immunology and Allergy, The Montreal General Hospital, Montreal, Québec H3G 1A4, Canada*

STANLEY D. SOMERFIELD • *Division of Clinical Immunology and Allergy, The Montreal General Hospital, Montreal, Québec H3G 1A4, Canada*

TERESA STELMASZYŃSKA • *Institute of Medical Biochemistry, Nicolaus Copernicus Academy of Medicine, 31-034 Kraków, Poland*

BARBARA STYRT • *Department of Medicine, Michigan State University, East Lansing, Michigan 48824–1317*

GAIL THOMAS • *Department of Medical Research and Laboratories, St. Margaret's Hospital for Women, Boston, Massachusetts 02125 and Department of Obstetrics and Gynecology, Tufts University School of Medicine, Boston, Massachusetts 02125*

ERIC E. TURNER • *Department of Biochemistry, University of Washington, Seattle, Washington 98195*

JACK P. UETRECHT • *Faculties of Pharmacy and Medicine, University of Toronto and Sunnybrook Hospital, Toronto, Ontario M5S 1A1, Canada*

ZSUZSA VARGA • *First Department of Medicine, University Medical School, 4012 Debrecen, Hungary*

HORN-WEN WANG • *Department of Maternal Fetal Medicine, St. Margaret's Hospital for Women, Boston, Massachusetts 02125 and Department of Obstetrics and Gynecology, Tufts University School of Medicine, Boston, Massachusetts 02125*

PETER A. WARD • *Department of Pathology, University of Michigan Medical School, Ann Arbor, Michigan 48109-0650*

JEFFREY S. WARREN • *Department of Pathology, University of Michigan Medical School, Ann Arbor, Michigan 48109-0650*

SIGMUND A. WEITZMAN • *Department of Medicine, Section of Hematology/Oncology and the Cancer Center, Northwestern University Medical School, Chicago, Illinois 60611*

GIULIANO ZABUCCHI • *Institute of General Pathology, University of Trieste, 34127 Trieste, Italy*

JAN MACIEJ ZGLICZYŃSKI • *Institute of Medical Biochemistry, Nicolaus Copernicus Academy of Medicine, 31-034 Kraków, Poland*

Preface

When phagocytes are exposed to a number of different stimuli, they undergo dramatic changes in the way they process oxygen. Oxygen uptake increases markedly, frequently more than 50-fold; the phagocytes begin to produce large quantities of superoxide and hydrogen peroxide; and they immediately begin to metabolize large amounts of glucose by way of the hexose monophosphate shunt. This series of changes has become known as the respiratory burst. It was first believed that the major function of this respiratory burst was to generate powerful antibacterial agents by the partial reduction of oxygen. It is becoming apparent that the respiratory burst has much wider application, and its physiological function in many different biological areas is clear. In this volume, we have attempted to bring together the work of experts who have published extensively on the involvement of the respiratory burst in different physiological functions.

In the first three chapters, Dr. Borregaard and Dr. Berton and co-workers and Dr. Roos and co-workers bring together what is known about the respiratory burst. They present up-to-date versions of the biochemical and metabolic activities associated with the burst. In Chapter 4, Dr. Styrt and Dr. Klempner discuss the respiratory burst as it affects cellular ion homeostasis. Dr. Cohen and Dr. Britigan (Chapter 5) present some interesting data on the competition between the respiratory burst and bacteria for oxygen. Dr. Dobrina and Dr. Patriarca (Chapter 6) describe the involvement of the respiratory burst in the damaging of endothelial cells. Dr. Zabucchi, Dr. Menegazzi, and Dr. Patriarca (Chapter 7) discuss the respiratory burst and the eosinophils, evaluating comparisons between the eosinophils and the neutrophil.

Dr. Sagone (Chapter 8) discusses the possible effects of the reactive oxygen species resulting from the respiratory burst on the functional capacity of lymphocytes. Dr. Somerfield and Dr. Skamene (Chapter 9) present data concerning the modulation of the respiratory burst by naturally occurring substances. In Chapter 10, Dr. O'Brien discusses the oxidants formed by the respiratory burst and their physiological involvement in the oxidative metabolism and activation of drugs, carcinogens, and xenobiotics. Drug-induced agranulocytosis and other effects mediated by peroxidases during the respiratory burst are discussed by Dr. Uetrecht in Chapter 11. Dr. Sagone (Chapter 12) discusses the capacity of granulocytes to metabolize xenobiotics and further discusses the possible clinical implication of this. Dr. Gordon and Dr. Weitzman (Chapter 13) review evidence implicating oxygen-derived free radicals in carcinogenesis.

Dr. Warren, Dr. Ward, and Dr. Johnson discuss in Chapter 14 how phagocytes generate oxygen-derived radicals and how they are important mediators of tissue damage. Dr. Zgliczyński and Dr. Stelmaszyńska (Chapter 15) discuss the influence of the respiratory burst on infected tissue.

In Chapter 16, Dr. Schopf presents data involving the respiratory burst of phagocytes in the inflammatory reaction in psoriasis. The special role of reactive oxygen species in diabetes mellitus is discussed by Dr. Nagy, Dr. Fülöp, Dr. Paragh, and Dr. Fóris in Chapter 17. Dr. Turner and Dr. Shapiro (Chapter 18) discuss the involvement of the respiratory burst in fertilization. The respiratory burst and the development of generalized atherosclerosis, a subject that is beginning to receive serious consideration, is presented by Dr. Paragh, Dr. Kovács, Dr. Nagy, Dr. Fóris, and Dr. Fülöp in Chapter 19. Aging and reactive oxygen species associated with the respiratory burst are discussed by Dr. Fülöp, Dr. Fóris, Dr. Nagy, and Dr. Leövey in Chapter 20. Dr. Cetrulo and colleagues (Chapter 21) present evidence that implicates the respiratory burst in the onset of human labor, preterm labor, and premature rupture of the membranes. In Chapter 22, Dr. Najjar discusses the biochemical and biological aspects of tuftsin, a powerful chemoattractant that can enhance the respiratory burst of neutrophils in the presence of other soluble or particulate stimuli. We are most grateful to all of the contributors for their contributions.

Anthony J. Sbarra
Robert R. Strauss

Boston and Pittsburgh

Contents

1. The Respiratory Burst: An Overview

Niels Borregaard

2. The Respiratory Burst of Phagocytes

Giorgio Berton, Stefano Dusi, and Paolo Bellavite

5. An Expanded View of the Phagocytic Respiratory Burst: Bacterial Competition for Oxygen and Its Stimulation by Host Factor(s)

Myron S. Cohen and Bradley E. Britigan

6. The Respiratory Burst and Endothelial Cells

Aldo Dobrina and Pierluigi Patriarca

7. The Respiratory Burst of Eosinophils

Giuliano Zabucchi, Renzo Menegazzi, and Pierluigi Patriarca

9. Modulation of the Respiratory Burst by Naturally Occurring Substances

Stanley D. Somerfield and Emil Skamene

10. Oxidants Formed by the Respiratory Burst: Their Physiological Role and Their Involvement in the Oxidative Metabolism and Activation of Drugs, Carcinogens, and Xenobiotics

Peter J. O'Brien

11. Drug-Induced Agranulocytosis and Other Effects Mediated by Peroxidases during the Respiratory Burst

Jack P. Uetrecht

12. The Respiratory Burst and the Metabolism of Drugs

Arthur L. Sagone, Jr.

13. The Respiratory Burst and Carcinogenesis

Leo I. Gordon and Sigmund A. Weitzman

14. The Respiratory Burst and Mechanisms of Oxygen Radical-Mediated Tissue Injury

Jeffrey S. Warren, Peter A. Ward, and Kent J. Johnson

15. The Respiratory Burst of Neutrophilic Granulocytes and Its Influence on Infected Tissues: Indirect Consequences

Jan Maciej Zgliczyński and Teresa Stelmaszyńska

18. The Respiratory Burst of Fertilization

Eric E. Turner and Bennett M. Shapiro

19. The Respiratory Burst and Atherosclerosis

George Paragh, Éva M. Kovács, József T. Nagy, Gabriella Fóris, and Tamás Fülöp, Jr.

20. The Respiratory Burst and Aging

Tamás Fülöp, Jr., Gabriella Fóris, József T. Nagy, Zsuzsa Varga, and András Leövey

21. The Respiratory Burst and the Onset of Human Labor, Preterm Labor, and Premature Rupture of the Membranes

Curtis L. Cetrulo, Anthony J. Sbarra, Anjan Chaudhury, Mark T. Peters, Charles J. Lockwood, Gail Thomas, Joseph L. Kennedy, Jr., Farid Louis, Chris J. Shakr, and Horn-Wen Wang

22. Tuftsin: Biochemical and Biological Aspects

Victor A. Najjar

The Respiratory Burst and Its Physiological Significance

1

The Respiratory Burst

An Overview

NIELS BORREGAARD

1. INTRODUCTION

The increased respiration of phagocytes during phagocytosis was discovered in 1933 by Baldridge and Gerard,[1] who measured a small but significant increase in the oxygen consumption by canine neutrophils during phagocytosis of bacteria. This enhanced metabolic response of the neutrophils was initially attributed to an increased generation of mitochondrial energy to meet the needs of the phagocytic process, and it was not realized until 1959 that this increase in respiration was due to nonmitochondrial activity. The studies by Sbarra and Karnovsky[2] clearly established that this respiratory burst of phagocytosis is insensitive to mitochondrial inhibitors and that the energy needed for phagocytosis by neutrophils is provided by glycolysis. It was later shown that consumption of ATP from a pool that is apparently not regained[3,4] also adds significantly to the energy expenditure during ingestion of particles by neutrophils. In macrophages, the energy for the phagocytic process is partially provided by creatine phosphate, which may ultimately be regained from oxidative metabolism.[5] Although different phagocytic cells may provide the extra energy needed for phagocytosis by activating different biochemical pathways, all phagocytes—neutrophils,[6,7] monocytes,[8,9] eosinophils,[10,11] and lung[12] and peritoneal macrophages[6,7,13]—mount a nonmitochondrial respiratory burst during phagocytosis due to activation of a unique biochemical entity, the membrane-bound NADPH oxidase. It has further been shown that the oxidase may be activated secondary to stimulation of the phagocytes with a wide variety of stimuli that act through binding to receptors other than the Fc and C3b receptors that mediate phagocytoses. These stimuli include *N*-formyl-methionyl-leucyl-phenylalanine (fMPL),[14] phorbol myristate acetate (PMA),[15] A 23187,[16] concanavalin A (Con A),[17] and others.[18–21] The significance of the respiratory burst of phagocy-

NIELS BORREGAARD • Department of Medicine and Hematology C, Gentofte Hospital, University of Copenhagen, DK-2900 Hellerup, Denmark.

TABLE I. Respiratory Burst Activities

Nonmitochondrial oxygen consumption
Carbon dioxide generation by the hexose monophosphate shunt
Generation of O_2^-, H_2O_2, $OH\cdot$, $^1O_2$, HOCl, chloramines
Chemiluminescence
Proton secretion
Depolarization

tosis was first fully understood when it became clear that a complete absence of this nonmitochondrial respiratory burst is the functional basis for the severe disease, chronic granulomatous disease (CGD),[22] in which the phagocytes are grossly incapable of killing and digesting phagocytosed micro-organisms.[23–25] Knowledge of CGD has since been invaluable in working out the biochemical basis for the respiratory burst and its effects on various models of microbial killing, cytotoxicity, and tissue damage. It was soon established that the primary product of the respiratory burst is the one-electron reduced species of oxygen, the superoxide anion, O_2^-.[26] At the pH and ionic conditions prevailing in the cell, superoxide will dismutate to H_2O_2 and O_2,[27,29] and it has been demonstrated that H_2O_2 is quantitatively generated from oxygen during the respiratory burst.[30,31] It was then established that the H_2O_2 generated during the respiratory burst and myeloperoxidase liberated from the azurophil granules constitute a very powerful microbicidal system that is primarily responsible for the oxygen-dependent bactericidal activity of the phagocytes.[32]

The respiratory burst was first defined as the increased respiration, i.e., O_2 consumption and CO_2 generation induced by phagocytosis. Later, other activities, kinetically and functionally related to the burst in O_2 consumption and absent from CGD phagocytes have been adopted under the term respiratory burst activities. These include the generation of the reduced oxygen species, O_2^-, H_2O_2, singlet oxygen,[33] and hydroxyl radicals.[34,35] More recently, the MPO-generated products HOCl[36] and chloramines[37] have been included. Finally, it has been realized that translocation of protons from the interior of the cell to the intracellular medium and the associated depolarization of the cells are intimately related to the respiratory burst[38–40] in a way that justifies the inclusion of these phenomena in the respiratory burst activities. (Table I).

2. BIOCHEMICAL ACTIVITIES ASSOCIATED WITH THE RESPIRATORY BURST

2.1. Origin of Electrons for the Respiratory Burst Oxidase

Despite initial disagreement about the nature of the ultimate electron donor for the respiratory burst oxidase,[41–44] it is now firmly established that the electrons are donated by NADPH.[45–57] The respiratory burst oxidase is thus an NADPH oxidase, as first claimed by Rossi and Zatti.[58]

This fits well with the observation that the hexose monophosphate shunt

(HMPS) is activated during the respiratory burst,[23] since the activity of the HMPS in phagocytes is normally governed by the availability of $NADP^+$,[59,60] originating in the NADPH oxidase or elsewhere. A respiratory quotient of 1 has been observed during the respiratory burst.[6,39] Although this usually indicates glucose oxidation in the Embden–Meyerhoff pathway and the citric acid (Krebs) cycle,[61] this does not seem to be the case in neutrophils.[2,3] Studies on the oxidation of glucose in neutrophils during phagocytosis using radiolabeled glucose indicate that most oxidation of glucose takes place in the HMPS,[62–64] but direct measurements of glucose utilization have been unable to account entirely for the stoichiometry of the CO_2 generation.[3] Nevertheless, a tight functional relationship exists between HMPS and the respiratory burst oxidase, since the absence of the oxidase, as in the classic CGD, is associated with absent activation of the HMPS during phagocytosis. This indicates that the normal stimulus for HMPS activation is the intracellular product of NADPH oxidase, $NADP^+$. Severe deficiency of glucose 6-phosphate dehydrogenase (G6PD), an enzyme in the HMPS, is associated with a functional defect in the respiratory burst of phagocytosis similar to the situation in CGD. This indicates that the HMPS is the main donor of electrons for the oxidase.[65,66] Thus, in normal phagocytes, the burst in CO_2 generation seems to result from the oxidation of glucose in the HMPS, the activity of which is regulated entirely by the respiratory burst oxidase.

2.2. Oxygen Consumption

The magnitude of the oxygen consumption elicited by stimulation of phagocytes is dependent on the stimulus used. In our hands, serum opsonized zymosan is the most potent activator of the burst and can elicit consumptions of oxygen

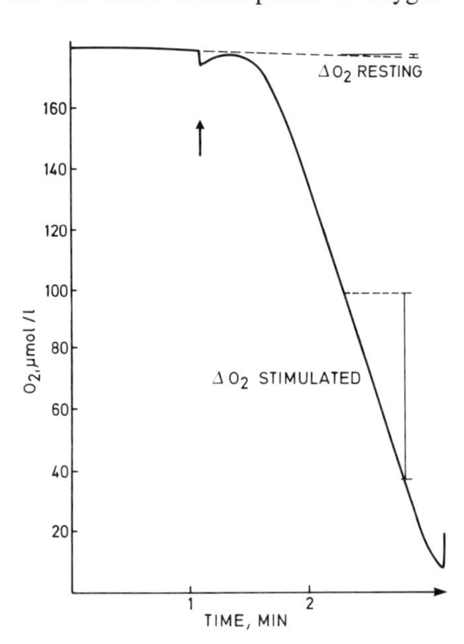

FIGURE 1. Respiratory burst of phagocytes. Polarographic tracing of oxygen consumption by isolated human neutrophils, 2×10^7 cells/ml. Arrow indicates addition of serum-opsonized zymosan.

ranging up to 5–6 fmoles/cell per min (Fig. 1). Measurements of oxygen consumption are performed in a closed compartment fitted with a Clark-type oxygen electrode. Some stimuli elicit a long-lasting burst that is nearly constant for 10–20 min,[1,2,7,15,67] whereas others, such as arachidonic acid[68] and fMLP,[16,99] result in bursts of considerably shorter duration. Although most data on the kinetics of the respiratory burst refer to a dynamic state, measured as the maximal rate of oxygen consumption (Fig. 1), this may be misleading in the evaluation of the magnitude of the burst, and in particular in the assessment of the quantity of reactive oxygen species generated, since the rate may be rapidly decreasing (see Section 4).

2.3. O_2^- Generation

It was first established by Babior et al.[26] that superoxide O_2^-, the one-electron-reduced species of oxygen species of oxygen, is generated by neutrophils during the respiratory burst. Ironically, Curnutte and Tauber[70] later showed that the stimulus initially used to demonstrate this is a particularly weak stimulus for O_2^- generation, and an unexplained discrepancy exists between the magnitude of the O_2 consumption and the O_2^- generation elicited by Latex particles.[70,71] This discrepancy may be partly explained by the methodology used. Superoxide anions are usually quantitated by their ability to reduce ferricytochrome *c* to ferrocytochrome *c*, which absorbs at 550 nm.[72] This reduction may be effectuated by many other redox species, and only the reduction that can be inhibited by the enzyme superoxide dismutase can be attributed to superoxide anions.

It should be kept in mind when assessing the kinetics of the respiratory burst that O_2 consumption is measured as the disappearence of O_2 from the entire medium, whereas only the O_2^- that reacts with cytochrome *c* will be recorded. For economic reasons, most assays of cytochrome *c* reduction are carried out at concentrations of cytochrome *c* at which the rate of cytochrome reduction by superoxide is a second-order reaction.[73] Furthermore, cytochrome *c* does not gain access to the interior of the cell.[74] It can therefore be questioned whether the cytochrome *c* will react with superoxide generated within the phagocytic vacuole.[74,75] Another complicating factor is that some of the products of the respiratory burst, in particular the products of the myeloperoxidase system, may reoxidize the reduced ferrocytochrome *c*, leading to an underestimation of the amount of cytochrome *c* reduced by superoxide.[76]

Most reports on the stoichiometry between O_2 consumption and O_2^- generation therefore suffer from the comparison of one activity that is measured quantitatively with another activity that may be underestimated by a factor that is often unknown. To circumvent this problem, Dri et al.[30] reasoned that the correct stoichiometry between O_2 consumption and superoxide generation could be calculated by measuring the ratio between the magnitude of the cytochrome *c* reduction and the decrease in O_2 consumption caused by the addition of cytochrome *c*. Using this approach, a stoichiometry between superoxide generation and oxygen consumption of 4 : 1 was established.[30] This is the ratio to be expected if all oxygen consumed is initially

TABLE II. Stoichiometry of the Respiratory Burst[a]

HMPS	$1G6P + 12NADP^+ + 6H_2O \rightarrow 6HCO_3^- + 12NADPH + 18H^+ + P_i$
NADPH oxidase	$12NADPH + 24O_2 \rightarrow 24O_2^- + 12NADP^+ + 12H^+$
O_2^- dismutation	$24H^+ + 24O_2^- \rightarrow 12H_2O_2 + 12O_2$
Catalase	$12H_2O_2 \rightarrow 6O_2 + 12H_2O$
Sum	$1G6P + 6O_2 \rightarrow 6HCO_3^- + 6H^+ + 6H_2O + P_i$

[a]G6P, glucose 6-phosphate.

reduced to superoxide and is eventually recovered as H_2O[77] (Table II). Others have found a stoichiometry of 2 : 1,[70] which is what should be expected if all O_2 is initially reduced to superoxide and recovered as H_2O_2 or as HOCl.[39]

2.4. Generation of H_2O_2

It follows from the reactivity of O_2^- that if this species is generated, it will rapidly dismutate to H_2O_2 and O_2 (Table II). The latter will reenter the medium. Although the rate constant for this dismutation reaction is pH dependent (8.5×10^7 $M^{-1}sec^{-1}$ at pH 4.8 and 4.5×10^5 M^{-1} sec^{-} at neutral pH[27]), the rate constant is sufficiently high at neutral pH to prevent O_2^- from accumulating during the respiratory burst. On the other hand, the dismutation is sufficiently slow to permit the detection of O_2^- by cytochrome *c* and to permit the interaction of O_2^- with other oxidase products (as discussed in Section 2.6), unless the steady-state O_2^- concentration is lowered enzymatically by the addition of superoxide dismutase (SOD), in which case the rate constant for the dismutation reaction is increased to 1.9×10^9 $M^{-1}sec^{-1}$.[78] Some indications have been put forward that the spontaneous dismutation of O_2^- may result in the generation of an excited form of oxygen, i.e., singlet oxygen $^1O_2$[79–81]:

$$2O_2^- + 2H^+ \rightarrow H_2O_2 + {}^1O_2$$

but this possibility is not supported by quantitative data.[82,83] Thus, whenever O_2^- is formed by phagocytes, H_2O_2 will be generated at one half the rate (Table II).

Generation of H_2O_2 by phagocytosing neutrophils was first demonstrated by Iyer et al.[29] Until the recent introduction of a H_2O_2-specific electrode,[31] these measurements were indirect, using peroxidation of a fluorescent probe to quantitate H_2O_2. Both these and the electrode method require that the metabolism of H_2O_2 by catalase or myeloperoxidase be blocked by azide, if the generation of H_2O_2 is to be quantitated.[84,85] However, the intracellular H_2O_2-scavenging system, the glutathione peroxidase system, is not blocked by azide, and the generation of H_2O_2 may thus be underestimated. Most reports on the stoichiometry between O_2 consumption and H_2O_2 generation give values between 2 : 1[70,84] and 1 : 1.[30,83] Theoretically, if the metabolism of H_2O_2 is completely blocked, the stoichiometry between O_2 consumption and H_2O_2 generation should be 1 : 1 (Table II).

Histochemical methods have been applied to verify the subcellular localization of H_2O_2 generation. These techniques have indicated that this process takes place at the plasma membrane and at the phagosomal membrane.[7,86]

2.5. Myeloperoxidase-Mediated Reactions

Myeloperoxidase (MPO) is a heme enzyme localized in the primary (azurophil) granules of neutrophils[87–89] and in a granule subset of monocytes,[90] whereas the enzyme disappears as monocytes differentiate into macrophages.[91,92] These may not be entirely without MPO, as this cationic protein has been shown to be taken up by macrophages from adjacent neutrophils by endocytosis or phagocytosis.[93] Myeloperoxidase released from the primary granules to the phagolysosomes is believed to generate the reactive oxygen species responsible for the oxygen-dependent microbicidal activity of the phagocytes,[32,94] as reviewed by Klebanoff.[28] The potency of this system is dependent on MPO as the catalyst and on H_2O_2 and a halide as reagents. Although thyroxine (T_4) and tri-iodothyronine (T_3) may be taken up by neutrophils and become deiodinated,[95] it is not known whether I^- is liberated to the phagosome during phagocytosis. The halide that normally prevails in the phagosome is therefore Cl^-, and the reactive MPO system will therefore be

$$H_2O_2 + Cl^- + MPO$$

A series of reactive species, such as Cl_2,[96] $OH\cdot$,[97] and $^1O_2$[98], has been claimed to be generated directly or indirectly by this system, but it has now been established that the major primary product is hypochlorous acid HOCl.[31,36,99] The generation of HOCl can be demonstrated by trapping the HOCl with taurine, which is converted to its chloramine by HOCl. The chloramine then further reacts with 5,5′-dithiobis(2-nitrobenzoic acid) (DTNB), which is oxidized to 5-thio-2-nitrobenzoic acid (TNB), with a resultant shift in absorption maximum from 412 nm to 350 nm. Using this method, it has been demonstrated that huge amounts of HOCl are generated during the respiratory burst in neutrophils[36] and monocytes[100] by the MPO system:

$$\text{MPO:} \quad H_2O_2 + Cl^- + H^+ \rightarrow HOCl + H_2O$$

It has been argued that $^1O_2$ may be another product of the MPO system. However, singlet oxygen has been shown to be a product of the MPO system only at extreme pH, and not to be generated in significant quantities by this pathway under conditions expected to occur in intact cells.[101]

2.6. Hydroxyl Radicals

As outlined above and as further substantiated recently,[102] O_2^- is the primary product of the respiratory burst oxidase, and H_2O_2 is a secondary product generated from the dismutation of O_2^-, and probably not originating directly from a two-electron step reduction of O_2.[104] The product of the three-electron step reduction of O_2, the extremely reactive hydroxyl radical OH·, has not been shown to generated by any electron transport chain directly. OH· may be formed in an iron-catalyzed reaction, the classic Haber–Weiss reaction[104] in which O_2^- and H_2O_2 are reactants:

$$O_2^- + Fe^{3+} \rightarrow Fe^{2+} + O_2$$

$$Fe^{2+} + H_2O_2 \rightarrow Fe^{3+} + OH\cdot + OH^-$$

Another mechanism, not requiring a metal catalyst, has been suggested to be operative in activated human neutrophils, by Okolow–Zubkowsky and Hill,[97] in which HOCl formed by the MPO system reacts with O_2^-:

$$HOCl + O_2^- \rightarrow OH\cdot + O_2 + Cl^-$$

Methods employed to detect and quantitate the generation of hydroxyl radicals by phagocytes include ethylene formation from methional[34] and methane formation from dimethyl sulfoxide.[105] The hydroxyl radical signals generated by these systems are inhibited by both catalase and SOD, indicating that the OH· originates in a Haber–Weiss-type reaction. It has been claimed, however, that these methods used to quantitate OH· generation are nonspecific and may detect products of the MPO system in addition to OH·.[28,106]

In stead, it has been suggested that demonstration of a characteristic electron spin resonance (ESR) signal from the spin trap 5,5-dimethylpyrroline-*N*-oxide (DMPO) is a more specific method for demonstrating the formation of OH·.[97,106,107] Using this method, the hydroxyl radical signal is inhibited by superoxide dismutase and hydroxyl radical scavengers but not by catalase or inhibitors of the MPO system. This argues against any of the two proposed mechanisms being operative in the generation of OH· in neutrophils. It was recently suggested that the DMPO–OH signal that reflects OH· interaction with DMPO might arise from the DMPO–OOH signal, which reflects interaction of O_2^- with DMPO, and it was questioned whether neutrophils generate hydroxyl radicals at all.[108]

Since $Fe^{2+/3+}$ is not freely available in the cells, lactoferrin liberated from specific granules of neutrophils has been suggested to enhance the generation of hydroxyl radicals by neutrophils.[109] Iron-saturated lactoferrin has been demonstrated to enhance the generation of OH· by O_2^--producing systems.[109] However, native lactoferrin is only approximately 10% iron saturated,[110] and the argument may be turned around that the 90% unsaturated lactoferrin may inhibit the formation of hydroxyl radicals by chelating any free Fe that might otherwise have participated in the Haber–Weiss reaction.

The evidence for the involvement of hydroxyl radicals in the mediation of respiratory burst-dependent processes relies largely on the use of scavengers such as mannitol, which, albeit interfering little with O_2^- and H_2O_2 themselves, cannot be considered discriminants for OH·.[28,111,112]

2.7. Chemiluminescence

The light emission associated with the respiratory burst of neutrophils was discovered by Allen et al.[33] Initially, this chemiluminescence was demonstrated without amplification. This activity shows the same characteristics as the oxygen consumption elicited during phagocytosis, i.e., absence from CGD neutrophils[113] and the initiation after a short lag period. Chemiluminescence has also been demonstrated to be generated by eosinophils,[114] macrophages,[115] and monocytes.[116] The unenhanced chemiluminescence emits light as a broad wavelength peak around 570 nm.[117] This indicates that decay of $^1O_2$ is not directly responsible for the chemiluminescence (the major peak from the decay of $^1O_2$ is at 1268 nm[118]), but this does not exclude that the energy liberated in the decay of $^1O_2$ is transferred to other compounds that will then emit light when returning to their ground states.[28,119] $^1O_2$ has been proposed to evolve by two main pathways, one independent of, and the other dependent on, MPO:

$$\text{I:} \quad 2H^+ + 2O_2^- \rightarrow {}^1O_2 + H_2O_2$$

$$\text{II:} \quad HOCl + H_2O_2 \rightarrow {}^1O_2 + H_2O + Cl^- + H^+$$

Although chemiluminescence response generated by stimulated phagocytes is a sensitive parameter of the respiratory burst, in particular when enhanced by luminol or lucinogen,[120] quantitative data do not support that $^1O_2$ itself is generated in appreciable amounts by phagocytes.[83,101]

2.8. Formation of Chloramines

It was shown by Segal et al.[77] that the oxygen consumed in the respiratory burst is ultimately recovered as H_2O. Although it is evident that reactive intermediates and derivatives are formed in significant quantities, these are all relatively short-lived as a consequence of their reactivity. However, Weiss et al.[37] demonstrated the generation of stable yet reactive oxidizing species, identified as *N*-chloramines, from HOCl and peptides released from the neutrophils during the respiratory burst. The quantity of chloramines generated probably depends on the incubation medium of the phagocytes. The stoichiometry of the generation of long-lived chloramines by neutrophils in serum has not yet been worked out.

2.9. Proton Transport during the Respiratory Burst

It has long been a dogma that the phagocytic vacuole is acidified during the process of phagocytosis (reviewed by Roos et al.[121]). In a heretic report by Segal et al.,[122] it was claimed however, that the phagocytic vacuole is acidified only if the respiratory burst is defective and that the primary effect of the respiratory burst is to alkalinize the phagocytic vacuole. Although this respiratory burst associated al-

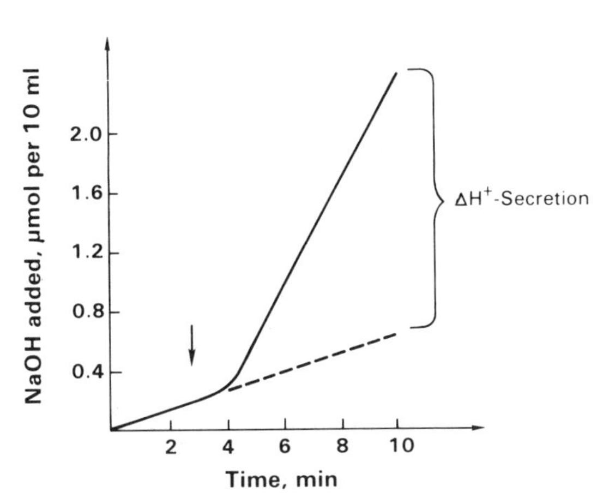

FIGURE 2. Proton secretion during phagocytosis. Human neutrophils, 5 × 10^6 cells/ml, were incubated in a isotonic 1 mM phosphate buffer. Arrow indicates addition of serum-opsonized zymosan. pH of the buffer was kept constant at 7.41 by titration with NaOH by an automated titration unit. (From Borregaard *et al.*[39])

kalinization of the phagocytic vacuole has been confirmed by others,[123] it is contradicted by the documentation that the overall effect of the respiratory burst is to elicit an extrusion of protons out of the cell and presumably into the phagocytic vacuole as well,[38,39,121] since the phagocytic vacuole is formed partly by an invagination of the plasma membrane. The biochemical pathways responsible for the H^+ generation have now been identified. In addition to a basal level of proton generation accounted for by the lactate production,[38] the phagocytes generate protons in a manner that corresponds precisely to the burst in oxygen consumption or superoxide generation (Fig. 2), i.e., the burst in proton generation is absent in CGD phagocytes and displays kinetics similar to that of the burst in oxygen consumption with a stoichiometry of 1 mole H^+ generated for each mole of O_2 consumed.[38,39] We have shown that the source of this H^+ is CO_2 generated as bicarbonate during respiratory burst.[39]

The extrusion of H^+ is dependent on an exchange with extracellular Na^+ and is not linked directly to the transport of electrons involved in O_2^- generation by the NADPH oxidase. This is demonstrated by the observation that H^+ secretion can be blocked by amiloride or by exchange of extracellular Na^+ with choline without resulting in a corresponding inhibition of the generation of O_2^-.[124–127] Na^+/H^+ antiporters are found in the plasma membrane of most mammalian cells and are not restricted to phagocytes.[128] The mechanism responsible for the activation of the Na^+/H^+ exchange during the respiratory burst of phagocytosis has not been worked out. The antiporter is activated when the intracellular pH decreases, but the cytosolic pH is reported to rise during activation of the respiratory burst.[126,127]

An additional H^+-translocating mechanism must exist, however, to account for the extracellular H^+ consumption that occurs during the dismutation of superoxide:

$$2O_2^- + 2H^+ \rightarrow H_2O_2 + O_2$$

This dismutation reaction takes place in the extracellular medium (or the phagocytic vacuole), as demonstrated by the observation that cytochrome *c* is reduced by the stimulated neutrophils, although the cytochrome does not enter intact cells.[75] The number of protons consumed in this reaction equals the number generated intracellularly during the respiratory burst less the protons generated from carbonic acid in the HMPS (see Table II). Thus, a proton-translocating mechanism coupled to the electron transport must exist in order to permit transport of protons and electrons in equal numbers across the plasma membrane during the respiratory burst. This mechanism has not been identified. It cannot be identical to the Na^+/H^+-exchange mechanism, but it could very well be a function of ubiquinone, as discussed in Section 3.1.

The overall reaction of the respiratory burst results in the generation of 1 mole H^+ per mole of O_2 consumed. This is very close to the figures observed.[39,39] It is possible, however, that a net H^+ consumption takes place within the phagocytic vacuole, since the generation of HOCl by MPO will result in a net proton consumption because of the high pK_a (7.5) of HOCl[129]:

$$H^+ + H_2O_2 + Cl^- \rightarrow HOCl + H_2O$$

In a confined environment such as the phagocytic vacuole, the pH may therefore rise as a result of MPO-mediated activities rather than as a result of the burst itself. This effect of MPO has been observed in purified NADPH oxidase preparations,[130] but the intravacuolar pH of MPO-deficient leukocytes or the effect of MPO inhibitors on the intravacuolar pH of normal cells was not investigated in the original reports.[122,123]

2.10. Depolarization

Neutrophils, like other mammalian cells, have a membrane potential with a slightly positive surface charge, i.e., a negative membrane potential. Stimulation of neutrophils by a wide variety of stimuli, e.g., A 23187, fMLP, PMA, immune complexes, and serum opsonized zymosan, elicits depolarization followed by repolarization after a variable length of time (5–10 min), as reviewed by Seligman and Gallin.[131] Depolarization of neutrophils is often measured as change in the fluorescence of the cyanine dyes di-*O*-$C_5(3)$ or di-*S*-$C_3(5)$, whose partition between the external medium and the cell is determined by the membrane potential and can be registered continuously by fluorescence spectroscopy. The distribution of the lipid-soluble cation triphenylmethyl phosphonium ($TPMP^+$) has also been used to measure membrane potentials, but the use of this compound does not permit continuous recordings.[131,132] The microelectrode technique has been applied to larger cells as monocytes and macrophages.[133,134]

Because depolarization is initiated shortly after stimulation, before the respiratory burst is fully activated,[40,132] depolarization has been considered a primary event in the activation of phagocytes, especially since the discovery that membrane potential changes were absent or grossly abnormal in neutrophils from CGD patients.[40,131,135] Several observations argue against this notion. First, CGD cells respond to activation with a normal rise in intracellular Ca^{2+} concentration.[136,137] This shows that the signal-transducing mechanism in neutrophils is operating normally, and as further evidence, the CGD cells are known to degranulate normally and to have a normal capacity for phagocytosis.[138] Thus, depolarization cannot be a signal-transducing mechanism for phagocyte activation. Second, normal cells can be made to depolarize by exchanging K^+ for Na^+ in the external medium, which does not by itself elicit a respiratory burst.[139,140] Third, the two forms of CGD, the autosomal-recessive and the X-linked forms, which result from complementary biochemical defects, share the same functional defect in membrane depolarization, as they share the same functional defect in the ability to mount a respiratory burst.[40,135] It is therefore most likely that membrane depolarization is an epiphenomenon of the respiratory burst, likely reflecting the ionic fluxes such as the Na^+/H^+ exchange or HCO_3^-/Cl^- exchange that take place as a consequence of the respiratory burst activity (see Section 2.9).

It has been observed that membrane depolarization is induced by very low concentrations of stimuli that apparently do not elicit a respiratory burst.[141,142] This may reflect that depolarization occurs simply as a consequence of only slight respiratory burst activity, although no proof of this has been presented. In support of this possibility, it should be noted that the concentration of the stimulus fMLP needed to activate Na^+/H^+ exchange in neutrophils is similar to the concentrations that produce membrane depolarization.[126,141,142]

3. THE NADPH OXIDASE

The key activity responsible for the respiratory burst is the NADPH oxidase. This entity is functionally defined as an NADPH-dependent superoxide dismutase-inhibitable cytochrome *c* reductase. Several authorities now agree that such an entity exists in activated phagocytes and has the two characteristic features[6,12,45,48–51,54–58,69]; i.e., the NADPH oxidase is virtually absent from unstimulated phagocytes and is absent from activated CGD phagocytes that are completely incapable of mounting a respiratory burst; whereas the latter still holds true, the former has been modified by the observation that a significant NADPH oxidase activity can be induced in subcellular fractions of unstimulated normal phagocytes by treatment with long-chain fatty acids, such as arachidonic acid and by sodium dodecyl sulfate (SDS).[143–147] It is very likely that this in vitro-activated NADPH oxidase is similar if not identical to the NADPH oxidase of activated phagocytes, since the K_m for NADPH is identical for the two activities, and since no NADPH oxidase activity can be induced in fractions from CGD phagocytes.[146]

Because of the prime importance of this "enzyme" for phagocyte function, much effort has been conducted to characterize and purify the NADPH oxidase activity of stimulated cells. The major obstacle against the successful achievement of this goal has been the instability of the enzyme, which is due either to autodestruction as a consequence of activity or to the loss of activity during purification.[45,48–51,54,57,149,149] The NADPH oxidase has been shown to have a requirement for FAD under some circumstances. This indicates that FAD is an integral part of the oxidase that may be lost during purification.[149–151] Recently, a protein has been purified that is claimed to be the NADPH oxidase.[54,57] This protein has the same K_m for NADPH as the NADPH oxidase of crude preparations of activated neutrophils. Its turnover rate of O_2^- is similar to that calculated for the NADPH oxidase. The protein eluted as a single band from molecular sieve chromatography and polyacrylamide gel electrophoresis under nondenaturing conditions, whereas three bands with molecular masses of 67,000, 48,000, and 32,000 M_r, respectively, were observed under denaturing conditions. This protein contains FAD but apparently no cytochrome. It is not unequivocally established whether this protein is identical to the NADPH oxidase operating in the cells or is just an active component. Several observations argue against the concept that the NADPH oxidase is a one-protein enzyme.

3.1. The NADPH Oxidase as an Electron-Transport Chain

The concept of a multicomponent electron-transport chain as the biochemical basis for the respiratory burst was first presented by Segal et al., who turned the focus to a *b* cytochrome of unique low midpoint potential in human phagocytes.[152–155] Although this cytochrome had been identified several years previously in horse neutrophils by Hattori[156] and in rabbit neutrophils by Shinagawa et al.,[157] these observations had generally been ignored. This cytochrome, b_{-245}, is not yet fully characterized despite several claims that it has been purified. The reported molecular weights range from 11,000 to 127,000.[158–163] The most recent report indicates that the *b* cytochrome is a heterodimer with subunits of 23,000 and 76,000–92,000 M_r.[164] Several observations indicated that this cytochrome participates in electron transport of direct importance for the respiratory burst.

First, the *b* cytochrome is reduced when normal neutrophils are stimulated during anaerobic conditions.[165,166] Although the rate of reduction of the cytochrome under anaerobic conditions is not compatible with the rate of electron flow through the respiratory burst oxidase during aerobic conditions,[167] this does not rule out the possibility that the *b* cytochrome normally transports electrons to O_2, since O_2 by itself has been shown to affect the rate of reduction of the cytochrome. By applying stop-flow spectrophotometry, Cross et al. were able to demonstrate that the rate of flow of electrons through the *b*-cytochrome component of an NADPH oxidase preparation was identical to the rate of O_2^- generation.[168,169] Second, this reduction of the *b* cytochrome has been found to be absent in neutrophils from all patients with CGD investigated (reviewed by Segal[170]), which included both the autosomal-recessive form, the X-linked form (X-CGD), and the form characterized by an abnormal K_m of the NADPH oxidase.[171–173] Since the defect in CGD neutrophils is strictly limited to the respiratory burst, these observations strongly indicate that the *b* cytochrome is involved in electron-transport linked to the respiratory burst.

The study of phagocytes from CGD patients has yielded more information about the nature of the NADPH oxidase because several genetically different forms of the disease exist. In most X-CGD cases, the *b* cytochrome is missing, judged from both the absence of a *b*-cytochrome spectrum in the cells[170–172,174] and the failure of anticytochrome *b* antibody to bind to the cells.[175] In some cases, an FAD flavoprotein is missing as well.[151,176–180] These observations support the concept of the NADPH oxidase as an electron-transport chain, of which one component is a *b* cytochrome and another is an FAD flavoprotein, indicating that the absence or defects in one or both components result in defects in the respiratory burst machinery, i.e., CGD. The interpretation of these straight observations has been somewhat obfuscated by the recent cloning of the gene responsible for classic X-linked CGD, since apparently no sequence coding for a *b* cytochrome was found, although the *b*-cytochrome spectrum was absent from the patient cells used to subtract background genetic material.[180] One explanation might be that the expression of the *b* cytochrome is dependent on the expression of some other component of the electron-transport chain, e.g., the FAD flavoprotein and that the primary genetic defect is in the latter. Another possibility is that the protein encoded for by this CGD gene may be the relevant *b* cytochrome and that the amino acid sequence, like the midpoint potential, is markedly different from that of other cytochromes. These possibilities should be resolved when the amino acid sequence of the purified *b* cytochrome is at hand.

The autosomal-recessive form of CGD has not generally been associated with lack of *b* cytochrome or FAD flavoprotein,[170,174,179,180,182] although a few families have been described in which the b cytochrome seems to be defective.[178,179] Most autosomal CGD patients are characterized by a defective phosphorylation of a 47,000-M_r protein.[183] This protein awaits further identification. It is not known whether the defective phosphorylation is due to a defect in the kinase that mediates this phosphorylation or to defects in the protein substrate. It has been claimed that the *b* cytochrome is phosphorylated,[162] but the evidence is not irrefutable, particularly because no agreement exists as to the molecular weight of the *b* cytochrome. Interestingly, the defects responsible for the autosomal and the X-linked forms of the disease have been shown to be complementary, since fusion of respiratory burst-defective monocytes from each of both types of CGD results in fused cells capable of mounting a respiratory burst.[184] This makes it unlikely but does not exclude the possibility that both types of CGD result from defects in the *b* cytochrome.

In addition to these structural indications that the NADPH oxidase is a multicomponent system, some functional indications have recently been presented. Several groups have demonstrated that NADPH oxidase activity can be induced in subcellular fractions of unstimulated phagocytes by combining a factor present in the cytosol with a factor present in the membrane fraction, in the presence of either arachidonate or SDS. Both the particulate and the soluble factor are essential for activity.[143–147]

Whereas these functional studies and the genetic evidence favor the concept of the NADPH oxidase as a multicomponent system, studies on isolated preparations of the NADPH oxidase from activated cells are conflicting. There is agreement that

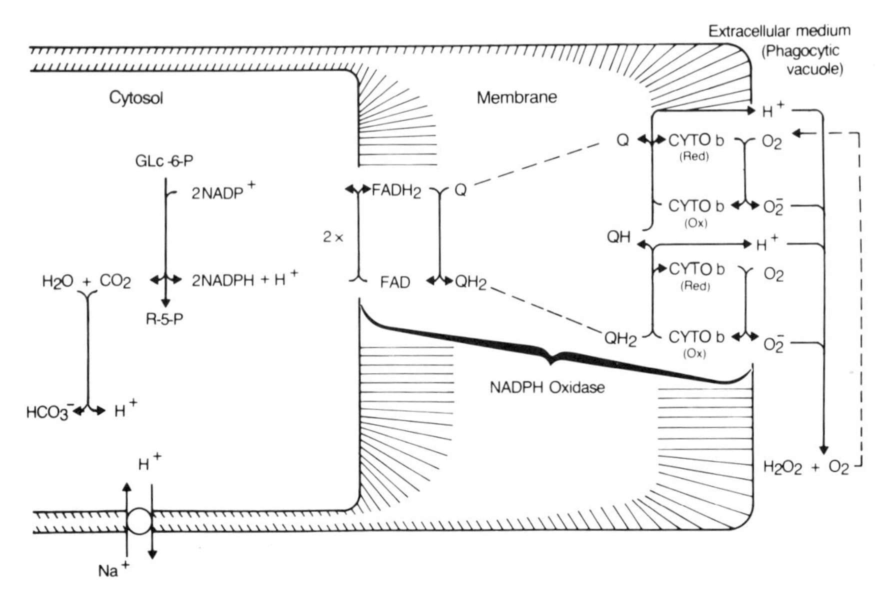

FIGURE 3. Model of NADPH oxidase as an electron-transport chain. Schematic representation of an activated neutrophil. Q, QH, QH_2: Oxidized, univalently reduced, and fully reduced ubiquinone, respectively. Glc-6-P (G6P), Glucose 6-phosphate; R-5-P, ribulose 5-phosphate.

the NADPH oxidase contains FAD, as evidenced by the following observations[45,50–52,54,57,149–151,162,170,176,185]: (1) the enzyme has a requirement for FAD, if Triton is used during solubilization of the enzyme; (2) substitution by flavins or flavenoids incapable of one-electron transfer abolishes the ability of the enzyme to generate O_2^-; and (3) FAD is present in most preparations of the NADPH oxidase. Although the *b* cytochrome has been identified in these preparations as well, in a molar ratio with FAD of 2 : 1 or 1 : 1, Glass et al.[57] were unable to confirm this in the most recent report on purified NADPH oxidase. By contrast, Bellavite et al.[186] found high concentrations of *b* cytochrome and little FAD in their most recent report on the purified oxidase.

One of the first indications that the NADPH oxidase is a multicomponent electron-transport chain was presented by Schneider and co-workers,[187] who identified ubiquinone-50 together with *b* cytochrome in mouse peritoneal leukocytes. These studies have been extended to human neutrophils, and it is reported that a ubiquinone is present in neutrophils together with the *b* cytochrome in a molar ratio of 1 : 1.[188,189] Furthermore, it was shown that this ubiquinone can be identified together with the *b* cytochrome in phagocytic vacuoles isolated by flotation, that the ubiquinone becomes reduced when the NADPH oxidase is stimulated, and finally that substitution by quinone inhibitors incapable of electron transport inhibits the respiratory burst, without inhibiting other aspects of activated neutrophil function. These observations all support the idea that ubiquinone-50 is involved in the electron transport of the respiratory burst.

These findings are somewhat contradicted by the inability of other groups to identify ubiquinone in relevant fractions from neutrophils,[154,193,194] although some workers have been able to confirm the findings of Schneider's group.[195–198] It is still an open question whether ubiquinone-50 participates in nonmitochondrial electron transport of phagocytes, although teleologically it would seem ideally suited to transport both electrons and protons across the membrane of the phagocyte, as depicted in the model of the NADPH oxidase as an electron-transport chain (Fig. 3).

3.2. Subcellular Localization of the NADPH Oxidase and Its Components

The NADPH oxidase products O_2^- and H_2O_2 have been demonstrated by histochemistry to be generated in the plasma membrane and the phagosomal membrane of phagocytes.[7,86,199] This indicates that the active oxidase is localized in these parts of the cell. Subcellular fractionation studies on activated neutrophils unequivocally support this possibility.[177,201–203] However, the subcellular localization of the electron-transport chain components, the *b* cytochrome, the FAD-flavoprotein, and possibly the ubiquinone, is not yet settled. We and others have found that most (approximately 80%) of the *b* cytochrome in human neutrophils is localized in the membrane of the specific granules.[177,188,189,204–207] Although some disagreement exist whether the *b* cytochrome may be localized in tertiary gelatinase-containing granules, we recently showed that that the cytochrome is localized in the specific granules.[208] On the basis of these findings, we have suggested that degranulation of specific granules is important for activation of the NADPH oxidase, by translocating electron-transport chain components from an intracellular store to the plasma membrane[177,204] (Fig. 4). This concept has been

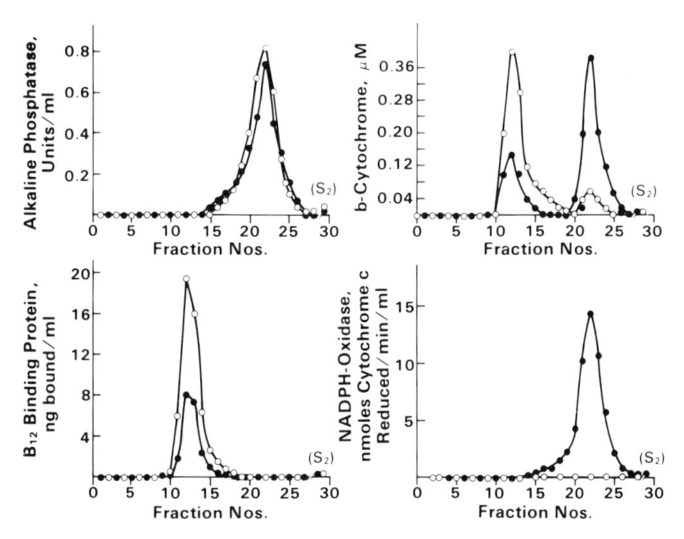

FIGURE 4. Subcellular localization of *b*-cytochrome and NADPH oxidase. Resting (closed symbols) and PMA-stimulated (open symbols) human neutrophils were disrupted by nitrogen cavitation and fractionated on Percoll density gradients. (From Borregaard and Tauber.[177])

further strengthened by the observation that the specific granules contain a FAD flavoprotein in a 1 : 2 molar ratio with the *b* cytochrome and that this flavoprotein is missing together with the *b* cytochrome in the CGD patients we have studied.[177] Others have been unable to find as clear-cut a pattern. This may be because different genetic forms of the disease have been studied or may be attributable to differences in the subcellular fractionation techniques employed.[205]

The finding that neutroplasts, which are granule-depleted neutrophils, are capable of generating a respiratory burst,[193,209,210] does not argue against the concept that an intracellular reservoir of electron-transport chain components exists that translocates to the plasma membrane and the phagosome membrane to function as an active NADPH oxidase, since the plasma membrane of neutroplasts has been shown to be enriched in granule membrane components.[211] The neutroplast model does, however, demonstrate that translocation of granule membrane components per se does not lead to activation of the NADPH oxidase. It should be noted that a few other laboratories have found most of the *b* cytochrome present in the plasma membrane of resting neutrophils.[153,212,213] This is probably explained by differences in cell handling and subcellular fractionation techniques.

The concept of recruitment of NADPH oxidase components from internal stores is now further supported by our identification of a novel granule subset in neutrophils.[214] The density of this granule subset is close to that of plasma membranes, and we have demonstrated that the majority of the 20% of the *b* cytochrome, previously thought to be in the plasma membrane, is localized in the membrane of these granules, which is completely translocated to the plasma membrane upon stimulation by even very low concentrations of soluble stimuli that cause little or no degranulation of specific granules.[215] Monocytes are known to contain similar easily mobilizable intracellular vesicles, but it is yet to be demonstrated whether these contain *b* cytochrome or FAD flavoprotein.[216]

3.3. Activation of the NADPH Oxidase

A wide variety of membrane perturbants have been shown to be capable of activating the respiratory burst (see Section 1). Although the signal transducing mechanism for this activation has not been yet worked out, there are some indications that activation of the NADPH oxidase involves phosphorylation of some component(s) of the NADPH oxidase, most likely by protein kinase C: (1) activation of neutrophils by several stimuli has been shown to be paralleled by translocation of protein kinase C from the cytosol to the membranes[217–219]; (2) PMA, the intracellular receptor of which is protein kinase C,[220,221] is a potent activator of the respiratory burst[15] and of NADPH oxidase activity[69,217,218]; (3) NADPH oxidase activity can be induced in subcellular fractions of unstimulated neutrophils by conditions that favor protein phosphorylation by protein kinase C[222]; (4) fluoride, which has properties both as a phosphoprotein phosphatase inhibitor and as *N*-regulatory protein activator, is capable of activating the respiratory burst[19]; (5) some inhibitors of protein kinase C are capable of blocking the activation of the respiratory burst[223]; (6) absent phosphorylation of a 47,000-M_r protein during PMA stimulation is associated with autosomal-recessive CGD, i.e., disability of the phagocytes to mount a respiratory burst[184]; and (7) a 31,500-M_r phosphorylated protein is claimed to copurify with NADPH oxidase activity during purification of the latter.[162,224,225] Albeit suggestive, these arguments are only indirect, and definite proof of the direct involvement of a kinase reaction, in particular of a protein kinase C-mediated reaction in activating the respiratory burst is lacking.

Several observations argue against the involvement of protein kinase C in the activation of the respiratory burst: (1) the activation of the respiratory burst is not inhibited by some inhibitors of protein kinase C[226,227], (2) the PMA analogue mezerine may activate the respiratory burst without causing any activation of protein kinase C,[228] and (3) that long-chain polyunsaturated fatty acids (e.g., arachidonic acid) can elicit immediate activation of the respiratory burst that is brought to an immediate stop if the stimulus is removed by albumin.[229] Although it has been shown that arachidonate may activate protein kinase C,[230] arachidonate is able to activate the NADPH oxidase in a cell-free system consisting of membranes and a cytosolic factor without the addition of ATP and without a requirement for Ca^{2+}. This makes it unlikely that this activation is mediated by protein kinase C,[145,146] unless protein kinase C has been proteolytically converted to a Ca^{2+}/phospholipid-insensitive form.

Instead, some indications are given that protease activity may be involved in the activation of the respiratory burst.[231] It should be kept in mind that proteins exist whose activity is changed dramatically by phosphorylation but for which the same functional changes are induced by limited proteolytic treatment.[232] It was recently shown that the serine protease calpain is bound to the plasma membrane of neutrophils together with protein kinase C during activation of the cells and that Ca^{2+} may induce a calpain-mediated proteolytic activation of protein kinase C, which then becomes insensitive to Ca^{2+} and phospholipid.[233–235] Evidence has been presented that different routes may lead to activation of the NADPH oxidase.[69]

4. KINETICS OF THE RESPIRATORY BURST

It is a feature of the respiratory burst that a lag period of 30–45 sec is needed before full activation of the burst is elicited.[39,67] This lag may be shortened by agents such as cytochalasins,[236] which augment the degranulation of stimulated neutrophils, but the lag is also present in neutroplasts,[209,210] indicating that it does not represent the time needed for degranulation. Rather, it may represent the time needed for signal transduction to result in phosphorylation of the NADPH oxidase. This is indicated by the observation that the lag period is absent when arachidonate is used as stimulus.[229] Arachidonate probably does not activate the NADPH oxidase via phosphorylation (see Section 3.3).

At the other end, the respiratory burst stops after variable periods of time, depending on the stimulus and the experimental conditions. In particular, if high concentrations of phagocytes are used, the supply of O_2 in the medium will be rapidly exhausted and the burst brought to a stop (see Fig. 1). If O_2 is still present, the respiration will decrease and eventually terminate. Although not definately proved, the main reason for the termination of the respiratory burst seems to be autoinactivation of the NADPH oxidase, since the activity of this enzyme that may be recovered from the phagocytes decreases with time after activation.[69,148]

The magnitude of the respiratory burst is also dependent on the stimulus. The respiratory burst is dependent on the delivery of electrons from the HMPS, but neither the flux of glucose to the HMPS nor the capacity of the enzyme system involved in the HMPS seems to be limiting for the magnitude of the burst, since only extensive defects in this system are reported to influence the magnitude of the respiratory burst (see Section 2.1).

By contrast, several observations indicate that the activity of the NADPH oxidase is rate limiting for the respiratory burst: (1) that priming of cells by applying one stimulus in suboptimal concentrations is followed by both an enhanced respiratory burst, as measured in intact cells, and an increased NADPH oxidase activity, as measured in membrane preparations of cells, when a second stimulus is applied[238]; and (2) that there is a CGD subset in which the respiratory burst is defective but not absent due to a decreased affinity of the NADPH oxidase for its substrate.[24,25,172,239]

Finally, it should be mentioned that activation of the respiratory burst is not an all-or-none phenomenon. This may be implicit in the observation that different stimuli activate the burst to different degrees, but the existence of subsets of phagocytes with different responsiveness toward various stimuli might partly explain these differences.[240] It has been established, however, that individual cells will respond in a graded fashion to stimulation by one stimulus, when different concentrations of the stimulus are applied.[241]

5. DEVELOPMENT OF THE ABILITY TO GENERATE A RESPIRATORY BURST

The ability of phagocytes to generate a respiratory burst is acquired during maturation of these in the bone marrow. Normal myeloid bone marrow cells become able to generate O_2^-, measured as reduction of nitroblue tetrazolium (NBT), only after maturation beyond the myelocyte stage.[242] This is supported by studies during differentiation of the promyelocytic HL-60 cell, although this cell line may not be a good model to study because of the lack of specific granules in the end-stage differentiated cells and their inability to achieve full capacity to generate a respiratory burst.[243,244] We have studied cells from chronic myeloid leukemia (CML) patients and shown that these cells acquire the ability to generate a respiratory burst during maturation at the same time that specific granules and b cytochrome appear. This takes place at the myelocyte/metamyelocyte stage.[245] Others have found that a final maturation of the functional capabilities of neutrophils is achieved as these leave the bone marrow.[246]

6. THE RESPIRATORY BURST IN NONPHAGOCYTIC CELLS

The ability to generate a significant respiratory burst seems to be a unique feature of phagocytes. Although this has also been claimed to be a feature of normal fibroblasts[247] and lymphocytes, in particular natural killer (NK) cells,[48] these observations have been refuted by others.[249–252] It has been established, however, that a considerable respiratory burst may be elicited in Epstein-Barr virus (EBV)-infected *b* lymphocytes.[253] This observation may be important in the further characterization of the gene(s) responsible for the respiratory burst. Apart from this infected nonphagocytic cell line, no normal mammalian nonphagocytic cells are known to have a respiratory burst apparatus similar to that present in the phagocytes. A striking similarity exists, however, between the respiratory burst elicited in phagocytes and the initial metabolic events that follow fertilization of sea urchin eggs.[254,255] This implies that the respiratory burst, in addition to the devastating effects mediated by the professional phagocytes, may be of more subtle importance for less differentiated cells.

ACKNOWLEDGMENTS. This work was supported by grants 12-5340 and 12-6134 from the Danish Medical Research Council, by grant 86-132 from the Danish Cancer Society, by grant 233-627 from the Danish Arthritis Foundation, and by Knud Højgaard's Fund. Mrs. Steffy Plinius is gratefully acknowledged for preparing the manuscript.

REFERENCES

1. Baldridge CW, Gerard RW: The extra respiration of phagocytosis. *Am J Physiol* **103:**235–236, 1933.
2. Sbarra AJ, Karnovsky ML: The biochemical basis of phagocytosis. I. Metabolic changes during the ingestion of particles by polymorphonuclear leukocytes. *J Biol Chem* **234:**1355–1362, 1959.
3. Borregaard N, Herlin T: Energy metabolism of human neutrophils during phagocytosis. *J Clin Invest* **70:**550–557, 1982.
4. Tauber AI, Roberts MF: ^{31}P NMR spectroscopy of phorbol-myristate-acetate stimulated polymorphonuclear human leukocytes. *FEBS Lett* **129:**105–108, 1981.
5. Loike JD, Kozler VF, Silverstein SC: Increased ATP and creatine phosphate turnover in phagocytosing mouse peritoneal macrophages. *J Biol Chem* **254:**9558–9564, 1979.
6. Rossi F, Romeo D, Patriarca P: Mechanism of phagocytosis-associated oxidative metabolism in polymorphonuclear leucocytes and macrophages. *J Reticuloendothel Soc* **12:**127–149, 1972.
7. Badwey JA, Curnutte JT, Robinson JM, et al: Comparative aspects of oxidative metabolism of neutrophils from human blood and guinea pig peritonea: Magnitude of the respiratory burst, dependence upon stimulating agents and localization of the oxidase. *J Cell Physiol* **105:**541–551, 1980.
8. Davis WC, Huber H, Douglas SD, et al: A defect in circulating mononuclear phagocytes in chronic granulomatous disease of childhood. *J Immunol* **101:**1093–1095, 1968.
9. Kragballe K, Borregaard N, Brandrup F, et al: Relation of monocyte and neutrophil oxidative metabolism to skin and oral lesions in carriers of chronic granulomatous disease. *Clin Exp Immunol* **43:**390–398, 1981.

10. Mickenberg, ID, Root RK, Wolf SM: Bactericidal and metabolic properties of human eosinophils. *Blood* **39:**67–80, 1972.
11. DeChatelet LR, Shirley PS, McPhail LC, et al: Oxidative metabolism of the human neutrophil. *Blood* **50:**525–535, 1977.
12. Hoffman M, Autor AP: Production of superoxide anion by an NADPH oxidase from rat pulmonary macrophages. *FEBS Lett* **121:**352–354, 1980.
13. Oren R, Farnham AE, Saito K, et al: Metabolic patterns in three types of phagocytic cells. *J Cell Biol* **17:**487–501, 1963.
14. Lehmeyer JE, Snyderman R, Johnston RB Jr: Stimulation of neutrophil oxidative metabolism by chemotactic peptides. Influence of calcium ion concentration and cytochalasin B and comparison with stimulation by phorbol myristate acetate. *Blood* **54:**35–45, 1979.
15. Repine JE, White JG, Clawson CC, et al: The influence of phorbol myristate acetate on oxygen consumption by polymorphonuclear leukocytes. *J Lab Med* **83:**911–920, 1974.
16. Becker EL, Sigman M, Oliver JM: Superoxide production induced in rabbit polymorphonuclear leukocytes by synthetic chemotactic peptides and A 23187. Nature of the receptor and the requirement for Ca^{2+}. *Am J Pathol* **95:**81–87, 1979.
17. Cohen HJ, Chovaniec ME, Wilson K, et al: Con-A-stimulated superoxide production by granulocytes: Reversible activation of NADPH oxidase. *Blood* **60:**1188–1194, 1982.
18. Lopez, AF, Williamson J, Gamble JR, et al: Recombinant human granulocyte–macrophage colony-stimulating factor stimulates in vitro mature human neutrophil and eosinophil function, surface receptor expression and survival. *J Clin Invest* **78:**1220–1228, 1986.
19. Curnutte JT, Babior BM, Karnovsky ML: Fluoride-mediated activation of the respiratory burst in human neutrophils, a reversible process. *J Clin Invest* **63:**637–647, 1979.
20. Abramson S, Hoffstein ST, Weissmann G: Superoxide anion generation by human neutrophils exposed to monosodium urate. *Arthritis Rheum* **25:**174–180, 1982.
21. Badwey JA, Curnutte JT, Karnovsky ML: Cis-polyunsaturated fatty acids induce high levels of superoxide production by human neutrophils. *J Biol Chem* **256:**12640–12643, 1981.
22. Bridges RA, Berendes H, Good RA: A fatal granulomatous disease of childhood. The clinical, pathological and laboratory findings of a new syndrome. *J Dis Child* **97:**387–408, 1959.
23. Holmes B, Page AR, Good RA: Studies on the metabolic activity of leukocytes from patients with a genetic abnormality of phagocyte function. *J Clin Invest* **46:**1422–1432, 1967.
24. Tauber AI, Borregaard N, Simons ER, et al: Chronic granulomatous disease. A syndrome of phagocyte oxidase deficiencies. *Medicine (Baltimore)* **62:**286–309, 1983.
25. Gallin JI, Fauci AS (eds): *Advances in Host Defense Mechanisms*. New York, Raven, 1983, vol 3: *Chronic granulomatous disease*.
26. Babior BM, Kipnes RS, Curnutte JT: Biological defense mechanisms. The production by leukocytes of superoxide, a potential bactericidal agent. *J Clin Invest* **52:**741–744, 1973.
27. Behar D, Czapski G, Rabani J, et al: The acid dissociation constant and decay kinetics of the perhydroxyl radical. *J Phys Chem* **74:**3209–3213, 1970.
28. Klebanoff SJ: Oxygen-dependent cytotoxic mechanisms of phagocytes, in Gallin JI, Fauci AS (eds): *Advances in Host Defense Mechanisms*. New York, Raven, 1983, vol. 1: *Phagocytic Cells*, p 111.
29. Iyer GYN, Islam DMF, Quastel JH: Biochemical aspects of phagocytosis. *Nature (Lond)* **192:**535–541, 1961.
30. Dri P, Bellavite P, Berton G, et al: Interrelationship between oxygen consumption, superoxide anion generation and hydrogen peroxide formation in phagocytosing guinea pig polymorphonuclear leukocytes. *Mol Cell Biochem* **23:**109–122, 1979.
31. Test ST, Weiss SJ: Quantitative and temporal characterization of the extracellular H_2O_2 pool generated by human neutrophils. *J Biol Chem* **259:**399–405, 1984.
32. Klebanoff SJ: Myeloperoxidase contribution to the microbicidal activity of intact leukocytes. *Science* **169:**1095–1097, 1970.
33. Allen RC, Stjernholm RL, Steele RH: Evidence for the generation of an electronic excitation

state(s) in human polymorphonuclear leukocytes and its participation in bactericidal activity. *Biochem Biophys Res Commun* **47**:679–684, 1972.
34. Tauber AI, Babior BM: Evidence for hydroxyl radical production by human neutrophils. *J Clin Invest* **60**:374–379, 1977.
35. Weiss SJ, Rustag PK, Lobuglio AF: Human granulocyte generation of hydroxyl radical. *J Exp Med* **147**:316–323, 1978.
36. Weiss SJ, Klein R, Slivka A, et al: Chlorination of taurine by human neutrophils: Evidence for hypochlorous acid generation. *J Clin Invest* **70**:598–607, 1982.
37. Weiss SJ, Lampert MB, Test ST: Long-lived oxidants generated by human neutrophils: Characterization and bioactivity. *Science* **222**:625–628, 1983.
38. VanZwieten R, Wever R, Hamers MN, et al: Extracellular proton release by stimulated neutrophils. *J Clin Invest* **68**:310–313, 1982.
39. Borregaard N, Schwarts JH, Tauber AI: Proton secretion by stimulated neutrophils. Significance of hexose monophosphate shunt activity as source of electrons and protons for the respiratory burst. *J Clin Invest* **74**:455–459, 1984.
40. Whitin JC, Chapman CE, Simons ER, et al: Correlation between membrane potential changes and superoxide production in human granulocytes stimulated by phorbol myristate acetate. Evidence for defective activation in chronic granulomatous disease. *J Biol Chem* **255**:1874–1878, 1980.
41. Evans WH, Karnovsky ML: The biochemical basis of phagocytosis. IV. Some aspects of carbohydrate metabolish during phagocytosis. *Biochemistry* **1**:159–166, 1962.
42. Baehner RL, Karnovsky ML: Deficiency of reduced nicotinamide-adenine dinucleotide oxidase in chronic granulomatous disease. *Science* **162**:1277–1279, 1968.
43. Segal AW, Peters TJ: Characterization of the enzyme defect in chronic granulomatous disease. *Lancet* **1**:1363–1365, 1976.
44. Badwey JA, Karnovsky ML: Production of superoxide and hydrogen peroxide by an NADH-oxidase in guinea pig polymorphonuclear leukocytes, modulation by nucleotides and divalent cations. *J Biol Chem* **254**:11530–11537, 1979.
45. Babior BM: The nature of the NADPH oxidase, in Gallin J I, Fauci AS (eds): *Advances in Host Defense Mechanisms.* New York, Raven, 1983, vol 3: *Chronic Granulomatous Disease,* p 91.
46. Goldstein IM, Cerquiera M, Lind S, et al: Evidence that the superoxide generating system of human leukocytes is associated with the cell surface. *J Clin Invest* **59**:249–254, 1977.
47. Hohn DC, Lehrer RI: NADPH oxidase deficiency in X-linked chronic granulomatous disease. *J Clin Invest* **55**:707–713, 1975.
48. Babior BM, Curnutte JT, McMurrich BJ: The particulate superoxide forming system from human neutrophils. Properties of the system and further evidence supporting its participation in the respiratory burst. *J Clin Invest* **58**:989–996, 1976.
49. Gabig TG, Kipnes RS, Babior BM: Solubilization of the O_2^--forming activity responsible for the respiratory burst in human neutrophils. *J Biol Chem* **253**:6663–6665, 1978.
50. Tauber AI, Goetzl EJ: Structural and catalytic properties of the solubilized superoxide-generating activity of human polymorphonuclear leukocytes. Solubilization, stabilization and partial characterization. *Biochemistry* **18**:5576–5584, 1979.
51. Gabig TG, Babior BM: The O_2^--forming oxidase responsible for the respiratory burst in human neutrophils. Properties of the solubilized enzyme. *J Biol Chem* **254**:9070–9074, 1979.
52. Wakeyama H, Takeshige K, Takeyanagi R, et al: Superoxide forming NADPH oxidase preparation of pig polymorphonuclear leukocytes. *Biochem J* **205**:593–601, 1982.
53. DeChatelet LR, Shirley PS: Pyridine nucleotide-dependent generation of hydrogen peroxide by a particulate fraction from human neutrophils. *J Immunol* **126**:1165–1169, 1981.
54. Market M, Glass GA, Babior BM: Respiratory burst oxidase from human neutrophils. Purification and some properties. *Proc Natl Acad Sci USA* **82**:3144–3148, 1985.
55. Suzuki H, Pabst MJ, Johnston RB Jr: Enhancement by Ca^{2+} or Mg^{2+} of catalytic activity of the superoxide-producing NADPH oxidase in membrane preparations of human neutrophils and monocytes. *J Biol Chem* **260**:3635–3639, 1985.

56. Umei T, Takeshige K, Minakami S: NADPH binding component of neutrophil superoxide generating oxidase. *J Biol Chem* **261:**5229–5232, 1986.
57. Glass GA, DeLise DM, DeTogni P, et al: The respiratory burst oxidase of human neutrophils. Further studies of the purified enzyme. *J Biol Chem* **261:**13247–13251, 1986.
58. Rossi F, Zatti M: Biochemical apects of phagocytosis in polymorphonuclear leukocytes. NADH and NADPH oxidation by the granules of resting and phagocytizing cells. *Experientia (Basel)* **20:**21–23, 1964.
59. Beck WS: The control of leukocyte glycolysis. *J Biol Chem* **232:**251–270, 1958.
60. Beck WS: Occurrence and control of the phosphogluconate oxidation pathway in normal and leukemic leukocytes. *J Biol Chem* **232:**271–283, 1958.
61. McGilvery RW: *Biochemistry: A Functional Approach.* Philadelphia, WB Saunders, 1970.
62. Stjernholm RL, Manac RC: Carbohydrate metablism in leukocytes. XIV. Regulation of pentose cycle activity and glycogen metabolism during phagocytosis, *J Reticuloendothel Soc* **8:**550–560, 1970.
63. Esmann V: The metabolism of (1-^{14}C)-, (2-^{14}C)-, (3,4-^{14}C)-, and (6-^{14}C)-glucose in normal and diabetic polymorphonuclear leukocytes and during phagocytosis, *Diabetologia* **4:**188–194, 1964.
64. McCall CE, Bass DA, Cousart S, et al: Enhancement of hexose uptake in human polymorphonuclear leukocytes by activated complement C5a, *Proc Natl Acad Sci USA* **76:**5896–5900, 1979.
65. Cooper MR, DeChatelet LR, McCall CE, et al: Complete deficiency of leukocyte glucose-6-phosphate dehydrogenase with defective bactericidal activity. *J Clin Invest* **51:**769–778, 1972.
66. Baehner R , Johnston RB Jr, Nathan DG: Comparative studies of the metabolic and bactericidal characteristics of severe glucose-6-phosphate dehydrogenase-deficient polymorphonuclear leukocytes and leukocytes from children with chronic granulomatous disease. *J Reticuloendothel Soc* **12:**150–169, 1972.
67. Weening RS, Roos D, Loos JA: Oxygen consumption of phagocytizing cells in human leukocyte and granulocyte preparations. A comparative study. *J Lab Clin Med* **83:**570–576, 1974.
68. Cohen HJ: Chovaniec ME, Takahashi K, et al: Activation of human granulocytes by arachidonic acid: Its use and limitations for investigating granulocyte functions. *Blood* **67:**1103–1109, 1986.
69. McPhail LC, Snyderman R: Activation of the respiratory burst enzyme in human polymorphonuclear leukocytes by chemoattractants and other soluble stimuli. Evidence that the same oxidase is activated by different transductional mechanisms. *J Clin Invest* **72:**192–200, 1983.
70. Curnutte JT, Tauber AI: Failure to detect superoxide in human neutrophils stimulated with latex particles. *Pediatr Res* **17:**281–284, 1983.
71. Segal AW, Meshulam T: Production of superoxide by neutrophils: A reappraisal. *FEBS Lett* **100:**27–32, 1979.
72. Van Gelder BF, Slater EC: The extinction coefficient of cytochrome c. *Biochim Biophys Acta* **58:**593–595, 1964.
73. Weening RS, Wever R, Roos D: Quantitative aspects of the production of superoxide radicals by phagocytizing human granulocytes. *J Lab Clin Med* **85:**245–252, 1975.
74. Root RK, Metcalf J: H_2O_2 release from human granulocytes during phagocytosis. Relationship to superoxide anion formation and cellular catabolism of H_2O_2: Studies with normal and cytochalasin b-treated cells. *J Clin Invest* **60:**1266–1279, 1977.
75. Roos D, Eckman CM, Yazdabaklıslı M, et al: Excretion of superoxide by phagocytes, measured with cytochrome c entrapped in resealed erythrocyte ghosts. *J Biol Chem* **259:**1770–1775, 1984.
76. Cuperus RA, Muijsers AO, Wever R: Myeloperoxidase from human leukocytes exerts superoxide dismutase activity; a pathway with compound III as intermediate. *Eur J Clin Invest* **16:**45A, 1986 (abst 254).
77. Segal AW, Clark J, Allison AC: Tracing the fate of oxygen consumed during phagocytosis by human neutrophils with $^{15}O_2$· *Clin Sci Mol Med* **55:**413–415, 1978.
78. Rabani J, Klug D, Fridovich I: Decay of the HO_2 and O_2 radicals catalyzed by superoxide dismutase. A pulse radiolytic investigation. *Israel J Chem* **10:**1095–1106, 1972.

79. Kahn AU: Singlet molecular oxygen from superoxide anion and sensitized fluorescence of organic molecules. *Science* **168:**476–477, 1970.
80. Kahn AU: Activated oxygen: Singlet molecular oxygen and superoxide anion. *Photochem Photobiol* **28:**615–627, 1978.
81. Mayeda EA, Bard AJ: Singlet oxygen. The suppression of its production in dismutation of superoxide ion by superoxide dismutase. *J Am Chem Soc* **96:**4023–4024, 1974.
82. Nilsson R, Kearns DR: Role of singlet oxygen in some chemiluminescence and enzyme oxidation reactions. *J Phys Chem* **78:**1681–1683, 1974.
83. Foote CS, Shook FC, Abakerli RA: Chemistry of superoxide anion. 4. Singlet oxygen is not a major product of dismutation. *J Am Chem Soc* **102:**2503–2504, 1980.
84. Homan-Muller JWT, Weening RS, Roos D: Production of hydrogen peroxide by phagocytizing human granulocytes. *J Lab Clin Med* **85:**198–207, 1975.
85. Root RK, Metcalf J, Oshino N, et al: H_2O_2 release from human granulocytes during phagocytosis. Documentation, quantitation, and some regulating factors. *J Clin Invest* **55:**945–955, 1975.
86. Briggs RT, Drath DB, Karnovsky ML, et al: Localization of NADH oxidase on the surface of human polymorphonuclear leukocytes by a new cytochemical method. *J Cell Biol* **67:**566–586, 1975.
87. Agner K: Verdoperoxidase. A ferment isolated from leukocytes. *Acta Physiol Scand* **2**(suppl 8)**:**1–62, 1941.
88. West BC, Rosenthal AS, Gelb NA, et al: Separation and characterization of human neutrophil granules. *Am J Pathol* **77:**41–66, 1974.
89. Spitznagel JK, Dalldorf FG, Leffell MS, et al: Character of azurophil and specific granules purified from human polymorphonuclear leukocytes. *Lab Invest* **30:**774–785, 1974.
90. Nichols BA, Bainton DF: Differentiation of human monocytes in bone marrow and blood: Sequential formation of two granule populations. *Lab Invest* **29:**27–40, 1973.
91. VanFurth R, Hirsch JG, Federko ME: Morphology and peroxidase cytochemistry of mouse promonocytes, monocytes and macrophages. *J Exp Med* **132:**794–812, 1970.
92. Bainton DF, Golde DW: Differentiation of macrophages from normal human bone marrow in liquid culture. Electron microscopy and cytochemistry. *J Clin Invest* **61:**1555–1569, 1978.
93. Heifets L, Imai K, Goren MB: Expression of peroxidase-dependent iodination by macrophages ingesting neutrophil debris. *J Reticuloendothel Soc* **28:**391–404, 1980.
94. Passo SA, Weiss SJ: Oxidative mechanisms utilized by human neutrophils to destroy *Escherichia coli*. *Blood* **63:**1361–1368, 1984.
95. Klebanoff SJ, Green WL: Degradation of thyroid hormones by phagocytosing human leukocytes. *J Clin Invest* **52:**60–72, 1963.
96. Zgcliczynski JM, Selveraj RJ, Paul BB, et al: Chlorination by the myeloperoxidase–H_2O_2-Cl/ antimicrobial system at acid and neutral pH. *Proc Soc Exp Biol Med* **154:**418–422, 1977.
97. Okolow-Zubkowska HJ, Hill HAO: An alternative mechanism for the production of hydroxyl radical by stimulated neutrophils, in Rossi F, Patriarca P (eds): *Advances in Experimental Medicine and Biology*. New York, Plenum, 1982, vol 141: *Biochemistry and Function of Phagocytes*, p 423.
98. Rosen H, Klebanoff SJ: Formation of singlet oxygen by the myeloperoxidase-mediated antimicrobial system. *J Biol Chem* **252:**4803–4810, 1977.
99. Agner K: Biological effects of hypochlorous acid formed by "MPO" peroxidation in the presence of chloride ion, in Åkesson A, Ehrenberg A (eds): *Structure and Function of Oxidation-Reduction Enzymes*. New York, Pergamon, 1972, vol 18, p 329.
100. Lampert MB, Weiss SJ: The chlorinating potential of the human monocyte. *Blood* **63:**645–651, 1983.
101. Kanofsky JR, Wright JR, Miles-Richardson GE, et al: Biochemical requirements for singlet oxygen production by purified human myeloperoxidase. *J Clin Invest* **74:**1489–1495, 1984.
102. Makino R, Tanaka T, Ilizuka T, et al: Stoichiometric conversion of oxygen to superoxide anion during the respiratory burst in neutrophils. Direct evidence by a new method for measurement of

superoxide anion with diacetyldeuteroheme-substituted horseradish peroxidase. *J Biol Chem* **261:**11444–11447, 1986.
103. Green TR, Wu DE: The NADPH : O_2 oxidoreductase of human neutrophils. Stoichiometry of univalent and divalent reduction of O_2. *J Biol Chem* **261:**6010–6015, 1986.
104. Haber F, Weiss J: The catalytic decomposition of hydrogen peroxide by iron salts. *Proc R Soc Lond Ser A* **147:**332–351, 1934.
105. Repine JE, Eaton JW, Anders MW, et al: Generation of hydroxyl radical by enzymes, chemicals and human phagocytes in vitro. Detection with the anti-inflammatory agent dimethyl sulfoxide. *J Clin Invest* **64:**1642–1651, 1979.
106. Rosen H. Klebanoff SJ: Hydroxyl radical generation by polymorphonuclear leukocytes measured by electron spin resonance spectroscopy. *J Clin Invest* **64:**1725–1729, 1979.
107. Green MR, Hill HAO, Okolow-Zubkowska MJ, et al: The production of hydroxyl and superoxide radicals by stimulated human neutrophils. Measurements by EPS spectroscopy. *FEBS Lett* **100:**23–26, 1979.
108. Britigan BE, Rosen GM, Chai Y, et al: Do human neutrophils make hydroxyl radical? Determination of free radicals generated by human neutrophils activated with a soluble and particulate stimulus using electronic paramagnetic resonance spectrometry. *J Biol Chem* **261:**4426–4431, 1986.
109. Ambruso DR, Johnston RB Jr: Lactoferrin enhances hydroxyl radical production by human neutrophils, neutrophil particulate fractions and an enzymatic generating system. *J Clin Invest* **67:**352–360, 1981.
110. Broxmeyer HE, Smithyman A, Eger RR, et al: Identification of lactoferrin as the granulocyte-derived inhibitor of colony-stimulating activity production. *J Exp Med* **148:**1052–1067, 1978.
111. Borregaard N, Kragballe K: Role of oxygen in antibody-mediated cytotoxocity mediated by monocytes and neutrophils. *J Clin Invest* **66:**676–683, 1980.
112. Newburger PE, Tauber AI: Heterogenous pathways of oxidizing radical production in human neutrophils and the HL-60 cell line. *Pediatr Res* **16:**856–860, 1982.
113. Johnston RD Jr, Keele BB Jr, Misra HP: The role of superoxide anion generation in phagocytic bactericidal activity. Studies with normal and chronic granulomatous disease leukocytes. *J Clin Invest* **55:**1357–1372, 1975.
114. Klebanoff SJ, Durak DT, Rosen H, et al: Functional studies on human peritoneal eosinophils. *Infect Immun* **17:**167–173, 1977.
115. Beall GD, Repine JE, Hoidal JR: Chemiluminescence by human alveolar macrophages: Stimulation with heat killed bacteria and phorbol myristate acetate. *Infect Immun* **17:**117–120, 1977.
116. Johnston RB Jr, Lehmeyer JE, Guthrie LA: Generation of superoxide anion and chemiluminescence by human monocytes during phagocytosis and on contact with surfacc bound immunoglobulin G. *J Exp Med* **143:**1551–1556, 1976.
117. Cheson BD, Christensen RL, Sperling R, et al: The origin of chemiluminescence of phagocytosing granulocytes. *J Clin Invest* **58:**789–796, 1976.
118. Browne RJ, Ogryzlo EA: The yield of singlet oxygen in the reaction of chlorine with hydrogen peroxide. *Can J Chem* **43:**2915–2916, 1965.
119. Andersen RB, Lint TF, Brendzel AM: Chemically shifted singlet oxygen spectrum. *Biochim Biophys Acta* **542:**527–536, 1978.
120. Allen RC: Chemiluminescence and the study of phagocyte redox metabolism, in Rossi F, Patriarca P (eds): *Advances in Experimental Medicine and Biology* New York, Plenum, 1982, vol 141: *Biochemistry and Function of Phagocytes*, p 411.
121. Roos D, Hamers MN, VanZweiten R, et al: Acidification of the phagocytic vacuole: A possible defect in chronic granulomatous disease, in Gallin JI, Fauci AS (eds): *Advances in Host Defense Mechanisms*. New York, Raven, 1983, vol 3: *Chronic Granulomatous Disease*, p 145.
122. Segal AW, Geisow M, Garcia R, et al: The respiratory burst of phagocytosis is associated with a rise in vacuolar pH. *Nature (Lond)* **290:**406–409, 1982.
123. Cech P, Lehrer RI: Phagolysosomal pH of human neutrophils. *Blood* **63:**88–95, 1984.

124. Wright J, Schwartz JH, Olson R, et al: Proton secretion by the sodium-hydrogen ion antiporter in the human neutrophil. *J Clin Invest* **77:**782–788, 1986.
125. Grinstein S, Furuya W: Characterization of the amiloride-sensitive Na^+-K^+ antiport of human neutrophils. *Am J Physiol* **250:**c283–c291, 1986.
126. Simchowitz L: Intracellular pH modulates the generation of superoxide radicals in human neutrophils. *J Clin Invest* **76:**1079–1089, 1985.
127. Grinstein, S, Furuya W: Amiloride-sensitive Na^+/H^+ exchange in human neutrophils: Mechanism of activation by chemotactic factors. *Biochem Biophys Res Commun* **122:**755–762, 1984.
128. Aronson PS: Kinetic properties of the plasma membrane Na–H exchanger. *Annu Rev Physiol* **47:**545–560, 1985.
129. Weiss SJ: in Ward A (ed): *Handbook of Inflammation.* New York, North-Holland, 1983, vol 4, p 37.
130. Gabig TG, Lefker BA, Ossanna PJ, et al: Proton stoichiometry associated with human neutrophil respiratory burst reactions. *J Biol Chem* **259:**13166–13171, 1984.
131. Seligman BE, Gallin JI: Abnormality in elecited membrane potential changes in neutrophils from patients with chronic granulomatous disease, in Gallin JI, Fauci AS (eds): *Advances in Host Defense Mechanisms.* New York, Raven, 1983, vol 3: *Chronic Granulomatous Disease,* p 195.
132. Korchak HM, Weissmann G: Changes in membrane potential of human granulocytes antecede the metabolic response to surface stimulation. *Proc Natl Acad Sci USA* **75:**3818–3822, 1978.
133. Gallin EK, Wiederhold ML, Lipsky PE, et al: Spontaneous and induced membrane hyperpolarization in macrophages. *J Cell Physiol* **86:**653–662, 1975.
134. Kuori J, Noa M, Diaz B: Hyperpolarization of rat peritoneal macrophages phagocytosing latex particles. *Nature (Lond)* **283:**868–869, 1980.
135. Seligman BE, Gallin JI: Use of lipophilic probes of membrane potential to assess human neutrophil activation. *J Clin Invest* **66:**493–503, 1980.
136. Herlin T, Borregaard N: Early changes in cyclic AMP and calcium efflux during phagocytosis by neutrophils from normals and patients with chronic granulomatous disease. *Immunology* **48:**17–28, 1983.
137. Lew DP, Wollheim C, Seger RA, et al: Cytosolic free calcium changes induced by chemotactic peptide in neutrophils from patients with chronic granulomatous disease. *Blood* **63:**231–233, 1984.
138. Voetman AA, Weening RS, Hamers MN, et al: Phagocytosing human neutrophils inactivate their own granular enzymes. *J Clin Invest* **67:**1541–1549, 1981.
139. Schimcovitz L, Spilberg I, Deweer P: Sodium and potassium fluxes and membrane potential of human neutrophils. Evidence for an electrogenic sodium pump. *J Gen Physiol* **79:**453–479, 1982.
140. DellaBianca V, Bellavite P, DeTogni P, et al: Studies on stimulus–response coupling in human neutrophils. I. Role of monovalent cations in the respiratory burst and secretory response to N-formyl methionylleucylphenylalanine. *Biochim Biophys Acta* **755:**497–505, 1983.
141. Seeds MC, Parce JW, Szejda P, et al: Independent stimulation of membrane potential changes and the oxidative metabolic burst in polymorphonuclear leukocytes. *Blood* **65:**233–240, 1985.
142. Gallin JI, Seligman BE, Fletcher MP: Dynamics of human neutrophil receptors for the chemoattractant fmet-leu-phe. *Agents Actions* **12**(suppl)**:**290–308, 1983.
143. Bromberg Y, Pick E: Unsaturated fatty acids stimulate NADPH-dependent superoxide production by cell-free system derived from macrophages. *Cell Immunol* **88:**213–221, 1984.
144. Heynemann RA, Vercauteren RE: Activation of a NADPH oxidase from horse polymorphonuclear leukocytes in a cell free system. *J Reticuloendothel Soc* **36:**751–759, 1984.
145. McPhail LC, Shirley PM, Clayton CC, et al: Activation of the respiratory burst enzyme from human neutrophils in a cell free system. *J Clin Invest* **75:**1735–1739, 1985.
146. Curnutte JT: Activation of human neutrophil nicotineamide adenine dinucleotidephosphate reduced (triphosphopyridine nucleotide reduced) oxidase by arachidonic acid in a cell-free system. *J Clin Invest* **75:**1740–1743, 1985.
147. Bromberg Y, Pick E: Activation of NADPH-dependent superoxide production in a cell-free system by sodium dodecyl sulfate. *J Biol Chem* **260:**13539–13545, 1985.

148. Jandl RC, Andre-Schwartz J, Borges-DuBois L, et al: Termination of the respiratory burst in human neutrophils. *J Clin Invest* **61:**1176–1185, 1977.
149. Babior BM, Kipnes RS: Superoxide-forming enzyme from human neutrophils: Evidence for a flavin requirement. *Blood* **50:**517–524, 1977.
150. Light DP, Walsh C, O'Callagan AM, et al: Characteristics of the co-factor requirements for the superoxide generating NADPH oxidase of human polymorphonuclear leukocytes. *Biochemistry* **20:**1468–1476, 1981.
151. Gabig TG: The NADPH-dependent O_2^- generating oxidase from human neutrophils. Identification of a flavoprotein component that is deficient in a patient with chronic granulomatous disease. *J Biol Chem* **258:**6352–6356, 1983.
152. Segal AW, Jones OTG: Novel cytochrome b system in phagocytic vacuoles of human granulocytes. *Nature (Lond)* **276:**515–517, 1978.
153. Segal AW, Jones OTG: The subcellular distribution and some properties of the cytochrome b component of the microbiocidal oxidase system of human neutrophils. *Biochem J* **182:**181–188, 1979.
154. Cross AR, Jones OTG, Harper AM, et al: Oxidation-reduction properties of the cytochrome b found in the plasma membrane fraction of human neutrophils. A possible oxidase in the respiratory burst. *Biochem J* **194:**559–606, 1981.
155. Segal AW, Garcia R, Goldstone AH, et al: Cytochrome b_{-245} of neutrophils is also present in human monocytes macrophages and eosinophils. *Biochem J* **196:**363–367, 1981.
156. Hattori H: Studies on the labile, stabile NADI oxidase and peroxidase staining reactions in the isolated particles of horse granulocytes. *Nagoya J Med Sci* **23:**362–378, 1961.
157. Shinagawa Y, Tanaka C, Teraoka A, et al: A new cytochrome in neutrophil granules of rabbit leukocytes. *J Biochem* **59:**623–624, 1966.
158. Harper AM, Dunne MJ, Segal AW: Purification of cytochrome b_{-245} from human neutrophils. *Biochem J* **219:**519–529, 1984.
159. Pember SO, Heyl BL, Kinkade JM, et al: Cytochrome b_{558} from (bovine) granulocytes. Partial purification from Triton X-114 extracts and properties of the isolated cytochrome. *J Biol Chem* **159:**10590–10596, 1984.
160. Harper AM, Chaplin MF, Segal AW: Cytochrome b_{-245} from human neutrophils is a glycoprotein. *Biochem J* **227:**783–788, 1985.
161. Lutters R, vanSchaik MLJ, vanZweiten R, et al: Purification and partial characterization of the b-cytochrome from human polymorphonuclear leukocytes. *J Biol Chem* **260:** 2237–2244, 1985.
162. Bellavite P, Papini E, Zeni E, et al: Studies on the nature and activation of O_2^- forming NADPH oxidase of leukocytes. Identification of a phosphorylated component of the active enzyme. *Free Rad Res Commun* **1:**11–29, 1985.
163. Ilizuka T, Kanagashaki S, Makino R, et al: Studies on the neutrophil b-cytochrome in situ by low temperature absorption spectroscopy. *J Biol Chem* **260:**12049–12053, 1985.
164. Segal AW: Absence of both cytochrome b_{-245} subunits from neutrophils in X-linked chronic granulomatous disease. *Nature (Lond)* **326:**88–91, 1987.
165. Segal AW, Jones OTG: Reduction and subsequent oxidation of a cytochrome b of human neutrophils after stimulation with phorbol myristate acetate. *Biochem Biophys Res Commun* **88:**130–134, 1979.
166. Borregaard N, Simons ER, Clark RA: Involvement of cytochrome b_{-245} in the respiratory burst of human neutrophils. *Infect Immun* **38:**1301–1303, 1982.
167. Gabig TG, Schervish EW, Santinger JT: Functional relationship of the cytochrome b to the superoxide generating oxidase of human neutrophils. *J Biol Chem* **257:**4114–4119, 1982.
168. Cross AR, Higson F, Jones OTG, et al: The enzymic reduction and kinetics of oxidation of cytochrome b_{-245} of neutrophils. *Biochem J* **204:**479–485, 1982.
169. Cross AR, Parkinson JF, Jones OTG: Mechanism of the superoxide-producing oxidase of neutrophils by NADPH. *Biochem J* **226:**881–884, 1985.
170. Segal AW: Chronic granulomatous disease: A model for studying the role of cytochrome b_{-245} in

health and disease, in Gallin JI, Fauci AS (eds): *Advances in Host Defense Mechanisms*. New York, Raven, 1983, vol 3: *Chronic Granulomatous Disease*, p 121.
171. Segal AW, Jones OTG: Absence of cytochrome b reduction in stimulated neutrophils from both female and male patients with chronic granulomatous disease. *FEBS Lett* **110:**111–114, 1980.
172. Seger RA, Tiefenauer L, Matsunaga T, et al: Chronic granulomatous disease due to granulocytes with abnormal NADPH oxidase activity and deficient cytochrome b. *Blood* **61:**423–428, 1983.
173. Borregaard N, Cross AR, Herlin T, et al: A variant form of X-linked chronic granulomatous disease with normal nitroblue tetrazoleum slide test and cytochrome b. *Eur J Clin Invest* **13:**243–247, 1983.
174. Segal AW, Cross AR, Garcia RC, et al: Absence of cytochrome b_{-245} in chronic granulomatous disease. A multicenter European evaluation of its incidence and relevance. *N Engl J Med* **308:**245–251, 1983.
175. Nakamura M, Murakami M, Koga T, et al: Monoclonal antibody 7D5 raised to cytochrome b_{558} of human neutrophils: Immunocytochemical detection of the antigen in peripheral phagocytes of normal subjects, patients with chronic granulomatous disease, and their carrier mothers. *Blood* **69:**1404–1408, 1987.
176. Cross AR, Jones OTG, Garcia R, et al: The association of FAD with the cytochrome b_{-245} of human neutrophils. *Biochem J* **208:**759–763, 1982.
177. Borregaard N, Tauber AI: Subcellular localization of the human neutrophil NADPH oxidase: b-cytochrome and associated flavoprotein. *J Biol Chem* **259:**47–52, 1984.
178. Weening R, Corbeel L, deBoer M, et al: Cytochrome b deficiency in an autosomal form of chronic granulomatous disease. A third form of chronic granulomatous disease recognized by monocyte hybridization. *J Clin Invest* **75:**915–920, 1985.
179. Ohno Y, Buescher ES, Roberts R, et al: Reevaluation of cytochrome b and flavin adenine dinucleotide in neutrophils from patients with chronic granulomatous disease, and description of a family with probable autosomal recessive inheritance of cytochrome b deficiency. *Blood* **67:**1132–1138, 1986.
180. Boehler M-C, Seger RA, Mouy R, et al: A study of 25 patients with chronic granulomatous disease: A new classification by correlating respiratory burst, cytochrome b and flavoprotein. *J Clin Immunol* **6:**136–145, 1986.
181. Royer-Pokora B, Kunkel LM, Monaco AP, et al: Cloning the gene for an inherited human disorder—chronic granulomatous disease—on the basis of its chromosomal location. *Nature (Lond)* **322:**32–38, 1986.
182. Borregaard N, Johansen KS, Taudorf E, et al: Cytochrome b is present in neutrophils from patients with chronic granulomatous disease. *Lancet* **1:**949–951, 1979.
183. Segal AW, Heyworth PG, Cockroft S, et al: Stimulated neutrophils from patients with autosomal recessive chronic granulomatous disease fail to phosphorylate a M_r-44000 protein. *Nature (Lond)* **316:**547–549, 1985.
184. Hamers MN, deBoer M, Meerhof LJ: Complementation in monocyte hybrids revealing genetic heterogeniety in chronic granulomatous disease. *Nature (Lond)* **307:**553–555, 1984.
185. Bellavite P, Cross AR, Serra MC, et al: The cytochrome b and flavin content and properties of the O_2^- forming NADPH oxidase solubilized from activated neutrophils. *Biochim Biophys Acta* **746:**40–47, 1983.
186. Bellavite P, Cassatella MA, Papini E: Presence of cytochrome b_{-245} in NADPH oxidase preparations from human neutrophils. *FEBS Lett* **199:**159–163, 1986.
187. Millard JA, Gerard KW, Schneider DL: The isolation from rat peritoneal leukocytes of plasma membrane enriched in alkaline phosphatase and a b-type cytochrome. *Biochem Biophys Res Commun* **90:**312–319, 1979.
188. Sloan EP, Crawford DR, Schneider DL: isolation of plasma membrane from human neutrophils and determination of cytochrome b and ubiquione content. *J Exp Med* **153:**1316–1328, 1981.
189. Mollinedo F, Schneider DL: Subcellular localization of cytochrome b and ubiquione in a tertiary

granule of resting human neutrophils and evidence for a proton pump ATPase. *J Biol Chem* **259:**7143–7150, 1984.
190. Crawford DR, Schneider DL: Ubiquione content and respiratory burst activity of latex filled phagolysosomes isolated from human neutrophils and evidence for the possible involvement of a third granule. *J Biol Chem* **258:**5363–5367, 1983.
191. Crawford DR, Schneider DL: Evidence that a quinone may be required for the production of superoxide and hydrogen peroxide in neutrophils. *Biochem Biophys Res Commun* **99:**1277–1286, 1981.
192. Crawford DR, Schneider DL: Identification of ubiquinone-50 in human neutrophils and its role in microbicidal events. *J Biol Chem* **257:**6662–6668, 1982.
193. Lutter R, vanZweiten RS, Hamers MN, et al: Cytochrome b, flavins and ubiquinone in enucleated human neutrophils (polymorphonuclear leukocytes cytoplasts). *J Biol Chem* **759:**9603–9606, 1984.
194. Cross AR, Jones OTG, Carcia RC, et al: The subcellular localization of ubiquinone in human neutrophils. *Biochem J* **216:**765–768, 1983.
195. Cunningham CC, DeChatelet LR, Spach PI, et al: Identification and quantitation of electron-transport components in human polymorphonuclear neutrophils. *Biochim Biophys Acta* **682:**430–435, 1982.
196. Lee JA: Histochemical identification of ubiquinone in neutrophil polymorphonuclear granules. *Exp Cell Biol* **54:**89–93, 1986.
197. Gabig TG, Lefker BA: Activation of the human neutrophil NADPH oxidase results in coupling of electron carrier function between ubiquinone-10 and cytochrome b_{559}. *J Biol Chem* **260:**3991–3995, 1985.
198. Bougonoux P, Bonvini E, Stevenson HC, et al: Identification of ubiquinone-50 as the major methylated nonpolar lipid in human monocytes. Regulation of its biosynthesis via methionine-dependent pathways and relationship to superoxide production. *J Biol Chem* **258:**4339–4344, 1983.
199. Ohno Y-I, Hirai K-I, Kanoh T, et al: Subcellular localization of H_2O_2 production in human neutrophils stimulated with particles and an effect of cytochalasin B on the cells. *Blood* **60:**253–260, 1982.
200. Ohno Y-I, Kirai K-I, Kanoh T, et al: Subcellular localization of hydrogen peroxide production in human polymorphonuclear leukocytes stimulated with lectins, phorbol myristate acetate, and digitonin: An electron microscopic study using $CeCl_3$. *Blood* **60:**1195–1202, 1982.
201. Dewald B, Baggiolini M, Curnutte JT, et al: Subcellular localization of the superoxide-forming enzyme in human neutrophils. *J Clin Invest* **63:**21–29, 1979.
202. Bellavite P, Serra MC, Davoli A, et al: Selective enrichment of NADPH oxidase activity in phagosomes from guinea pig polymorphonuclear leukocytes. *Inflammation* **6:**21–29, 1982.
203. Tsunawaki S, Kaneda M, Kakinuma K: Activation of guinea pig polymorphonuclear leukocytes with soluble stimulators leads to nonrandom distribution of NADPH oxidase in the plasma membrane. *J Biochem* **4:**655–664, 1983.
204. Borregaard N, Heiple JM, Simons ER, et al: Subcellular localization of the b-cytochrome component of the human neutrophil microbicidal oxidase: Translocation during activation. *J Cell Biol* **97:**52–61, 1983.
205. Ohno Y, Seligman B, Gallin JI: Cytochrome b translocation to human neutrophil plasma membranes and superoxide release. Differential effects of N-formylmethionylleucylphenylalanine, phorbol myristate acetate and A 23187. *J Biol Chem* **260:**2409–2414, 1985.
206. Parkost CA, Cochrane CG, Schmitt M, et al: Regulation of the oxidative response of human granulocytes to chemoattractants. No evidence for stimulated traffic of redox enzymes between endo and plasma membranes. *J Biol Chem* **260:**6541–6547, 1985.
207. Higson FK, Durbin L, Paulotsky N, et al: Studies on cytochrome b_{-245} translocation in the PMA stimulated of the human neutrophil NADPH oxidase. *J Immunol* **135:**519–524, 1985.
208. Borregaard N, Bjerrum O: Subcellular localization of cytochrom b_{-245}: The specific granules are the intracellular store. *Eur J Clin Invest* **17:**A70, 1987 (abst).

209. Roos D, Voetman AA, Meerhof LJ: Functional activity of enucleated human polymorphonuclear leukocytes. *J Cell Biol* **97:**368–377, 1983.
210. Korchak HM, Roos D, Giedd KN, et al: Granulocytes without degranulation: Neutrophil function in granulocyte-depleted cytoplasts. *Proc Natl Acad Sci USA* **80:**4968–4972, 1983.
211. Petrequin P, Todd RF III, Smolen JE, et al: Expression of specific granule markers on the cell surface of neutrophil cytoplasts. *Blood* **67:**1119–1125, 1986.
212. Garcia RC, Segal AW: Changes in the subcellular distribution of the cytochrome b_{-245} on stimulation of human neutrophils. *Biochem J* **219:**233–242, 1984.
213. Yamaguchi T, Kaneda M, Kakinuma K: Is cytochrome b_{558} translocated to the plasma membrane of granulocytes during the activation of neutrophils? *J Biochem* **99:**953–959, 1986.
214. Borregaard N, Miller LJ, Springer TA: Chemoattractant regulated mobilization of a novel intracellular compartment in human neutrophils. *Science* **237:** 1204–1206, 1987.
215. Bjerrum O, Borregaard N: Dual granule localization of the dormant NADPH oxidase and cytochrome b_{559} in human neutrophils: The structural basis for a graduated activation of the microbicidal respiratory burst, *J Immunol*, in press.
216. Miller LJ, Bainton DF, Borregaard N, et al: Stimulated mobilization of monocyte Mac-1 and p150,95 adhesion proteins from a latent intracellular site to the cell surface. *J Clin Invest* **80:** 535–544, 1987.
217. Wolfson M, McPhail LC, Nasrallah VN, et al: Phorbol myristate acetate mediates redistribution of protein kinase C in human neutrophils: Potential role in the activation of the respiratory burst enzyme. *J Immunol* **135:**2057–2162, 1985.
218. Pike MC, Jakoi L, McPhail LC, et al: Chemoattractant-mediated stimulation of the respiratory burst in human polymorphonuclear leukocytes may require appearance of protein kinase activity in the cells' particulate fraction. *Blood* **67:**909–913, 1986.
219. Pontremoli S, Melloni E, Michetti M, et al: Differential mechanisms of translocation of protein kinase C to plasma membranes in activated human neutrophils. *Biochem Biophys Res Commun* **136:**228–234, 1986.
220. Castagni M, Takai Y, Kabuchi K, et al: Direct activation of calcium-activated phospholipid-dependent protein kinase by tumor promoting phorbol esters. *J Biol Chem* **257:**7847–7851, 1981.
221. Leach KL, James ML, Blumberg PM: Characterization of a specific phorbol ester aporeceptor in mouse brain cytosol. *Proc Natl Acad Sci USA* **80:**4208–4212, 1983.
222. Cox JA, Jeng AY, Sharkey NA, et al: Activation of the human neutrophil nicotinamide dinucleotid phosphate (NADPH) oxidase by protein kinase C. *J Clin Invest* **76:**1932–1938, 1985.
223. Wilson E, Olcott MC, Bell RM, et al: Inhibition of the oxidative burst in human neutrophils by sphingoid long-chain bases. Role of protein kinase C in activation of the burst. *J Biol Chem* **261:**12616–12623, 1986.
224. Gennaro R, Florio C, Romeo D: Activation of protein kinase C in neutrophil cytoplasts. Localization of protein substrates and possible relationship with stimulus–response coupling. *FEBS Lett* **180:**185–190, 1985.
225. Papini E, Grezeskowiak M, Bellavite P, et al: Protein kinase C phosphorylates a component of the NADPH oxidase of neutrophils. *FEBS Lett* **190:**204–208, 1985.
226. Craig G, McPhail LC, Marfat A, et al: Role of protein kinase C in stimulation of human polymorphonuclear leukocyte oxidative metabolism by various agonists. Differential effects of a novel protein kinase inhibitor. *J Clin Invest* **77:**61–65, 1986.
227. Wright CD, Hoffman MD: The protein kinase C inhitors H-7 and H-9 fail to inhibit human neutrophil activation. *Biochem Biophys Res Commun* **135:**749–755, 1986.
228. Balazovich KJ, Smolen JE, Boxer LA: Ca^{2+} and phospholipid-dependent protein kinase (protein kinase C) activity is not necessarily required for secretion by human neutrophils. *Blood* **68:**810–817, 1986.
229. Badway JA, Curnutte JT, Robinson JM: Effect of free fatty acids on release of superoxide and on changes of shape by human neutrophils reversibility by albumin. *J Biol Chem* **259:**7870–7877, 1984.

230. McPhail LC, Clayton CC, Snyderman R: A potential second messenger role for unsaturated fatty acids: Activation of a Ca^{2+}-dependent protein kinase. *Science* **224:**622–625, 1984.
231. Kitagawa S, Takaku F: Role of serine proteases in superoxide production by human neutrophils, monocytes and basophils, in Rossi F, Patriarca P (eds): *Advances in Experimental Medicine and Biology.* New York, Plenum, 1982, vol 141: *Biochemistry and Function of Phagocytes,* p 441.
232. Huston RB, Krebs EG: Activation of skeletal muscle phosphorylase kinase by Ca^{2+}. II. Identification of the kinase activating factor as a proteolytic enzyme. *Biochemistry* **7:**2116–2122, 1968.
233. Melloni E, Pontremoli S, Michetti M, et al: The involvement of calpain in the activation of protein kinase C in neutrophils stimulated by phorbol myristate acetate. *J Biol Chem* **261:**4101–4105, 1986.
234. Pontremoli S, Melloni E, Michetti M, et al: Biochemical response in activated human neutrophils mediated by protein kinase C and a Ca^{2+}-requiring proteinase. *J Biol Chem* **261:**8309–8318, 1986.
235. Melloni E, Pentremoli S, Salamino F, et al: ATP induces the release of neutral serine proteinase and enhances the production of superoxide anion in membranes from phorbol ester-activated neutrophils. *J Biol Chem* **261:**11437–11439, 1986.
236. Badwey JA, Curnutte JT, Berder CB: Cytochalasin E diminishes the lag phase in the release of superoxide by human neutrophils. *Biochem Biophys Res Commun* **106:**170–174, 1982.
237. Tsan M-F: Phorbol myristate acetate induced neutrophil autotoxicity. *J Cell Physiol* **105:**327–334, 1980.
238. McPhail LC, Snyderman R: Mechanisms of regulating the respiratory burst. *Contemp Top Immunobiol* **14:**247–281, 1984.
239. Lew PD, Southwick FS, Stossel TP, et al: A variant of chronic granulomatous disease. Deficient oxidative metabolism due to a low-affinity NADPH oxidase. *N Engl J Med* **305:**1329–1333, 1981.
240. Gallin JI: Human neutrophil heterogeneity exists, but is it meaningful? *Blood* **63:**977–983, 1984.
241. Bass DA, Parce JW, DeChatelet LA: Flow cytometric studies of oxidative product formation by neutrophils: A graded response to membrane stimulation. *J Immunol* **130:**1910–1917, 1983.
242. Zakhirey B, Root RK: Development of oxidase activity by human bone marrow granulocytes. *Blood* **54:**429–439, 1979.
243. Roperts PJ, Cross AR, Jones OTG, et al: Development of cytochrome b and an active oxidase system in association with maturation of a human promyelocytic (HL-60) cell line. *J Cell Biol* **95:**720–726, 1982.
244. Newburger PE, Speir C, Barry N, et al: Development of the superoxide generating system during differentiation of HL-60 cell human promyelocytic cell line. *J Biol Chem* **259:**3771–3776, 1984.
245. Segel EK, Ellegaard J, Borregaard N: Development of the phagocytic and cidal capacity during maturation of myeloid cells. Studies on cells from patients with chronic myelogenous leukemia. *Br J Haematol* **67:**3–10, 1987.
246. Berkow RL, Dodsen RW: Purification and functional evaluation of mature neutrophils from human bone marrow. *Blood* **68:**853–860, 1986.
247. Fikrig SM, Smitnick EM, Suniharalingham K, et al: Fibroblast nitroblue tetrazoleum test and in-utero diagnosis of chronic granulomatous disease. *Lancet* **1:**818–819, 1980.
248. Roder JC, Helfand SL, Vermeister J, et al: Oxygen intermediates are triggered early in the cytolytic pathway of human NK-cells. *Nature (Lond)* **298:**569–572, 1982.
249. Seger R, Steinmann B: Prenatal diagnosis of chronic granulomatous disease: Unreliability of fibroblast nitroblue tetrazoleum test. *Lancet* **1:**1260, 1981.
250. Nathan CF, Mercer-Smith JA, Desantis NM, et al: Role of oxygen in T-cell mediated cytolysis. *J Immunol* **129:**2164–2171, 1982.
251. Kay HD, Smith DL, Sullivan G, et al: Evidence for a nonoxidative mechanism of human natural killer (NK) cell cytotoxicity by using mononuclear effector cells from healthy donors and from patients with chronic granulomatous disease. *J Immunol* **131:**1784–1788, 1983.
252. El-Hag A, Clark RA: Intact natural killer activity in chronic granulomatous disease: Evidence against an oxygen-dependent cytotoxic mechanism. *J Immunol* **132:**569–570, 1984.
253. Volkman DJ, Buescher ES, Gallin JI, et al: B cell lines as models for inherited phagocyte diseases:

abnormal superoxide generation in chronic granulomatous disease and giant granules in Chediac–Higashi syndrome. *J Immunol* **133:**3006–3009, 1984.

254. Foerder CA, Klebanoff SJ, Shapiro BM: Hydrogen peroxide production, chemiluminescence and the respiratory burst of fertilization interrelated events in early sea urchine development. *Proc Natl Acad Sci USA* **75:**3183–3187, 1978.
255. Klebanoff SJ, Foerder CA, Eddy EM, et al: Metabolic similarities between fertilization and phagocytosis. Conservation of a peroxidatic mechanism. *J Exp Med* **149:**938–953, 1979.

2

The Respiratory Burst of Phagocytes

GIORGIO BERTON, STEFANO DUSI, and PAOLO BELLAVITE

Studies on the oxygen metabolism of phagocytes began in 1933 with the observation that during phagoytosis, white blood cells (WBC) undergo a striking increase in oxygen consumption.[1,2] These observations did not attract much attention until the 1950s, when the metabolic activities of neutrophils during phagocytosis of foreign matter were reinvestigated.[3–7] The key initial observations that led to further research in the field were that (1) stimulated oxygen consumption is accompanied by enhanced catabolism of glucose through the hexose monophosphate shunt,[4,6] (2) inhibitors of mitochondrial oxidative respiration do not block oxygen consumption,[5,6] and (3) hydrogen peroxide is produced in parallel.[8,9]

During the past 30 years, the oxygen metabolism of phagocytes has been investigated in detail. Table I presents a provisional outline of the main observations of particular relevance in directing further research. Several important aspects are still unknown. However, some relevant conclusions have been reached on the significance of this peculiar cell function in the whole organism. These can be summarized as follows: (1) oxygen metabolism of the phagocyte represents an important mechanism of natural defense against invasion of microorganisms; (2) eukaryotic cells, including tumor cells, are also killed by products derived from the metabolism of oxygen; (3) abnormal or unrestrained stimulation of this function in phagocytes is responsible for the tissue damage associated with some important human diseases; and (4) in vivo modulation of phagocytes oxygen metabolism during the development of immunity mediated by cells of the T lineage is important in the acquired resistance against intracellular pathogens and tumor cell growth.

This chapter reviews some of main aspects of the respiratory burst of phagocytes at the cellular and the biochemical level. The interested reader can obtain a comprehensive view of this function in other detailed reviews that have appeared during the past few years.[38–43]

GIORGIO BERTON, STEFANO DUSI, and PAOLO BELLAVITE • Institute of General Pathology, University of Verona, 37134 Verona, Italy.

TABLE I. Milestones in the Understanding of the Respiratory Burst of Phagocytes

Year(s)	Milestone	Reference
1933	Discovery that leukocytes enhance their oxygen consumption during phagocytosis	(1,2)
1956–1961	Start of the reinvestigation of the metabolism of oxygen in leukocytes and its definition as insensitive to mitochondrial inhibitors	(3,5–8)
1957–1959	Description of the activation of the hexose monophosphate shunt during phagocytosis	(4,6)
1961–1968	Identification of H_2O_2 as a product of the respiratory burst	(8,9)
1967–1970	Description of the H_2O_2-myeloperoxidase–halide system as a powerful microbicidal system	(10–12)
1963–1964	Characterization of NADH and NADPH oxidases as the enzymatic basis of the respiratory burst	(13–16)
1973	Demonstration that superoxide anion is produced by stimulated neutrophils	(17)
1959–1975	Description of the chronic granulomatous disease as a defect of the respiratory burst and subsequently of the NADPH oxidase activity	(18–21)
1979	Demonstration that the NADPH oxidase is located in the plasma membrane	(22)
1971–1980	Demonstration that macrophages can be activated to produce higher amounts of toxic oxygen products	(23–27)
	This is correlated with microbicidal and cytotoxic activities	(28–29)
1966–1978	Discovery and subsequent reappraisal of cytochrome *b* as a unique phagocyte chromophore	(30–31)
	Discovery of its absence in CGD patients	(32,33)
1979	First attempts to solubilize and purify the NADPH oxidase	(34,35)
1984	Activation of the NADPH oxidase in a cell-free system by arachidonic acid	(36,37)

1. FEATURES OF THE RESPIRATORY BURST OF PHAGOCYTES

The pivotal studies on the events following phagocytosis cited in Table I led to the concept of respiratory burst to define the stimulation of the oxygen metabolism of phagocytes. This term is now used to describe a series of interconnected events, which include an increase in a nonmitochondrial oxygen consumption; the production of highly reactive oxygen species such as superoxide anion (O_2^-), hydrogen peroxide (H_2O_2) and other oxygen radicals; and the oxidation of glucose through the hexose monophosphate shunt.

This form of respiratory burst can be evoked in different types of phagocytic cells, i.e., in polymorphonuclear neutrophils and eosinophils, in circulating monocytes, and in tissue macrophages. Attempts to demonstrate the existance of a respiratory burst in other immune or inflammatory cells, e.g., cytotoxic T lymphocytes, natural killer (NK) cells, and platelets, produced controversial results.[44–46]

The respiratory burst is triggered by a wide and heterogeneous group of molecules capable of interacting with the phagocyte surface. Early studies showed that not only is phagocytosis able to stimulate oxygen consumption but that soluble agents have this capability as well.[47–50] The respiratory burst can be triggered by (1) stimuli that interact with specific surface receptors, e.g., chemotactic peptides, immunoglobulin G (IgG), complement factors, leukotriene; (2) aggregation of surface molecules with antibodies and lectins; (3) degradation of plasma membrane phospholipids with exogenous phospholipase C; (4) unsaturated fatty acids and detergents; (5) elevation of cytosolic free calcium by ionophores; (6) stimulation of protein kinase C by phorbol diesters and diacylglycerol analogues; and (7) dissociation of G-protein subunits by fluoride. This simple list of stimuli and their different ways of interacting with the phagocyte gives an idea of the complexity of the mechanism(s) of activation of the respiratory burst. This point is addressed in some detail later on.

The respiratory burst is an impressive phenomenon with regard to the rapidity of its occurrence and intensity. Activation of the oxygen metabolism can be observed after a few seconds from the addition of an appropriate stimulus. In 1 min, one million stimulated cells can consume up to 5–10 nmoles of oxygen, something like 50–100 times more than resting cells. The intensity of the respiratory burst varies, depending on the cell type, the cell sources, and the animal species. Tissue macrophages are highly heterogeneous in their response and can be modulated by bacterial and cell-derived factors to metabolize oxygen in response to appropriate stimuli with a severalfold higher intensity.[51–53] The time course of the respiratory burst varies according to the stimulant used. For example, phagocytosable particles induce a progressive increase of respiration that accompanies the engulfing act and ceases when phagocytosis is completed,[15] chemotactic peptides trigger an almost instantaneous response that lasts only a few minutes, and phorbol diesters cause progressive irreversible activation that lasts several minutes in neutrophils and even 1 hr in macrophages. Stimulation of the respiratory burst can be reversed by removing the stimulus after washing the cells[54] or by adding an excess of a competitor for the same receptor.[55] Also, oxidative inactivation of the stimulus has been described.[56] Once deactivated, the respiratory burst can be reactivated by a second stimulus,[56,57] but desensitization of the system may occur. Both selective, i.e., due to receptor downregulation or uncoupling[58–60] and nonselective desensitization have been reported.[59–62]

TABLE II. Methods of Assaying the Respiratory Burst of Phagocytes

Method	Description
Oxygen consumption	Clark-type oxygen electrode
Hydrogen peroxide production	Fluorimetric measurement of peroxidase-catalyzed oxidation of scopoletin or of homovanillic acid; photometric detection of peroxidase spectrum; electron microscopic detection of cerium deposits
Superoxide anion production	Spectrophotometric measurement of cytochrome *c* reduction; microscopic examination of formazan formation in cells incubated in the presence of nitroblue of tetrazolium (NBT test)
Light production	Chemiluminescence detection with luminometer
Hexose monophosphate pathway activity	Production of $^{14}CO_2$ from [1-^{14}C]glucose

The set of phenomena included in the concept of respiratory burst can be studied by different well-established techniques based essentially on the measure of the total oxygen consumed, of the products of the oxygen reduction, and of the amount of glucose oxidized through the hexose monophosphate shunt. Table II lists the most frequently used methods in the study of the respiratory burst of phagocytes.

2. ENZYMATIC BASIS OF THE RESPIRATORY BURST

2.1. Major Properties of NADPH Oxidase, the O_2^--Generating System of Phagocytes

Studies on the enzymatic basis of the respiratory burst began during the 1960s in the laboratories of Karnovsky, Quastel, and Rossi. The matter soon attracted the attention of several groups of investigators and still remains the most challenging problem of the cell biology of phagocytes. It is now generally agreed that, as clearly proposed by Rossi and Zatti[15] initially, the enzymatic activity that accounts for the respiratory burst is a NADPH oxidase. This oxidase is apparently the only system present in eukaryotic cells able to perform exclusively a monovalent reduction of oxygen according to the reaction

$$\text{NADPH} + 2\text{O}_2 + \text{H}^+ \rightarrow \text{NADP}^+ + 2\text{O}_2^- + 2\text{H}^+$$

The O_2^- formed in this reaction can then give rise to other products known as reactive intermediates of oxygen reduction or toxic oxygen products such as hydrogen peroxide and hydroxyl radical. Hydrogen peroxide can derive from the spontaneous or the superoxide dismutase-catalyzed dismutation of O_2^-:

$$2\text{O}_2^- + 2\text{H}^+ \rightarrow \text{H}_2\text{O}_2 + \text{O}_2$$

The formation of hydroxyl radical can derive from a secondary Haber–Weiss reaction, in which O_2^- reduces H_2O_2, probably with the obligatory catalysis of Fe^{2+} chelates:

$$\text{O}_2^- + \text{H}_2\text{O}_2 \rightarrow \text{OH}\cdot + \text{OH}^- + \text{O}_2$$

Other nonenzymatic chain reactions have been reported to occur at acidic pH and high NADPH concentrations and with the catalysis of Mn^{2+} ions[63–65]:

$$\text{NADPH} + \text{O}_2^- + \text{H}^+ \rightarrow \text{NADP}\cdot + \text{H}_2\text{O}_2$$

$$\text{NADP}\cdot + \text{O}_2 \rightarrow \text{NADP}^+ + \text{O}_2^- + \text{H}^+$$

Among the set of enzymatic reactions set in motion by the NADPH oxidase, it is logical to include those linked to the hexose monophosphate shunt as well. Oxidation of NADPH to $NADP^+$ decreases by threefold the ratio between the reduced and the oxidized form of pyridine nucleotides in phagocytosing neutrophils,[66] thereby activating the glucose 6-phosphate dehydrogenase (G6PD) to catalyze the reaction

$$NADP^+ + \text{glucose 6-phosphate} \rightarrow NADPH + \text{ribulose 5-phosphate} + H^+ + \tfrac{1}{2}CO_2$$

Other pathways of activation of the hexose monophosphate shunt are those linked to the degradation (through the glutathione cycle) of the H_2O_2 formed that enters the cytoplasmic space:

$$H_2O_2 + 2GSH \rightarrow 2H_2O + GSSG$$

$$GSSG + NADPH + H^+ \rightarrow 2GSH + NADP^+$$

The main properties of the NADPH oxidase as an enzymatic system can be summarized as follows:

1. It is undetectable in resting cells and becomes active after exposure of the cells to appropriate stimuli. After stimulation for a few minutes at 37°C, cooling of the cells, and washing off of the stimulus, the activity can be easily assayed in disrupted cells. Subcellular fractionation studies have demonstrated that the activity copurifies with the plasma membrane and is apparently enriched in the phagosomal membrane.[22,67]
2. The optimum pH for the activity is 7.0–7.5.
3. Apparent K_m values are 0.015–0.08 for NADPH and 0.01–0.03 mM for O_2.
4. The activity is optimal in the presence of FAD, phospholipids, and low (0.02–0.05%) concentrations of detergents, if assayed in membrane preparations.
5. The activity is insensitive to inhibitors of oxidative phosphorylation (CN^-, antimycin A, rotenone) and inhibited by sulfhydryl-reactive agents (*p*-chloromercuribenzoate, mersalyl, *N*-ethyl maleimide), FAD analogues (5-carba-5-deaza FAD), pyridine nucleotide analogues (cibacron blue), metal chelating agents (bathophenanthroline sulfonate), and relatively high concentrations of detergents.
6. Appropriate concentrations of arachidonic acid[36] and sodium dodecyl sulfate (SDS)[68] can activate the superoxide-forming system in disrupted cells or in a reconstituted mixture composed of membranes or specific granules[69] plus cytosol.

2.2. Activity of the NADPH Oxidase in Relationship to the Respiratory Burst of Intact Phagocytes

The problem of the adequacy of the NADPH oxidase to account for the respiratory burst of intact phagocytes has been addressed by different investigators. Evidence has been provided that the activity of the oxidase in membrane fractions is directly proportional to the respiration of intact cells. We have indeed shown that both the NADPH-dependent O_2^--generating system and the NADPH-generating system, i.e., NADPH produced by the activity of the hexose monophosphate shunt, are competent to support the respiratory burst of intact phagocytes.[70]

A relevant aspect of the activity of the NADPH oxidase in relationship to the respiratory burst of the intact cell has recently emerged from studies on the molecular basis of the phenomenon of macrophage activation. Bacterial or cell-derived factors are well known to upmodulate the respiratory burst of tissue macrophages in response to appropriate stimuli; this phenomenon is directly linked to the acquirance of resistance against intracellular pathogens.[51–53] In macrophages activated by in vivo injection of bacteria, bacterial derivatives or other agents the NADPH oxidase was shown to display alteration in its kinetic properties with a striking 5- to 10-fold decrease in its K_m.[71–73] Significantly, a similar alteration was observed with human macrophages activated in vitro with γ-interferon (IFN_γ),[74] now considered the major macrophage activating factor derived from T lymphocytes. Furthermore, the inhibition of macrophage respiratory burst by factor(s) released by tumor cells is accompanied by an enhancement of the K_m of the NADPH oxidase.[75] These results substantiate the notion that the intrinsic capability of phagocytes to metabolize oxygen in response to environmental stimuli can be modulated by modifications of the affinity of the NADPH oxidase for its substrate.

3. NATURE OF NADPH OXIDASE

3.1. An O_2^--Forming Electron-Transfer Chain

Although the properties of the NADPH oxidase as an enzymatic system capable of reducing the oxygen molecule monovalently are relatively well defined, its molecular structure is far from being clarified.

The bulk of evidence accumulated so far indicates that the O_2^--generating system of phagocytes is formed by various components arranged as an electron-transport chain. The available data favor the following model of transport of reducing equivalents from NADPH to oxygen:

$$\text{NADPH} \rightarrow \text{flavoprotein} \rightarrow \text{cytochrome } b_{-245} \rightarrow O_2$$

The participation of a flavoprotein as an NADPH-oxidizing component is suggested by various elements:

1. A flavoprotein is required for theoretical reasons because, whatever the electron acceptor (i.e., a cytochrome *b* or the oxygen itself), an intermediate electron carrier capable of transferring electrons from a two-electron donor (e.g., NADPH) to a one-electron acceptor would be obligatory.
2. A flavin cofactor in the form of FAD (but not other cofactors, e.g., FMN, riboflavin, ADP, AMP) was shown to be required to optimize NADPH oxidase activity of preparations solubilized with detergents.[76]

3. FAD analogues, e.g., 5-carba-5-deaza-FAD, inhibit the NADPH oxidase activity at micromolar concentrations.[77]
4. The addition of NADPH to preparations of solubilized oxidase in anaerobic conditions causes a reduction in FAD that is selective[78] or accompanied by cytochrome *b* reduction,[79] whether flavoproteins and cytochrome *b* are resolved by purification procedures or not, respectively.
5. Electron-spin resonance (ESR) studies showed a NADPH-dependent formation of flavin free radicals with characteristics of neutral semiquinone (FADH·) in membrane preparations obtained from stimulated neutrophils.[80]

Studies during the past 10 years have found a *b*-like cytochrome to be the most substantial component of the O_2^--forming system of phagocytes. This hemoprotein was discovered during the 1960s and later defined as the major chromophore of neutrophils and other types of phagocytes. It was shown to have a peculiar peak of absorbance in the reduced state at 558 nm and a low midpoint potential of −245 mV.[81] Several lines of evidence suggest that cytochrome b_{558} is a primary component of the O_2^--forming system:

1. Cytochrome b_{558} is the major chromophore of phagocytes, including macrophages that contain mitochondrial and endoplasmic reticulum *b* cytochromes[92]; and its low midpoint potential is suitable for its role as an electron donor to oxygen, since the midpoint potential of the couple O_2/O_2^- is −330 mV.
2. During myelocyte differentiation, cytochrome *b* is expressed in parallel with the acquisition of the ability to undergo a respiratory burst.[83]
3. When neutrophils[84] or macrophages[82] are stimulated in an anaerobic environment, the cytochrome *b* is reduced and can then be rapidly reoxidized upon readdition of oxygen.
4. Cytochrome *b* behaves as a typical component of a terminal oxidase, as demonstrated by its ability to bind carbon monoxide[85]; it must be pointed out, however, that this finding was not confirmed by others.[86]
5. In anerobiosis, the addition of NADPH to membrane preparations or solubilized NADPH oxidase obtained from stimulated neutrophils is accompanied by a reduction in cytochrome *b*.[79]

Any hypothesis on the nature of the O_2^--forming activity of phagocytes must be compatible with the observations made on the cells of patients affected by chronic granulomatous disease (CGD). This genetic disease includes several variants transmitted either with an X-linked or autosomal-recessive mechanism. The common trait of all the described CGD syndromes is that in the homozygous form, the phagocytes are unable to undergo a respiratory burst in response to appropriate stimuli.

Phagocytes of CGD patients have been studied extensively to challenge the model of electron transfer from NADPH to oxygen described above. Several observations would indeed substantiate that model:

1. Phagocytes of patients affected by the X-linked form of CGD (X-CGD) are cytochrome *b* negative[32,33]; in the autosomal-recessive variant of the disease, stimulation of the cell in anaerobiosis is not accompanied by reduction of cytochrome *b*.[87]
2. A deficiency of flavoproteins or of FAD has been reported to occur in some CGD variants.[88,89]
3. Hybridization of monocytes of cytochrome b_{558}-negative (e.g., b_{-558}) CGD patients with monocytes of patients with a variant cytochrome b_{558}-positive form (e.g., b_{558} reconstitutes the O_2^--forming activity, suggesting that the assembly of two or more components are required for this activity.[90]

Experimental observations supporting the model in which both a flavoprotein and the cytochrome b_{558} are essential components of the O_2^--generating system are contradictory about the ratio between the two components. Either ratios of one flavoprotein for one cytochrome *b*[88] or one flavoprotein for 5–20 cytochrome *b*[91,92] have been reported (see Rossi et al.[93] for further discussion).

Although supported by most investigators in the field, there are grounds to conceive of models of electron transfer to oxygen that are different from that described. First, the participation of quinones[94,95] mediating electron transfer from a flavoprotein to cytochrome *b* can not be excluded. Claims have also been made that a flavoprotein[96] or an unidentified electron carrier[97] can transfer electrons directly to oxygen.

3.2. Molecular Structure of NADPH Oxidase

The structure of discrete components constituting the NADPH oxidase is still a matter of controversy. Attempts have been made in several laboratories to purify the oxidase but, up to now the results obtained are more contradictory than confirmatory of those of other investigators.

A major impediment to the purification of the oxidase is its extreme lability on detergent extraction and on manipulation by usual purification methods. Nonetheless, preparations of NADPH oxidase activity still conserving the O_2^--forming ability have been obtained in different laboratories by chromatographic techniques. The molecular weights of this preparation on gel-filtration chromatography ranged between 150,000 and 1,000,000. Electrophoresis of these preparations in polyacrylamide gels under denaturating conditions showed the presence of different bands with major components of 87,000[98] and 32,000.[99,100] With other methods of purification, active NADPH oxidase preparations gave, at electrophoretic analysis, components of 70,000,[101] 65,000,[97] or three major bands with molecular weights of 67,000, 48,000, and 32,000.[96,102]

To avoid some of the difficulties in purifying preparations conserving NADPH oxidase activity, different groups of investigators undertook the task of purifying to homogeneity the cytochrome b_{558}. The rationale behind this approach is the possibility of detecting the spectrum of the reduced cytochrome *b* in preparations in which the use of strong detergents has inactivated NADPH oxidase activity. Investigations performed at different laboratories have resulted in purification from human, pig, and bovine neutrophils of materials containing up to 20 nmoles cytochrome *b*/mg proteins, i.e., enriched about 200 times with respect to isolated membranes. Under nondenaturating conditions, this activity showed molecular weights of about 200,000. Electrophoretic analysis under denaturating conditions gave different molecular species. The reported molecular weights of major proteins present in partially purified preparations of cytochrome b are as follows: (1) 68,000–78,000 M_r with conversion to 55,000 M_r after carbohydrate cleavage[103]; (2) 11,000, 12,000, and 14,000 M_r[104]; (3) 127,000 M_r[105]; (4) 31,500 M_r with minor species of 90,000–100,000 M_r[100]; (5) 22,000 and 91,000 M_r with conversion of this last to 48,000 M_r after carbohydrate cleavage[106]; and (6) 23,000 and 78,000–90,000 M_r species, which are absent in X-CGD.[107]

Recent results have provided extensive and definitive evidence that the cytochrome b of human neutrophils is a heterodimer composed of two polypeptides of 91,000–92,000 and 22,000–23,000 M_r and that the former of this component is heavily glycosylated and can be converted to a lower molecular weight component of 50,000–55,000 M_r after removal of N-linked sugars.[108–110]

The difficulties encountered to purify to homogeneity discrete components of the O_2^--generating system are certainly one of the major reasons of the current uncertainty about the real nature of the NADPH oxidase. The most important new methodologies potentially able to overcome these difficulties, i.e. the isolation of monoclonal antibodies and gene–cloning techniques, do not appear to have been exploited at the level to push the field a step forward. Certainly, it should be only a question of time. Monoclonal antibodies against a partially purified oxidase preparation have been useful in identifying a novel proteolipid that is possibly involved in the mechanism of activation of the respiratory burst.[111] Elegant studies have permitted the cloning of a gene coding for the synthesis of a protein missing in one and abnormally transcribed in three patients with the X-CGD[112]; recently it has been shown that this is the high molecular weight component of the heterodimeric cytochrome *b* complex (see above).[108,109]

4. MECHANISMS OF ACTIVATION OF THE RESPIRATORY BURST

The molecular events underlying the shift of the phagocytic cell from a quiescent state to that of a cell that actively metabolizes oxygen can be addressed in two different ways: (1) to define single discrete events that ensue from the perturbation of the plasma membrane with appropriate stimuli in order to establish a possible causal relationship between one of these and the eventual activation of the NADPH oxidase; and (2) to face the problem of the actual molecular modification of the NADPH oxidase, which is responsible for its conversion to a highly efficient system of electron transport.

Needless to say, the available information derives mainly from studies performed on the basis of the first way to address this central issue of the respiratory burst of phagocytes. In line with the burst of studies on mechanisms of transmembrane signaling in eukaryotic cells, the past few years have provided new insights into the molecular events that precede or accompany activation of the respiratory burst. These matters have been treated in detail in recent reviews.[43,113] What should be underscored is that the overwhelming evidence favors the view that specific receptors enable a GTP-binding protein to activate a phospholipase C that hydrolyzes, in a presumably selective way, phosphoinositides with formation of secondary messengers involved in elevating intracellular free calcium (inositol 3- and 4-phosphates) and in activating protein kinase C (diacylglycerol). Evidence has been presented, however, that activation of the respiratory burst can be completely dissociated from the phosphoinositide turnover[114]; therefore, it is not necessarily dependent on a unique and stereotyped sequence of events.

The molecular alteration(s) responsible for the conversion of the O_2^--generating system to an active state are substantially conjectural and are based on indirect evidence. A comprehensive list of these alterations would include (1) a covalent, post-translational modification of NADPH oxidase components by phosphorylation, acylation, or methylation reactions; (2) a modification of the lipid environment in which the oxidase is embedded; and (3) the assembly of components of the electron-transfer chain, which can be either directly dependent on crosslinking or secondary to the above-mentioned modifications of proteins or lipids.

Evidence has been accumulated during the past few years that phosphorylation reactions are indirectly or directly linked to the activation of the NADPH oxidase. As originally shown by Schneider et al.,[115] activation of the respiratory burst of phagocytes is accompanied by the phosphorylation of several cytosolic and membrane proteins. The phosphorylation of three of these proteins shows a good correlation indeed with the activation of the respiratory burst in terms of dose dependence on the stimulus, kinetics, and effect of inhibitors.[116,117] Of these proteins, 64,000, 50,000, and 21,000 M_r, only the last has been identified, probably the light chain of myosin. Also unknown is the role of a protein of 44,000 to 48,000 M_r, which is phosphorylated in normal neutrophils but not in the neutrophils of patients with the autosomal-recessive form of CGD.[118] Since both the amount of cytochrome b_{558} and of flavoproteins in this CGD variant is normal, this protein is more likely involved in the activation mechanism rather than a component of the NADPH oxidase.

Ground for the role of phosphorylation reactions in the activation of the respiratory burst would also derive from the observation that the NADPH oxidase activity of neutrophils membranes can be stimulated by added purified brain protein kinase C in the presence of the appropriate cofactors necessary to sustain its activity.[119] However, to our knowledge, this results has not yet been confirmed by other investigators.

The only possible evidence of phosphorylation of a component of the NADPH oxidase was obtained in our laboratory and derives from the observation that the

32,000-M_r protein that copurifies with the oxidase and constitutes a major species of highly purified preparations of cytochrome *b* is markedly phosphorylated upon cell stimulation.[100,120] Furthermore, in a cell-free system, phosphorylation of the 32,000-M_r polypeptide was obtained in the presence of calcium, phosphatidylserine, and PMA, indicating that it is due to protein kinase C.[121]

In conclusion, it should be stated that the role of phosphorylation as an essential phenomenon in the conversion of the NADPH oxidase to its active state cannot be established until the oxidase components are defined more precisely. Until then, it can be cautiously stated that activation of the respiratory burst is accompanied by phosphorylation reactions for which the real significance on the response of the cell is unknown.

It has long been known that perturbation of the lipid environment of the phagocyte surface can activate the respiratory burst of phagocytes. Displacement of membrane cholesterol by saponin is an efficient stimulus of the respiratory burst.[48,49] Treatment of neutrophils with exogenous phospholipase C is also effective.[50] Finally, unsaturated fatty acids trigger the respiratory burst.[122] In the light of more recent data, the effect of phospholipase C can be explained with the production of diacylglycerol, the endogenous activator of protein kinase C; fatty acids have also been shown to trigger phosphorylation reactions.[123] However, the generation of secondary messengers by all the maneuvers that alter the lipid environment are far from conclusive. For example, the effect of saponin has not been reinvestigated, and the stimulatory activity of fatty acid is correlated with their capability to alter membrane fluidity.[124] Furthermore, one must face the intriguing evidence that arachidonic acid and even sodium dodecyl sulfate (SDS) are able to activate the NADPH oxidase in a cell-free system in conditions in which protein kinase C activity is not favored.[36–37,68,125] It is not yet possible to conclude whether activation of phospholipases by different types of stimuli converge in altering the lipid environment in which the NADPH oxidase is embedded and this causes its conversion to an active state. The perturbation of membrane lipids could indeed cause conformational modification of the oxidase which mimicks some more subtle modification operative in the response of the cell to physiologic environmental simuli.

Independent of the mechanisms with which this would occur, there are some indications that the electron-transport chain could be assembled at the moment of stimulation. Early studies showed that nontoxic doses of glutaraldehyde inhibited the activation of the respiratory burst only if added before the stimulus, but not afterward.[126] More recently, another bifunctional crosslinking agent, *N*-hydroxysuccinimide, has been shown to have the same effect.[127] Significantly, cleavage of the crosslinker restores the neutrophil's response, indicating that aggregation of surface molecules plays indeed a role in stimulation of the respiratory burst. However, it is unknown whether the transmembrane signaling or the assembly of oxidase components is affected by the inhibition of the lateral diffusion of surface molecules. Evidence has also been presented that part of cytochrome b_{558} translocates from granules to the plasma membrane during activation of the respiratory burst[128] and appears to be incorporated in the phagocytic vacuoles membrane.[129]

Contradictory results have been obtained, however, with studies of localization[130] and also with the behavior of cytoplasts, which are devoid of granules but normally activatable.[131]

We think it reasonable to conclude that the plasma membrane pool of cytochrome *b* is sufficient to trigger the respiratory burst and that translocation of the intracellular pool is a concomitant event that might increase the efficiency or the duration of the oxidase activity. No direct evidence exists to show that assembly of the NADPH oxidase components are a necessary step in the activation of the respiratory burst.

5. BIOLOGICAL SIGNIFICANCE OF THE RESPIRATORY BURST OF PHAGOCYTES

Products derived from the activation of the phagocyte respiratory burst are now universally known as important mediators of host reactions. The most extensive knowledge on their role has been obtained by means of studies on the mechanism of killing of different pathogens. Products derived from the reduction of oxygen by the NADPH oxidase are part of a powerful microbicidal system,[39] and the ability of macrophages to kill intracellular pathogens is correlated with their oxygen metabolism.[132] Significantly, this last capability has been shown to be finely modulated during the development of cell-mediated immunity. IFN_{γ} produced by activated T lymphocytes activates macrophage oxygen metabolism[133] and microbial killing in parallel.[134] More recently, it has been shown that members of the cytokine family, like granulcoyte–macrophage colony-stimulating factor and IFN_{γ} itself, are able to enhance the respiratory burst to environmental stimuli in neutrophils as well.[135–138] The modulation of phagocyte respiratory burst by cytokines produced by DNA technologies will certainly be exploited as a means of enhancing host resistance toward pathogens.

Several observations indicate that products of the respiratory burst can also have the ability, by themselves or in concert with other factors, to damage eukaryotic cells. The hydrogen peroxide–halide peroxidase system has long been recognized as a mediator of cytotoxicity.[139,140] Good correlation has been found between the ability of neutrophils and different populations of macrophages to undergo a respiratory burst and to lyse tumor cells in parallel.[29] Finally, the action of both neutrophils and macrophages as effector cells in antibody-dependent cell-mediated cytotoxicity has been shown to be mediated as well by molecules produced during activation of the respiratory burst.[141,142] These data encourage studies on the effect of in vivo modulation of phagocytes respiratory burst by IFN_{γ} and other cytokines in controlling tumor cell growth.

Cell damage consequent to activation of phagocytes respiratory burst can conceivably accompany inflammatory processes where such complement factors as C5a, leukotriene, platelet-activating factor, or immobilized immune complexes are present. In the clinical situation known as adult respiratory distress syndrome (ARDS) or shock lung, C5a generated during passage of blood through membranes used for extracorporeal haemodialyis, or during burns, trauma, endotoxin intoxication causes aggregation of neutrophils; the activation of the respiratory burst that ensues is responsible for damage to endothelial cells, followed by extravasation of blood constituents.[143,144]

Products of the respiratory burst can have even more far-reaching effects on the development and regulation of the inflammatory process than that caused by induction of cell injury. Evidence has in fact, been presented to show that these products can affect other cell functions, such as platelet aggregation and secretion, endothelial cell permeability, mast cell secretion. Fibrous proteins, hyaluronic acid, and proteoglycans of connective tissues, enzymes, and protein inhibitors can also be targets of oxygen metabolites undergoing structural and functional alteration.

ACKNOWLEDGMENTS. This work was supported by grants 86.00478.04 from the National Research Council of Italy (C.N.R.), Gruppo Immunologia, and by Ministero Pubblica Istruzione. We are very grateful to Professor Filippo Rossi for his helpful suggestions in the preparation of this chapter and for having introduced us into the research on the respiratory burst of phagocytes.

REFERENCES

1. Baldridge CW, Gerard RW: The extra respiration of phagocytosis. *Am J Physiol* **103:**235–236, 1933.
2. Ado AD: Uber den Verlauf der oxidativen und glykolytischen Prozesse in den Leukocyten des entzundeten Gewebes wahrend der Phagocytose. *Z. Ges. Exp. Med.* **87:**473–480, 1933.
3. Stahelin H, Suter E, Karnovsky ML: Studies on the interaction between phagocytes and tubercle bacilli. I. Observations on the metabolism of guinea pig leukocytes and influence of phagocytosis. *J Exp Med* **104:**121–136, 1956.
4. Stahelin H, Karnovsky ML, Farnham AE, Suter E: Studies on the interaction between phagocytes and tubercle bacilli. III. Some metabolic effects in guinea pigs associated with infection with tubercle bacilli. *J Exp Med* **105:**265–277, 1957.
5. Becker H, Munder G, Fischer H: Uber den Leukocytenstoffwechsel bei der Phagocytose. *Hoppe Seylers Z Physiol Chem* **313:**266–275, 1958.
6. Sbarra AJ, Karnovsky ML: The biochemical basis of phagocytosis. I. Metabolic changes during the ingestion of particles by polymorphonuclear leukocytes. *J Biol Chem* **234:**1355–1362, 1959.
7. Cohn ZA, Morse SI: Functional and metabolic properties of polymorphonuclear leucocytes. I. Observations on the requirements and consequences of particle ingestion. *J Exp Med* **111:**667–687, 1960.
8. Iyer GYN, Islam MF, Quastel JH: Biochemical aspects of phagocytosis. *Nature (Lond)* **192:**535–541, 1961.
9. Paul B, Sbarra AJ: The role of the phagocytes in host–parasite interactions. XIII. The direct quantitative estimation of H_2O_2 in phagocytizing cells. *Biochim Biophys Acta* **156:**168–178, 1968.
10. Klebanoff SJ: A peroxidase-mediated anti-microbial system in leukocytes. *J Clin Invest* **46:**1078, 1967.
11. Klebanoff SJ: Iodination of bacteria: A bactericidal mechanism. *J Exp Med* **126:**1063–1078, 1967.
12. Strauss RR, Paul BB, Jacobs AA, Sbarra AJ: Role of the phagocyte in host–parasite interactions.

XXII. H_2O_2-dependent decarboxylation and deamination by myeloperoxidase and its relationship to antimicrobial activity. *J Reticuloendothel Soc* **7**:754–761, 1970.
13. Iyer GYN, Quastel JH: NADPH and NADH oxidation by guinea pig polymorphonuclear leucocytes. *Can J Biochem* **41**:427–434, 1963.
14. Cagan RH, Karnovsky ML: Enzymatic basis of the respiratory stimulation during phagocytosis. *Nature (Lond)* **204**:255–257, 1964.
15. Rossi F, Zatti M: Changes in the metabolic pattern of polymorphonuclear leukocytes during phagocytosis. *Br J Exp Pathol* **45**:548–559, 1964.
16. Rossi F, Romeo D, Patriarca P: Mechanism of phagocytosis-associated oxidative metabolism in polymorphonuclear leukocytes and macrophages. *J Reticuloendothel Soc* **12**:127–149, 1972.
17. Babior BM, Kipnes RS, Curnutte JT: Biological defense mechanisms. The production by leukocytes of superoxide, a potential bactericidal agent. *J Clin Invest* **52**:741–744, 1973.
18. Bridges RA, Berendes H, Good RA: A fatal granulomatous disease of childhood. The clinical, pathological, and laboratory features of a new syndrome. *J Dis Child* **97**:387–408, 1959.
19. Holmes B, Page AR, Good RA: Studies of the metabolic activity of leukocytes from patients with a genetic abnormality of phagocytic function. *J Clin Invest* **46**:1422–1432, 1967.
20. Curnutte JT, Whitten DM, Babior BM: Defective superoxide formation by granulocytes from patients with chronic granulomatous disease. *N Engl J Med* **290**:593–597, 1974.
21. Hohn DC, Lehrer RI: NADPH oxidase deficiency in X-linked chronic granulomatous disease. *J Clin Invest* **55**:707–713, 1975.
22. Dewald B, Baggiolini M, Curnutte JT, Babior BM: Subcellular localization of the superoxide-forming enzyme in human neutrophils. *J Clin Invest* **63**:21–29, 1979.
23. Nathan CF, Karnovsky ML, David JR: Alterations of macrophage functions by mediators from lymphocytes. *J Exp Med* **133**:1356–1376, 1971.
24. Karnovsky ML, Lazdins J, Drath D, Harper A: Biochemical characteristics of activated macrophages. *Ann NY Acad Sci* **256**:266–274, 1975.
25. Rossi F, Zabucchi G, Romeo D: Metabolism of phagocytosing mononuclear phagocytes, in Van Furth R (ed): *Mononuclear Phagocytes in Immunity, Infection and Pathology*. Oxford, Blackwell Scientific, 1975, p 441.
26. Nathan CF, Root RK: Hydrogen peroxide release from mouse peritoneal macrophages: Dependence on sequential activation and triggering. *J Exp Med* **146**:1648–1662, 1977.
27. Johnston RB, Godzik CA, Cohn ZA: Increased superoxide anion production by immunologically activated and chemically elicited macrophages. *J Exp Med* **148**:115–127, 1978.
28. Nathan C, Nogueira N, Juangbhanich et al: Activation of macrophages in vivo and in vitro. Correlation between hydrogen peroxide release and killing of *Trypanosoma cruzi*. *J Exp Med* **149**:1056–1068, 1979.
29. Nathan CF, Brukner LM, Silverstein SC, Cohn ZA: Extracellular cytolysis by activated macrophages and granulocytes. *J Exp Med* **149**:84–99, 1979.
30. Shinagawa Y, Tanaka C, Teraoka A, Shinagawa Y: A new cytochrome in neutrophilic granules of rabbit leukocytes. *J Biochem* **59**:622–624, 1966.
31. Segal AW, Jones OTG: Novel cytochrome b system in phagocytic vacuoles of human granulocytes. *Nature (Lond)* **276**:515–517, 1978.
32. Segal AW, Jones OTG, Webster D, Allison AC: Absence of a newly described cytochrome b from neutrophils of patients with chronic grenulomatous disease. *Lancet* **2**:446–449, 1978.
33. Segal AW, Cross AR, Gárcia RC, et al: Absence of cytochrome b-245 in chronic granulomatous disease. A multicenter European evaluation of its incidence and relevance. *N Engl J Med* **308**:245–251, 1983.
34. Tauber AI, Goetzl EJ: Structural and catalytic properties of the solubilized superoxide-generating activity of human polymorphonuclear leukocytes. Solubilization, stabilization in solution and partial characterization. *Biochemistry* **18**:5576–5584, 1979.
35. Gabig TG, Babior BM: The O_2^- forming oxidase responsible for the respiratory burst in human neutrophils. Properties of the solubilized enzyme. *J Biol Chem* **254**:9070–9074, 1979.

36. Bromberg Y, Pick E: Unsaturated fatty acids stimulate NADPH-dependent superoxide production by cell-free system derived from macrophages. *Cell Immunol* **88:**213–221, 1984.
37. Heyneman RA, Vercauteren RE: Activation of a NADPH oxidase from horse polymorpholuclear leukocytes in a cell-free system. *J Leukocyte Biol* **36:**751–759, 1984.
38. Sbarra AJ, Selvaray RJ, Paul BB, et al: Biochemical, functional, and structural aspects of phagocytosis. *Int Rev Exp Pathol* **16:**249–271, 1976.
39. Klebanoff SJ, Clark RA: *The Neutrophil: Function and Clinical Disorders.* Amsterdam, North-Holland, 1978.
40. Rossi F, Patriarca P, Romeo D: Metabolic changes accompanying phagocytosis, in Sbarra AJ, Strauss R (eds): *The Reticuloendothelial System. A Comprehensive Treatise.* New York, Plenum, 1980, vol 2, p 153.
41. Badwey JA, Karnovsky ML: Active oxygen species and the functions of phagocytic cells. *Annu Rev Biochem* **49:**695–726, 1980.
42. Babior BM: Oxidants from phagocytes: Agents of defense and destruction. *Blood* **64:**959–966, 1984.
43. Rossi F: The O_2^- forming NADPH oxidase of the phagocytes: Nature, mechanisms of activation and function. *Biochim Biophys Acta* **853:**65–89, 1986.
44. Roder JC, Helfand SL, Werkmeister J, et al: Oxygen intermediates are triggered early in the cytolytic pathway of human NK cells. *Nature (Lond)* **298:**569–572, 1982.
45. Volkman DJ, Buescher ES, Gallin JI, Fauci AS: B cell lines as models for inherited phagocytic diseases: Abnormal superoxide generation in chronic granulomatous disease and giant granules in Chediak–Higashi syndrome. *J Immunol* **133:**3006–3009, 1984.
46. Nathan CF, Mercer-Smith JA, Desantis NM, Palladino M: Role of oxygen in T cell-mediated cytolysis. *J Immunol* **129:**2164–2171, 1982.
47. Strauss BS, Stetson CA: Studies on the effect of certain macromolecular substances on the respiratory activity of the leukocytes of peripheral blood. *J Exp Med* **112:**653–669, 1960.
48. Zatti M, Rossi F: Relationship between glycolysis and respiration in surfactant-treated leucocytes. *Biochim Biophys Acta* **148:**553–555, 1967.
49. Graham RC, Karnovsky MJ, Shafer AW, et al: Metabolic and morphological observations on the effect of surface-active agents on leukocytes. *J Cell Biol* **32:**629–647, 1967.
50. Patriarca P, Zatti M, Cramer R, Rossi F: Stimulation of the respiration of polymorphonuclear leucocytes by phospholipase C. *Life Sci* **9:**841–849, 1970.
51. Johnston RB Jr: Oxygen metabolism and the microbicidal activity of macrophages. *Fed Proc* **37:**2759–2764, 1978.
52. Nathan CF: Regulation of macrophage oxidative metabolism and parasitic activity, in Van Furth R (ed): *Mononuclear Phagocytes. Characteristics, Physiology and Function.* Dordrecht, Martinus Niijhoff, 1985, p 411.
53. Berton G, Gordon S: Role of the plasmamembrane in the regulation of superoxide anion release by macrophages, in Van Furth R (ed): *Mononuclear Phagocytes. Characteristics, Physiology and Function.* Dordrecht, Martinus Niijhoff, 1985, p 435.
54. Curnutte JT, Babior BM, Karnovsky ML: Fluoride-mediated activation of the respiratory burst in human neutrophils. A reversible process. *J Clin Invest* **63:**637–647, 1979.
55. Romeo D, Zabucchi G, Rossi F: Reversible metabolic stimulation of polymorphonuclear leukocytes and macrophages by concanavalin A. *Nature (Lond)* **243:**111–112, 1973.
56. Rossi F, De Togni P, Bellavite P, et al: Relationship between the binding of N-Formyl-Methionyl-Leucyl-Phenylalanine and the respiratory response in human neutrophils. *Biochem Biophys Acta* **758:**168–175, 1983.
57. Cohen HJ, Chovaniec ME, Wilson MK, Newburger PE: Con-A-stimulated superoxide production by granulocytes: Reversible activation of NADPH oxidase. *Blood* **60:**1188–1194, 1982.
58. Simkowitz L, Atkinson JP, Spilberg I: Stimulus-specific deactivation of chemotactic factor-induced cyclic AMP response and superoxide generation by human neutrophils. *J Clin Invest* **66:**736–747, 1980.

59. Berton G, Gordon S: Desensitization of macrophages to stimuli which induce secretion of superoxide anion. Down-regulation of receptors for phorbol myristate acetate. *Eur J Immunol* **13:**620–627, 1983.
60. Berton G, Gordon S: Modulation of macrophage mannosyl-specific receptors by cultivation on immobilised zymosan. Effects on superoxide release and phagocytosis. *Immunology* **49:**705–715, 1983.
61. De Togni P, Cabrini G, Di Virgilio F: Cyclic AMP inhibition of fMet-Leu-Phe-dependent metabolic responses in human neutrophils is not due to its effects on cytosolic Ca^{2+}. *Biochem J* **224:**629–635, 1984.
62. De Togni P, Bellavite P, Della Bianca V, et al: Intensity and kinetics of the respiratory burst of human neutrophils in relation to receptor occupancy and rate of occupation by formyl-methionyl-leucyl-phenylalanine. *Biochim Biophys Acta* **838:**12–22, 1985.
63. Patriarca P, Dri P, Kakinuma K, et al: Studies on the mechanism of metabolic stimulation in polymorphonuclear leukocytes during phagocytosis. I. Evidence for superoxide involvement in the oxidation of $NADPH_2$. *Biochim Biophys Acta* **385:**380–386, 1975.
64. Curnutte JT, Karnovsky ML, Babior BM: Manganese-dependent NADPH oxidation by granulocyte particles. Role of superoxide and nonphysiological nature of the manganese requirement. *J Clin Invest* **57:**1059–1067, 1976.
65. Bellavite P, Berton G, Dri P: Studies on the NADPH oxidation by subcellular particles from phagocytosing polymorphonuclear leucocytes. Evidence for the involvement of three mechanisms. *Biochim Biophys Acta* **591:**434–444, 1980.
66. Zatti M, Rossi F: Early changes of hexose monophosphate pathway activity and of NADPH oxidation in phagocytizing leukocytes. *Biochim Biophys Acta* **99:**557–561, 1965.
67. Bellavite P, Serra MC, Davoli A, Rossi F: Selective enrichment of NADPH oxidase activity in phagosomes from guinea pig polymorphonuclear leukocytes. *Inflammation* **6:**21–29, 1982.
68. Bromberg Y, Pick E: Activation of NADPH dependent superoxide production in a cell-free system by sodium dodecyl sulfate. *J Biol Chem* **260:**13539–13545, 1985.
69. Clark RA, Leidal KG, Pearson DW, Nauseef WM: NADPH oxidase of human neutrophils: Subcellular localization and characterization of an arachidonate-activatable superoxide-generating system. *J Biol Chem* **262:**4065–4074, 1987.
70. Berton G, Bellavite P, Dri P, et al: The enzyme responsible for the respiratory burst in elicited guinea pig peritoneal macrophages. *J Pathol* **136:**273–279, 1982.
71. Sasada M, Pabst JM, Johnston RB: Activation of mouse peritoneal macrophages by lipopolysaccharide alters the kinetic parameters of the superoxide-producing NADPH oxidase. *J Biol Chem* **258:**9631–9635, 1983.
72. Tsunawaki S, Nathan CF: Enzymatic basis of macrophage activation. Kinetic analysis of superoxide production in lysates of resident and activated mouse peritoneal macrophages and granulocytes. *J Biol Chem* **259:**4305–4312, 1984.
73. Berton G, Cassatella M, Cabrini G, Rossi F: Activation of mouse macrophages causes no change in expression and function of phorbol diester receptors, but is accompanied by alterations in the activity and kinetic parameters of NADPH oxidase. *Immunology* **54:**371–379, 1985.
74. Cassatella M, Della Bianca V, Berton G, Rossi F: Activation by gamma interferon of human macrophages capability to produce toxic oxygen molecules is accompanied by decreased K_m of the superoxide generating NADPH oxidase. *Biochem Biophys Res Commun* **132:**908–914, 1985.
75. Tsunawaki S, Nathan CF: Macrophage deactivation. Altered kinetic properties of superoxide-producing enzyme after exposure to tumor cell-conditioned medium. *J Exp Med* **164:**1319–1331, 1986.
76. Babior BM, Kipnes RS: Superoxide-forming enzyme from human neutrophils: Evidence for a flavin requirement. *Blood* **50:**517–524, 1977.
77. Light DR, Walsh C, O'Callaghan AM, et al: Characteristics of the cofactor requirements for the superoxide-generating NADPH oxidase of human polymorphonuclear leukocytes. *Biochemistry* **20:**1468–1476, 1981.

78. Gabig TG, Lefker BA: Catalytic properties of the resolved flavoprotein and cytochrome b components of the NADPH dependent O_2^--generating oxidase from human neutrophils. *Biochem Biophys Res Commun* **118:**430–436, 1984.
79. Cross AR, Parkinson JF, Jones OTG: The superoxide-generating oxidase of leucocytes. NADPH-dependent reduction of flavin and cytochrome b in solubilized preparations. *Biochem J* **223:**337–344, 1984.
80. Kakinuma K, Kaneda M, Chiba T, Ohnishi T: Electron spin resonance studies on a flavoprotein in neutrophil plasma membranes. Redox potentials of the flavin and its participation in NADPH oxidase. *J Biol Chem* **261:**9426–9432, 1986.
81. Cross AR, Jones OTG, Harper AM, Segal AW: Oxidation-reduction properties of the cytochrome b found in the plasmamembrane fraction of human neutrophils. A possible oxidase in the respiratory burst. *Biochem J* **194:**599–606, 1981.
82. Berton G, Cassatella M, Bellavite P, Rossi F: Molecular basis of macrophage activation. Expression of the low potential cytochrome b and its reduction upon cell stimulation in activated macrophages. *J Immunol* **136:**1393–1396, 1986.
83. Roberts PJ, Cross AR, Jones OTG, Segal AW: Development of cytochrome b and active oxidase system in association with maturation of a human promyelocytic (HL-60) cell line. *J Cell Biol* **95:**720–726, 1982.
84. Segal AW, Jones OTG: Reduction and subsequent oxidation of a cytochrome b of human neutrophils after stimulation with phorbol myristate acetate. *Biochem Biophys Res Commun* **88:**130–134, 1979.
85. Cross AR, Higson FK, Jones OTG, et al: The enzymic reduction and kinetics of oxidation of cytochrome b-245 of neutrophils. *Biochem J* **204:**479–485, 1982.
86. Iizuka T, Kanegasaki S, Makino R, Ishimura Y: Studies on neutrophil b-type cytochrome in situ by low temperature absorption spectroscopy. *J Biol Chem* **260:**12049–12053, 1985.
87. Segal AW, Jones OTG: Absence of cytochrome b reduction from both femele and male patients with chronic granulomatous disease. *FEBS Lett* **110:**111–114, 1980.
88. Cross AR, Jones OTG, Garcia R, Segal AW: The association of FAD with the cytochrome b-245 of human neutrophils. *Biochem J* **208:**759–763, 1982.
89. Gabig TG: The NADPH-dependent O_2^- generating oxidase from human neutrophils. Identification of a flavoprotein component that is deficient in a patient with chronic granulomatous disease. *J Biol Chem* **258:**6352–6356, 1983.
90. Hamers MN, de Boer M, Meerhof LJ, et al: Complementation in monocyte hybrids revealing genetic heterogeneity in chronic granulomatous disease. *Nature (Lond)* **307:**553–555, 1984.
91. Bellavite P, Jones OTG, Cross AR, et al: Composition of partially purified NADPH oxidase from pig neutrophils. *Biochem J* **223:**639–648, 1984.
92. Berton G, Papini E, Cassatella M, et al: Partial purification of the superoxide-generating system of macrophages. possible association of the NADPH oxidase activity with a low potential cytochrome b. *Biochim Biophys Acta* **810:**164–173, 1985.
93. Rossi F, Bellavite P, Serra MC, Papini E: Characterization of phagocyte NADPH oxidase, in Van Furth R (ed): *Mononuclear Phagocytes. Characteristics, Physiology and Function.* Boston, Martinus Nijhoff, 1985, p 423.
94. Crawford DR, Schneider DL: Evidence that a quinone may be required for the production of superoxide and hydrogen peroxide in neutrophils. *Biochem Biophys Res Commun* **99:**1277–1286, 1981.
95. Gabig TG, Lefker BA: Activation of the human neutrophil NADPH oxidase results in coupling of electron carrier function between ubiquinone-10 and cytochrome b_{559}. *J Biol Chem* **260:**3991–3995, 1985.
96. Glass GA, DeLisle DM, De Togni P, et al: The respiratory burst oxidase of human neutrophils. Further studies of the purified enzyme. *J Biol Chem* **261:**13247–13251, 1986.
97. Doussiere J, Vignais PV: Purification and properties of O_2^- generating oxidase from bovine polymorphonuclear neutrophils. *Biochemistry* **24:**7231–7239, 1985.

98. Sakane F, Takahashi K, Koyama Y: Purification and characterization of a membrane-bound NADPH–cytochrome c reductase capable of catalyzing menadione-dependent O_2^- formation in guinea pig polymorphonuclear leukocytes. *J Biochem. (Tokyo)* **96:**671–678, 1984.
99. Serra MC, Bellavite P, Davoli A, et al: Isolation from neutrophil membranes of a complex containing active NADPH oxidase and cytochrome b-245. *Biochim Biophys Acta* **788:**138–146, 1984.
100. Bellavite P, Papini E, Zeni L, et al: Studies on the nature and activation of O_2^- forming NADPH oxidase of leukocytes. Identification of a phosphorylated component of the active enzyme. *Free Rad Res Comm.* **1:**11–29, 1985.
101. Tamoto K, Washida N, Yukishige K, et al: Electrophoretic isolation of a membrane-bound NADPH oxidase from guinea-pig polymorphonuclear leukocytes. *Biochim Biophys Acta* **732:**569–578, 1983.
102. Markert M, Glass GA, Babior BM: Respiratory burst oxidase from human neutrophils: Purification and some properties. *Proc Natl Acad Sci USA* **82:**3144–3148, 1985.
103. Harper AM, Chaplin MF, Segal AW: Cytochrome b-245 from human neutrophils is a glycoprotein. *Biochem J* **227:**783–788, 1985.
104. Pember SO, Heyl BL, Kinkade JM Jr, Lambeth JD: Cytochrome b558 from (bovine) granulocytes. Partial purification from Triton X-114 extracts and properties of the isolated cytochrome. *J Biol Chem* **259:**10590–10595, 1984.
105. Lutter R, van Schaik MLJ, van Zwieten R, et al: Purification and partial characterization of the b-type cytochrome from human polymorphonuclear leukocytes. *J Biol Chem* **260:**2237–2244, 1985.
106. Parkos CA, Allen RA, Cochrane CG, Jesaitis AJ: Characterization of purified cytochrome b559 from the plasma membrane of stimulated human granulocytes. *J Cell Biol* **103:**1908, 1986 (abst).
107. Segal AW: Absence of both cytochrome b-245 subunits from neutrophils in X-linked chronic granulomatous disease. *Nature (Lond)* **326:**88–91, 1987.
108. Dinauer MC, Orkin SH, Brown R, Jesaitis AJ, Parkos CA: The glycoprotein encoded by the X-linked chronic granulomatous disease locus is a component of the neutrophil cytochrome b complex. *Nature* **327:**717–720, 1987.
109. Teahan C, Rowe P, Parker P, et al: The X–linked chronic granulomatous disease gene codes for the β-chain of cytochrome b_{-245}. *Nature* **327:**720–721, 1987.
110. Parkos CA, Allen RA, Cochrane CG, Jesaitis AJ: Purified cytochrome b from human granulocyte plasma membrane is comprised of two polypeptides with relative molecular weights of 91,000 and 22,000. *J Clin Invest* **80:**732–742, 1987.
111. Berton G, Rosen H, Ezekowitz RAB, et al: Monoclonal antibodies to a particulate superoxide-forming system stimulate a respiratory burst in intact guinea pig neutrophils. *Proc Natl Acad Sci USA* **83:**4002 4006, 1986.
112. Royer-Pokora B, Kunkel LM, Monaco AP, et al: Cloning the gene for an inherited human disorder—chronic granulomatous disease—on the basis of its chromosomal location. *Nature (Lond)* **322:**32–38, 1986.
113. McPhail LC, Snyderman R: Mechanisms of regulating the respiratory burst in leukocytes, in Snyderman R (ed): *Regulation of Leukocyte Function* New York, Plenum, 1984, p 247.
114. Grzeskowiak M, Della Bianca V, Cassatella M, Rossi F: Complete dissociation between the activation of phosphoinositide turnover and of NADPH oxidase by Formyl-methionyl-leucyl-phenylalanine in human neutrophils depleted of Ca^{2+} and primed by subtreshold doses of phorbol myristate acetate. *Biochem Biophys Res Commun* **135:**785–794, 1986.
115. Schneider C, Zanetti M, Romeo D: Surface-reactive stimuli selectively increase protein phosphorylation in human neutrophils. *FEBS Lett* **127:**4–8, 1981.
116. Irita K, Takeshige K, Minakami S: Protein phosphorylation in intact pig leukocytes. *Biochim Biophys Acta* **805:**44–52, 1984.
117. Ohtsuka T, Okamuar N, Ishibashi S: Involvement of protein kinase C in the phosphorylation of 46 kDa proteins which are phosphorylated in parallel with activation of NADPH oxidase in intact guinea pig polymorphonuclear leukocytes. *Biochim Biophys Acta* **888:**332–337, 1986.

118. Segal AW, Heyworth PG, Cockroft S, Barrowman MM: Stimulated neutrophils from patients with autosomal recessive chronic granulomatous disease fail to phosphorylate a M_r-44,000 protein. *Nature (Lond)* **316:**547–549, 1985.
119. Cox JA, Jeng JA, Sharkey NA, et al: Activation of the human neutrophil nicotinamide adenine dinucleotide phosphate (NADPH)-oxidase by protein kinase C. *J Clin Invest* **76:**1932–1938, 1985.
120. Rossi F, Bellavite P, Papini E: Respiratory response of phagocytes: Terminal NADPH oxidase and the mechanisms of its activation, in *Biochemistry of Macrophages. Ciba Foundation Symposium* 118. London, Pitman, 1986, p 172–195.
121. Papini E, Grzeskowiak M, Bellavite P, Rossi F: Protein kinase C phosphorylates a component of NADPH oxidase of neutrophils. *FEBS Lett* **190:**204–208, 1985.
122. Kakinuma K: Effects of fatty acids on the oxidative metabolism of leukocytes. *Biochim Biophys Acta* **348:**76–85, 1974.
123. McPhail LC, Clayton CC, Snyderman R: A potential second messenger role for unsaturated fatty acids: Activation of Ca^{2+}-dependent protein kinase C. *Science* **224:**622–624, 1984.
124. Badwey JA, Curnutte JT, Robinson JM, et al: Effects of free fatty acids on release of superoxide and on change of shape by human neutrophils: Reversibility by albumin. *J Biol Chem* **259:**7870–7877, 1984.
125. Tauber AI, Cox JA, Yeng AY, Blumberg PM: Subcellular activation of the human NADPH-oxidase by arachidonic acid and sodium dodecyl sulfate (SDS) is independent of protein kinase C. *Clin Res* **34:**664, 1986 (abst).
126. Romeo D, Zabucchi G, Rossi F: Surface modulation of oxidative metabolism of polymorphonuclear leucocytes, in Rossi F, Patriarca PL, Romeo D (eds): *Movement, Metabolism and Bactericidal Mechanisms of Phagocytes.* Padua, Piccin Medical Books, 1977, p 153.
127. Aviram I, Simons ER, Babior BM: Reversible blockade of the respiratory burst in human neutrophils by a cleavable cross-linking agent. *J Biol Chem* **259:**306–311, 1984.
128. Borregaard N, Heiple JM, Simons ER, Clark RA: Subcellular localization of the b-cytochrome component of the human neutrophil microbicidal oxidase: Translocation during activation. *J Cell Biol* **97:**52–61, 1983.
129. Segal AW, Jones OTG: Rapid incorporation of the human neutrophil plasmamembrane cytochrome b into phagocytic vacuoles. *Biochem Biophys Res Commun* **92:**710–715, 1980.
130. Yamaguchi T, Kaneda M, Kakinuma K: Is cytochrome b558 translocated into the plasmamembrane from granules during the activation of neutrophils? *J Biochem* **99:**953–959, 1986.
131. Lutter R, van Zwieten R, Weening RS, et al: Cytochrome b, flavins, and ubiquinone-50 in enucleated human neutrophils (polymorphonuclear leukocyte cytoplasts). *J Biol Chem* **259:**9603–9606, 1984.
132. Murray HW, Cohn ZA: Macrophage oxygen-dependent antimicrobial activity. III. Enhanced oxidative metabolism as an expression of macrophage activation. *J Exp Med* **152:**1596–1609, 1980.
133. Nathan CF, Murray HW, Wiebe ME, Rubin BY: Identification of interferon gamma as the lymphokine that activates human macrophages oxidative metabolism and antimicrobial activity. *J Exp Med* **158:**670–679, 1983.
134. Nathan CF; Interferon-gamma and macrophage activation in cell-mediated immunity, in Steinman RM and North RJ (eds): *Mechanisms of Host Resistance to Infectious Agents, Tumors and Allografts.* New York, Rockefeller University Press, 1986, p 165.
135. Weisbart RH, Golde DW, Clark SC, et al: Human granulocyte–macrophage colony-stimulating factor is a neutrophil activator. *Nature (Lond)* **314:**361–363, 1985.
136. Berton G, Zeni L, Cassatella MA, Rossi F: Gamma interferon is able to enhance the oxidative metabolism of human neutrophils. *Biochem Biophys Res Commun* **138:**1276–1782, 1986.
137. Cassatella MA, Cappelli R, Della Bianca V, et al: Interferon–gamma activates human neutrophil oxygen metabolism and exocytosis. *Immunology,* **63:**499–506, 1988.
138. Ezekowitz RAB, Orkin SH, Newburger PE: Recombinant interferon gamma augments phagocyte superoxide production and X-chronic granulomatous disease gene expression in X-linked variant chronic granulomatous disease. *J Clin Invest* **80:**1009–1016, 1987.

139. Edelson PJ, Cohn ZA: Peroxidase-mediated mammalian cell cytotoxicity. *J Exp Med* **138:**318–323, 1973.
140. Clark RA, Klebanoff SJ, Einstein A, Fefer A: Peroxidase–H_2O_2–halide system: Cytotoxicity effect on mammalian tumor cells. *Blood* **45:**161–170, 1975.
141. Clark RA, Klebanoff SJ: Studies on the mechanism of antibody-dependent polymorphonuclear leukocyte-mediated cytotoxicity. *J Immunol* **119:**1413–1414, 1977.
142. Nathan C, Cohn Z: Role of oxygen-dependent mechanisms in antibody-induced lysis of tumor cells by activated macrophages. *J Exp Med* **152:**198–208, 1980.
143. Jacobs HS, Craddock PR, Hammerschmidt DE, Moldow CF: Complement-induced granulocyte aggregation. An unsuspected mechanism of disease. *N Engl J Med* **302:**789–794, 1980.
144. Fantone JC, Ward PA: Mechanisms of lung parenchymal injury. *Am Rev Respir Dis* **130:**484–491, 1984.

3

The Respiratory Burst and the NADPH Oxidase of Phagocytic Leukocytes

DIRK ROOS, RENÉ LUTTER, and BEN G. J. M. BOLSCHER

1. INTRODUCTION

When phagocytic leukocytes are activated to kill microorganisms, these cells respond with considerable changes in their cellular metabolism and structure.[1] Most marked is a 20- to 30-fold increase in the oxygen consumption (i.e., the respiratory burst) that is not sensitive to inhibitors of the mitochondrial respiration.[2] The extra oxygen consumption reflects the action of a phagocyte-specific oxidase, responsible for the generation of reduced oxygen species.[1] This enzyme is localized in the plasma membrane and—after phagocytosis—in the phagosomal membranes of the cell.[3,4]

Superoxide ($O_2^- \cdot$) and hydrogen peroxide (H_2O_2) are the main products of this oxidase.[5,6] Although $O_2^- \cdot$ and H_2O_2 together might give rise to other, more reactive agents, such as hydroxyl radicals and singlet oxygen, the actual production of the latter oxygen species has not been supported by much experimental evidence.[7] Indeed, in a recent paper by Britigan et al.[8] describing the use of spin traps for the detection of free radicals, it was concluded that neutrophils do not generate hydroxyl radicals.

Whereas hydrogen peroxide is a potent oxidizing and microbicidal agent, superoxide is not. However, hydrogen peroxide,[9] and possibly also superoxide,[10] are consumed by the heme-containing enzyme myeloperoxidase. As a result, potent microbicidal hypohalites are generated, such as hypochlorous acid.

Although much research has been focused on the composition of the oxidase, no unifying model is yet available for this enzyme (system). Clearly, research in this field has been troubled by the instability and the low recoveries of the oxidase

DIRK ROOS, RENÉ LUTTER, and BEN G. J. M. BOLSCHER • Central Laboratory of the Netherlands Red Cross Blood Transfusion Service and Laboratory of Experimental and Clinical Immunology, University of Amsterdam, 1006 AK Amsterdam, The Netherlands.

activity in both particulate and solubilized oxidase preparations[11–13] (see also Section 3).

Several detailed reviews on the oxidase in general and its putative composition have recently appeared.[14,15] This chapter discusses problems related to the composition and enzymatic properties of subcellular oxidase preparations. Because studies on phagocytes with an abberant oxidase activity have provided useful information about the involvement of some of the putative oxidase components,[14] we refer to these studies wherever appropriate.

2. INTRACELLULAR KILLING OF MICROORGANISMS

The action of phagocytes against microorganisms can be described by three distinct processes: (1) a directed movement of the phagocyte to the microorganism (chemotaxis), followed by (2) the engulfment of the microorganism by the phagocyte (phagocytosis), and (3) the intracellular killing of the microorganism.[1,14] The latter process depends on two independent but supporting mechanisms, i.e., the generation of oxygen (-derived) microbicidal products and the release of microbicidal proteins from the granules into the developing phagosome.[1] This is illustrated in Fig. 1.

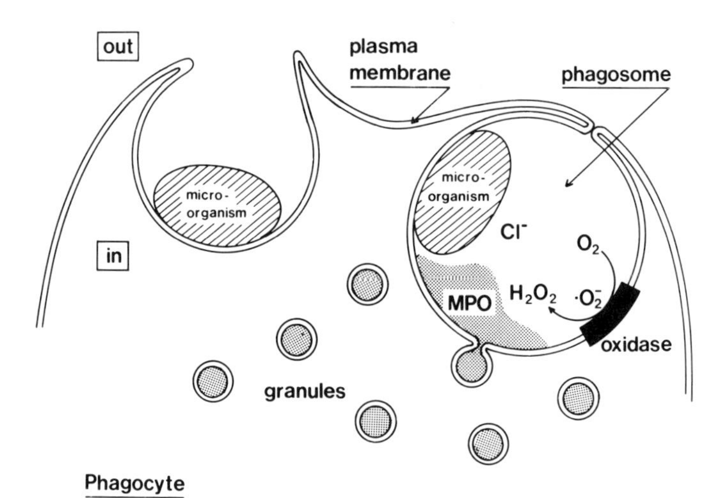

FIGURE 1. Schematic overview of reactions involved in the intracellular killing of microorganisms by phagocytic leukocytes. Microorganisms attach to the phagocyte surface, starting three independent processes: engulfment of the microorganism by pseudopods of the phagocyte (phagocytosis), fusion of intracellular granules with the phagosomal membrane (degranulation), and activation of the NADPH oxidase. Within the phagosome, superoxide is dismutated to hydrogen peroxide, which may then react with a halide (e.g., chloride) to a hypohalite (e.g., hypochlorous acid) in a reaction catalyzed by myeloperoxidase. Together with other microbicidal products and proteins, these agents kill the ingested microorganism.

2.1. Production of Reduced Oxygen Species

Superoxide production is usually quantitated by optical spectroscopy, employing the superoxide-dependent reduction of ferricytochrome *c*.[6] Superoxide dismutase (SOD) is an enzyme that specifically scavenges superoxide. The addition of SOD to the reaction mixture will therefore prevent the reduction of cytochrome *c* by superoxide, thus allowing the determination of the superoxide-specific reduction of cytochrome *c*. Hydrogen-peroxide production is measured directly with a polarographic method[16] or by following the peroxidase-mediated oxidation of a substrate.[17]

Measurement of both superoxide and hydrogen peroxide production by activated phagocytes suggests that the conversion of oxygen to superoxide and then to hydrogen peroxide is quantitative; at least under experimental conditions. In a typical series of experiments conducted in our laboratory, superoxide accounted for 93% and hydrogen peroxide for 87% of the increment in oxygen consumption during the respiratory burst of human neutrophils.

The evidence in support of the production of superoxide by stimulated phagocytes is unambiguous.[7] The ability of activated phagocytes to reduce ferricytochrome *c* enclosed in resealed erythrocyte ghosts and in an SOD-sensitive manner indicates that superoxide is actually released by activated phagocytes.[18] Hydrogen peroxide, detectable in the extracellular medium of activated cells, can be derived from either superoxide via a dismutation reaction (1) or from a direct divalent reduction of oxygen by the oxidase (2).

$$O_2 \xrightarrow{e^-} O_2^- \xrightarrow[pK\,=\,4.8]{H^+} HO_2$$

$$\begin{aligned} &HO_2 + O_2^- \xrightarrow[\text{(spontaneous or catalyzed by SOD)}]{\text{dismutation}} HO_2^- + O_2 \\ &HO_2^- + H^+ \xrightarrow[pK\,=\,11.7]{} H_2O_2 \end{aligned} \qquad (1)$$

$$O_2 \xrightarrow{e^-} O_2^- \xrightarrow{e^-} O_2^{2-} \xrightarrow{H^+} HO_2^- \xrightarrow{H^+} H_2O_2 \qquad (2)$$

Until recently, there was no definite evidence in favor of either of the two pathways leading to the generation of hydrogen peroxide, basically because no method was available for the independent and simultaneous measurement of superoxide and hydrogen peroxide. Makino et al.[19] used a diacetyl deuteroheme-substituted horseradish peroxidase (HRP) that can trap both superoxide and hydrogen peroxide; this leads to the formation of compound III or compound II of the peroxidase, respectively. Both compounds are stable and, moreover, are easily distinguished by optical spectroscopy, permitting the simultaneous quantification of both superoxide and hydrogen peroxide produced by activated phagocytes. The results of this study clearly indicate that the primary product of the oxidase in phorbol myristate acetate (PMA)-stimulated neutrophils is superoxide. However, the oxidase in subcellular fractions seems to be capable of both univalent and divalent reduction of oxygen, with a stoichiometry depending on the assay conditions.[20]

2.2. Fusion of Granules with the Developing Phagosome

During the formation of the phagosome, intracellular granules migrate to, and fuse with, the phagosome membrane; in this way, granules release their contents into the lumen of the phagosome. There are three different types of granules in human neutrophilic granulocytes: azurophilic granules (lysosomelike), specific granules, and the recently recognized tertiary granules.[21,22] The specific granules presumably play a role during the early stages of microbicidal activity, just before the actual attachment of the microorganism to the phagocyte.[23] The function of the tertiary granules in the microbicidal activity is unknown. The azurophilic granules are essential for some of the intraphagosomal activities. Myeloperoxidase, a heme-containing enzyme, is a constituent of azurophilic granules. At the expense of hydrogen peroxide, and possibly of superoxide as well,[10] highly microbicidal hypohalites are produced by myeloperoxidase. The azurophilic granules also contain various proteolytic enzymes and acid hydrolases that contribute to the killing and degradation of the engulfed microorganisms.[24]

Fusion of granules with the plasma membrane or with phagosomal membranes may serve another function, however. Fractionation of broken resting cells by centrifugation on a density gradient has demonstrated a dual localization of cytochrome b_{558}, one of the best characterized putative components of the oxidase (see Section 4.1). A minor portion of cytochrome b_{558} is detectable in the fraction containing the plasma membranes, while the majority is found in the fraction containing the specific granules. Similar fractionation of activated cells has indicated substantial translocation of cytochrome b_{558} from its intracellular site to the plasma membrane.[25] This disappearance of cytochrome b_{558} from its intracellular pool coincides with the depletion of marker enzymes for specific granules from the same fraction. It must be emphasized, however, that the intracellular pool of cytochrome b_{558} has by no means been identified as the specific granules, let alone as the membranes of these granules. So far, there is only circumstantial evidence that this concept may be correct.[25]

Borregaard et al.[25] proposed that fusion of the membranes of these granules with the plasma membrane might be a means of activating the oxidase, for example, by assembly of a complete and therefore active enzyme. However, a number of subsequent studies have established that although translocation of cytochrome b_{558} might be a relevant step in the activation of the oxidase, additional phenomena (e.g., phosphorylation) are required.[26–29] Translocation of oxidase components might be important in providing a functional oxidase enzyme at the cell surface. Translocation of specific granules, and therefore fusion of its membranes with the plasma membrane, seems an early step in the activation of the microbicidal process.[23] Possibly, degranulation of specific granules prepares the phagocyte's membrane for the forthcoming demand on functional oxidase components.

3. OXIDASE PREPARATIONS

Several attempts have been made to isolate an active oxidase preparation from neutrophilic granulocytes. Because the oxidase activity is only manifest after activation, the cells have to be stimulated before an active oxidase preparation can be isolated. Although several groups have recently discovered that the oxidase in subcellular fractions derived from resting phagocytes can also be activated,[30–33] such material has not been yet used to isolate active oxidase preparations. The oxidase activity in subcellular particulate preparations derived from activated cells is associated with the nonsoluble fraction. Therefore, to permit further purification of the oxidase, it is essential to extract the oxidase with detergent(s).

3.1. Properties of Particulate and Detergent-Treated Particulate Preparations

Some published data on the oxidase activity, i.e., the superoxide production in particulate and detergent-treated particulate preparations from human neutrophils, are collected in Table I. Particulate preparations derived from activated, and subsequently broken, human neutrophilic granulocytes can account for only 10–30% of

TABLE I. Comparison of Superoxide Production in Detergent-Treated Particulate Fractions Obtained in Various Laboratories[a,b]

Particulate fraction from activated	Additions	O_2^- production (nmoles/min × equiv. of 10^6 cells)	Reference
Control cells	None	0.85–1.10	34–36
	TX-100, FAD, PE	2.6	34–36
Control cells	None	1.5	12
	DOC	4.3	12
	DOC	7.3	37
CGD cells	None	<0.02	34,38
	TX-100, FAD	0.035	34,38

[a] Superoxide production was measured as a NADPH-dependent SOD-inhibitable cytochrome-*c* reduction[15]; 50 μg protein in a particulate preparation is equivalent to 10^6 cells.[36]

[b] CGD, chronic granulomatous disease; DOC, deoxycholate; FAD, flavin adenine dinucleotide; PE, phosphatidyl ethanolamine.

the superoxide-generating activity in intact neutrophils. Moreover, the specific activity in particulate preparations is significantly lower than in intact cells. These data have been confirmed with measurements of oxygen consumption and hydrogen-peroxide production in our laboratory (M. J. Pabst, unpublished observations).

Generally, it has been found that the oxidase prefers NADPH over NADH, with K_m values of about 30 μM and 500–800 μM, respectively. The oxidase is therefore usually referred to as an NADPH oxidase.

The addition of detergents to particulate preparations has considerable effects on the oxidase activity. Babior and Kipnes[34] noted that low concentrations of Triton X-100, a nonionic detergent, completely abolished the superoxide production in particulate preparations. Almost complete restoration of this activity was found upon addition of FAD (K_a FAD: 61 nM).[13,36] The addition of phosphatidyl ethanolamine resulted in a further twofold increase in the superoxide production in the detergent-treated FAD-reconstituted particulate fraction[36] (see Table I). Deoxycholate has also been used to solubilize the oxidase activity, first by Tauber et al.[11,12] and subsequently in a modified procedure by Gabig.[37] A critical deoxycholate concentration in a very narrow concentration range added to a particulate neutrophil preparation increased the oxidase activity. Neither FAD nor phosphatidyl ethanolamine augmented this oxidase activity, in contrast to the results obtained with Triton X-100-treated preparations[11] (see Table I).

An important criterion of specificity in isolating the oxidase activity from (human) neutrophilic phagocytes is the absence of this activity in identically treated preparations from activated chronic granulomatous disease (CGD) cells or nonactivated normal cells. CGD is a syndrome in which the phagocytes are unable to produce reactive oxygen species. Therefore, stimulated CGD phagocytes show no, or a significantly decreased, SOD-inhibitable cytochrome-*c* reductase activity and hydrogen peroxide production. As a consequence, CGD phagocytes are unable to kill certain ingested microorganisms, although such cells are still perfectly capable of phagocytosis and degranulation.[14]

Particulate preparations obtained from activated CGD cells hardly produce any superoxide, as shown in Table I.[38] The slightly higher activities observed in particulate preparations from nonactivated normal cells (data not shown) might be due to the isolation and subsequent treatment of the cells. This may have caused modest activation of the oxidase.

The increase in the specific activity of particulate preparations upon the addition of detergents and, in the presence of Triton X-100, of FAD and phosphatidyl ethanolamine, requires some comment. Although the SOD-inhibitable cytochrome-*c* reductase assay is specific for superoxide production by the phagocyte oxidase, it remains to be established whether the values obtained are representative for the actual superoxide production by the intact cells. In other words, the increase in the specific activity by detergent addition to particulate preparations might be due to nonenzymatic reactions. In a recent paper by Green and Pratt,[39] in which either cytochrome *c* or acetylated cytochrome *c* was used to measure the superoxide production of particulate preparations, cytochrome *c* was found to enhance superoxide production. This might especially apply to the superoxide measurement of detergent-treated particulate preparations.

Another possible mechanism explaining the higher specific activities in detergent-treated particulate preparations involves the fact that free reduced redox mediators, such as FAD and quinones, can reduce oxygen to superoxide.[40,41] Furthermore, it has been reported that free flavins are able to uncouple the enzymatic activity of flavoproteins, for example, of salicylate hydroxylase (E.C. 1.14.13.1).[42] Similar effects have also been described for other flavoenzymes.[43,44] This uncoupling has been explained by an exchange of reducing equivalents via bound flavins to free flavins.[42] Thus, NADPH provides reducing equivalents to the oxidase, possibly to the putative flavoprotein of the oxidase (see Section 4.2). These reducing equivalents might subsequently be transmitted to free endogenous or exogenous flavins. The reduced flavin might then reduce oxygen to superoxide. Moreover, it has been observed that (flavin) semiquinone radicals can give rise to a substantial SOD-inhibitable reduction of ferricytochrome *c*.[45] Semiquinone radicals in the presence of cytochrome *c* that hardly generate any superoxide may instead reduce cytochrome *c* directly. However, the addition of SOD will remove the traces of superoxide produced and will displace the overall equilibrium, resulting in significant additional H_2O_2 formation. In preparations derived from CGD cells and nonstimulated cells, a (partial) flow of electrons through the nonfunctional oxidase cannot occur, and therefore the nonenzymatic interaction of flavins will not take place in such preparations.

Attempts at further purification of the oxidase have been troubled by the instability of the enzymatic activity and the resistance to solubilization. The superoxide production in detergent-treated particulate preparations decays rapidly, with a half-life (t-½) as short as 30–60 min at room temperature.[12,13] The activity in deoxycholate-treated preparations can be stabilized to some extent by certain additives (ethylene glycol, glycerol, and dimethyl sulfoxide).[11] Tauber and Goetzl[11] observed similar rates of deterioration of oxidase activity in detergent-treated and detergent-free particulate preparations, which argues against a relationship between the action of the detergents and the decrease in activity.

These investigators[11] have further shown that an optimal deoxycholate concentration of 0.25% (w/v) solubilized only up to 13% of the oxidase activity in a particulate preparation free of other detergents. Better results were obtained with Triton X-100-treated particulate preparations, in which 50% of the oxidase activity was regarded solubilized, because it had run through an XM300 ultrafiltration filter.[36] Such filters retain globular molecules with a molecular mass equal to or larger than 300,000 daltons.

3.2. Properties of Solubilized Oxidase Preparations

Several groups of investigators have reported the partial purification of active solubilized oxidase preparations (for a recent review, see Rossi[15]). Common features of these preparations are again low recovery and instable enzymatic activity, similar to the features of detergent-treated particulate preparations. There is disagreement as to the molecular composition of the NADPH oxidase; i.e., some preparations contain FAD but no cytochrome b_{558},[46,47] whereas other preparations, obtained in a similar way, contain cytochrome b_{558} but hardly any FAD.[48–50] Preparations without either of these chromophores,[51] or with both,[52] have also been reported, as well as preparations that contain FAD and cytochrome b_{558}, as well as ubiquinone.[53] Whether these redox components are involved in the NADPH oxidase activity of phagocytes (discussed in Section 4) certainly cannot be deduced from the above-mentioned reports. Possibly, undetectable amounts of any of these prostetic groups are sufficient to permit specific but nonenzymatic superoxide production.

One important point in this respect is the assay used to measure the oxidase activity. This is usually limited to the SOD-inhibitable ferricytochrome-*c* reduction. Apart from the problems mentioned in Section 3.1, it should also be realized that detergent treatment and solubilization of the oxidase may have caused structural changes that result in facilitated electron transport to ferricytochrome *c*. Indications for such changes were found in our laboratory: several deoxycholate-treated preparations consumed oxygen and generated superoxide but did not produce hydrogen peroxide, even in the presence of SOD and azide. This is hard to understand, because H_2O_2 would be expected to be formed whenever $O_2^{-}\cdot$ is produced. Possibly, only minute amounts of $O_2^{-}\cdot$ are produced, leading to an amplified, but still SOD-sensitive, reduction of cytochrome *c* via nonspecific side reactions (see Section 3.1). In our opinion, H_2O_2 generation should therefore be included in oxidase activity tests.

Another problem is the specificity of the NADPH oxidase activity. Because the yields are low, it is impossible to obtain a solubilized oxidase preparation from small amounts of blood. Consequently, the specificity of the oxidase preparation cannot be checked by showing absence of activity in a preparation derived from CGD cells. In fact, many investigators use animal blood as the starting material, adding another problem to this complicated research area. However, it is generally found that oxidase preparations derived from resting cells are devoid of enzymatic activity, the K_m for NADPH is always far below that for NADH, and the SOD-independent substrate reduction is usually low. Therefore, the oxidase activity in most preparations is probably at least qualitatively related to the NADPH oxidase in intact phagocytes.

4. THE NADPH OXIDASE SYSTEM

In analogy to the mitochondrial respiratory chain, the NADPH oxidase of phagocytic leukocytes has been postulated to be composed of several components in a short electron-transfer chain. This idea is based on the possible involvement of a flavoprotein, ubiquinone, and cytochrome b_{558} in the oxidase activity. From genetic studies, it is known that at least three proteins are involved in the expression of oxidase activity in intact cells.[54] However, these proteins are not necessarily structural components of the oxidase system, because proteins of the oxidase activation machinery and proteins involved in post-translational events are also required for oxidase activity. Biochemically, one might expect more than one component in the NADPH oxidase system to dissect the two-electron transfer from NADPH to the system into a one-electron transfer from the system to oxygen. In the following sections, the arguments for and against the involvement of cytochrome b_{558}, a flavoprotein, and ubiquinone in the NADPH oxidase system are critically evaluated. Moreover, a brief survey of the attempts to characterize these components and their possible function in the oxidase enzyme is presented.

4.1. Cytochrome b_{558}

Studies on CGD neutrophils have contributed significantly to the identification of at least one component involved in the oxidase activity. Using optical spectroscopy, Segal and co-workers[55,56] observed the presence of a *b* heme in normal neutrophils that was missing in the neutrophils from CGD patients with the X-chromosome-linked form of the disease. Moreover, these investigators showed that activation of normal neutrophils under anaerobic conditions resulted in the reduction of the *b* heme. Subsequent gassing with oxygen caused reoxidation of the heme.[57] This last result indicates that the *b*-heme moiety in activated neutrophils functions in an electron-transport chain with oxygen. In accordance with the accepted nomenclature for cytochromes, this hemoprotein has been tentatively named cytochrome b_{558} after the position of the α-band at 558 nm in the heme absorbance spectrum. In the literature the name cytochrome b_{-245} is also used, referring to its extremely low redox midpoint potential.

4.1.1. Involvement in the NADPH Oxidase System

Various reports have appeared that confirm the correlation between the absence of the cytochrome-b_{558} absorbance spectrum in neutrophils from X-linked CGD patients and the inability of these cells to produce reactive oxygen species.[58,59] Studies on the oxidase activity at the single-cell level in relatives of these male patients reveal a mosaic distribution of normal and defective neutrophils in female carriers of the disease.[59] This distribution is explained by the lyonization phenomenon, i.e., the random inactivation of either the normal or the affected X-chromosome, providing additional evidence for the X-linked nature of this specific defect. In a recent study[59] with carriers of the X-linked type of CGD, a close correlation was observed between the percentage of normal (oxidase-active) cells and the cytochrome-b_{558} content expressed as percentage of the content in normal cells. Figure 2 presents our own results, which confirm the published findings. This strict correlation suggests that only those cells that contain (a normal amount of) cytochrome b_{558} are able to generate reduced oxygen species.

A similar correlation between the content of cytochrome b_{558} and the oxidase

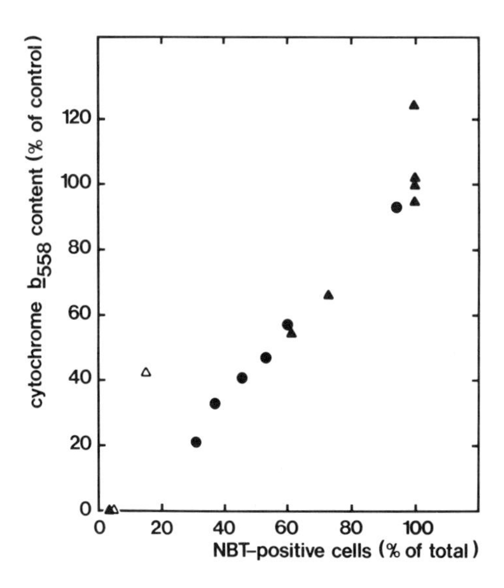

FIGURE 2. Correlation between the relative amount of cytochrome b_{558} and the percentage of oxidase-active cells in carriers of the X-linked form of chronic granulomatous disease. Oxidase activity was determined with the nitroblue-tetrazolium slide test[60] and cytochrome b_{558} was measured, as described.[54] The cytochrome-b_{558} content in normal neutrophils is 4.4 ± 0.7 pmoles cytochrome b_{558} per 10^6 neutrophils (mean ± SD; $N = 13$). The correlation coefficient calculated from values for 15 carriers is 0.90. (●) Mothers; (▲) sisters or maternal grandmothers; (△) possible heterozygotes.

activity has been observed with neutrophils from so-called X-linked partially cytochrome-*b*-deficient CGD patients.[61,62] The cells from these patients contain only 10–30% of the normal amount of cytochrome b_{558} and show a similar deficiency of the respiratory burst (R. S. Weening and D. Roos, unpublished observations).

Finally, Iizuka et al.[63] found an inhibitory effect of heme-iron-binding agents, such as pyridine and imidazole, on the oxidase activity in intact neutrophils, together with a spectral change in cytochrome b_{558}. Both the oxidase inhibition and the spectral change were reversible, again suggesting the involvement of cytochrome b_{558} in the NADPH oxidase system.

Some arguments against the participation of cytochrome b_{558} in the oxidase activity have also been put forward. One is the slow reducibility of the heme under anaerobic conditions, in cells and in particulate cell fractions. In the presence of NADPH, only 38% of the total cytochrome-b_{558} content is reduced within 30 min.[12,52] However, Cross et al.[64] provided evidence that the reduction of cytochrome b_{558} proceeds at maximal speed only in the presence of oxygen. Their calculated turnover values can now account for the enzymatic reduction of oxygen by cytochrome b_{558} with a velocity similar to the rate of superoxide generation in a partially purified oxidase preparation.[46,64]

Another counterargument for a role of cytochrome b_{558} in the oxidase activity is the existence of partially purified oxidase preparations with SOD-inhibitable NADPH-cytochrome-*c* reductase activity, but without the apparent presence of cytochrome *b*. However, as discussed in Section 3, this activity does not necessarily represent the (complete) NADPH oxidase activity of intact cells; moreover, trace amounts of cytochrome *b* might be sufficient to catalyze the transfer of electrons to oxygen.

4.1.2. Molecular Composition

Thus, cytochrome b_{558} is now widely accepted as a component of the NADPH oxidase of phagocytic leukocytes. As such, it is by far the best characterized component of this system. Because cytochrome b_{558} shows an easily recognizable light absorbance spectrum, several attempts have been made to purify and characterize this protein. A number of difficulties have troubled these efforts. Cytochrome b_{558} is an integral membrane protein requiring detergents for its solubilization and subsequent purification. It is also a protein that is poorly detectable with the usual protein staining procedures. Moreover, identification on sodium dodecyl sulfate (SDS) gels is hampered by the loss of the heme group from the protein in the presence of SDS. Finally, it is apparently a poor immunogen, because only very recently a precipitating polyclonal antibody against cytochrome b_{558} has been described.[65]

Table II summarizes some of the properties of cytochrome b_{558} reported by several research groups. Initially, general disagreement existed as to the molecular weight, especially when determined by sodium dodecyl sulfate–polyacrylamide gel electrophoresis (SDS–PAGE). The two most recent reports,[68,70] however, agree in that they describe two apparent subunits of the protein: one that appears as a smear on the gel between 76,000 and 92,000 M_r and one as a sharp band at about 23,000 M_r. These findings have been confirmed by our group (B. G. J. M. Bolscher and D. Roos, unpublished observations). The identification of the two proteins as true subunits of cytochrome b_{558} depends on a number of independent criteria: copurification of both proteins with the heme of the cytochrome through affinity and gel-filtration chromatography,[68,70] and apparent absence of both proteins in the neutrophils from X-linked CGD patients,[68,70,72] hydrodynamic properties of the complex,[65] cross-linking of both proteins into a complex with M_r 120,000–135,000,[65,70] and precipitation of the complex by an antibody against the 23,000-dalton subunit.[65] Which of the two subunits carries the heme is unknown.

A clue for the answer to this last question may be derived from the work of

TABLE II. Comparison of Some Properties of Cytochrome b_{558} Protein Reported by Several Research Groups[a]

Investigators	M_r determination		Glycoprotein
	Gel filtration	SDS–PAGE	
Segal et al.[66–68]	500,000–600,000	76,000–92,000, 23,000	Yes
Pember et al.[69]	n.d.	14,000, 12,000, 11,000	n.d.
Bellavite et al.[48]	n.d.	32,000	n.d.
Parkos et al.[65,70]	100,000–135,000[b]	91,000, 22,000	Yes
Lutter et al.[71]	235,000	127,000	Yes
Bolscher et al. (unpublished data)	235,000	60,000–90,000, 23,000	Yes

[a]n.d., not determined.
[b]Corrected for the contribution of detergents.

Orkin's group.[72,73] These investigators have cloned a gene on the X-chromosome that is abnormal in X-linked CGD. The transcript of the gene is expressed in hematopoietic cells of the phagocytic lineage and was absent or structurally abnormal in four X-CGD patients. However, the gene encodes a protein that appears different from cytochrome b_{558}, because its predicted amino acid composition is different from that derived from purified cytochrome-b_{558} preparations.[67,71] Moreover, the protein encoded by the X-CGD gene does not show an evident heme-binding site. Royer-Pokora et al.[73] hypothesized that this protein might form a complex with cytochrome b_{558}, essential for the intracellular stability of the cytochrome. The absence of the heme signal in X-linked CGD cells might therefore be secondary to the absence or abnormal structure of the predicted X-CGD protein. Orkin's group has subsequently succeeded in raising polyclonal antisera against a synthetic peptide with an amino acid sequence derived from the complementary DNA (cDNA) sequence of the X-CGD gene and against a fusion protein produced in *E. coli*. These antibodies were then shown to react with the 90,000-dalton component of purified cytochrome b_{558}.[72] This would identify the high-molecular-mass subunit of cytochrome b_{558} as the X-CGD protein, in accordance with the predicted molecular mass of 54,000 daltons of the latter[74] versus 48,000–50,000 daltons for the deglycosylated high-molecular-mass subunit in cytochrome-b_{558} preparations.[70] A positive identification of both proteins was obtained by Teahan et al.[74] who sequenced 43 amino acids at the N terminus of the high M_r subunit of cytochrome b_{558} and found homology with the X-CGD nucleotide sequence.

4.1.3. Function

From the pyridine hemochrome spectrum, one protoheme IX per polypeptide was calculated.[71] Considerably higher molar absorbance coefficients were obtained from purified as compared with unpurified cytochrome b_{558}.[71] This has important implications for the calculation of the specific heme content of purified preparations, a fact that is not always recognized.[68] The midpoint potential of cytochrome *b* is sufficiently low (−220 to −250 mV) to permit reduction of oxygen to superoxide by electrons donated by the cytochrome. We found this midpoint potential to be even lower in purified preparations.[71]

Whether the heme moiety mediates the electron transfer to oxygen, i.e., whether cytochrome b_{558} acts as a terminal oxidase, is a matter of dispute. This has been investigated with carbon monoxide (CO)-binding studies. CO reacts with reduced heme groups similarly to the action of oxygen and induces a pronounced shift in the position of the absorbance peaks of the reduced hemoprotein. In cytochrome b_{558}, however, CO induces only a minor blue shift of 1–2 nm in the position of the Soret peak.[71,75–77] Moreover, relatively long incubation times are required to complete CO binding with the heme in cytochrome b_{558} (Lutter et al.,[71] but see also Cross et al.[76]). Treatment of the cytochrome-b_{558}–CO complex with oxygen results in the removal of CO and the concomitant recovery of oxidized cytochrome b_{558}, a process with similar slow kinetics as those of the binding of CO.[71] Each of these reactions can be measured in intact cells as well as with crude or purified cytochrome-b_{558} preparations. Cross et al.[76,78] found faster reactions between CO and cytochrome b_{558}, but similar slow reactions with butyl-isocyanide, another heme ligand.[52] The binding of CO to the heme in cytochrome b_{558} is agreed on by most investigators but, because of its slow kinetics and minor effect on the spectral characteristics, the relevance of this phenomenon to a possible direct heme–oxygen interaction is doubtful. It should also be noted that neither CO nor azide or cyanide inhibits the respiratory burst in activated neutrophils.[2,77] Thus, even if these ligands bind to the heme moiety of cytochrome b_{558}, they do not inhibit the transfer of electrons from the oxidase to oxygen.

4.2. A Flavoprotein

Glucose 6-phosphate dehydrogenase (G6PD) is the first enzyme in the hexose monophosphate (HMP) shunt. The HMP shunt supplies the neutrophil with NADPH. It has been found that neutrophils with a drastic reduction in the G6PD activity (i.e., with an activity of less than 5% of normal) also exhibit a decreased respiratory burst.[79,80] Therefore, it has been postulated that NADPH provides the oxidase with reducing equivalents for the reduction of oxygen. This idea has been strengthened by the low K_m for NADPH in particulate and solubilized oxidase preparations[36,48,51,81] and by the inhibition of the oxidase activity by NADPH analogues.[82,83] Therefore, it is now generally accepted that the superoxide-generating enzyme in phagocytic leukocytes is an NADPH oxidase. In analogy to other NADPH-dependent enzymes,[84,85] it has been proposed that a flavin moiety in the oxidase accepts the electrons from NADPH. Support for flavin involvement in the oxidase activity comes from a number of observations.

4.2.1. Involvement in the NADPH Oxidase System

Babior and Kipnes[34] found that addition of Triton X-100 to a particulate fraction of activated neutrophils completely destroys the NADPH-dependent superoxide generation of these preparations (see Section 3). This activity was restored by addition of FAD. Studies by Light et al.[12] confirmed the FAD dependency in Triton X-100-treated oxidase preparations. These investigators also observed that FAD analogues (in contrast to FAD) capable of only two-electron transfer reactions did not restore the oxidase activity.

Cross et al.[52] showed that FAD present in a particulate oxidase preparation was reduced by NADPH under anaerobic conditions and reoxidized after admittance of oxygen to the preparation. This finding indicates that, like cytochrome b_{558}, a flavin is present that functions in a redox chain with oxygen (but not necessarily in the same chain). These investigators also showed that the addition of diphenylene iodonium, an inhibitor of the NADPH oxidase, prevents the anaerobic reduction of both FAD and cytochrome b_{558}.[86]

Similarly, Gabig and Lefker[87] found that separation of an active oxidase preparation into fractions containing either the flavin(s) or cytochrome b_{558} results in loss of oxidase activity. In the flavin-containing fraction, FAD was still reduced by NADPH under anaerobic conditions and reoxidized by oxygen. This fraction no longer generated superoxide, but it did reduce artificial electron acceptors in an NADPH-dependent reaction. The cytochrome-b_{558}-containing fraction did not generate superoxide either, and the heme was no longer reduced by NADPH under anaerobic conditions. After reduction by dithionite, however, cytochrome b_{558} was again rapidly reoxidized by oxygen. Apparently, reconstitution experiments with the two fractions failed. These experiments suggest a flavin involvement in the oxidase activity, in a sequence NADPH $\rightarrow$ flavin $\rightarrow$ cytochrome $b_{558} \rightarrow O_2$.

With electron-spin resonance (ESR) studies, Kakinuma et al.[88] found that anaerobic redox titrations with NADPH of plasma membrane fractions from pig neutrophils resulted in flavin semiquinone formation only when the cells from which the membranes had been prepared were prestimulated to activate the oxidase. About 50% of the flavins in the membranes showed a midpoint potential at pH 7 of about -280 mV. This E_m value lies between that of cytochrome b_{558} (E_m -245 mV) and that of $NADP^+$/NADPH (E_m -320 mV), suggesting a two-electron transfer from NADPH to FAD and a one-electron transfer from FAD to cytochrome b_{558}. Unfortunately, the relationship of these findings to the NADPH oxidase was not unequivocally established.

In a more recent report, Kakinuma et al.[47] solubilized the NADPH oxidase from pig neutrophil membranes with octyl glucoside and found it to be acidic (pI 5.0) and mainly of 67,000 dalton molecular mass, in accordance with two other reports.[51,82] Kakinuma's preparation contained FAD but no cytochrome, and she speculated[47] that perhaps the NADPH oxidase flavoprotein obtained by her might form a complex with the basic cytochrome b_{558} in situ. Thus, Kakinuma proposes that the flavoprotein is only one component of the total oxidase complex.

4.2.2. Deficiencies in Chronic Granulomatous Disease

A number of reports in the literature claim deficiencies in the content of noncovalently bound FAD in neutrophils or subcellular fractions from CGD patients. Cross et al.[89] found only one half the normal FAD content in neutrophil plasma membranes of three X-linked CGD patients. By contrast, these investigators found normal amounts of FAD in membranes derived from neutrophils of patients with the common autosomal form of CGD. Neutrophils from two different male CGD patients with an unidentified genetic background were analyzed for FAD by Gabig.[35] A particulate fraction of the neutrophils from one of these patients contained only 8% of the normal FAD content, and preparations from the other patient contained a normal amount of FAD. Another study by the same group[90] has shown a flavin deficiency (30% of normal) in the cells from two other CGD patients, one with the X-linked form and one presumably with an autosomal form, and in cells from the mother of the X-linked CGD patient. Borregaard and Tauber[91] described the presence of 50–60% of normal FAD values in extracts of total cells from two genetically undefined CGD patients. This last study, in contrast to that by Cross et al.,[89] indicated that the flavoprotein relevant for the oxidase activity was not localized in the plasma membrane but in a fraction containing the specific granules. Borregaard and Tauber[91] found no abnormalities in the FAD content of the fraction containing the plasma membranes or in the cytosolic fraction. More recently, Ohno et al.[92] published a survey of 30 CGD patients. In 18 patients with the autosomal form of the disease, a normal amount of FAD was found in a particulate fraction of the neutrophils. In 4 of the 11 patients with the X-linked form of the disease (X-CGD), decreased amounts of FAD were found; however, because of the considerable variation in the FAD levels per individual, these results are difficult to interpret. Bohler et al.[62] investigated 25 CGD patients. FAD was measured in a cholate extract of a membrane fraction that showed NADPH oxidase activity in normal cells. In three patients with the autosomal form of the disease, the amount of FAD was normal. In four of the 18 patients with the X-linked cytochrome-*b*-negative form of the disease, the flavin spectrum appeared normal but the amount was decreased. In four other patients with this type of CGD, the spectrum was abnormal.

Our group has also examined the FAD content of cells from various CGD patients. Because it seemed clear from previous studies that FAD deficiencies might

TABLE III. Contents of Noncovalently Bound FAD in Cytoplast Preparations from Normal Donors and from Patients with Chronic Granulomatous Disease

Donor	Flavin content[a] (pmoles per 10^6 cytoplasts)	
	Membranes	Cytoplasm
X-linked form of CGD		
Patient I	0.36	0.34
Patient II	0.25	0.30
Common autosomal form of CGD		
Patient I	0.30	0.37
Patient II	0.35	0.48
Patient III	0.52	0.39
Rare autosomal form of CGD[b]		
Father (carrier)	0.43	0.30
Mother (carrier)	0.51	0.28
Patient I	0.39	0.37
Patient II	0.51	0.37
Patient III	0.39	0.27
Sister (carrier)	0.38	0.26
Controls (N = 7)	0.45 ± 0.09 (Mean ± SD)	0.42 ± 0.09 (Mean ± SD)

[a]For experimental conditions, see Lutter *et al.*[27]
[b]From Weening *et al.*[54]

be restricted to certain types of CGD, we used only cells from patients who had been genetically classified by complementation studies with fused monocytes.[54,93] Because some studies had indicated that an FAD deficiency might be confined to certain cell fractions, we used neutrophil cytoplasts, i.e., enucleated and granule-depleted neutrophils. These cytoplasts can still be triggered to express oxidase activity to a degree comparable to that of intact neutrophils[94] but obviously contain less flavoproteins irrelevant to the oxidase system. We found normal FAD levels in the cytoplast fractions from all CGD patients studied (Table III).

Obviously, more extensive studies with large well-defined groups of CGD patients are needed before conclusions about a possible FAD deficiency can be drawn. In addition, only noncovalently bound flavins have been studied so far, although covalently bound flavins might, in principle, also be relevant for the oxidase activity.

4.2.3. Molecular Weight

Several attempts have been made to determine the molecular weight of the oxidase protein that accepts the reducing equivalents from NADPH. Umei et al.[82] used 2,3-dialdehyde NADPH and found this compound to inhibit the oxidase activity in a solubilized preparation from pig neutrophils, in competition with NADPH. After subsequent treatment with cyanoboro[^{3}H]hydride, these investigators observed a specific labeling of a 66,000-M_r protein in their preparation. The labeling was prevented by excess NADPH or pretreatment of the preparation with *p*-chloromercuribenzoate, but apparently no attempts were made to study the effect of neutrophil stimulation on the labeling or to investigate the labeling in preparations from CGD cells.

As mentioned in Section 4.2.1, Kakinuma et al.[47,88] purified a flavoprotein from pig neutrophils with NADPH oxidase activity; this protein has a molecular weight of 67,000.

Doussière et al.[83] used *N*-4-azido-2-nitrophenyl aminobutyryl $NADP^+$ (arylazido $NADP^+$), a photoactivatable $NADP^+$ analogue that binds covalently to $NADP^+$- or NADPH-converting proteins. Both arylazido $NADP^+$ and arylazido NADPH inhibited the oxidase activity in particulate and solubilized oxidase preparations from bovine neutrophils, in competition with NADPH. Several proteins were photolabeled by these compounds. The photolabeling of one of these proteins, with a molecular weight of 65,000, was decreased when the oxidase preparation was derived from nonstimulated neutrophils or when Cibacron Blue or mersalyl (oxidase inhibitors) were added together with arylazido NADP(H). During the oxidase purification procedure, the 65,000-M_r protein emerged as the preferentially photolabeled protein. Because these investigators had previously found this 65,000-M_r protein to be the major constituent of their purified oxidase preparations,[51] they suggested that this protein acts both as an NADPH dehydrogenase and as an oxygen reductase to generate O_2^-.

Cross and Jones[86] used diphenylene [^{125}I]iodonium, an inhibitor of oxidase activity, to label components of a solubilized oxidase preparation from pig neutrophils. A protein of 45,000 M_r was preferentially labeled; this reaction was inhibited by NADPH. No difference was found in labeling between oxidase prepared from stimulated or unstimulated cells. CGD cells were not studied, but the investigators suggested that this 45,000-M_r protein might be identical to a protein of similar molecular weight that is phosphoryated upon stimulation of normal neutrophils but not upon stimulation of autosomal CGD cells.[95] Moreover, in rat peritoneal macrophages, an additional protein of 70,000 M_r was labeled.[96] The notion that either of these proteins is a flavoprotein lacks experimental support, however.

Taken together, these results suggest that the flavoprotein involved in the NADPH oxidase may have a molecular mass of about 67,000 daltons.

4.3. Ubiquinone-50

In view of the possible involvement of both a flavoprotein and cytochrome b_{558} in the oxidase activity, it has been postulated that the NADPH oxidase is a short electron-transfer chain. In analogy to the mitochondrial respiratory chain, the electron-transfer chain of the oxidase might contain ubiquinone. Several investigators have therefore searched for ubiquinones in oxidase preparations from neutrophils.

Schneider and co-workers[97] were the first to present evidence for the presence in neutrophils of a lipid with quinonelike properties, in a later report identified as ubiquinone-50.[98] Moreover, the stimulatory and inhibitory effects of quinone analogues on the oxidase activity of neutrophils suggested to these authors the requirement of a quinone in this activity.[99] The presence of a nonmitochondrial quinone in neutrophils was confirmed by Cunningham et al.,[100] although in considerably lower amounts than those reported by Schneider's group.[101] Cross et al.[76] analyzed plasma membranes of neutrophils, which are expected to contain the oxidase, but failed to detect ubiquinone. This work was criticized for its presumably inefficient extraction procedure for ubiquinone.[100] Moreover, Schneider et al.[101] suggested a subcellular localization for ubiquinone in a compartment different from the plasma membrane, recently specified as the tertiary granules.[102]

Earlier work by our group has indicated that ubiquinones do not play a role in the respiratory burst of activated neutrophils.[103] Spectroscopic measurements indicated that 2-*n*-heptyl-hydroxyquinoline-*N*-oxide (HQNO), an inhibitor of quinone-dependent electron-transfer reactions, had no effect on the reduction of cytochrome b_{558} in cells stimulated under anaerobic conditions. Moreover, ultraviolet (UV) radiation, which destroys quinones, did not affect the respiratory burst of neutrophils. Even with the use of an efficient extraction procedure for quinones we have been unable to detect any ubiquinone in either intact neutrophils or in neutrophil cytoplasts.[27] We thus excluded a role for ubiquinone in the oxidase activity. However, we noticed that platelets, which constitute a significant contamination in many neutrophil preparations, are a rich source of ubiquinones. Subfractionation studies by Cross et al.[104] revealed the presence of ubiquinone only in mitochondrial fractions derived from their neutrophil preparation, again calling into question a function for ubiquinone in the respiratory burst of neutrophils.

In a more recent report, Gabig and Lefker[53] described the presence of FAD, ubiquinone, and cytochrome b_{558} in a particulate oxidase preparation from human neutrophils, in a molar ratio of about 1.3 : 1 : 2. Addition of NADPH under anaerobic conditions to a preparation derived from unstimulated cells resulted in reduction of FAD and ubiquinone; similar studies with a preparation from stimulated cells showed reduction of all three components. The investigators conclude that activation of the respiratory burst in neutrophils results in linkage of electron-carrier function between endogenous ubiquinone and cytochrome b_{558}. However, their neutrophil preparations contained "less than one platelet per neutrophil," which may have yielded sufficient ubiquinone to explain their results. Indeed, there is no experimental evidence that ubiquinone forms part of an *intrinsic* electron-transfer chain in neutrophils. Moreover, these results are in contrast to those obtained by Kakinuma et al.,[88] who found that the FAD in oxidase preparations from unstimulated cells was not reduced by NADPH. Possibly, exogenous ubiquinone in the preparations of Gabig and Lefker[53] mediated a nonspecific interaction between NADPH and a flavoprotein.

5. CONCLUSIONS

The NADPH oxidase of phagocytic leukocytes is an important defense system against invading microorganisms. This enzyme mediates the transfer of electrons from NADPH to oxygen, with superoxide ($O_2^- \cdot$) as the sole product. Hydrogen peroxide is produced via dismutation of superoxide; there are no indications for the formation of singlet oxygen or hydroxyl radicals by phagocytes. (This does not exclude the formation of these reactive oxygen products in inflammatory loci via, e.g., extraphagocytic iron-catalyzed reactions.) The NADPH oxidase is inactive in resting cells; during phagocytosis, it is activated into a highly effective antimicrobial system. The resulting increase in cell respiration is called the respiratory burst. During activation, translocation of oxidase components to the plasma membrane and the phagosomal membrane takes place; the function of this process is unknown.

The purification and characterization of NADPH oxidase have been hampered by several problems: instability of the enzymatic activity in particulate or solubilized cell fractions, resistance to solubilization, low recovery of enzymatic activity, and a lack of unambiguous assay systems. The commonly employed SOD-sensitive ferricytochrome-*c* reduction, although specific for superoxide, may not reflect the true enzymatic generation of superoxide. We recommend the additional assay of hydrogen peroxide generation.

Three components have been proposed to participate in the NADPH oxidase system: cytochrome b_{558}, a flavoprotein, and ubiquinone-50. Of these, cytochrome b_{558} is widely accepted as a true oxidase component and is by far the best characterized. This protein consists of two subunits, one heavily glycosylated chain that is absent or abnormal in X-linked chronic granulomatous disease, and one smaller subunit. Which of these two subunits carries the heme is unknown. The function of the heme group is probably the transfer of electrons to oxygen, although studies with carbon monoxide have not provided evidence for this concept.

Indications for the involvement of a flavoprotein are mainly derived from anaerobic treatment of oxidase preparations with NADPH, which leads to reduction of cytochrome b_{558} as well as reduction of a flavoprotein. Dissection of these two phenomena by chemical or physical means indicates a sequence NADPH → flavoprotein → cytochrome b_{558} → oxygen. Flavin deficiencies in X-linked chronic granulomatous disease have not been established beyond any doubt. The molecular weight of the flavoprotein is probably around 67,000; identity with other proteins involved in the respiratory burst phenomenon lacks experimental proof.

Most likely, ubiquinone-50 does not take part in the NADPH oxidase of intact phagocytes. Indications to this effect are probably caused by contamination of phagocyte suspensions with platelets. Complete understanding of the NADPH oxidase structure requires purification of oxidase components and reconstitution of an active oxidase system.

REFERENCES

1. Klebanoff SJ, Clark RA: *The Neutrophil: Function and Clinical Disorders*. Amsterdam, Elsevier/North-Holland Biomedical, 1978.
2. Sbarra AJ, Karnovsky ML: The biochemical basis of phagocytosis. I. Metabolic changes during the ingestion of particles by polymorphonuclear leukocytes. *J Biol Chem* **234:**1355–1362, 1959.
3. Dewald B, Baggiolini M, Curnutte JT, et al: Subcellular localization of the superoxide-forming enzyme in human neutrophils. *J Clin Invest* **63:**21–29, 1979.
4. Briggs RT, Drath DB, Karnovsky ML, et al: Localization of NADH oxidase on the surface of human polymorphonuclear leukocytes by a new cytochemical method. *J Cell Biol* **67:**566–586, 1975.
5. Iyer GYN, Islam MF, Quastel JH: Biochemical aspects of phagocytosis. *Nature (Lond)* **192:**535–541, 1961.
6. Babior BM, Kipnes RS, Curnutte JT: Biological defense mechanisms: The production by leukocytes of superoxide, a potential bactericidal agent. *J Clin Invest* **52:**741–744, 1973.
7. Hamers MN, Roos D: Oxidative stress in human neutrophilic granulocytes: Host defence and self defence, in Sies H (ed): *Oxidative Stress*. London, Academic, 1985, p 351.
8. Britigan BE, Rosen GM, Chai Y, et al: Do human neutrophils make hydroxyl radical? *J Biol Chem* **261:**4426–4431, 1986.
9. Klebanoff SJ: Myeloperoxidase–halide–hydrogen peroxide anti-bacterial system. *J Bacteriol* **95:**2131–2138, 1968.
10. Cuperus RA, Muijsers AO, Wever R: The superoxide dismutase activity of myeloperoxidase; formation of Compound III. *Biochim Biophys Acta* **871:**78–84, 1986.
11. Tauber AI, Goetzl EJ: Structural and catalytic properties of the solubilized superoxide-generating activity of human polymorphonuclear leukocytes. Solubilization, stabilization in solution, and partial characterization. *Biochemistry* **18:**5576–5584, 1979.
12. Light DR, Walsh C, O'Callaghan AM, et al: Characteristics of the cofactor requirements for the superoxide-generating NADPH oxidase of human polymorphonuclear leukocytes. *Biochemistry* **20:**1468–1476, 1981.
13. Babior BM, Peters WA: The O_2^--producing enzyme of human neutrophils. Further properties. *J Biol Chem* **256:**2321–2323, 1981.

14. Tauber AI, Borregaard N, Simons E, et al: Chronic granulomatous disease: A syndrome of phagocyte deficiencies. *Medicine (Baltimore)* **62:**286–309, 1983.
15. Rossi F: The O_2^{-}·-forming NADPH oxidase of the phagocyte: Nature, mechanisms of activation and function. *Biochim Biophys Acta* **853:**65–89, 1986.
16. Test ST, Weiss SJ: Quantitative and temporal characterization of the extracellular H_2O_2 pool generated by human neutrophils. *J Biol Chem* **259:**399–405, 1984.
17. Ruch W, Cooper PH, Baggiolini M: Assay of H_2O_2 production by macrophages and neutrophils with homovanillic acid and horse radish peroxidase. *J Immunol Methods* **63:**347–357, 1983.
18. Roos D, Eckmann CM, Yazdanbakhsh M, et al: Excretion of superoxide by phagocytes measured with cytochrome-c entrapped in resealed erythrocyte ghosts. *J Biol Chem* **259:**1770–1775, 1984.
19. Makino R, Tanaka T, Iizuka T, et al: Stoichiometric conversion of oxygen to superoxide anion during the respiratory burst in neutrophils. *J Biol Chem* **261:**11444–11447, 1986.
20. Green TR, Wu DE: The NADPH : O_2 oxidoreductase of human neutrophils. Stoichiometry of univalent and divalent reduction of O_2. *J Biol Chem* **261:**6010–6015, 1986.
21. Bainton DF: Sequential degranulation of the two types of polymorphonuclear leukocyte granules during phagocytosis of microorganisms. *J Cell Biol* **58:**249–264, 1973.
22. Dewald B, Bretz U, Baggiolini M: Release of gelatinase from a novel secretory compartment of human neutrophils. *J Clin Invest* **70:**518–525, 1982.
23. Gallin JI: Neutrophil specific granules: A fuse that ignites the inflammatory response. *Clin Res* **32:**320–328, 1984.
24. Wright DG: The neutrophil as a secretory organ of the host defense, in Gallin JI, Fauci AS (eds): *Advances in Host Defense Mechanisms.* New York, Raven, 1982, vol 1: *Phagocytic Cells,* p 75.
25. Borregaard N, Heiple JN, Simons ER, et al: Subcellular localization of the b-cytochrome component of the human neutrophil microbicidal oxidase: Translocation during activation. *J Cell Biol* **97:**52–61, 1983.
26. Garcia RC, Segal AW: Changes in the subcellular distribution of the cytochrome b_{-245} on stimulation of human neutrophils. *Biochem J* **219:**233–242, 1984.
27. Lutter R, van Zwieten R, Weening R, et al: Cytochrome b, flavins, and ubiquinone-50 in enucleated human neutrophils (PMN cytoplasts). *J Biol Chem* **259:**9603–9606, 1984.
28. Ohno Y, Seligmann BE, Gallin JI: Cytochrome *b* translocation to human neutrophil plasma membranes and superoxide release. *J Biol Chem* **268:**2409–2414, 1985.
29. Parkos CA, Cochrane CG, Schmitt M, et al: Regulation of the oxidative response of human granulocytes to chemoattractants. *J Biol Chem* **260:**6541–6547, 1985.
30. Heyneman RA, Vercauteren RE: Activation of a NADPH oxidase from horse polymorphonuclear leukocytes in a cell-free system. *J Leukocyte Biol* **36:**751–759, 1984.
31. McPhail LC, Shirley PS, Clayton CC, et al: Activation of the respiratory burst enzyme from human neutrophils in a cell-free system. *J Clin Invest* **75:**1735–1739, 1985.
32. Curnutte JJ: Activation of human neutrophil nicotinamide adenine dinucleotide phosphate, reduced (triphosphopyridine nucleotide, reduced) oxidase by arachidonic acid in a cell-free system. *J Clin Invest* **75:**1740–1743, 1985.
33. Bromberg Y, Pick E: Activation of NADPH-dependent superoxide production in a cell-free system by sodium dodecyl sulphate. *J Biol Chem* **260:**13539–13545, 1985.
34. Babior BM, Kipnes RS: Superoxide-forming enzyme from human neutrophils: Evidence for a flavin requirement. *Blood* **50:**517–524, 1977.
35. Gabig TG: The NADPH-dependent O_2^{-}-generating oxidase from human neutrophils. Identification of a flavoprotein component that is deficient in a patient with chronic granulomatous disease. *J Biol Chem* **258:**6352–6356, 1983.
36. Gabig TG, Babior BM: The O_2^{-}-forming oxidase responsible for the respiratory burst in human neutrophils. Properties of the solubilized enzyme. *J Biol Chem* **254:**9070–9074, 1979.
37. Gabig TG, Schervish EW, Santinga JT: Functional relationship of the cytochrome *b* to the superoxide-generating oxidase of human neutrophils. *J Biol Chem* **257:**4114–4119, 1982.

38. Curnutte JT, Kipnes RS, Babior BM: Defect in pyridine nucleotide dependent superoxide production by a particulate fraction from the granulocytes of patients with chronic granulomatous disease. *N Engl J Med* **293:**628–632, 1975.
39. Green TR, Pratt KL: A reassessment of the product specificity of the NADPH : O_2 oxidoreductase of human neutrophils. *Biochem Biophys Res Commun* **142:**213–220, 1987.
40. Misra HP, Fridovich I: The univalent reduction of oxygen by reduced flavins and quinones. *J Biol Chem* **247:**188–192, 1971.
41. Michelson AM: Chemical production of superoxide anions by reaction between riboflavin, oxygen and reduced nicotinamide adenine dinucleotide. Specificity of test for $O_2^- \cdot$, in Michelson AM, McCord JM, Fridovich I (eds): *Superoxide and Superoxide Dismutase*. London, Academic, 1977, p 87.
42. Tu SC, Romero FA, Wang LH: Uncoupling of the substrate monooxygenation and reduced pyridine nucleotide oxidation activities of salicylate hydroxylase by flavins. *Arch Biochem Biophys* **209:**423–432, 1981.
43. Kishore GM, Snell EE: Reactivity of an FAD-dependent oxygenase with free flavins: A new mode of uncoupling in flavoprotein oxygenases. *Biochem Biophys Res Commun* **87:**518–523, 1979.
44. Grover TA, Piette LH: Influence of flavin addition and removal on the formation of superoxide by NADPH-cytochrome P-450 reductase: A spin-trap study. *Arch Biochem Biophys* **212:**105–114, 1981.
45. Winterbourn CC: Cytochrome c reduction by semiquinone radicals can be indirectly inhibited by superoxide dismutase. *Arch Biochem Biophys* **209:**159–167, 1981.
46. Glass GA, DeLisle DM, DeTogni P, et al: The respiratory burst oxidase of human neutrophils. Further studies of the purified enzyme. *J Biol Chem* **261:**13247–13251, 1986.
47. Kakinuma K, Fukuhara Y, Kaneda M: The respiratory burst of neutrophils. Separation of an FAD enzyme and its characterization. *J Biol Chem* **262:**12316–12322, 1987.
48. Bellavite, P, Jones OTG, Cross AR, et al: Composition of partially purified NADPH oxidase from pig neutrophils. *Biochem J* **223:**639–648, 1984.
49. Berton G, Papini E, Cassatella MA, et al: Partial purification of the superoxidase-generating system of macrophages. Possible association of the NADPH oxidase activity with a low-potential (−247 mV) cytochrome *b*. *Biochim Biophys Acta* **810:**164–173, 1985.
50. Bellavite P, Cassatella MA, Papini E, et al: Presence of cytochrome b_{-245} in NADPH oxidase preparations from human neutrophils. *FEBS Lett* **199:**159–163, 1986.
51. Doussière J, Vignais PV: Purification and properties of an $O_2^- \cdot$-generating oxidase from bovine polymorphonuclear neutrophils. *Biochemistry* **24:**7231–7239, 1985.
52. Cross AR, Parkinson JF, Jones OTG: The superoxide-generating oxidase of leucocytes. NADPH-dependent reduction of flavin and cytochrome *b* in solubilized preparations. *Biochem J* **223:**337–344, 1984.
53. Gabig TG, Lefker BA: Activation of the human neutrophil NADPH oxidase results in coupling of electron carrier function between ubiquinone-10 and cytochrome b_{559}. *J Biol Chem* **260:**3991–3995, 1985.
54. Weening RS, Corbeel L, de Boer M, et al: Cytochrome *b* deficiency in an autosomal form of chronic granulomatous disease. A third form of chronic granulomatous disease recognized by monocyte hybridization. *J Clin Invest* **75:**915–920, 1985.
55. Segal AW, Jones OTG: Novel cytochrome *b* system in phagocytic vacuoles of human granulocytes. *Nature (Lond)* **276:**515–517, 1978.
56. Segal AW, Jones OTG, Webster D, et al: Absence of a newly described cytochrome *b* from neutrophils of patients with chronic granulomatous disease. *Lancet* **2:**446–449, 1978.
57. Segal AW, Jones OTG: Reduction and subsequent oxidation of a cytochrome *b* of human neutrophils after stimulation with phorbol myristate acetate. *Biochem Biophys Res Commun* **88:**130–134, 1979.
58. Borregaard N, Johansen KS, Esmann V: Quantitation of superoxide production in human poly-

morphonuclear leukocytes from normals and three types of chronic granulomatous disease. *Biochem Biophys Res Commun* **90:**214–219, 1979.

59. Segal AW, Cross AR, Garcia RC, et al: Absence of cytochrome b_{-245} in chronic granulomatous disease: A multicenter European evaluation of its incidence and relevance. *N Engl J Med* **308:**245–251, 1983.
60. Meerhof LJ, Roos D: Heterogeneity in chronic granulomatous disease detected with an improved nitroblue tetrazolium slide test. *J Leukocyte Biol* **39:**699–711, 1986.
61. Borregaard N, Cross AR, Herlin T, et al: A variant form of X-linked chronic granulomatous disease with normal nitroblue tetrazolium slide test and cytochrome *b*. *Eur J Clin Invest* **13:**243–247, 1983.
62. Bohler M-C, Seger RA, Mouy R, et al: A study of 25 patients with chronic granulomatous disease: A new classification by correlating respiratory burst, cytochrome *b* and flavoprotein. *J Clin Immunol* **6:**136–145, 1986.
63. Iizuka T, Kanegasaki S, Makino R, et al: Pyridine and imidazole reversibly inhibit the respiratory burst in porcine and human neutrophils: Evidence for the involvement of cytochrome b_{558} in the reaction. *Biochem Biophys Res Commun* **130:**621–626, 1985.
64. Cross AR, Parkinson JF, Jones OTG: Mechanism of the superoxide producing oxidase neutrophils. O_2 is necessary for the fast reduction of cytochrome b_{-245} by NADPH. *Biochem J* **226:**881–884, 1985.
65. Parkos CA, Allen RA, Cochrane CG, et al: The quaternary structure of the plasma membrane *b*-type cytochrome of human granulocytes. *Biochim Biophys Acta* **932:**71–83, 1988.
66. Harper AM, Dunne MJ, Segal AW: Purification of cytochrome b_{-245} from human neutrophils. *Biochem J* **219:**519–527, 1984.
67. Harper AM, Chaplin MF, Segal AW: Cytochrome b_{-245} from human neutrophils is a glycoprotein. *Biochem J* **227:**783–788, 1985.
68. Segal AW: Absence of both cytochrome b_{-245} subunits from neutrophils in X-linked chronic granulomatous disease. *Nature (Lond)* **326:**88–91, 1987.
69. Pember SO, Heyl BL, Kinkade JM, et al: Cytochrome b_{558} from (bovine) granulocytes. Partial purification from Triton X-114 extracts and properties of the isolated cytochrome. *J Biol Chem* **259:**10590–10595, 1984.
70. Parkos CA, Allen RA, Cochrane CG, et al: Purified cytochrome *b* from human granulocyte plasma membrane is comprised of two polypeptides with relative molecular weights of 91,000 and 22,000. *J Clin Invest* **80:**732–742, 1987.
71. Lutter R, van Schaik MLJ, van Zwieten R, et al: Purification and partial characterization of the *b*-type cytochrome from human polymorphonuclear leukocytes. *J Biol Chem* **260:**2237–2244, 1985.
72. Dinauer MC, Orkin SH, Brown R, et al: The glycoprotein encoded by the X-linked chronic granulomatous disease locus is a component of the neutrophil cytochrome *b* complex. *Nature* **327:**717–720, 1987.
73. Royer-Pokora B, Kunkel LM, Monaco AP, et al: Cloning the gene for an inherited human disorder—chronic granulomatous disease—on the basis of its chromosomal localization. *Nature (Lond)* **322:**32–38, 1986.
74. Teahan C, Rowe P, Parker P, et al: The X-linked chronic granulomatous disease gene codes for the β-chain of cytochrome b_{-245}. *Nature* **327:**720–721, 1987.
75. Shinagawa Y, Tanaka C, Teraoka A, et al: A new cytochrome in neutrophilic granules of rabbit leukocyte. *J Biochem (Tokyo)* **59:**622–624, 1966.
76. Cross AR, Jones OTG, Harper AM, et al: Oxidation-reduction properties of the cytochrome *b* found in the plasma membrane fraction of human neutrophils. A possible oxidase in the respiratory burst. *Biochem J* **194:**599–606, 1981.
77. Morel F, Vignais PV: Examination of the oxidase function of the *b*-type cytochrome in human polymorphonuclear leucocytes. *Biochim Biophys Acta* **764:**213–225, 1984.

78. Cross AR, Higson FK, Jones OTG, et al: The enzymatic reduction and kinetics of oxidation of cytochrome b_{-245} of neutrophils. *Biochem J* **204:**479–485, 1982.
79. Cooper MR, DeChatelet LR, McCall CE, et al: Complete deficiency of leukocyte glucose-6-phosphate dehydrogenase with defective bactericidal activity. *J Clin Invest* **51:**769–778, 1972.
80. Baehner RL, Johnston RB, Nathan DC: Comperative study of the metabolic and bactericidal characteristics of severly glucose-6-phosphate dehydrogenase-deficient polymorphonuclear leukocytes and leukocytes from children with chronic granulomatous disease. *J Reticuloendothel Soc* **12:**150–169, 1972.
81. Babior BM, Curnutte JT, Kipnes RS: Pyridine nucleotide-dependent superoxide production by a cell-free system from human granulocytes. *J Clin Invest* **56:**1035–1042, 1975.
82. Umei T, Takeshige K, Minakami S: NADPH binding component of neutrophil superoxide-generating oxidase. *J Biol Chem* **261:**5229–5232, 1986.
83. Doussière J, Laporte F, Vignais PV: Photolabeling of a $O_2^{-}\cdot$-generating protein in bovine polymorphonuclear neurophils by an arylazido $NADP^{+}$ analog. *Biochem Biophys Res Commun* **139:**85–93, 1986.
84. Singer TP, Edmondson DE: Flavoproteins. Overview, in Fleischer S, Packer L (eds): *Methods in Enzymology.* New York, Academic, 1978, vol 53: *Biomembranes. Part D. Biological Oxidations, Mitochondrial and Microbial Systems*, p 397.
85. Walsh C: Scope of chemical redox transformations catalyzed by flavoenzymes, in Massey V, Williams CH (eds): *Flavins and Flavoproteins.* Amsterdam, Elsevier, 1982, p 121.
86. Cross AR, Jones OTG: The effect of the inhibitor diphenylene iodonium on the superoxide-generating system of neutrophils. *Biochem J* **237:**111–116, 1986.
87. Gabig TG, Lefker BA: Catalytic properties of the resolved flavoprotein and cytochrome *b* components of the NADPH dependent $O_2^{-}\cdot$ generating oxidase from human neutrophils. *Biochem Biophys Res Commun* **118:**430–436, 1984.
88. Kakinuma K, Kaneda M, Chiba T, et al: Electron spin resonance studies on a flavoprotein in neutrophil plasma membranes. Redox potentials of the flavin and its participation in NADPH oxidase. *J Biol Chem* **261:**9426–9432, 1986.
89. Cross AR, Jones OTG, Garcia R, et al: The association of FAD with the cytochrome b_{-245} of human neutrophils. *Biochem J* **253:**759–763, 1982.
90. Gabig TG, Lefker BA: X-linked recessive and apparent autosomal recessive chronic granulomatous disease: FAD and cytochrome b_{559} content of neutrophil oxidase fractions from patients and family members. *Blood* **62**(suppl 1)**:**80, 1983 (abst).
91. Borregaard N, Tauber AI: Subcellular localizations of the human neutrophil NADPH oxidase. *b*-Cytochrome and associated flavoprotein. *J Biol Chem* **259:**47–52, 1984.
92. Ohno Y, Buescher ES, Roberts R, et al: Reevaluation of cytochrome *b* and flavine adenine dinucleotide in neutrophils from patients with chronic granulomatous disease and description of a family with probable autosomal recessive inheritance of cytochrome *b* deficiency. *Blood* **67:**1132–1138, 1986.
93. Hamers MN, de Boer M, Meerhof LJ, et al: Complementation in monocyte hybrids revealing genetic heterogeneity in chronic granulomatous disease. *Nature (Lond)* **307:**553–555, 1984.
94. Roos D, Voetman AA, Meerhof LJ: Functional activity of enucleated human polymorphonuclear leukocytes. *J Cell Biol* **97:**368–377, 1983.
95. Segal AW, Heyworth PG, Cockcroft S, et al: Stimulated neutrophils from patients with autosomal recessive chronic granulomatous disease fail to phosphorylate a M_r 44,000 protein. *Nature (Lond)* **316:**547–549, 1985.
96. Hancock JT, Jones OTG: The inhibition by diphenyleneiodonium and its analogues of superoxide generation by macrophages. *Biochem J* **242:**103–107, 1987.
97. Millard JA, Gerard KW, Schneider DL: The isolation from rat peritoneal leukocytes of plasma membrane enriched in alkaline phosphatase and a *b*-type cytochrome. *Biochem Biophys Res Commun* **90:**312–319, 1979.

98. Crawford DR, Schneider DL: Identification of ubiquinone-50 in human neutrophils and its role in microbicidal events. *J Biol Chem* **257:**6662–6668, 1982.
99. Crawford DR, Schneider DL: Evidence that a quinone may be required for the production of superoxide and hydrogen peroxide in neutrophils. *Biochem Biophys Res Commun* **99:**1277–1286, 1981.
100. Cunningham CC, DeChatelet LR, Spach PI, et al: Identification and quantitation of electron-transport components in human polymorphonuclear neutrophils. *Biochim Biophys Acta* **682:**430–435, 1982.
101. Sloan EP, Crawford DR, Schneider DL: Isolation of plasma membrane from human neutrophils and determination of cytochrome *b* and quinone content. *J Exp Med* **153:**1316–1328, 1981.
102. Crawford DR, Schneider DL: Ubiquinone content and respiratory burst activity of latex-filled phagolysosomes isolated from human neutrophils and evidence for the probable involvement of a third granule. *J Biol Chem* **258:**5363–5367, 1983.
103. Hamers MN, Lutter R, van Zwieten R, et al: Characterization of the $O_2^-/H_2O_2^-$ generating system in human neutrophils, in Bors W, Saran M, Tait D (eds): *Oxygen Radicals in Chemistry and Biology*. Berlin, Walter de Gruyter, 1984, p 843.
104. Cross AR, Jones OTG, Garcia R, et al: The subscellular localization of ubiquinone in human neutrophils. *Biochem J* **216:**765–768, 1983.

4

The Respiratory Burst and Cellular Ion Homeostasis

BARBARA STYRT and MARK S. KLEMPNER

1. INTRODUCTION

The respiratory burst of phagocytic cells uses electron transfer to produce a series of highly reactive toxic oxygen metabolites. This process is accompanied by a complex pattern of ion redistribution between the cell and its environment and among subcellular compartments. The relationships between ion fluxes and the respiratory burst have been under intensive investigation in recent years, spurred by the rapidly progressing elucidation of the NADPH oxidase complex responsible for the burst and by the introduction of several fluorescent indicators facilitating the study of intracellular ion transients.

This chapter briefly reviews the ionic events associated with phagocyte activation and their relationship to the respiratory burst. Emphasis is placed on the neutrophil, which has been subjected to more exhaustive scrutiny in recent literature, but macrophage studies are also discussed where pertinent information is available.

2. ION DISTRIBUTION IN THE RESTING NEUTROPHIL

Like other mammalian cells. the neutrophil at rest maintains transmembrane ion gradients. These gradients are accounted for both by differential membrane permeability to different ions and by the operation of energy-requiring pumps in the plasma membrane and in organelle membranes.

BARBARA STYRT • Department of Medicine, Michigan State University, East Lansing, Michigan 48824-1317. MARK S. KLEMPNER • Department of Medicine, Tufts–New England Medical Center, Boston, Massachusetts 02111.

2.1. Calcium

Measurement of cytosol free calcium in the neutrophil has been greatly facilitated by the recent development of fluorescent chelating agents—most notably Quin 2[1] and Fura 2,[2] the fluorescence of which can be used to quantitate the local ionized calcium concentration. Neutrophils and macrophages maintain a cytosolic-free calcium concentration of approximately 100–200 nM,[3–5] some four orders of magnitude lower than serum ionized calcium.[6]

Maintenance of this large gradient is an active process. In addition to membrane binding and local chelation by calcium-binding proteins, mechanisms for lowering cytosol free calcium include extrusion from the cell and sequestration within organelles. Calcium is transported outward across the plasma membrane by an adenosine triphosphate (ATP)-dependent calcium pump[7–9] similar to those described in other cell types.[10] The lysosome also has an ATP-dependent calcium pump that transports calcium into the organelle.[11] The relative importance of organelle sequestration may be judged from the fact that the mitochondria can take up calcium to maintain free calcium outside the organelles at appoximately 600 nM, while the nonmitochondrial pool of organelles can regulate ambient calcium at approximately 200 nM.[12]

Calcium levels within organelles have not been fully elucidated, partly because the use of the fluorescent indicators is limited by their cytoplasmic distribution. A high steady-state calcium concentration in the granules would be expected on the basis of their active calcium uptake[11] and on the analogy of platelet secretory granules, which have an estimated free calcium of 12 μM.[13] While preliminary evidence (Styrt and Klempner, unpublished data) suggests that total calcium in the neutrophil granules is in fact high relative to the whole cell, accurate measurement of free calcium within the organelles awaits improvements in methodology.

2.2. Hydrogen

Studies of pH distribution in the neutrophil indicate that the largest gradients are across organelle membranes rather than across the plasma membrane. At physiological external pH, intracellular pH in the neutrophil has been estimated at 7.2–7.3.[14–18] These measurements are derived from equilibration of the weak acid [^{14}C]5,5-dimethyloxazolidine-2,4-dione (DMO) or from loading of cells with fluorescein derivatives with a pH-dependent fluorescence spectrum. Both methods used acidic indicators that should be excluded from highly acid intracellular compartments and should thus primarily report the cytosolic pH; they agree in estimating a pH that is acidic relative to the extracellular milieu but only very slightly so.

Measurement of intracellular pH of the neutrophil by equilibration of a weak base, [^{14}C]trimethylamine (TMA), yields a lower intracellular pH value of 6.35.[15] Since a basic indicator would be expected to accumulate in intracellular acidic compartments, this discrepancy is consistent with the assumption that neutrophil cytosol has a pH only slightly lower than that of the external milieu, but that the cell has a heterogeneous pH distribution with organelle pools of much greater acidity. Studies with isolated neutrophil granules support this hypothesis and suggest that, as in lysosomes of other mammalian cells[19–22] and in other granule types, such as platelet dense granules,[23] mast cell granules,[24] and chromaffin granules,[25] the intragranular compartment is intensely acidic. Thus, neutrophil granules from disrupted cells maintain a 2-pH-unit transmembrane gradient with an internal pH of approximately 5.6.[26] Furthermore, a weakly basic fluorescent indicator, 9-aminoacridine, co-localizes with granule markers after incubation with whole cells,[27] suggesting again that granule sequestration is the most likely explanation for the low intracellular pH values obtained with TMA. Amino acid distribution studies have suggested that the azurophil granule is the predominant acidic granule compartment in the neutrophil.[28]

The fluorescence spectrum of fluoresceinated dextran has been used to estimate intralysosomal pH in macrophages.[29–31] Highly acid values (~4.8) have been obtained similar to those calculated for isolated neutrophil granules by weak base distribution. Thus, divergent methodologic approaches lead to the consistent conclusion that phagocytic cells have a cytosol pH just slightly lower than their physiologic surroundings but also contain a lysosomal granule compartment of very low pH.

Maintenance of the lysosomal pH gradient may be attributable to a lysosomal membrane proton ATPase, again in agreement with studies using other cell types.[32–37] The presence of highly charged impermeant proteins within the lysosome may also contribute to the pH gradient by the creation of a Donnan equilibrium.[38]

2.3. Sodium, Potassium, and the Membrane Potential

The neutrophil maintains an inside-negative electrical potential at the plasma membrane. The magnitude of this potential has been variously reported at values ranging from −20 to −100 mV.[39–42] The variability probably results both from differences in techniques of measurement and from different procedures for cell isolation and storage.[41] Maintenance of this potential has been ascribed to both potassium and sodium gradients and to operation of an electrogenic Na-K ATPase.[41,42] Macrophages have been reported as having a potassium-channel-determined membrane potential on the order of −80 mV.[43]

There is evidence for the maintenance of an electrical potential across intracellular membranes as well. An inside-negative membrane potential has been reported in rat liver lysosomes, varying from −9 to −121 mV according to the method of measurement[44]; this was diminished by agents that collapse the transmembrane proton gradient, suggesting that the two gradients are interrelated. However, both the rat liver lysosomal proton ATPase[44] and chromaffin granule protein ATPase[45] are reported as operating to increase rather than decrease the internal potential of the organelle.

In isolated neutrophil lysosomes we have measured an inside-negative membrane potential of -34 ± 3 mV using the distribution of tetraphenylmethylphosphonium ion and -104 ± 5 mV using tetraphenylphosphonium (unpublished results). Potential could not be reliably calculated using thiocyanate anion, suggesting that it is excluded from the lysosome as would be expected with an inside-negative potential. This range of results is compatible with those obtained in other cell types (e.g., -9 ± 10 mV using thiocyanate versus -121 ± 6 using tetraphenylphosphonium in rat liver lysosomes[44]) and supports the hypothesis that the net result of neutrophil lysosomal ion transport is not electroneutral.

3. IONIC EVENTS DURING ACTIVATION OF THE RESPIRATORY BURST

Activation of the respiratory burst triggers a series of ionic shifts, notably a rise in cytosol calcium and a biphasic change in cytosol pH. While the phenomenology of changes in ion transport is rapidly being elucidated, the causal links between ion transport and the respiratory burst remain somewhat conjectural.

3.1. Calcium

The relationship between activation of the respiratory burst and cellular calcium redistribution is complex and varies according to the pathway of activation. For the discussion of this relationship, it is useful to take one well-studied pathway as a prototype, and to consider alternative mechanisms in comparison with it.

3.1.1. Receptor-Mediated Activation by fmet-leu-phe

The formylated peptide fmet-leu-phe, a potent chemotactic stimulus for neutrophils, has also been intensively investigated as a stimulus for the respiratory burst. It is likely to be a physiologic means of stimulation as it belongs to a class of formylated peptides produced by bacteria such as *Escherichia coli*[46] and presumably serves to modulate neutrophil responses to bacteria in vivo. Two receptor types of differing affinity for fmet-leu-phe have been identified on the neutrophil surface,[47] and a pool of latent receptors in the neutrophil specific granules can be expressed on the surface after stimulation.[48]

After stimulation with fmet-leu-phe, neutrophil cytosol free calcium rises to several times its resting value over a very short time course (often <1 min to peak values ranging from ~ 300 nM to micromolar levels).[35,49,50] Influx of calcium across the membrane does take place during activation[51–54] and presumably contributes to the rise in cytosol calcium, although efflux is also activated[41,55] and may eventually predominate.[41,56] The rise is diminished but is not abolished by chelation of extracellular calcium, indicating that it is partly due to mobilization of calcium from intracellular stores[57,58]; this could represent displacement of membrane-bound calcium[59,60] and/or release from organelle sequestration. The inability of granule-depleted cytoplasts to raise cytosol calcium upon stimulation in low calcium media[60] supports the hypothesis that neutrophil granules serve as a site of releasable calcium stores comparable to sarcoplasmic reticulum[61] and macrophage endoplasmic reticulum.[62,63] However, activation of the respiratory burst by stimuli of this type appears to be partially dependent on extracellular calcium[50,60] and is correlated with exogenous calcium uptake,[52] suggesting that calcium influx across the plasma membrane may be necessary for maximal NADPH oxidase assembly, or that the presence of external calcium may be necessary for maintenance of intracellular stores.[64]

The calcium dependency of the respiratory burst can be explained as a complex positive feedback system. It is suggested that fmet-leu-phe binding to its membrane receptor is coupled to a guanosine triphosphate (GTP)-binding protein and that this coupling activates a phosphoinositidase that catalyzes cleavage of phospholipids to produce inositol triphosphate (IP3) and diacylglycerol.[65] Diacylglycerol activates protein kinase C (PKC),[66] while IP3 raises cytosol free calcium (presumably by releasing calcium from a nonmitochondrial intracellular pool),[67] and elevated calcium in turn promotes phospholipid metabolism and PKC activation.[68–70] PKC is thought to trigger the respiratory burst by phosphorylating a protein or proteins involved in assembly of the NADPH oxidase complex.[71]

Evidence for the involvement of a GTP-binding protein is provided by GTP-stimulated phosphoinositide hydrolysis[72] and by studies with pertussis toxin.[73–75] This toxin acts by ADP-ribosylating a class of guanine nucleotide regulatory proteins. Pretreatment of neutrophils with pertussis toxin blocks both the rise in cytosol calcium induced by fmet-leu-phe and the functional activation of the cell, suggesting that an intact G protein is required both for the calcium transient and for activation of the respiratory burst. The G protein in turn may serve this function by lowering the calcium concentration required for activation of phospholipase C at the neutrophil membrane.[65]

Several of the subsequent steps in the activation sequence are supported by experimental evidence from studies with isolated components. IP3 releases nonmitochondrial calcium stores in detergent-permeabilized cells,[12,63] while synthetic diacyglycerol can activate protein kinase C and stimulate the respiratory burst.[76] The local calcium concentration not only contributes to PKC activation but also modulates association of PKC with the neutrophil membrane.[69]

3.1.2. Alternative Pathways of Activation

This sequence of events provides a conceptual framework within which the interactions of other respiratory burst stimuli with calcium fluxes can be addressed. The phorbol esters are potent activators of the respiratory burst that do not raise cytosol calcium and, indeed, stimulate efflux of calcium from cells.[77–79] Their capacity to activate the burst can be explained on the grounds that they directly stimulate redistribution and activation of protein kinase C and thus mimic the physiologic function of diacylglycerol.[80,81] Calcium nevertheless appears to be of contributory importance to their function, since verapamil[82] inhibits superoxide production stimulated by phorbol myristate acetate (PMA). A calcium/calmodulin-dependent kinase might also be involved in their function, since calmodulin antagonists inhibit some phorbol ester-mediated functions in other cell systems.[83]

Another means of bypassing the fmet-leu-phe receptor-mediated activation sequence is illustrated by the calcium ionophore A23187, which equilibrates calcium across the membrane and may activate PKC directly via this large rise in cytosol calcium.[84] Another ionophore, ionomycin, also equilibrates calcium efficiently but is a far weaker stimulus for the respiratory burst.[4] It may be relevant that A23187 stimulates neutrophil arachidonate metabolism,[85] with the potential for secondary stimulation by leukotriene products.

Other physiologic stimuli also appear to trigger the cell by calcium-dependent pathways. Staphylococcal α-toxin is reported to stimulate leukotriene B_4 production by forming pores in the neutrophil membrane that increase passive permeability to calcium (and nonselectively to other small molecules).[86] Opsonized particles, upon phagocytosis by neutrophils, produce a rise in cytosol free calcium,[87] which is most pronounced in the periphagosomal area[88] and might be expected to contribute to NADPH oxidase activation in the phagosomal membrane. Neutrophil spreading on surfaces is likewise accompanied by a cytosolic calcium increase.[89]

The complement fragment C5a[90] and leukotriene B_4[91,92] both appear to stimulate the respiratory burst by a sequence of events that includes a rise in cytosol free calcium, although the mechanism has not been fully elucidated. Leukotriene B_4 reportedly does not stimulate PKC mobilization[84,93] but does mediate IP3 production.[94] It may involve a GTP-binding protein, since the rise in cytosol calcium can be inhibited by pertussis toxin.[73,94] Platelet activating factor produces a calcium rise at low concentrations and phosphoinositide breakdown at higher concentrations.[95]

Several qualifications must be added to this model. First, the respiratory burst is not activated by agents that produce small or transient rises in cytosol calcium, including low doses of fmet-leu-phe[49] or ionomycin[4] or the antibody 7C3.[96] This suggests that there is a threshold in concentration and duration for efficacy of the calcium rise in producing NADPH oxidase activation and that in most settings, the rise in cytosol calcium acts in conjunction with another stimulatory pathway. Second, there is evidence that PKC activation is not a necessary final common pathway for NADPH oxidase assembly.[97,98] Superoxide production by neutrophils stimulated by fmet-leu-phe or C5a is not significantly decreased in the presence of a newly developed PKC inhibitor,[99] suggesting that these agents can promote phosphorylation of proteins in the NADPH oxidase complex via alternative pathways such as a calcium/calmodulin-dependent kinase.

Thus, it appears that triggering of the respiratory burst via NADPH oxidase activation takes place via multiple different pathways and that it can be initiated by multiple stimuli, some of which activate more than one pathway. Studies in this area are proceeding rapidly and may soon produce a more coherent picture of the process of signal transduction. It may be said that an increase in cytosol free calcium is intimately involved in this process; that this calcium may be derived from organelle stores, from membrane-bound stores, and from the external media; and that there are alternative mechanisms of activation which may be mediated by a large calcium rise alone, by calcium-independent enzyme pathways alone, or by an interaction between calcium-dependent and -independent processes. This complex system of backup and feedback effects may serve the function of ensuring an adequate host response to a wide variety of situations. However, there is little evidence that occurrence of the respiratory burst itself has a feedback function in the regulation of cytosolic calcium (discussed for other ion transport systems, below). Although an attenuation of membrane calcium mobilization has been reported,[100] the cytosol calcium response to fmet-leu-phe is normal in cells from patients with chronic granulomatous disease (CGD), which do not mount a respiratory burst.[5]

3.1.3. Microenvironmental Calcium after NADPH Oxidase Activation

Although the respiratory burst enzyme can be activated by both calcium-dependent and -independent mechanisms, its activity after assembly may also be modulated by calcium. In phagocytic vesicles from rabbit macrophages, superoxide production is reportedly inhibited by calcium.[101] However, in NADPH-oxidase-enriched preparations from stimulated neutrophils, activity is impaired by EDTA[102] and enhanced by calcium.[103] The production of chemiluminescence by the myeloperoxidase–hydrogen peroxide–halide system or following generation of superoxide appears to be independent of ambient calcium,[104,105] suggesting that the production of other oxygen metabolites is unaffected. Also, the microbicidal effects of the myeloperoxidase–hydrogen peroxide–halide system are not inhibited by EDTA.[106]

When a phagocyte fulfills its host-defense function by ingesting bacteria, a new intracellular compartment—the phagosome—is formed. As calcium is mobilized into the cytosol and out of intracellular organelles, these transport processes must also affect calcium concentrations in the phagosome, which is the target site for activity of the products of the respiratory burst. However, little is known about actual calcium levels in the phagosome. Both the plasma membrane calcium pump and the lysosomal membrane calcium pump should transport calcium into the phagosome, leading to an anticipated calcium concentration greater than that in the cytosol. By contrast, studies using a strain of *Yersinia pestis* whose enzyme production is modulated by ambient calcium indicate that free calcium is less than 100 μM in the macrophage phagosome[107]—at least an order of magnitude lower than serum concentrations. The process of cellular activation may produce changes in calcium transport at the phagosomal membrane, which remain to be defined.

3.2. Hydrogen

Studies of pH in three distinct compartments have been useful in defining the redistribution of hydrogen ions during phagocyte activation. Cytosol pH has been measured using the fluorescence spectrum of fluorescein derivatives[18] or the partitioning of radiolabeled weak acids[17]; lysosomal pH has been monitored by the distribution of weak bases[28]; and the pH within the phagocytic vacuole has been assessed using ingested particles labeled with pH indicators.[108–112] The relationship between the respiratory burst and local pH changes appears to be reciprocal, with a maximal respiratory burst taking place only under optimal pH conditions and with the products of the respiratory burst, in turn, modifying local pH. This section deals with observed intracellular pH changes during normal activation and with information derived from abnormalities of the respiratory burst and/or of pH regulation.

3.2.1. Intracellular pH Changes during Activation

Changes in cytoplasmic pH have been assessed after treatment of neutrophils with several different stimuli. The chemotactic peptide fmet-leu-phe has been repeatedly reported to produce a net alkalinization[14,17,113]; in some accounts, this is preceded by a smaller transient acidification. The changes measured are usually no more than a few tenths of a pH unit. One study[114] has reported macrophage cytoplasmic acidification (0.14 pH units) after treatment with formylnorleucyl leucyl phenylalanine, followed by recovery to near baseline values.

A biphasic response of acidification followed by alkalinization has also been reported after stimulation by calcium ionophore A23187, suggesting that calcium flux can lead directly to local pH change.[115] This study must be interpreted cautiously, however, since the indicator used was 9-aminoacridine, which is distributed preferentially in the neutrophil granules rather than in the cytosol, and stimulatory concentrations of A23187 might be expected to interfere with the fluorescence of the indicator.

The effect of phorbol esters on cytosol pH has been the subject of varying reports. One group of investigators has found a monophasic pH rise after PMA stimulation,[115] while another group, using TPA as the stimulus,[18] reports acidification followed by partial or complete recovery. These discrepancies could reflect either methodological differences or differing mechanisms of action; however, there has been little to suggest that the phorbol esters vary significantly among themselves in their modes of activation.

3.2.2. Changes in pH in the Phagocytic Vacuole

Interactions between pH and NADPH oxidase are particularly important in the phagolysosome, the new intracellular compartment created by phagocyte ingestion of bacteria. The phagocytic vacuole is formed by invagination of the plasma membrane to enclose a pocket of the external media; it presumably begins at the same pH as the external environment. This begins to change rapidly during the period in which lysosomal contents are discharged into the phagosome, and the respiratory burst is activated in the phagosomal membrane.

The initial event is a transient rise in pH to approximately 7.5 to 7.8.[110,112,117]

This phase lasts a few minutes, encompassing the period of maximal activity of the respiratory burst. Thus, the phagosomal pH is at an optimal level for NADPH oxidase activity during the period of intensive oxidative metabolism.

The respiratory burst may in fact be the direct cause of this transient phagosomal alkalization. After superoxide is produced in the NADPH oxidase-catalyzed reduction of oxygen, it dismutates to hydrogen peroxide via the reaction

$$2O_2^- + 2H^+ \rightarrow H_2O_2 + O_2$$

with a net consumption of protons.[118] When the H^+ produced during the initial reaction

$$NADPH + 2O_2 \rightarrow NADP^+ + 2O_2^- + H^+$$

is taken into account, the overall result would still be a drop in free proton concentration.[114] This is therefore a tenable explanation for phagosomal alkalinization.

Following this initial rise, there is a more gradual fall in pH reaching a level of 5.6 to 6.[110,112] Multiple sources of protons for this acidification have been proposed, and the actual sequence of events is not fully understood but is probably multifactorial.

One major candidate for a source of phagolysosomal acidification is a vacuolar membrane proton ATPase, which would translocate hydrogen ions from the cytosol into the phagosome. Analogous proton ATPases have been described in a variety of cells ranging from calf brain[120] to turtle bladder.[121] They are present in the membrane of endocytic vesicles of macrophages and fibroblasts[122,123] and apparently are not derived from lysosomal membrane ATPases, since endosomal acidification can be detected before discharge of lysosomal contents into the phagosome.[124]

Another potential source of phagosomal acid is the local production of lactate by cellular metabolism.[125] This may well be a contributory factor but is not quantitatively adequate to account for all the protons released by stimulated neutrophils.[126] Carbonic anhydrase activity has also been proposed as a mechanism for phagolysosomal acidification,[127] but confirmation of its importance is lacking, and carbonic anhydrase inhibitors were not found to block extracellular release of protons attributed to the hexose monophosphate shunt.[128]

Phagosome-lysosome fusion may itself be an important factor in acidification of the phagocytic vacuole. Even if endosomal acidification in other cell types begins before such fusion is detected, the acidification phase in the neutrophil phagosome takes place largely after phagosome–lysosome fusion.[129] The low pH contents of the lysosome,[26–31] especially large nonpermeant molecules, may represent a reservoir of acid substances whose discharge into the phagolysosome is instrumental in lowering the ambient pH during the microbicidal process.

Finally, it remains questionable whether the respiratory burst itself contributes to phagolysosomal acidification. The production of superoxide does generate excess protons that could produce a net acidification of the microenvironment. However, dismutation of superoxide to hydrogen peroxide should involve consumption of these excess hydrogen ions. Therefore, the net contribution of NADPH oxidase activity to the acidification phase of phagolysosomal pH change may depend on whether superoxide is used by other pathways before its conversion to hydrogen peroxide.

3.2.3. Regulation of pH in the Absence of the Respiratory Burst

Since pH changes occur during activation of the respiratory burst and the respiratory burst may itself influence cellular pH homeostasis, determination of cause-and-effect relationships between the two processes is a complex and problematic task. Some insight into this issue can be derived from studies in CGD cells, which lack the respiratory burst but have normal phagocytic responses. Both intracellular and intraphagosomal pH regulation are reportedly abnormal in CGD.

In a study of intracellular pH changes in response to the phorbol ester TPA,[18] CGD cells demonstrated sustained intracellular alkalinization after stimulation as compared with transient but monophasic acidification in normal control cells. This alkalinization was sensitive to the inhibition of Na^+/H^+ exchange by amiloride and was attributed to direct activation of a sodium–hydrogen exchanger without metabolic generation of acid. These findings would imply that the NADPH oxidase system is itself an important source of acid in the initial decline of poststimulatory cytosol pH.

Regulation of phagosomal pH in CGD is also reportedly abnormal, although different investigators have reported conflicting findings.[110,130] The suggested abnormality is a lack of the initial phagosomal alkalinization, with a normal or supranormal acidification phase. The failure of alkalinization would suggest that dismutation of superoxide to hydrogen peroxide is responsible for the initial rise in pH. The preservation of the acidification phase indicates that the respiratory burst is not necessary for phagosomal acidification, although it cannot be entirely ruled out as a contributory factor.

3.2.4. Experimental Perturbation of pH Regulation

Additional information on the relationship between pH homeostasis and the respiratory burst has come from in vitro treatment of cells with agents that disturb pH regulation. Of particular interest in this context is amiloride, which blocks sodium–hydrogen exchange. There is increasing evidence that the phagocyte plasma membrane contains a sodium–hydrogen exchanger and that stimulation of the cell activates this exchanger, resulting in a net extrusion of hydrogen ions from the cell in exchange for sodium from the environment.[131–133] This process appears to account for intracellular alkalinization in response to fmet-leu-phe. Treatment with amiloride inhibits cytosolic alkalinization, PKC activation,[134] and superoxide generation,[132] suggesting that the processes are interdependent at this step.

Another inhibitor that affects both neutrophil functional responses and pH modulation in response to fmet-leu-phe is pertussis toxin, which acts by modifying guanine nucleotide regulatory proteins. The toxin reportedly blocks both phases of intracellular pH change,[115] suggesting that coupling of the fmet-leu-phe–receptor complex to a GTP binding protein is important to initiation of the poststimulatory pH changes.

The relationship between intracellular pH and the respiratory burst has also been studied via direct manipulation of intracellular pH. Cells that have been alkalinized with ammonium chloride or acidified with CO_2, and then washed and stimulated during the period of overshoot to acid and alkaline pH respectively, display respectively inhibited and enhanced superoxide generation.[113] This superoxide response correlates with cytosol pH at the time of stimulation but, since the pH status is rapidly changing at the time of study, the causal relationship between pH and superoxide production is not clear. The creation of an abnormal transmembrane pH gradient was not sufficient in these studies to generate a respiratory burst, but rather appeared to modulate the cellular response to another stimulus.

In studying the relationship between lysosomal pH and cellular function, we have examined the effect on the respiratory burst of weak bases that diminish the pH gradient between lysosome and cytosol.[135] These agents inhibit superoxide production in response to diverse stimuli (PMA, fmet-leu-phe, A23187, NaF, and opsonized zymosan), with a dose response that correlates with their effects on lysosomal pH. By the use of particulate preparations from stimulated cells, pretreatment with these agents was shown to inhibit activation of the NADPH oxidase complex, leading to activation of decreased amounts of enzyme with normal kinetic properties. This suggests that existence of intracellular pH gradients may be important to normal assembly of the enzyme complex in stimulated cells. Once assembled, the NADPH oxidase is active over a broad pH range with a pH optimum in the region of neutrality.[136,137] Its activity is not significantly affected by treatment with lysosomotropic weak bases either after stimulation of the cells or after disruption and isolation of the particulate fraction.[135]

The consequence of phagolysosomal acidification is a shift of local pH away from the optimum for NADPH oxidase activity. However, the acid pH that is attained enhances the activity of many of the lysosomal enzymes discharged into the phagosome. In particular, it is close to the pH optimum for the myeloperoxidase–hydrogen peroxide–halide microbicidal system, one of the more important pathways by which the products of the respiratory burst actually kill bacteria.[138,139] When lysosomal components are studied in combination with hydrogen peroxide, killing of bacteria is clearly enhanced by dropping the pH from the early phagosome level of 7.0–7.5 to the late phagolysosome level of 5.5.[140] Thus, acidification of the phagolysosome modifies the local pH from a level that is optimal for generation of toxic oxygen metabolites to an environment more favorable for their microbicidal activity.

3.3. Sodium, Potassium, and the Membrane Potential

The membrane potential maintained by neutrophils is linked to sodium–potassium balance and is impaired when Na–K ATPase activity is blocked by ouabain. The potential is thought to be a function largely of the potassium gradient, although it has been suggested that the sodium gradient may be more important.[41,42]

Somewhat more is known about this issue in the macrophage, whose resting membrane potential is apparently maintained by voltage-dependent K^+ channels.[43] The presence of these channels varies with duration of cell culture and adherence, suggesting that they may be an important element in cellular maturation and function.

Macrophages can be depolarized by stimulation, with a response involving both sodium and potassium flux. The macrophage Fc receptor has been shown to be a cation channel, and ligand binding increases permeability to potassium and sodium.[43] Both the increased permeability and consequent current fluctuations have been demonstrated in isolated Fc receptors incorporated into lipid bilayers. The channel has a low permeability to calcium, but its activity is modified by changes in calcium concentration on the side opposite the ligand binding site, suggesting a role for cytosol calcium changes in modulation of the membrane potential.

Neutrophils undergo transient depolarization in response to diverse stimuli, including fmet-leu-phe, PMA, and calcium ionophore A23187.[39,40,141,142] Depolarization is accompanied by sodium influx followed by efflux, and by potassium influx.[143,144] This may involve alteration of Na/K ATPase activity. However, both fmet-leu-phe and phorbol esters have also been reported to activate a Na^+/H^+ antiport in the plasma membrane.[131–133] This may account for both sodium influx and cytosol alkalinization. The antiport is sensitive to amiloride and can also be activated by acid loading or hyperosmolarity. Since the latter interventions do not stimulate the cell, activation of the antiport is not an adequate signal for neutrophil functional responses.

The relationship between membrane depolarization and cellular function may be relatively specific. Depolarization has been linked to the respiratory burst by multiple investigators using a variety of stimuli and of membrane potential indicators. Its importance for other cellular activities such as degranulation remains to be proved.

Again, the evidence for specific relevance of membrane depolarization to the respiratory burst comes from studies in genetically abnormal cells. Neutrophils from patients with Chediak–Higashi syndrome, which have morphologically abnormal granules and degranulate abnormally, have a depolarization response that is normal in configuration but slightly delayed in its time course.[40] By contrast, CGD cells do not exhibit the changes in membrane potential elicited in normal cells by either fmet-leu-phe or phorbol esters.[40,141] These cells have a normal resting membrane potential and a normal response to the potassium ionophore valinomycin, suggesting that the defect cannot be explained simply on the basis of abnormal membrane permeability to monovalent cations. Thus, it is possible that the component of the NADPH oxidase complex missing in CGD (e.g., cytochrome *b*)[145] is necessary for normal modulation of the membrane potential in response to cellular activation.

Depolarization has been shown to precede the detectible appearance of products of the respiratory burst. The time elapsed between evidence of membrane depolarization by fluorescent methods and detection of superoxide is 20–30 sec, making depolarization one of the earliest events measured in the activation sequence.[39]

This early occurrence, association with multiple stimuli with different mechanisms of action, and abnormality in CGD suggest that depolarization could be the initial postreceptor signal for initiation of the respiratory burst. However, depolarization is not itself adequate to produce a respiratory burst. Depolarization using ouabain, potassium plus valinomycin, or gramicidin does not lead to increases in oxidative metabolism.[146,147] Thus, if changes in membrane potential are a signal for NADPH oxidase activation, they probably act in concert with another transduction mechanism.

It is questionable whether the respiratory burst can be activated in the absence of depolarization. While a modest amount of oxidative metabolism has been reported in neutrophils stimulated with fmet-leu-phe without depolarization,[142] this effect was highly dependent on the method of cell isolation and monitoring of membrane potential was discontinuous so that transient events could have been missed. It would seem at least that membrane depolarization is an early event in the vast majority of instances in which the respiratory burst is triggered, and may participate together with other ion transients (such as calcium) in signal transduction for NADPH oxidase activation in the neutrophil. The abnormality of depolarization in CGD cells would suggest that components of the NADPH oxidase complex may themselves be essential for normal depolarization, if lack of an element such as cytochrome *b* is sufficient to produce the CGD phenotype. The mechanism for this reciprocal influence remains to be defined.

3.4. Other Ion Systems

Homeostatic mechanisms for calcium, pH, and the potassium/sodium dependent membrane potential appear to be of major importance for normal triggering of the respiratory burst. A number of other ionic transport and binding systems may be relevant to the respiratory burst but have not been as extensively investigated or are important only in certain special settings. We shall touch briefly on the evidence for a handful of other ion systems as modulators of the respiratory burst.

3.4.1. Magnesium

Whereas magnesium is important for the ATP-dependent transport of calcium across intracellular membranes[7–10] and thus has a supporting role in many of the calcium events discussed, its full importance in intracellular events is only gradually being elucidated. There is evidence that magnesium can potentiate the activity of the assembled NADPH oxidase complex.[103] The NADPH oxidase is inactivated by removal of divalent cations by resin binding or EDTA chelation. Activity can be restored by either calcium or magnesium, but not by other divalent cations. Kinetic analysis indicated that magnesium does not interact with the substrate (NADPH) but interacts with some component of the oxidase system, with a K_d value of 3 μM, so as to enhance its stability and activity. This suggests that either calcium or magnesium is a normal constituent of the activated enzyme complex. However, magnesium could not substitute for calcium in restoring the respiratory burst activity in calcium-depleted cells.[58]

3.4.2. Selenium

Selenium is a cofactor for activity of glutathione peroxidase, which catalyzes one of the pathways for detoxification of hydrogen peroxide. Animals fed a selenium-deficient diet have low concentrations of granulocyte selenium, low glutathione peroxidase activity, and a defective respiratory burst.[148–150] The mechanism for this effect has been explored in selenium-deficient rats.[148] In this model, the initial rate of triggering of the respiratory burst has been found to be normal. However, NADPH oxidase activity declines more rapidly after stimulation than in cells from selenium-replete animals, and the respiratory burst is subnormal in cells from selenium-deficient animals treated with hydrogen peroxide. These findings support the hypothesis that selenium-deficient cells have normal initiation of the respiratory burst but that the hydrogen peroxide they produce damages the NADPH oxidase, producing an inability to sustain the burst. Chemiluminescence of cells from selenium-deficient goats can be restored by in vitro addition of selenium, suggesting that glutathione peroxidase is present in these cells but inactive due to the absent cofactor, and capable of activation by ion repletion.

3.4.3. Other Cations

Inhibition of the respiratory burst has been demonstrated after in vitro incubation of neutrophils with mercury, silver, or copper[151]; however, copper-deficient cattle reportedly have impaired NBT reduction by neutrophils.[150] Lead is reported to decrease erythrocyte superoxide dismutase[152]—this, if generalized to other tissues, could enhance susceptibility to damage by products of the respiratory burst. Gold inhibits the respiratory burst in its therapeutic organic forms[153] but apparently not as the free cation.[151] Zinc, manganese, and copper inhibit the activity of the NADPH oxidase in particulate fractions prepared from stimulated neutrophils.[103]

3.4.4. Fluoride

Fluoride has been cited as an activator of the respiratory burst. This halide anion is well delineated as a stimulus of neutrophil superoxide production and appears to trigger the respiratory burst selectively out of proportion to its effect on other neutrophil activities, such as degranulation.[154] Its mechanism of action and its potential physiologic role remain to be fully defined. Fluoride is known to interact with G proteins and could therefore trigger the respiratory burst by a pathway similar to the action of fmet-leu-phe. A rise in cytosol calcium has been demonstrated in fluoride-stimulated cells, which would be compatible with this hypothesis.[155] However, unlike the effect of fmet-leu-phe, fluoride activation was not inhibited by pertussis toxin. If fluoride activity is mediated by a G protein, this would suggest that it either interacts with a different G protein from that coupled to the fmet-leu-phe receptor or is able to activate the same G protein by a pathway that is not affected by pertussis toxin ADP ribosylation.

3.4.5. Other Anions

Many of the studies of cation flux during neutrophil activation have disregarded potential contributions of concomitant anion transport. Studies of so-called anion channel blockers have revealed an interesting selective effect on neutrophil function, with inhibition of degranulation but not of the respiratory burst.[156,157] Thus, it would appear that NADPH oxidase can be activated without a major requirement for anion flux sensitive to these agents.

The supply of inorganic anions in the phagosome may, on the other hand, be quite important for optimal antimicrobial effectiveness of the respiratory burst. Halide anions such as chloride and iodide are important for bacterial killing by the myeloperoxidase–hydrogen peroxide–halide system.[138,139] Chlorination of amines may also be important to the production of longer-lived antimicrobial agents in the phagosomal milieu.[158–161]

4. CONCLUSIONS

Redistribution of ions is one of the major events in initiation and maintenance of the phagocytic respiratory burst. Membrane depolarization and elevation of cytosol calcium are among the earliest steps in the usual activation sequence leading to activation of NADPH oxidase. Although they usually operate concurrently with additional excitatory events, and although alternative pathways exist for activating NADPH oxidase by other mechanisms, these ion fluxes are evidently of major importance to the role of phagocyte oxidative metabolism in host defense.

After activation of NADPH oxidase, further ionic events contribute to the optimal microbicidal function of the products of the respiratory burst. Acidification of the phagocytic vacuole may be partially mediated by superoxide production and leads to an environment suitable for activity of the myeloperoxidase–hydrogen peroxide–halide system. Calcium may continue to play a modulatory role the significance of which is gradually being elucidated.

Several issues of potentially great importance to the ionic milieu of the respiratory burst are just beginning to be investigated. These include the possible role of calcium/calmodulin-dependent protein kinase in NADPH oxidase activation, the precise mechanism of membrane depolarization and its effect on the oxidase, and calcium regulation in the phagolysosome. Investigation in this field is evolving so fast that a review at this point is necessarily incomplete, and much of the information presented here may soon require reevaluation as new data accumulate. Ongoing studies may bring order as well as added detail to what is now a tantalizing, rapidly growing, but highly complex field of investigation.

REFERENCES

1. Tsien RY, Pozzan T, Rink TJ: Calcium homeostasis in intact lymphocytes: cytoplasmic free calcium monitored with a new, intracellularly trapped fluorescent indicator. *J Cell Biol* **94:**325–334, 1982.
2. Grynkiewicz G, Poenie M, Tsien RY: A new generation of Ca^{2+} indicators with greatly improved fluorescence properties. *J Biol Chem* **260:**3440–3450, 1985.
3. White JR, Naccache PH, Molski TFP, et al: Direct demonstration of increased intracellular concentration of free calcium in rabbit and human neutrophils following stimulation by chemotactic factor. *Biochem Biophys Res Commun* **113:**44–50, 1983.
4. Pozzan T, Lew DP, Wollheim CB, Tsien RY: Is cytosolic ionized calcium regulating neutrophil activation? *Science* **221:**1413–1415, 1983.
5. Lew DP, Wollheim CB, Seger RA, Pozzan T: Cytosolic free calcium changes induced by chemotactic peptide in neutrophils from patients with chronic granulomatous disease. *Blood* **63:**231–233, 1984.
6. Young DS: Implementation of SI Units for clinical laboratory data. *Ann Intern Med* **106:**114–129, 1987.
7. Volpi, M, Naccache PH, Sha'afi RI: Calcium transport in inside-out membrane vesicles prepared from rabbit neutrophils. *J Biol Chem* **258:**4153–4158, 1983.
8. Ochs DL, Reed PW: ATP-dependent calcium transport in plasma membrane vesicles from neutrophil leukocytes. *J Biol Chem* **258:**10116–10122, 1983.
9. Lagast H, Lew PD, Waldvogel FA: Adenosine triphosphase-dependent calcium pump in the plasma membrane of guinea pig and human neutrophils. *J Clin Invest* **73:**107–115, 1984.
10. Carafoli E, Zurini M: The Ca^{2+}-pumping ATPase of plasma membranes. *Biochim Biophys Acta* **683:**279–301, 1982.
11. Klempner MS: An adenosine triphosphate-dependent calcium uptake pump in human neutrophil lysosomes. *J Clin Invest* **76:**303–310, 1985.
12. Prentki M, Wollheim CB, Lew PD: Ca^{2+} homeostasis in permeabilized human neutrophils. *J Biol Chem* **259:**13777–13782, 1984.
13. Grinstein S, Furuya W, VanderMeulen J, Hancock RGV: The total and free concentrations of Ca^{2+} and Mg^{2+} inside platelet secretory granules. *J Biol Chem* **258:**14774–14777, 1983.
14. Molski TFP, Naccache PH, Volpi M, et al: Specific modulation of the intracellular pH of rabbit neutrophils by chemotactic factors. *Biochem Biophys Res Commun* **94:**508–514, 1980.
15. Simchowitz L, Roos A: Regulation of intracellular pH in human neutrophils. *J Gen Physiol* **85:**443–470, 1985.
16. Grinstein S, Elder B, Furuya W: Phorbol ester-induced changes of cytoplasmic pH in neutrophils: Role of exocytosis in Na^{+}–H^{+} exchange. *Am J Physiol* **248:**C379–C386, 1985.
17. Simchowitz L: Chemotactic factor-induced activation of Na^{+}/H^{+} exchange in human neutrophils. II. Intracellular pH changes. *J Biol Chem* **260:**13248–13255, 1985.
18. Grinstein S, Furuya W, Bigger WD: Cytoplasmic pH regulation in normal and abnormal neutrophils. *J Biol Chem.* **261:**512–514, 1986.
19. Goldman R, Rottenberg H: Ion distribution in lysosomal suspensions. *FEBS Lett* **33:**233–238, 1973.
20. Reijngoud DJ, Tager JM: Measurement of intralysosomal pH. *Biochim Biophys Acta* **297:**174–178, 1973.
21. Reijngoud DJ, Tager TM: The permeability properties of the lysosomal membrane. *Biochim Biophys Acta* **472:**419–449, 1977.
22. Reijngoud DJ, Oud DS, Tager JM: Effect of ionophores on intralysosomal pH. *Biochim Biophys Acta* **448:**303–313, 1976.

23. Grinstein S, Furuya W: The electrochemical H^+ gradient of platelet secretory alpha-granules. *J Biol Chem* **258:**7876–7882, 1983.
24. Johnson RG, Carty SE, Fingerhood BJ, Scarpa A: The internal pH of mast cell granules. *FEBS Lett* **120:**75–78, 1980.
25. Salama G, Johnson RG, Scarpa A: Spectrophotometric measurements of transmembrane potential and pH gradients in chromaffin granules. *J Gen Physiol* **75:**109–140, 1980.
26. Styrt B, Klempner MS: Internal pH of human neutrophil lysosomes. *FEBS Lett* **149:**113–116, 1982.
27. Styrt B, Johnson PC, Klempner MS: Differential lysis of plasma membranes and granules of human neutrophils by digitonin. *Tissue Cell* **17:**793–800, 1985.
28. Ransom JT, Reeves JP: Accumulation of amino acids within intracellular lysosomes of rat polymorphonuclear leukocytes incubated with amino acid methyl esters. *J Biol Chem* **258:**9270–9275, 1983.
29. Ohkuma S, Poole B: Fluorescence probe measurement of the intralysosomal pH in living cells and the perturbation of pH by various agents. *Proc Natl Acad Sci USA* **75:**3327–3331, 1978.
30. Poole B, Ohkuma S: Effect of weak bases on the intralysosomal pH in mouse peritoneal macrophages. *J Cell Biol* **90:**665–669, 1981.
31. Geisow MJ: Fluorescein conjugates as indicators of subcellular pH. *Exp Cell Res* **150:**29–35, 1984.
32. Mego JL: The ATP-dependent proton pump in lysosome membranes. *FEBS Lett* **107:**113–115, 1979.
33. Reeves JP, Reames T: ATP stimulates amino acid accumulation by lysosomes incubated with amino acid methyl esters. *J Biol Chem* **256:**6047–6053, 1981.
34. Schneider DL: ATP-dependent acidification of intact and disrupted lysosomes. *J Biol Chem* **256:**3858–3864, 1981.
35. Moriyama Y, Takano T, Ohkuma S: Acridine orange as a fluorescent probe for lysosomal proton pump. *J Biochem* **92:**1333–1336, 1982.
36. Ohkuma S, Moriyama Y, Takano T: Identification and characterization of a proton pump on lysosomes by fluorescein isothiocyanate-dextran fluorescence. *Proc Natl Acad Sci USA* **79:**2758–2762, 1982.
37. Schneider DL: ATP-dependent acidification of membrane vesicles isolated from purified rat liver lysosomes. *J Biol Chem* **258:**1833–1838, 1983.
38. Hollemans M, Donker-Koopman, W, Tager JM: A critical examination of the evidence for an Mg ATP-dependent proton pump in rat liver lysosomes. *Biochim Biophys Acta* **603:**171–177, 1980.
39. Korchak HM, Weissman G: Changes in membrane potential of human granulocytes antecedes the metabolic responses to surface stimulation. *Proc Natl Acad Sci USA* **75:**3818–3822, 1978.
40. Seligmann BE, Gallin JI: Use of lipophilic probes of membrane potential to assess human neutrophil activation. *J Clin Invest* **66:**493–503, 1980.
41. Gallin JI, Seligmann BE: Neutrophil chemoattractant f-met-leu-phe receptor expression and ionic events following activation, in Snyderman R (ed): *Contemporary Topics in Immunobiology*. New York, Plenum, 1984, vol 14: *Regulation of Leukocyte Function*, pp 83–108.
42. Bashford CL, Pasternak CA: Plasma membrane potential of neutrophils generated by the Na^+ pump. *Biochim Biophys Acta* **817:**174–180, 1985.
43. Gallin EK: Ionic channels in leukocytes. *J Leukocyte Biol* **39:**241–254, 1986.
44. Harikumar P, Reeves JP: The lysosomal proton pump is electrogenic. *J Biol Chem* **258:**10403–10410, 1983.
45. Apps DK: Proton-translocating ATPase of chromaffin granule membranes. *Fed Proc* **41:**2775–2780, 1982.
46. Marasco WA, Phan SH, Krutzsch H, Showell HJ, Feltner DE, Nairn R, Becker EL, Ward PA: Purification and identification of formyl-methionyl-leucyl-phenylalamine as the major peptide neutrophil chemotactic factor produced by *Escherichia coli*. *J Biol Chem* **259:**5430–5439, 1984.
47. Snyderman R, Pike MC: Chemoattractant receptors on phagocytic cells. *Annu Rev Immunol* **2:**257–281, 1984.

48. Gallin JI: Neutrophil specific granules: A fuse that ignites the inflammatory response. *Clin Res* **32:**320–328, 1984.
49. Korchak HM, Vienne K, Rutherford LE, Wilkenfeld C, Filkenstein MC, Weissman G: Stimulus response coupling in the human neutrophil. II. Temporal analysis of changes in cytosolic calcium and calcium efflux. *J Biol Chem* **259:**4076–4082, 1984.
50. Lew PD, Wollheim CB, Waldvogel FA, Pozzan T: Modulation of cytosolic-free calcium transients by changes in intracellular calcium-buffering capacity: Correlation with exocytosis and O_2^- production in human neutrophils. *J Cell Biol* **99:**1212–1220, 1984.
51. Naccache PH, Showell HJ, Becker EL, Sha'afi RI: Changes in ionic movements across rabbit polymorphonuclear leukocyte membranes during lysosomal enzyme release. *J Cell Biol* **75:**635–649, 1977.
52. Korchak HM, Rutherford LE, Weissman G: Stimulus response coupling in the human neutrophil. I. Kinetic analysis of changes in calcium permeability. *J Biol Chem* **259:**4070–4075, 1984.
53. Rossi, F, della Bianca V, Grzeskowiak M, et al: Relationships between phosphoionositide metabolism, Ca^{2+} changes and respiratory burst in formyl-methionyl-leucyl-phenylalanine-stimulated human neutrophils. *FEBS Lett* **181:**253–258, 1985.
54. Strauss RG, Snyder EL: Uptake of extracellular calcium by neonatal neutrophils. *J Leukocyte Biol* **37:**423–429, 1985.
55. Chandler D, Meusel G, Schumaker E, Stapleton C: FMLP-induced enzyme release from neutrophils: A role for intracellular calcium. *Am J Physiol* **245:**C196–C202, 1983.
56. Styrt B, Klempner MS: Cytoskeletal inhibitors and calcium transport: A link between degranulation and calcium efflux from stimulated neutrophils. *Clin Res* **35:**493A, 1987 (abst).
57. Sklar LA, Oades ZG: Signal transduction and ligand-receptor dynamics in the neutrophil. Ca^{2+} modulation and restoration. *J Biol Chem* **260:**11468–11475, 1985.
58. Nakagawara M, Takeshige K, Sumimoto H, et al: Superoxide release and intracellular free calcium of calcium-depleted human neutrophils stimulated by *N*-formyl-methionyl-leucyl-phenylalanine. *Biochim Biophys Acta* **805:**97–103, 1984.
59. Naccache PH, Volpi M, Showell HJ, et al: Chemotactic factor-induced release of membrane calcium in rabbit neutrophils. *Science* **203:**461–463, 1979.
60. Torres M, Coates TD: Neutrophil cytoplasts: Relationships of superoxide release and calcium pools. *Blood* **64:**891–895, 1984.
61. Shoshar-Barmatz V: Chemical modification of sarcoplasmic reticulum. *Biochem J* **240:**509–517, 1986.
62. Hirata M, Hamachi T, Hashimoto T, et al: Ca^{2+} release in the endoplasmic reticulum of guinea pig peritoneal macrophages. *J Biochem* **94:**1155–1163, 1983.
63. Hirata M, Sasaguri T, Hamachi T, et al: Irreversible inhibition of Ca^{2+} release in saponin-treated macrophages by the photoaffinity derivative of inositol-1,4,5-triphosphate. *Nature (Lond)* **317:**723–725, 1985.
64. Gabler WL, Creamer HR, Bullock WW: Role of extracellular calcium in neutrophil responsiveness to chemotactic tripeptides. *Inflammation* **10:**281–292, 1986.
65. Snyderman R, Smith, CD, Verghese MW: Model for leukocyte regulation by chemoattractant receptors: Roles of a guanine nucleotide regulatory protein and polyphosphoinositide metabolism. *J Leukocyte Biol* **40:**785–800, 1986.
66. Nishizuka Y: The role of protein kinase C in cell surface signal transduction and tumor promotion. *Nature (Lond)* **308:**693–698, 1984.
67. Berridge MJ, Irvine RF: Inositol triphosphate, a novel second messenger in cellular signal transduction. *Nature (Lond)* **312:**315–321, 1984.
68. Berridge MJ: A novel cellular signaling system based on the integration of phospholipid and calcium metabolism. *Calcium Cell Function* **III:**1–35, 1982.
69. Melloni E, Pontremoli S, Michetti M, et al: Binding of protein kinase C to neutrophil membranes in the presence of Ca^{2+} and its activation by a Ca^{2+}-requiring proteinase. *Proc Natl Acad Sci USA* **82:**6435–6439, 1985.

70. Cockcroft S: The dependence on Ca^{2+} of the guanine-nucleotide-activated polyphosphoinositide phosphodiesterase in neutrophil plasma membrane. *Biochem J* **240:**503–507, 1986.
71. Papini E, Grzeskowiak M, Bellavite P, Rossi F: Protein kinase C phosphorylates a component of NADPH oxidase of neutrophils. *FEBS Lett* **190:**204–208, 1985.
72. Barrowman MM, Cockcroft S, Gomperts BD: Two roles for guanine nucleotides in the stimulus-secretion sequence of neutrophils. *Nature (Lond)* **319:**504–507, 1986.
73. Molski TFP, Naccache PH, Marsh ML, et al: Pertussis toxin inhibits the rise in the intracellular concentration of free calcium that is induced by chemotactic factors in rabbit neutrophils: Possible role of the "G proteins" in calcium mobilization. *Biochem Biophys Res Commun* **124:**644–650, 1984.
74. Goldman DW, Chang FH, Gifford LA, et al: Pertussus toxin inhibition of chemotactic factor-induced calcium mobilization and function in human polymorphonuclear leukocytes. *J Exp Med* **162:**145–156, 1985.
75. Krause KH, Schlegel W, Wollheim CB, et al: Chemotactic peptide activation of human neutrophils and HL-60 cells. *J Clin Invest* **76:**1348–1354, 1985.
76. Fujita I, Irita K, Takeshige K, Minakami S: Diacylglycerol, 1-oleoyl-2-acetyl-glycerol, stimulates superoxide generation from human neutrophils. *Biochem Biophys Res Commun* **120:**318–324, 1984.
77. Lagast H, Pozzan T, Waldvogel FA, Lew DP: Phorbol myristate acetate stimulates ATP-dependent calcium transport by the plasma membrane of neutrophils. *J Clin Invest* **73:**878–883, 1984.
78. Rickard JE, Sheterline P: Evidence that phorbol ester interferes with stimulated Ca^{2+} redistribution by activating Ca^{2+} efflux in neutrophil leucocytes. *Biochem J* **231:**623–628, 1985.
79. della Bianca V, Grzeskowiak M, Cassatella MA, et al: Phorbol 12, myristate 13, acetate potentiates the respiratory burst while it inhibits phosphoinositide hydrolysis and calcium mobilization by formyl-methionyl-leucyl-phenylalanine in human neutrophils. *Biochem Biophys Res Commun* **135:**556–565, 1986.
80. diVirgilio F, Lew DP, Pozzan T: Protein kinase C activation of physiological processes in human neutrophils at vanishingly small cytosolic Ca^{2+} levels. *Nature (Lond)* **310:**691–693, 1984.
81. Wolfson M, McPhail LC, Nasrallah VN, Snyderman R: Phorbol myristate acetate mediates redistribution of protein kinase C in human neutrophils: Potential role in the activation of the respiratory burst enzyme. *J Immunol* **135:**2057–2062, 1985.
82. della Bianca V, Grzeskowiak M, de Togni P, et al: Inhibition by verapamil of neutrophil responses to formylmethionyl-leucylphenylalanine and phorbol myristate acetate. Mechanisms involving Ca^{2+} changes, cyclic AMP and protein kinase C. Biochim Biophys Acta **845:**223–236, 1985.
83. Froscio M, Guy GR, Murray AW: Calmodulin inhibitors modify cell surface changes triggered by a tumor promoter. *Biochem Biophys Res Commun* **98:**829–835, 1981.
84. Nishihira J, McPhail LC, O'Flaherty JT: Stimulus-dependent mobilization of protein kinase C. *Biochem Biophys Res Commun* **134:**587–594, 1986.
85. Walsh CE, Waite BM, Thomas MJ, de Chatelet LR: Release and metabolism of arachidonic acid in human neutrophils. *J Biol Chem* **256:**7228–7234, 1981.
86. Suttorp N, Seeger W, Zucker-Reimann J, et al: Mechanism of leukotriene generation in polymorphonuclear leukocytes by staphylococcal alpha-toxin. *Infect Immun* **55:**104–110, 1987.
87. Lew DP, Andersson T, Hed J, et al: Ca^{2+}-dependent and Ca^{2+}-independent phagocytosis in human neutrophils. *Nature (Lond)* **315:**509–511, 1985.
88. Sawyer DW, Sullivan JA, Mandell GL: Intracellular free calcium localization in neutrophils during phagocytosis. *Science* **230:**663–666, 1985.
89. Kurskal BA, Shak S, Maxfield FR: Spreading of human neutrophils is immediately preceded by a large increase in cytoplasmic free calcium. *Proc Natl Acad Sci USA* **83:**2919–2923, 1986.
90. Gennaro R, Pozzan T, Romeo D: Monitoring of cytosolic free Ca^{2+} in C5a-stimulated neutrophils: Loss of receptor-modulated Ca^{2+} stores and Ca^{2+} uptake in granule-free cytoplasts. *Proc Natl Acad Sci USA* **81:**1416–1420, 1984.
91. Lew PD, Dayer JM, Wollheim CB, Pozzan T: Effect of leukotriene B_4, prostaglandin E_2 and

arachidonic acid on cytosolic free calcium in human neutrophils. *FEBS Lett* **166:**44–48, 1984.
92. Goldman DW, Gifford LA, Olson DM, Goetzl EJ: Transduction by leukotriene B_4 receptors of increases in cytosolic calcium in human polymorphonuclear leukocytes. *J Immunol* **135:**525–530, 1985.
93. Hansson A, Serhan CN, Haeggstrom J, et al: Activation of protein kinase C by lipoxin A and other eicosanoids. Intracellular action of oxygenation products of arachidonic acid. *Biochem Biophys Res Commun* **134:**1215–1222, 1986.
94. Andersson T, Schlegel W, Monod A, et al: Leukotriene B_4 stimulation of phagocytes results in the formation of inositol 1,4,5-trisphospinate. *Biochem J* **240:**333–340, 1986.
95. Naccache PH, Molski MM, Volpi M, et al: Biochemical events associated with the stimulation of rabbit neutrophils by platelet-activating factor. *J Leukocyte Biol* **40:**533–548, 1986.
96. Apfeldorf WJ, Melnick DA, Meshulam T, et al: A transient rise in intracellular free calcium is not a sufficient stimulus for respiratory burst activity in human polymorphonuclear leukocytes. *Biochem Biophys Res Commun* **132:**674–680, 1985.
97. Cooke E, Hallett M: The role of C-kinase in the physiological activation of the neutrophil kinase. *Biochem J* **232:**323–327, 1985.
98. Balazovich KJ, Smolen JE, Boxer LA: Ca^{2+} and phospholipid-dependent protein kinase (protein kinase C) activity is not necessarily required for secretion by human neutrophils. *Blood* **68:**810–817, 1986.
99. Gerard C, McPhail LC, Marfat A, et al: Role of protein kinases in stimulation of human polymorphonuclear leukocyte oxidative metabolism by various agonists. *J Clin Invest* **77:**61–65, 1986.
100. Sullivan GW, Donowitz GR, Sullivan JA, Mandell GL: Interrelationships of polymorphonuclear neutrophil membrane-bound calcium, membrane potential, and chemiluminescence: Studies in single living cells. *Blood* **64:**1184–1192, 1984.
101. Lew DP, Stossel TP: Effect of calcium on superoxide production by phagocytic vesicles from rabbit alveolar macrophages. *J Clin Invest* **67:**1–9, 1981.
102. Babior BM, Peters WA: The O_2-producing enzyme of human neutrophils. *J Biol Chem* **256:**2321–2323, 1981.
103. Suzuki H, Pabst MJ, Johnston RB Jr: Enhancement by Ca^{2+} or Mg^{2+} of catalytic activity of the superoxide-producing NADPH oxidase in membrane fractions of human neutrophils and monocytes. *J Biol Chem* **260:**3635–3639, 1985.
104. de Chatelet LR, Shirley PS: Chemiluminescence of human neutrophils induced by soluble stimuli: Effect of divalent cations. *Infect Immun* **35:**206–212, 1982.
105. Dahlgren C, Briheim G: Comparison between the luminol-dependent chemiluminescence of polymorphonuclear leukocytes and of the myeloperoxidase-HOOH system: Influence of pH, cations and protein. *Photochem Photobiol* **41:**605–610, 1985.
106. Klebanoff SJ: The iron–H_2O_2–iodide cytotoxic system. *J Exp Med* **156:**1262–1267, 1982.
107. Pollack CR, Straley SC, Klempner MS: Probing the phagolysosomal environment of human macrophages with a Ca^{2+}-responsive operon fusion in *Yersinia pestis*. *Nature (Lond)* **322:**834–836, 1986.
108. Mandell GL: Intraphagolysosomal pH of human polymorphonuclear neutrophils. *Proc Soc Exp Biol Med* **134:**447–449, 1970.
109. Jensen MS, Bainton DF: Temporal changes in pH within the phagocytic vacuole of the polymorphonuclear leukocyte. *J Cell Biol* **56:**379–388, 1973.
110. Segal AW, Geisow M, Garcia R, et al: The respiratory burst of phagocytic cells is associated with a rise in vacuolar pH. *Nature (Lond)* **290:**406–409, 1981.
111. Bassoe CF, Lairum OD, Glette J, et al: Simultaneous measurement of phagocytosis and phagosomal pH by flow cytometry: Role of polymorphonuclear neutrophilic leukocyte granules in phagosome acidification. *Cytometry* **4:**254–262, 1983.
112. Cech P, Lehrer RI: Phagolysosomal pH of human neutrophils. *Blood* **63:**88–95, 1984.
113. Simchowitz L; Intracellular pH modulates the generation of superoxide radicals by human neutrophils. *J Clin Invest* **76:**1079–1089, 1985.

114. Holian A, Diamond MS, Daniel RP: Intracellular pH changes in alveolar macrophages associated with O_2^- production. *Clin Res* **31**:365A, 1983.
115. Satoh M, Nanri H, Takeshige K, Minakami S: Pertussis toxin inhibits intracellular pH changes in human neutrophils stimulated by *N*-formyl-methionyl-leucyl-phenylalanine. *Biochem Biophys Res Commun* **131**:64–69, 1985.
116. Weisman SJ, Punzo A, Ford C, Sha'afi RI: Intracellular pH changes during neutrophil activation: Na^+–H^+ antiport. *Clin Res* **34**:473A, 1986.
117. Geisow MJ, Evans WH: pH in the endosome. *Exp Cell Res* **150**:36–46, 1984.
118. Babior BM: Oxidants from phagocytes: Agents of defense and destruction. *Blood* **64**:959–966, 1984.
119. Gabig TG, Lefker BA, Ossanna PJ, Weiss SJ: Proton stoichiometry associated with human neutrophil respiratory burst reactions. *J Biol Chem* **259**:13166–13171, 1984.
120. Forgac M, Cantley L, Wiedenmann B, et al: Clathrin-coated vesicles contain an ATP-dependent proton pump. *Proc Natl Acad Sci USA* **80**:1300–1303, 1983.
121. Cannon C, van Adelsberg J, Kelly S, al-Awqati Q: Carbon-dioxide-induced exocytotic insertion of H^+ pumps in turtle bladder luminal membrane: Role of cell pH and calcium. *Nature (Lond)* **314**:443–446, 1985.
122. Galloway CJ, Dean GE, Marsh M, et al: Acidification of macrophage and fibroblast endocytic vesicles *in vitro*. *Proc Natl Acad Sci USA* **80**:3334–3338, 1983.
123. Yamashiro DJ, Fluss SR, Maxfield FR: Acidification of endocytic vesicles by an ATP-dependent proton pump. *J Cell Biol* **97**:929–934, 1983.
124. McNeil PL, Tanasugarn L, Meigs JB, Taylor DL: Acidification of phagosomes is initiated before lysosomal enzyme activity is detected. *J Cell Biol* **97**:692–702, 1983.
125. Kakinuma K: Metabolic control and intracellular pH during phagocytosis by polymorphonuclear leukocytes. *J Biochem* **68**:177–185, 1970.
126. van Zwieten R, Wever R, Hamers MN, et al: Extracellular proton release by stimulated neutrophils. *J Clin Invest* **68**:310–313, 1981.
127. Cline M: Mechanism of acidification of the human leukocyte phagocytic vacuole. *Clin Res* **21**:595A, 1973.
128. Borregaard N, Schwartz JH, Tauber AI: Proton secretion by stimulated neutrophils. *J Clin Invest* **74**:455–459, 1984.
129. Segal AW, Dorling J, Coade S: Kinetics of fusion of the cytoplasmic granules with phagocytic vacuoles in human polymorphonuclear leukocytes. *J Cell Biol* **85**:42–59, 1980.
130. Roos D, Hamers MN, van Zwieten R, Weening RS: Acidification of the phagocytic vacuole: A possible defect in chronic granulomatous disease?, in Gallin J. I. (ed): *Advances in Host Defense Mechanisms,* New York, Raven, 1983, vol 3, pp 145–193.
131. Grinstein S, Furuya W: Amiloride-sensitive Na^+/H^+ exchange in human neutrophils: mechanism of activation by chemotactic factors. *Biochem Biophys Res Commun* **122**:755–762, 1984.
132. Wright J, Schwartz JH, Olson R, et al: Proton secretion by the sodium/hydrogen ion antiporter in the human neutrophil. *J Clin Invest* **77**:782–788, 1986.
133. Grinstein S, Furuya W, Cragoe EJ Jr: Volume changes in activated human neutrophils: the role of Na^+/H^+ exchange. *J Cell Physiol* **128**:33–40, 1986.
134. Besterman JM, May WS Jr, Levine H III, et al: Amiloride inhibits phorbol ester-stimulated Na^+/H^+ exchange and protein kinase C. *J Biol Chem* **260**:1155–1159, 1985.
135. Styrt B, Klempner MS: Inhibition of neutrophil oxidative metabolism by lysosomotropic weak bases. *Blood* **67**:334–342, 1986.
136. Babior BM: The nature of the NADPH oxidase. in Gallin JI, Fauci AS (eds): *Advances in Host Defense Mechanisms,* Raven, New York, 1983, vol 3, pp 91–118.
137. Gabig G, Babior BM: The O_2-forming oxidase responsible for the respiratory burst in human neutrophils. *J Biol Chem* **254**:9070–9074, 1979.
138. Klebanoff SJ: Myeloperoxidase–halide–hydrogen peroxide antibacterial system. *J Bacteriol* **95**:2131–2138, 1968.

139. Hamers MN, Sips HJ: Molecular mechanism of the bactericidal action of myeloperoxidase–H_2O_2–halide. *Adv Exp Med Biol* **141:**151–159, 1982.
140. Styrt B, Klempner MS: Interaction of pH, lysosomal contents, and oxygen metabolites in neutrophil microbicidal activity against *S. aureus* and *E. coli. Clin Res* **34:**534A, 1986.
141. Whitin JC, Chapman CE, Simons ER, et al: Correlation between membrane potential changes and superoxide production in human granulocytes stimulated by phorbol myristate acetate. *J Biol Chem* **255:**1874–1878, 1980.
142. Seeds MC, Parce JW, Szejda P, Bass DA: Independent stimulation of membrane potential changes and the oxidative metabolic burst in polymorphonuclear leukocytes. *Blood* **65:**233–240, 1985.
143. Naccache PH, Showell HJ, Becker EL, Sha'afi RI: Transport of sodium, potassium, and calcium across rabbit polymorphonuclear leukocyte membranes. *J Cell Biol* **73:**428–444, 1977.
144. Simchowitz L: Chemotactic factor-induced activation of Na^+/H^+ exchange in human neutrophils. *J Biol Chem* **260:**13237–13247, 1985.
145. Segal AW, Cross AR, Garcia RC, et al: Absence of cytochrome b-245 in chronic granulomatous disease. *N Engl J Med* **308:**245–251, 1983.
146. della Bianca V, Bellavite P, de Togni P, et al: Studies on stimulus-response coupling in human neutrophils. I. Role of monovalent cations in the respiratory and secretory response to *N*-formylmethionylleucylphenylalanine. *Biochim Biophys Acta* **755:**497–505, 1983.
147. Kitagawa S, Johnston RBJ: Relationship between membrane potential changes and superoxide-releasing capacity in resident and activated mouse peritoneal macrophages. *J Immunol* **135:**3417–3423, 1985.
148. Baker SS, Cohen HJ: Altered oxidative metabolism in selenium-deficient rat granulocytes. *J Immunol* **130:**2856–2860, 1983.
149. Aziz SS, Klesius PH, Frandsen JC: Effects of selenium on polymorphonuclear leukocyte function in goats. *Am J Vet Res* **45:**1715–1718, 1984.
150. Arthur JR, Boyne R: Superoxide dismutase and glutathione peroxidase activities in neutrophils from selenium deficient and copper deficient cattle. *Life Sci* **36:**1569–1575, 1985.
151. Malamud D, Dietrich SA, Shapiro IM: Low levels of mercury inhibit the respiratory burst in human polymorphonuclear leukocytes. *Biochem Biophys Res Commun* **128:**1145–1151, 1985.
152. Mylroie AA, Collins H, Umbles C, Kyle J: Erythrocyte superoxide dismutase and other parameters of copper status in rats ingesting lead acetate. *Toxicol Appl Pharmacol* **82:**512–520, 1986.
153. Davis T, Johnston C: Effects of gold compounds on function of phagocytic cells. *Inflammation* **10:**311–320, 1986.
154. Curnutte JT, Babior BM, Karnovsky ML: Fluoride-mediated activation of the respiratory burst in human neutrophils. *J Clin Invest* **63:**637–647, 1979.
155. Strnad CF, Wong K: Calcium mobilization in fluoride activated human neutrophils. *Biochem Biophys Res Commun* **133:**161–167, 1985.
156. Korchak HM, Smolen JE, Eisenstat BA, et al: Activation of the neutrophil: The role of anion fluxes in neutrophil secretion. *Ann NY Acad Sci* **358:**346–347, 1980.
157. Korchak HM, Eisenstat BA, Hoffstein ST, et al: Anion channel blockers inhibit lysosomal enzyme secretion from human neutrophils without affecting generation of superoxide anion. *Proc Natl Acad Sci USA* **77:**2721–2725, 1980.
158. Foote CS, Goyne TE, Lehrer RI: Assessment of chlorination by human neutrophils. *Nature (Lond)* **301:**715–716, 1983.
159. Weiss SJ, Lampert MB, Test ST: Long-lived oxidants generated by human neutrophils: Characterization and bioactivity. *Science* **222:**625–628, 1983.
160. Grisham MB, Jefferson MM, Telton DF, Thomas EL: Chlorination of endogenous amines by isolated neutrophils. *J Biol Chem* **259:**10404–10413, 1984.
161. Test ST, Lampert MB, Ossanna PJ, et al: Generation of nitrogen–chlorine oxidants by human phagocytes. *J Clin Invest* **74:**1341–1349, 1984.

5

An Expanded View of the Phagocytic Respiratory Burst

Bacterial Competition for Oxygen and Its Stimulation by Host Factor(s)

MYRON S. COHEN and BRADLEY E. BRITIGAN

1. INTRODUCTION

Baldridge and Gerard in 1933[1] made the seminal observation that leukocyte consumption of ambient O_2 increases markedly during the early stages of phagocytosis. Over the ensuing 43 years remarkable progress was made in characterizing the mechanisms and ramifications of this phenomenon, identified by Babior as the respiratory burst.[2] Consumption of O_2 by neutrophils and monocytes during phagocytosis leads to the purposeful formation of superoxide, a unique capacity of phagocytic cells. Production of superoxide and subsequent formation of other reactive oxygen intermediates (e.g., H_2O_2 and OCl^-) is critical to the microbicidal activity of these cells.[3] This is graphically demonstrated by the natural history of patients with chronic granulomatous disease (CGD) of childhood, whose phagocytes lack the capacity to develop a respiratory burst.[3,4]

1.1. Microbial Strategies to Resist Oxygen-Dependent Killing

Most human pathogens are recognized by the host during some stage of the infection process. In order to survive and cause clinical disease, the pathogen must

MYRON S. COHEN • Departments of Medicine and Microbiology and Immunology, Division of Infectious Diseases, University of North Carolina, Chapel Hill, North Carolina 27514. BRADLEY E. BRITIGAN • Department of Medicine, University of North Carolina, Chapel Hill, North Carolina 27514.

be able to evade multiple host defenses, including phagocytosis and exposure to resulting toxic oxygen reduction products. Bacteria are extremely adaptive, and a remarkable variety of mechanisms by which different species avoid phagocyte mediated killing has been described, which include interference with chemotaxis, attachment, ingestion, and the microbicidal process.[5] For organisms unable to avoid phagocytosis, interference with the formation or activity of phagocyte-derived oxygen-reduction products would provide an additional means of survival.

Although ingested by host phagocytes, several pathogens fail to trigger respiratory burst activity. Intrastrain variations in this characteristic appear to correlate with virulence. A virulent strain of *Salmonella typhi* is ingested to a similar extent as a nonvirulent strain but induces less stimulation of neutrophil O_2 consumption and chemiluminescence.[6,7] *Toxoplasma gondii* induces minimal respiratory burst activity in cultured monocytes or mouse peritoneal macrophages.[8] Formation of reactive oxygen intermediates appears critical for the microbicidal activity of peritoneal macrophages against *T. gondii.*[8] Similar inhibition of (or failure to stimulate) the neutrophil respiratory burst may follow ingestion of *Brucella abortus.*[9]

An additional strategy described for some microbial pathogens involves enzymatic inactivation of toxic oxygen reduction products. Mandell[10] studied multiple strains of *Staphylococcus aureus* that varied in their concentration of endogenous superoxide dismutase (SOD) or catalase. In vitro ingestion of high and low catalase strains by neutrophils was similar, but the high catalase strains were much more resistant to neutrophil-mediated killing. Virulence in a mouse model correlated with catalase but not SOD activity. Similar results have been obtained using *Listeria moncytogenes.*[11]

Beaman and colleagues[12] described the isolation of a unique SOD that appears to be secreted by a strain of *Nocardia asteroides* resistant to killing by neutrophils in vitro.[12] Release of this enzyme in a phagocytic vacuole would limit the accumulation of superoxide. This strain also possesses a high concentration of catalase, which could protect the organism from resulting hydrogen peroxide formation.

Several pathogens have been shown to interfere with lysosome–phagosome fusion.[5] Failure of degranulation to occur could limit OCl^- formation[13] by preventing release of granule-bound myeloperoxidase, as well as interfering with oxygen-independent microbicidal activity.[14]

Formation of oxygen reduction products by phagocytic cells requires the availability of molecular O_2. Cells incubated under anaerobic conditions are unable to generate a respiratory burst and are limited in their ability to kill some (but not all) pathogens.[15] Aerobic and microaerophilic microorganisms consume O_2 as well, using it as a terminal electron acceptor.[16] O_2 availability is limited at many tissue sites, suggesting that bacterial pathogens and phagocytic cells could be in direct competition for this substrate. We have investigated the dynamics of microbial metabolism during neutrophil phagocytosis with the hypothesis that such metabolism contributes to microbial resistance to oxygen-dependent neutrophil microbicidal activity. This chapter summarizes work that demonstrates in vitro competition between human neutrophils and several bacterial species for molecular O_2 and characterizes a host factor(s) that stimulates bacterial metabolism, altering the dynamics of this neutrophil–bacterial competition. Implications of these observations for bacterial pathogenesis and neutrophil microbicidal defense are also discussed.

2. BACTERIAL COMPETITION FOR OXYGEN

During evaluation of the postphagocytic respiratory burst, as reflected by the rate of disappearance of O_2 from a 1-ml vol of Hanks balanced salt solution (HBSS) in a Clark-type electrode, we noted that the bacteria we were examining (*Neisseria gonorrhoeae*, 10^8/ml) consumed more O_2 than did the phagocytes (5×10^6/ml).[17] Furthermore, addition of serum to the system had a remarkable effect on the rate of bacterial O_2 consumption, causing an immediate three- to fourfold increase (Fig. 1). Our initial experiments were conducted with *N. gonorrhoeae*, but similar results were obtained with *Staph. aureus* and *Escherichia coli*.[18] O_2 consumption rates were significantly increased when bacteria were exposed to 1% serum, and a maximal effect was demonstrated at a 20% (v/v) serum concentration.[17] Thus, in vitro experiments using oxygen consumption as a measure of the neutrophil respiratory burst during their interaction with live bacteria may be incorrectly interpreted. Increases in the rate of oxygen consumption seen with the addition of serum have generally been attributed to increased neutrophil stimulation by opsonization of bacteria or release of C5a rather than to bacterial oxygen consumption. Serum

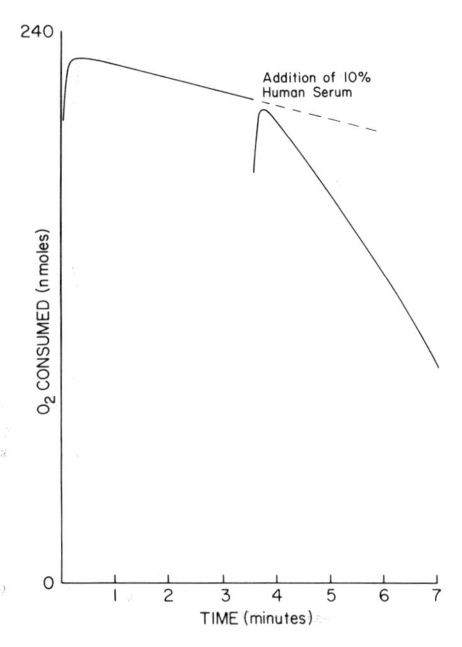

FIGURE 1. Typical tracing taken from a Clark O_2 electrode demonstrating the effect of 10% normal pooled serum on the O_2 consumption rate of *Neisseria gonorrhoeae*. Gonococci (10^8/ml) suspended in HBSS have a small steady rate of O_2 consumption as shown in the first part of the tracing. Addition of 10% serum leads to an immediate increase in the bacterial O_2 consumption rate, which continues until all O_2 is used (end point not shown). (From Cohen and Cooney.[17])

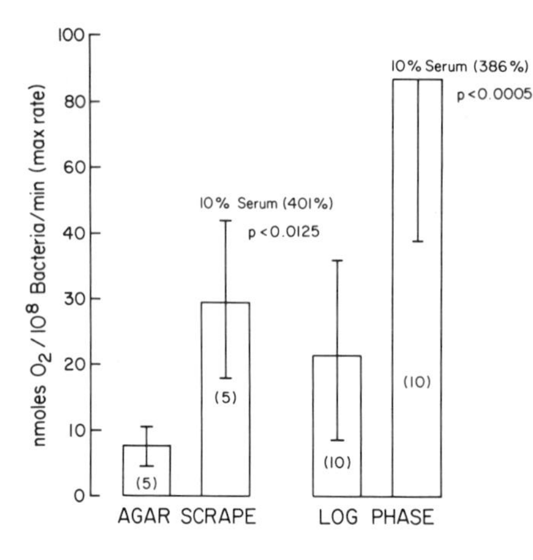

FIGURE 2. Effect of 10% serum on the maximal O_2 consumption rate of gonococci (10^8/ml) in HBSS that had been scraped from agar or grown to log phase. Columns represent the means and brackets the standard deviations of the maximal O_2 consumption rates observed over 5 min in a Clark O_2 electrode. The number of experiments is noted in parenthesis. Percentage increases in O_2 consumption above control and p values calculated by Student's *t*-test are also given. (From Cohen and Cooney.[17])

increases other parameters of bacterial metabolism to a similar extent, including glucose utilization, [^{14}C]adenine incorporation, and growth rate.[17,19] Serum-induced stimulation of bacterial metabolism is independent of the growth phase of the organism (Fig. 2) or its susceptibility to antibody–complement-dependent serum microbicidal activity. Characterization of the serum factor(s) responsible for this phenomena is addressed below.

These results suggested that stimulated bacterial O_2 consumption produces an anaerobic environment. To examine this possibility, we incubated several important bacterial pathogens in HBSS containing the indicator resazurin, which turns colorless when Eh decreases below −42 mV.[18] In the presence of 10% serum at 37°C, each organism (10^8/ml) turned the indicator colorless within 15 min (Table I). Even in the absence of serum, *E. coli* generated anaerobic conditions.[18] These results are

TABLE I. Ability of Bacteria to Create an Anaerobic Environment

Organism	Time until resazurin colorless (min)[a]	
	10% normal serum	10% dialyzed serum
Neisseria gonorrhoeae	7.4 ± 0.5	No change[b]
Staphylococcus aureus	13.3 ± 1.9	No change[b]
Escherichia coli	15.1 ± 1.3	17.2 ± 1.2

[a]Results are expressed as the minutes necessary (mean ± SEM, $N = 3$) for 10^8 organisms in 1 ml HBSS + 10% normal or dialyzed serum to turn the resazurin indicator colorless (anaerobic conditions, Eh < −42 mV) while vertically suspended in 12 × 75-mm glass tubes at 37°C in a shaker bath apparatus (72 strokes/min).

[b]Neither *N. gonorrhoeae* nor *Staph. aureus* turned the resazurin colorless while suspended in 10% dialyzed serum during a 30-min period of observation. (From Britigan et *al.*[18])

consistent with the disappearance of O_2 from a Clark electrode chamber (Fig. 3). To explore these results further in terms of their relevance to in vitro experiments assessing phagocyte microbicidal activity, the buffer Eh was determined during incubation of gonococci (10^8/ml) in the presence of 10% serum in a rotating end-over-end assay[20] or shaker bath system,[21] both of which are commonly used. The rotating system remained aerobic, presumably because O_2 at the air–fluid interface was continually mixed throughout the buffer. Likewise, conditions remained aerobic when bacteria were incubated in the shaker apparatus at a 30° angle at 240 strokes/min.[18] However, when the organisms were shaken vertically at the same rate or at 30° at a slower rate (72 strokes/min) Eh approached −42 mV. Although

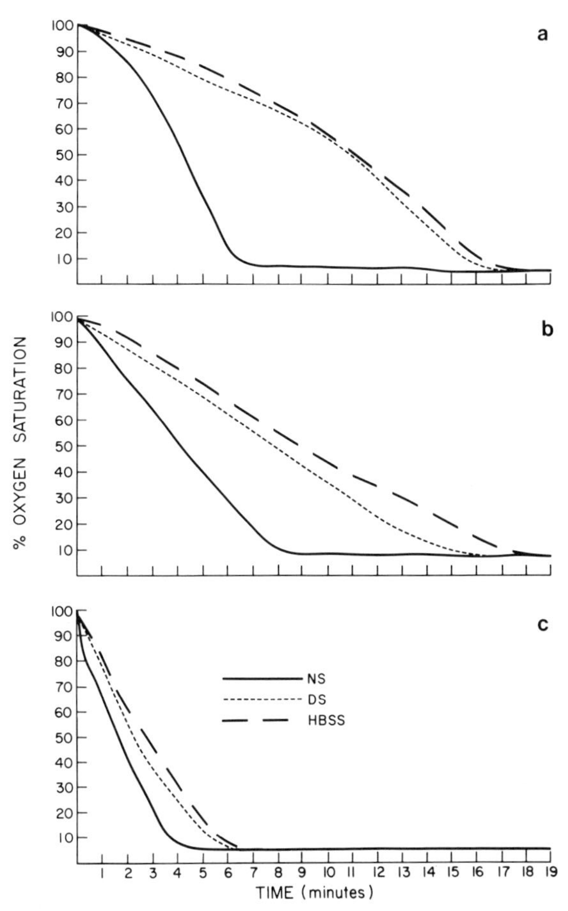

FIGURE 3. Representative Clark O_2 electrode tracing ($N = 3$–4) of O_2 consumption rates of (a) *Neisseria gonorrhoeae,* (b) *Staphylococcus aureus,* and (c) *Escherichia coli* suspended in HBSS, HBSS plus 10% dialyzed serum (DS), or HBSS plus 10% normal pooled serum (NS). For all three organisms, 10% normal serum induced a significant increase in O_2 consumption. (From Britigan et al.[18])

these experimental conditions may not be directly comparable to all experimental systems used to study bacterial–neutrophil interactions, they underscore the need to ensure continuous aeration of buffers containing metabolically active bacteria. Several published studies related to phagocyte microbicidal activity may have inadvertently been conducted under anaerobic conditions!

To explore further the relevance of these observations, bacteria were incubated in serum for brief periods before the addition of neutrophils. Phorbol myristate acetate (PMA), 100 ng/ml, was added simultaneously with neutrophils to ensure maximal stimulation.[22] Subsequent neutrophil formation of reactive oxygen intermediates was determined by on-line measurement of luminol-dependent luminescence (LDL).[18] Control experiments showed that all LDL detected was of neutrophil origin. PMA-stimulated neutrophils suspended in 10% serum in which bacteria (10^8/ml) had been allowed to preincubate for 5 min were unable to generate LDL (Fig. 4). The magnitude of bacterial inhibition of neutrophil LDL could be correlated to both the duration of bacterial preincubation (Fig. 5) and the number of organisms employed (Fig. 6). Inhibition of LDL was demonstrable when $>10^6$ gonococci/ml were employed. Results varied somewhat with bacterial species employed. The absolute concentration of microorganisms present rather than the particle/cell ratio was the most critical determinant in this system: significant inhibition of LDL could be observed at a 1 : 1 bacteria to cell ratio as long as 10^6 or more organisms were employed. A variety of control experiments demonstrated that bacteria did not interfere with the assay itself.[18] To offer more direct evidence of the ability of bacteria to limit neutrophil formation of reactive oxygen intermediates, we measured neutrophil hydrogen peroxide formation as assessed by the oxidation of scopoletin in the presence of horseradish peroxidase.[23] As serum interferes with this assay, we used *E. coli,* whose resting O_2 consumption is of sufficient magnitude that serum stimulation is not required to permit competition for O_2. Neutrophils (5×10^6/ml) suspended in HBSS in which *E. coli* had been allowed to preincubate for

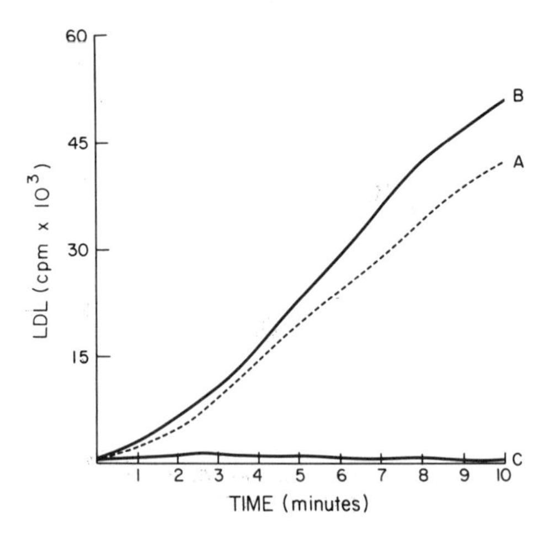

FIGURE 4. Representative tracing of luminol-dependent luminescence taken directly from the stripchart recorder attached to the luminometer. Neutrophils (10^5 ml) were stimulated with PMA while suspended in (A) HBSS; (B) HBSS plus 10% dialyzed serum in which *Neisseria gonorrhoeaea* had preincubated 5 min prior to neutrophil addition, and (C) HBSS plus 10% normal serum, in which *N. gonorrhoeae* had preincubated for 5 min as well. Time 0 is the time of neutrophil and PMA addition. Substitution of *S. aureus* for gonococci yielded similar results ($N = 3$). (From Britigan and Cohen.[18])

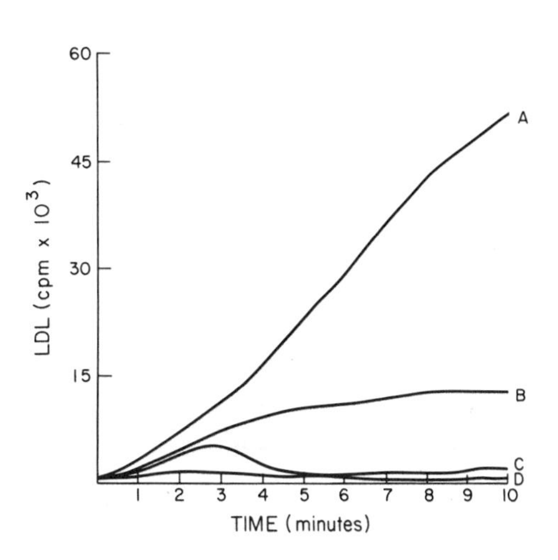

FIGURE 5. Tracing demonstrating the effect of time of bacterial preincubation in 10% serum (prior to neutrophil addition) on subsequent inhibition of PMA stimulated neutrophil LDL ($N = 5$). Neutrophils (10^5/ml) were stimulated with PMA while suspended in (A) HBSS plus 10% dialyzed serum and *Neisseria gonorrhoeae* preincubated for 5 min (control), or HBSS plus 10% normal serum container *N. gonorrhoeae* added at (B) 0, (C) 1, or (D) 5 min before the addition of neutrophils. Inhibition of neutrophil LDL increases as a function of duration of bacterial preincubation. Time 0 is the time of addition of neutrophils and PMA. Data were similar regardless of bacterial strain employed. (From Britigan and Cohen.[18])

5 min (100/1 bacteria to cell ratio) failed to generate H_2O_2 when stimulated with PMA[18] (Fig. 7). Simultaneous addition of neutrophils, bacteria, and PMA resulted in a normal rate of H_2O_2 production for 2 min, after which further H_2O_2 formation was not detected. Control experiments showed no bacterial interference with the detection system itself.

These observations suggested that bacteria are able to consume O_2 at a rate great enough to create conditions sufficiently anaerobic to prevent O_2 reduction. To explore this hypothesis further, the importance of ongoing bacterial metabolism was examined. Using the LDL system, bacterial preincubation was performed as described earlier except that 1 mM KCN was included in the reaction mixture. At this concentration, KCN completely inhibits gonococcal O_2 consumption[18,24] but has a

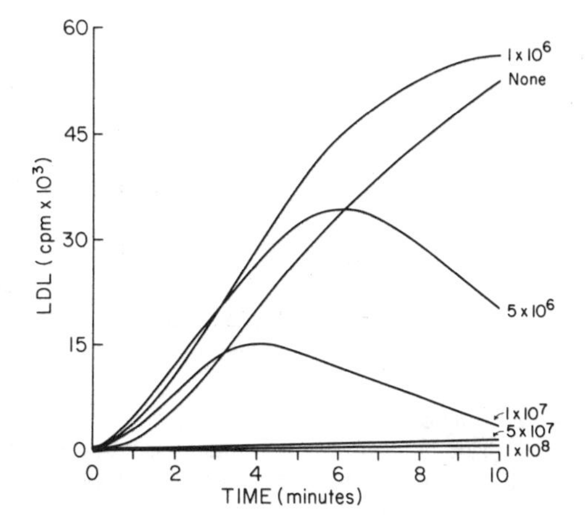

FIGURE 6. Tracings reflecting inhibition of neutrophil LDL by increasing *Neisseria gonorrhoeae* concentration; 10^5 neutrophils were stimulated with PMA while suspended in HBSS plus 10% serum, in which varying concentrations of *N. gonorrhoeaea* had incubated at 37°C for 5 min prior to addition of neutrophils ($N = 2$). Neutrophil LDL in the absence of bacteria (designated none) served as control. Results with *Escherichia coli* were similar ($N = 3$), except 0.5 log fewer organisms induced equivalent LDL inhibition. (From Britigan and Cohen.[18])

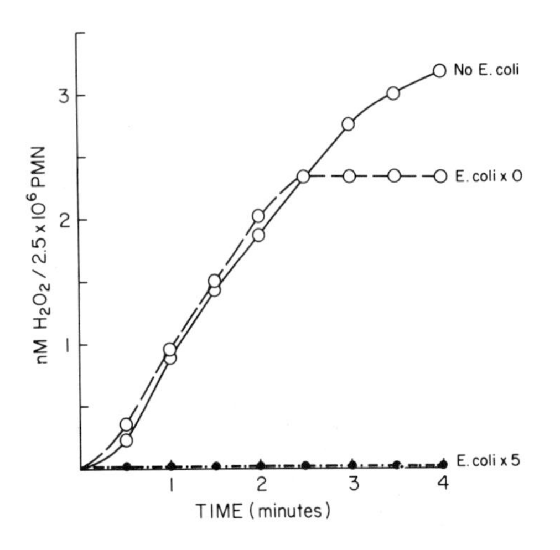

FIGURE 7. Tracings ($N = 2$) of mean H_2O_2 production following PMA stimulation of neutrophils (2.5 × 10^6/ml) while suspended in HBSS (no *Escherichia coli*); HBSS containing *E. coli* (10^8/ml), which had incubated for 5 min at 37°C prior to neutrophil and PMA addition (*E. coli* × 5); and HBSS, in which the same concentration of *E. coli* was added simultaneously to neutrophils and PMA (*E. coli* × 0). (From Britigan and Cohen.[18])

minimal effect on neutrophil LDL. Suppression of bacterial metabolism prevented the organism's ability to inhibit PMA-stimulated LDL (Fig. 8). Subsequently, gonococci (or *E. coli*) were preincubated for 5 min in the absence of KCN after which the system was reaerated; reaeration restored the system's capacity to support PMA-stimulated neutrophil LDL only transiently (Fig. 8). However, addition of KCN after the bacterial preincubation period allowed formation of LDL equivalent to control cells. This surprising result is important for several reasons. First, it eliminates the possibility that the bacteria released a substance into the buffer, which interfered with neutrophil O_2 reduction and/or LDL.[25] Second, it eliminates the possibility that the organisms depleted the buffer of another substrate (i.e., glucose), which is critical for maintenance of the neutrophil respiratory burst. Finally, it shows that competition for O_2 between bacteria and PMNS is dynamic:

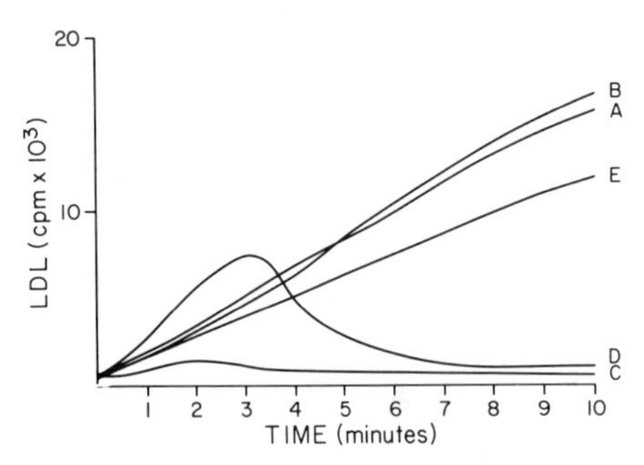

FIGURE 8. Tracings examining the effect of inhibition of bacterial metabolism with KCN or reaeration of the experimental system on bacterial inhibition of neutrophil LDL. Neutrophils (10^5/ml) and PMA were added to HBSS plus 10% dialyzed serum (A) or HBSS plus 10% normal serum following 5 min preincubation of *Escheria coli* (10^8/ml) in the presence (B) or absence (C) of 1 mM KCN. Conditions for tracing D were identical to C, except that the system was reaerated at the completion of bacterial preincubation. Tracing E was generated under condition identical to D, except that KCN was added at the time of reaeration. Addition of KCN after bacterial preincubation without reaeration yielded a curve identical to E. Time 0 is the time of addition of neutrophils and PMA. Similar results were obtained with *S. aureus* and *Neisseria gonorrhoeae.* (From Britigan and Cohen.[18])

even at low oxygen tensions if bacterial respiration is abrogated neutrophils are able to use the remaining oxygen to form reactive oxygen intermediates.

There is some discrepancy regarding the extent to which O_2 tensions must be reduced in order to interfere with the capacity of neutrophils to generate O_2 reduction products. Gabig and Babior[26] reported that neutrophil superoxide production (ferricytochrome C reduction) was not effected until O_2 concentrations decreased below 2.4 μM. By contrast, using LDL as a detection system for neutrophil O_2 reduction, Edwards et al.[27] reported that neutrophil LDL started to decrease when O_2 concentration was <120 μM and was $<20\%$ of control when the O_2 concentration dropped below 10 μM.

Oxygen tensions in the latter range have been reported to occur in experimental abscess cavities.[28] They are probably also encountered clinically in abscesses containing a mixture of aerobic and anaerobic pathogens. Failure of neutrophils to eliminate bacteria from an abscess cavity may relate to the ability of aerobic and microaerophilic organisms to limit the O_2 available to the phagocyte, leaving only oxygen-independent systems available.[29] Utilization of O_2 by aerobic pathogens may also contribute to the proliferation of anaerobic organisms within these abscesses.[29]

2.1. Limitations of Interpretation

To characterize this competition further, we must calculate the respective K_m for O_2 of the enzyme systems (bacterial and neutrophil) in competition for this substrate. In the case of neutrophils, O_2 is the terminal acceptor of a single electron transferred through an NADPH-dependent oxidase system, the features of which are described in detail elsewhere in this volume. Using rat neutrophils, Edwards et al.[30] showed that the K_m of this system for O_2 was approximately 9.6 μM for unstimulated and 3.7 μM for stimulated whole cells. V_{max} for respiration increased 51.7% with stimulation. However, this study was limited by use of whole cells and the insensitivity of conventional O_2 electrodes. Aerotolerant bacteria use O_2 as a terminal electron acceptor,[16] where a variety of substrates can act as electron donors. In comparing bacterial K_m with that of the neutrophil oxidase system, a major theoretical problem surrounds determination of the correct denominator for comparison: membrane volumes, protein concentrations, ratio of whole bacteria to neutrophils, and other factors. On the basis of work with animal models, a bacteria to neutrophil ratio of 100–1000/1 seems reasonable. The K_m for O_2 of *Klebsiella* aerogenes may be below 2 μM[16] (concentration not given), but this has not been calculated for most pathogenic bacteria.

The relevance of the buffers employed in this and other systems to in vivo conditions has yet to be determined. In the circulation, O_2 is tightly bound to hemoglobin until capillary delivery to tissue; the relative ability of the bacteria and neutrophils to extract O_2 from hemoglobin could well be important. Experimental conditions that would allow this competition to occur have not been developed. Furthermore, conclusions regarding bacterial–neutrophil competition for O_2 may not be applicable to conditions occurring when bacteria have been ingested. Available evidence suggest that neutrophil reduction of O_2 is by a membrane bound oxidase (allowed to use extracellular substrates) with transfer of superoxide into the phagosome; the rate of diffusion of O_2 into the phagosome itself is not known.

3. CHARACTERIZATION AND PURIFICATION OF SERUM FACTOR(S) THAT ENHANCE BACTERIAL COMPETITION FOR OXYGEN

Our work has suggested that bacterial recognition of one or more host factors allows competition for O_2. These factors have been best characterized in serum and myeloid cells but appear to be present in vaginal secretion and ascitic fluid as well.[31] Three different human pathogens (*N. gonorrhoeae, E. coli,* and *S. aureus*), have been evaluated. *N. gonorrhoeae* is of particular interest because it is strictly limited to a human reservoir and its pheno- and genotypic adaptations to this milieu have been characterized extensively.[32,33] Several aspects of this organism's metabolism have been studied.[24] Furthermore, gonococci appear to survive at sites of infection marked by significant inflammatory response indicating in vivo some degree of resistance to neutrophil microbicidal activity.

The stimulatory capacity of human serum is removed by overnight dialysis with membrane exclusion tubing of defined pore size against HBSS suggesting the critical factor(s) to be $<1000\ M_r$.[19] Activity is then demonstrable in the resulting dialysate. Activity persists following heating at 56° for 30 min, eliminating complement as a contributing factor.[17] Further characterization of the factor (summarized in Table II) showed it is not removed by lipid extraction or boiling. Following exposure to living bacteria, stimulatory capacity is lost, suggesting its utilization or inactivation.

Since *N. gonorrhoeae* is fastidious in its growth requirements, it initially seemed logical that serum contained gonococcal growth cofactors missing from

TABLE II. Characteristics of Serum Bacterial Metabolic Stimulatory Factor(s)

$<1000\ M_r$[a]
Heat stable[a]
Stimulates gonococcal electron transport proximal to cytochrome *c*[a]
Activity lost after induction of metabolic stimulation
Factors initially eliminated by add-back experiments: glucose, Ca^{2+}, Mg^{2+}, HCO_3^-, PO_4^{2-}, lactate, pyruvate, amino acids, ATP, AMP, NADPH, NADH, malate, glutathione, fatty acids, erythrocyte lysate
Active for *Neisseria gonorrhoeae, Staphylococcus aureus,* and *Escherichia coli*
Recoverable in the dialysate
Similar elution profile for serum- and cell-derived factor(s) using Sephadex and DEAE columns[a]

[a]These parameters are also true for the stimulatory factor(s) recovered from mammalian cells.

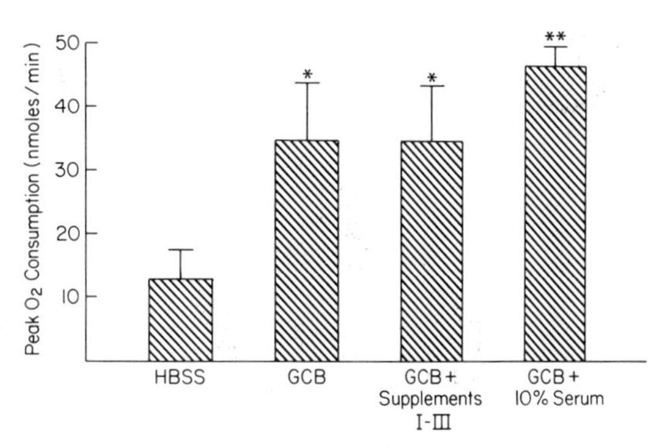

FIGURE 9. Comparison of maximal O_2 consumption rates over 5 min of *Neisseria gonorrhoeae* suspended in HBSS, gonococcal base broth (GCB), GCB plus Kellogg defined supplements 1–3, or GCB plus 10% normal pooled serum. Results are the mean ±SEM of four separate experiments done in duplicate. * = significant ($p < 0.05$) relative to HBSS control and ** = significant increase ($p < 0.01$) relative to HBSS, GCB, and GCB plus Kellogg supplements. (From Britigan et al.[19])

HBSS. However, addition of serum to optimal in vitro growth media [gonococcal base broth (GCB) containing Kellogg-defined supplements I-III, GCB (Ref. 34)] resulted in stimulation of O_2 consumption greater than that detected in GCB alone[19] (Fig. 9). Several approaches to characterize the factor(s) responsible for this stimulation were undertaken.

We hypothesized initially that the serum factor(s) was one of a number of substrates known to be important to gonococcal metabolism[16,24] and present in human serum. Consequently, we used our ability to remove the stimulatory factor(s) from serum by dialysis to pursue an add-back strategy. A variety of substances were added alone or in combination to the postdialysis retentate at concentrations present in serum.[35] In addition, all alterations in electrolyte concentrations resulting from the dialysis procedure were corrected. These maneuvers failed to restore stimulatory capacity to the retentate (see Table II).[19]

Serum stimulation of gonococcal glucose metabolism results in formation of reducing equivalents, leading to increased electron transport and O_2 consumption.[16,24] *N. gonorrhoeae* possesses a number of pathways for glucose catabolism including the pentose pathway, Entner–Doudoroff, and Embden–Meyerhof/TCA cycle.[24] In the absence of serum, the organism relies primarily on a combination of the pentose and Entner–Doudoroff cycles.[24] In the presence of 10% serum, a threefold increase over baseline activity in each pathway was detected,[19] precluding identification of this factor as a critical agent in a single glucose cycle. These results suggest, however, that the factor does not primarily stimulate oxygenation of an organic substrate.[36]

Oxygen is primarily used by gonococci as a terminal electron acceptor. Electrons may enter the system at a number of different points (Fig. 10). KCN, antimycin A, and amytal inhibit O_2 consumption (electron transport) resulting from interaction of gonococci with serum, suggesting serum stimulation results from addition of electrons to the transport system at a point proximal to coenzyme Q. In these experiments, several recognized electron donors used as a positive control did

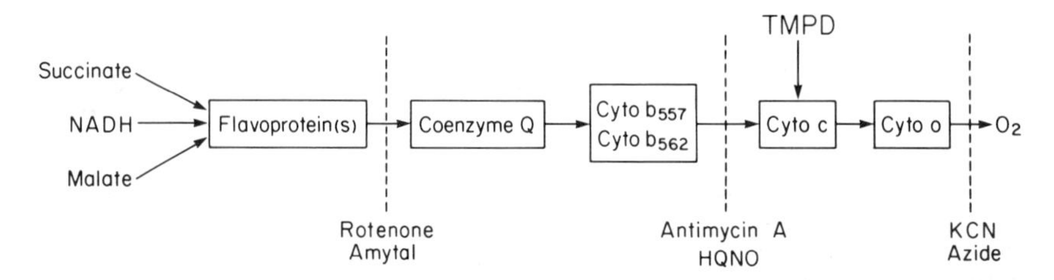

FIGURE 10. Diagram of proposed pathway of electron transport in *Neisseria gonorrhoeae* in which O_2 acts as the terminal electron acceptor. Points of blockage of electron transport for a number of respiratory inhibitors are noted. All these compounds inhibit serum-induced stimulation of gonococcal electron transport. This indicates that serum must lead to addition of electrons to the system proximal to coenzyme Q.

not stimulate intact organisms, presumably because the bacterial membrane is impermeable. Since serum stimulates whole organisms (as well as membrane preparations), it seems unlikely that the factor we have isolated was one of the known electron donors examined. Experiments are ongoing to explore the possibility that increased glucose transport resulting from exposure to this serum factor(s) contributes to enhanced gonococcal metabolism.

Given our inability to determine the identity of the serum stimulatory factor(s) indirectly, we have purified this material. Application of serum to a G-15 Sephadex column yields an elution profile with stimulatory activity in fractions resolved immediately prior to and overlapping the salt containing fraction. Application of these active G-15 fractions to a DEAE column with subsequent elution with Tris buffer (pH 8.2) containing increasing concentrations of NaCl results in recovery of activity in the initial fractions collected. The active fractions corresponded to a small OD 280 absorbance peak. However the stability of the factor(s) to acid hydrolysis as well as data presented in the next section suggest that stimulatory activity does not result from the presence of a peptide.

4. INTERACTION WITH MYELOID CELLS INCREASES BACTERIAL OXYGEN CONSUMPTION

Serum-induced stimulation of bacterial oxygen utilization may confound interpretation of experiments designed to measure stimulation of phagocyte respiratory burst activity by live bacteria in the presence of serum. Even in the absence of serum, however, the contribution of bacterial metabolism to the process needs to be kept in mind. Many organisms, including *N. gonorrhoeae*, stimulate neutrophil respiratory burst activity in the absence of serum,[37] the extent of which may vary from strain to strain, depending on the expression of different gonococcal surface structures.[38] In a previous study, we noted that the O_2 consumption rate of a suspension of gonococci and neutrophils was significantly higher than the sum of

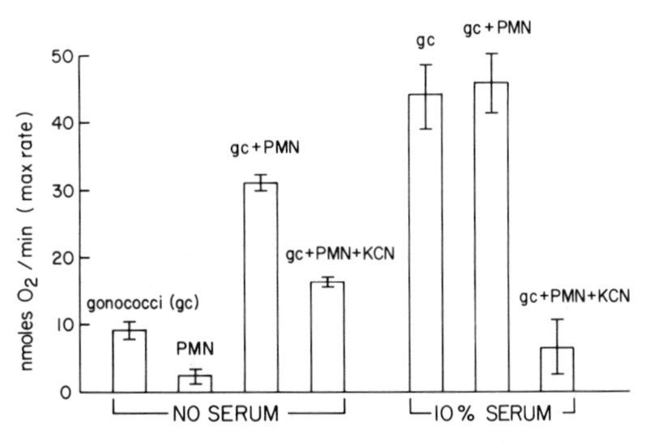

FIGURE 11. Relative contribution of *Neisseria gonorrhoeae* and neutrophils to O_2 consumption when incubated simultaneously in the presence and absence of 10% serum. Columns designate the mean ±SEM of these experiments performed in duplicate. KCN blocks gonococcal O_2 consumption but has little effect on that of stimulated neutrophils. In the presence of serum, gonococcal O_2 consumption is so great that determination of neutrophil O_2 consumption is impossible in the absence of KCN. Of note, even in the absence of serum the sum of gonococcal (gc) and stimulated neutrophil (gc + PMN + KCN) O_2 consumption does not account for all the O_2 consumed when the two (gc + PMN) are incubated simultaneously. This suggests that gonococcal O_2 consumption may also increase in response to neutrophils. (From Cohen and Cooney.[17])

the component rates.[17] In this study, the neutrophil component was measured by adding 1 mM KCN (Fig. 11). These data suggested that neutrophils were less effectively stimulated by metabolically inactive gonococci[39] or that neutrophil exposure also stimulated gonococcal O_2 consumption. We recently showed that even in the absence of serum bacterial (gonococcal) oxygen metabolism is enhanced by interaction with myeloid cells.[40]

We measured the effect of interaction of gonococci with the human promyelocytic HL-60 cell line. In their undifferentiated form HL-60 cells are unable to undergo a respiratory burst.[41] Exposure to HL-60 cells immediately increased gonococcal O_2 consumption two- to threefold (Fig. 12). In the presence of KCN (or

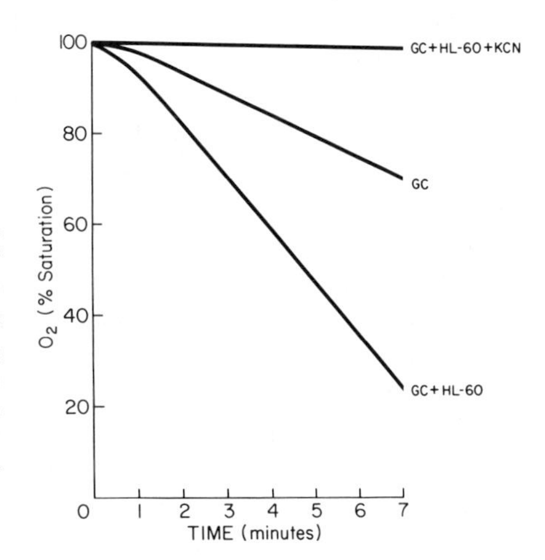

FIGURE 12. Representative ($N = 6$) Clark O_2 electrode tracing demonstrating the effect of undifferentiated HL-60 cells (5×10^6/ml) on the O_2 consumption rate of *Neisseria gonorrhoeae* suspended in HBSS. Gonococcal O_2 consumption increases two- to three-fold in the presence of the HL-60 cells. No O_2 consumption was detected with the addition of 1 mM KCN to the HL-60 gonococcal mixture or when HL-60 cells were incubated in the absence of gonococci. (From Britigan et al.[40])

with HL-60 cells alone), no oxygen was consumed, indicating that bacterial O_2 utilization was responsible for the effect. Subsequent experiments have shown that bacterial stimulation does not correlate with direct interaction with the eukaryotic cell membrane.[40] Rather, neutrophils and HL-60 cells appear to release a factor(s) into the culture supernatant that stimulates gonococcal O_2 utilization. Characterization of this cell-derived factor(s) has showed it to be similar to that found in serum: (1) stimulatory activity in media is removed by overnight dialysis in 1000-M_m membrane exclusion tubing; (2) KCN, antimycin A, and amytal block the ability of gonococci to respond to the cell-derived factor(s); and (3) the elution profile of stimulatory activity seen with G-15 Sephadex and DEAE column chromatography of this material is similar to that obtained with serum.

One possible explanation for the accumulation of a stimulating factor in supernatant following incubation of myeloid cells was that it was released as a consequence of cellular metabolism. Consistent with this hypothesis, no stimulatory activity was detectable in supernatant if cells were incubated at 4°C rather than 37°C.[40] Subsequently it was shown that NaF an inhibitor of cellular glucose metabolism had a similar effect while the protein synthesis inhibitor cyclohexamide failed to block acquisition of stimulatory activity.[40]

These data suggested the possibility that lactate, which is the principal product of resting neutrophil glucose catabolism[42] could be responsible for enhanced gonococcal metabolism. Both L(+) and D(−) lactate were shown to increase gonococcal O_2 consumption two fold with maximal activity observed at concentrations of 1–2 mg/dl.[40] Similar concentrations of lactate were detectable in supernatant following neutrophil incubation at 37°. Minimal lactate was detected if neutrophil incubation was performed at 4° or in the presence of NaF. The level of lactate present in supernatant closely paralleled its stimulatory activity. Lactate was rapidly depleted if gonococci were added to the supernatant.[40] The G-15 elution profile of lactate was identical to that of the serum and phagocyte derived stimulatory factors.[40] The G-15 active fractions derived from serum and HL-60 cells contained 10 and 4 mg/dl of lactate, respectively.[40]

These data indicate that lactate is responsible for serum and phagocyte induced stimulation of gonococcal metabolic activity. Addition of L-lactate to dialyzed serum partially restored its ability to stimulate gonococcal O_2 consumption (238% of control). However, this increase was significantly less than that observed (313%) with 10% normal serum suggesting that other factors may contribute to the serum-mediated effect.[40] Our previous failure to identify lactate as a serum stimulatory factor may have resulted from the fact that lactate was "added back" in conjunction with pyruvate which inhibits lactate oxidation by some organisms.[43]

The ability of gonococci to metabolize lactate via both L(+) and D(−) lactate dehydrogenase associated with the organism's inner membrane[44] is well recognized.[45] Mammalian lactate is predominantly L(+) implicating gonococcal L(+) LDH as most important in response to serum and phagocytes. Lactate is readily available at sites important to gonococcal infection. Once ingested within a neutrophil phagosome, gonococcal access to most physiologic sources of lactate would disappear. However, it seems likely that lactate derived from neutrophil glucose catabolism would be available.[46] The location of gonococcal LDH,[44] its persistent activity over a wide pH range, and its resistance to chemical and temperature stress[45] suggest it could remain active within the phagosome. Such ongoing bacterial metabolism could conceivably lead to a relatively anaerobic phagosomal environment, possibly limiting the ability of the neutrophil to generate microbicidal O_2 reduction products.

5. SUMMARY AND CONCLUSIONS

The work presented in this chapter is designed to shed new light on the dynamics of oxygen utilization resulting from the interaction of bacteria with phagocytic cells. The adaptability of bacterial pathogens such as *N. gonorrhoeae* to its human host has been previously emphasized.[32,33] Within this context, it seems clear that bacteria have superior genetic flexibility. It should not be surprising that strict or common human pathogens have adapted to use a variety of host factors for the purposes of growth and metabolism. However, from an investigational point of view, these phenomena deserve special emphasis. During the postphagocytic respiratory burst, it is clear that for many bacterial species microbial O_2 consumption is greater than that of the phagocyte. Furthermore, bacterial metabolism appears to be driven by one or more host factors, ultimately resulting in intense and dynamic competition for oxygen.

These results are of particular importance to investigators interested in phagocyte microbicidal mechanisms. This field has been roughly divided into O_2-dependent and independent defenses.[3,47] The severe illness of patients with defective O_2 reduction (CGD) has emphasized the importance of these metabolites for many microbes.[3,4] However, recent work has documented the potency of high- and low-molecular-weight catonic granule proteins that kill susceptible microbes independent of O_2.[47–51] Our observations have a variety of implications for oxygen-independent microbicidal mechanisms. First, some microbicidal experiments purported to show oxygen-dependent killing may have inadvertently been conducted under anaerobic conditions. Second, these observations have particular relevance for *N. gonorrhoeae*. Several lines of evidence indicate that gonococci are killed effectively by O_2 independent mechanisms,[37,49,51] implying the importance of this type of phagocyte system for an organism able to eliminate O_2 from the environment. In the case of gonococci, even some neutrophil cationic binding which are presumed to act by an O_2 independent mechanism appear to be less microbicidal under anaerobic conditions.[51] Third, bacterial metabolic activity appears to be important for the action of defensins, low-molecular-weight proteins that can kill a wide variety of pathogens.[48] Defensins appear to work only against metabolically active *E. coli* (R. Lehrer, personal communication). These observations add yet another layer of complexity to our understanding of phagocyte microbicidal systems (Fig. 13).

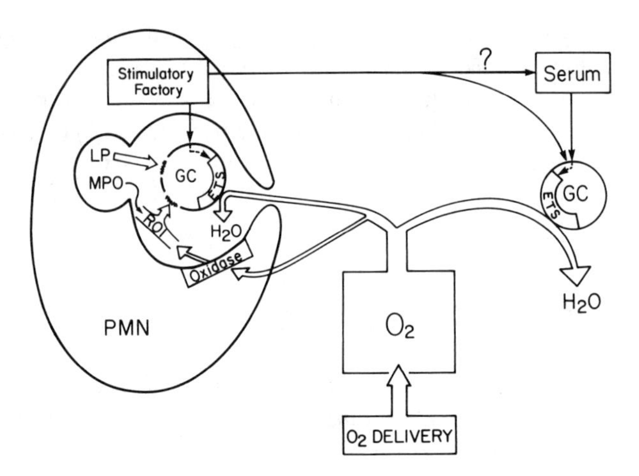

FIGURE 13. Summary of the interaction of bacteria with neutrophils. Oxygen-independent and -dependent microbicidal mechanisms must function in the face of bacterial metabolism and host factors that influence it. LP, lysosomal proteins of which cationic proteins such as defensins, bacterial permeability increasing (BPI) protein, and cathepsin G may be of particular relevance. MPO, myeloperoxidase; ETS, bacterial electron-transport system; roi, reactive oxygen intermediates; oxidase, neutrophil NADPH-dependent oxidase; gc, gonococci.

It cannot be argued that bacteria are consuming O_2 purposefully so as to make it unavailable to phagocytes, regardless of the derivation and nature of the stimulatory factor(s). Indeed, bacterial O_2 consumption is invariably linked to growth.[16] Survival advantages are almost certainly teleologically independent of the metabolic responses we have described. However, we believe that these observations justify an expanded definition of the respiratory burst that focuses on competition for oxygen between the bacteria and the phagocyte. Exploring this competition should help define the primal microbicidal process.[3,47]

ACKNOWLEDGMENTS. The authors want to thank Elizabeth Craig for her help with the preparation of this chapter, and Dr. P. F. Sparling for his comments related to this work. This study was supported by grants AI15036-07 and 2T32AI07001-09 from the National Institutes of Health. Bradley E. Britigan is the recipient of the Burroughs Wellcome Fellowship of the Infectious Disease Society of America.

REFERENCES

1. Baldridge CW, Gerard RW: The extra respiration of phagocytosis. *Am J Physiol* **103:**235–236, 1933.
2. Babior BM: Oxygen-dependent microbial killing. *N Engl J Med* **298:**659–668, 1978.
3. Root RK, Cohen MS: The microbial mechanisms of human neutrophils and eosinophils. *Rev Infect Dis* **3:**565–598, 1981.

4. Tauber AI, Borregaard N, Simons E, et al: Chronic granulomatous disease: A syndrome of phagocyte oxidase deficiencies. *Medicine (Baltimore)* **62:**286–308, 1983.
5. Densen P, Mandell G: Phagocyte strategy versus microbial tactics. *Rev Infect Dis* **2:**817–838, 1980.
6. Miller RM, Carbus J, Hornick RB: Lack of enhanced oxygen consumption by polymorphonuclear leukocytes on phagocytosis of virulent *Salmonella typhi. Science* **175:**1010–1011, 1972.
7. Kossack RE, Guerrant RL, Densen P, et al: Diminished neutrophil oxidative metabolism after phagocytosis of virulent *Salmonella typhi. Infect Immun* **31:**674–679, 1981.
8. Wilson CB, Tsai V, Remington JS: Failure to trigger the oxidative metabolic burst by normal macrophages: Possible mechanisms for survival of intracellular pathogens. *J Exp Med* **151:**328–346, 1980.
9. Kreutzer DL, Dreyfus LA, Robertson DC: Interaction of polymorphonuclear leukocytes with smooth and rough strains of *Brucella abortus. Infect Immun* **23:**737–742, 1979.
10. Mandell GL: Catalase, superoxide dismutase and virulence of *Staphylococcus aureus:* In vitro and in vivo studies with emphasis on staphylococcal-leukocyte interaction. *J Clin Invest* **55:**561–566, 1979.
11. Welch DF, Sword CP, Brehm S, et al: Relationship between superoxide dismutase and pathogenic mechanisms of *Listeria monocytogenes. Infect Immun* **23:**863–879, 1982.
12. Beaman BL, Scates SM, Moring E, et al: Purification and properties of a unique superoxide dismutase from *Nocardia asteroides. J Biol Chem* **258:**91–96, 1983.
13. Klebanoff S: Myeloperoxidase, halide-hydrogen peroxide antibacterial system. *J Bacteriol* **95:**2131–2138, 1968.
14. Jones TC, Hirsch JG: The interaction between *Toxoplasma gondii* and mammalian cells. II. The absence of lysosomal fusion with phagocytic vacuoles containing living parasites. *J Exp Med* **136:**1173–1194, 1972.
15. Mandell G: Bactericidal activity of aerobic and anaerobic polymorphonuclear neutrophils. *Infect Immun* **9:**337–341, 1974.
16. Harrison DEF: Regulation of respiration rate in growing bacteria. *Adv Microbiol Physiol* **14:**243–313, 1976.
17. Cohen MS, Cooney MH: A bacterial respiratory burst: Stimulation of *Neisseria gonorrhoeae* by human serum. *J Infect Dis* **150:**49–56, 1984.
18. Britigan BE, Cohen MS: Effects of human serum on bacterial competition with neutrophils for molecular oxygen. *Infect Immun* **52:**657–663, 1986.
19. Britigan BE, Chai Y, Cohen MS: Effects of human serum on the growth and metabolism of *Neisseria gonorrhoeae:* An alternative view of serum. *Infect Immun* **50:**738–744, 1985.
20. Root RK, Rosenthal AS, Balestra DJ: Abnormal bactericidal, metabolic and lysosomal fuctions of Chediak-Higashi syndrome leukocytes. *J Clin Invest* **51:**649–665, 1972.
21. Verhoef J, Peterson PR, Sabath DL, et al: Kinetics of staphylococcal opsonization, attachment, ingestion, and killing by human polymorphornuclear leukocytes using [^{3}H]-thymidine labeled bacteria. *J Immunol Methods* **14:**303–311, 1977.
22. DeChatelet LR, Shirley PS, Johnston RB: Effect of phorbol myristate acetate on the oxidative metabolism of polymorphonuclear leukocytes. *Blood* **47:**545–554, 1976.
23. Root RK, Metcalf J, Oshino N, et al: H_2O_2 release from human granulocytes during phagocytosis. 1. Documentation, quantitation, and some regulating factors. *J Clin Invest* **55:**945–955, 1975.
24. Morse SA: The biology of the gonococcus. *CRC Crit Rev Microbiol* **7:**93–189, 1978.
25. Rotstein OD, Pruett TL, Fiegel VD, et al: Succinic acid, a metabolic by-product of *Bacteroides* species, inhibits polymorphonuclear leukocyte function. *Infect Immun* **48:**402–408, 1985.
26. Gabrig TC, Bearman SI, Babior BM: Effects of oxygen tension and pH on the respiratory burst of human neutrophils. *Blood* **53:**1133–1139, 1979.
27. Edwards SN, Hallett MB, Campbell AK: Oxygen-radical production during inflammation may be limited by oxygen concentration. *Biochem J* **217:**851–854, 1984.
28. Hays RC, Mandel GL: pO_2, pH, and redox potential of experimental abscesses. *Proc Soc Exp Biol Med* **147:**29–30, 1974.

29. Rotstein OD, Pruett TL, Simmons RL: Mechanisms of microbial synergy in polymicrobial surgical infections. *Rev Infect Dis* **7:**151–170, 1985.
30. Edwards SW, Hallett MB, Lloyd D, et al: Decrease in apparent K_m for oxygen after stimulation of respiration of rat polymorphonuclear leukocytes. *FEBS Lett* **161:**60–64, 1983.
31. Cohen MS, Black JR, Proctor RA, et al: Host defence and the vaginal mucosa: A reevaluation. *Scand J Urol Nephrol* **86**(suppl)**:**13–22, 1985.
32. Britigan BE, Cohen MS, Sparling PF: Gonococcal infection: A model of molecular pathogenesis. *N Engl J Med* **312:**1683–1694, 1985.
33. Cannon JG, Sparling PF: The genetics of the gonococcus. *Annu Rev Microbiol* **38:**111–133, 1984.
34. Kellogg DS, Peacock WL Jr, Deacon WE, et al: *Neisseria gonorrhoeae.* I. Virulence genetically linked to clonal variation. *J Bacteriol* **85:**1274–1279, 1963.
35. Diam K: *Documenta Geigy Scientific Tables.* Ardsley, New York, Geigy Pharmaceuticals, 1962.
36. Hughes DE, Wimpenny JWT: Oxygen metabolism by micro-organisms. *Adv Microbiol Physiol* **3:**197–232, 1969.
37. Rest, RF, Fischer SH, Ingham ZZ, et al: Interaction of *Neisseria gonorrhoeae* with human neutrophils: Effects of serum and gonococcal opacity on phagocyte killing and chemiluminescence. *Infect Immun* **36:**737–744, 1982.
38. Rest RF, Lee N, Bowden C: Stimulation of human leukocytes by protein II+ gonococci is mediated by lectin-like gonococcal components. *Infect Immun* **50:**116–122, 1985.
39. DeChatelet LR, Mullikin DS, Shirley PS, et al: Phagocytosis of live versus heat-killed bacteria by human polymorphonuclear leukocytes. *Infect Immun* **10:**25–29, 1974.
40. Britigan BE, Klapper D, Svendsen T, Cohen MS: Phagocyte-derived lactate stimulates oxygen consumption by *Neisseria gonorrhoeae:* An unrecognized aspect of the oxygen metabolism of phagocytosis. *J Clin Invest* **81:**318–324, 1988.
41. Harris P, Ralph P: Human leukemic models of myelomonocytic development: A review of the HL-60 and U937 cell lines. *J Leukocyte Biol* **37:**407–422, 1985.
42. Munroe JF, Shipp JC: Glucose metabolism in leucocytes from patients with diabetes mellitus, with and without hypercholesterolemia. *Diabetes* **14:**584–590, 1965.
43. Garvie, EI: Bacterial lactate dehydrogenase. *Microbiol Rev* **44:**106–139, 1980.
44. Johnston KH, Gotschlich EC: Isolation and characterization of the outer membrane of *Neisseria gonorrhoeae. J Bacteriol* **119:**250–257, 1974.
45. Barron ESG, Hastings AB: Studies on biological oxidations II. The oxidation of lactic acid by α-hydroxyoxidase and its metabolism. *J Biol Chem* **100:**155–182, 1933.
46. Van Zwieten RR, Wever MN, Hamers RS, et al: Extracellular proton release by stimulated neutrophils. *J Clin Invest* **63:**310–313, 1981.
47. Elsbach P, Weiss J: A reevaluation of the roles of O_2-dependent and O_2-independent microbicidal systems of phagocytes. *Rev Infect Dis* **5:**843–853, 1983.
48. Ganz T, Selsted ME, Harwig SL, et al: Defensins. Natural peptide antibiotics of human neutrophils. *J Clin Invest* **76:**1427–1435, 1985.
49. Rest RF: Killing of *Neisseria gonorrhoeae* by human polymorphonuclear neutrophil granule extracts. *Infect Immun* **25:**574–579, 1979.
50. Modrzakowski MC, Spitznagel JK: Bactericidal activity of fractionated granule contents from human polymorphonuclear leukocytes: Antagonism of granule cationic proteins by lipopolysaccharide. *Infect Immun* **25:**597–602, 1979.
51. Casey SG, Shafer WM, Spitznagel JK: Anaerobiosis increases resistance of *Neisseria gonorrhoeae* to O_2-independent antimicrobial proteins from human polymorphonuclear granulocytes. *Infect Immun* **47:**401–407, 1985.

6

The Respiratory Burst and Endothelial Cells

ALDO DOBRINA and PIERLUIGI PATRIARCA

1. INTRODUCTION

Neutrophils are the first-line defense of the host against foreign invaders once they have reached the tissues. When at work, however, the neutrophils behave as a double-edged sword. In fact, the reactive oxygen intermediates (ROI) generated by the respiratory burst and the granular components released during degranulation contribute to killing of ingested microorganisms on one side, but, on the other they may damage tissues in both their cellular and extracellular components. Several cell types, including the neutrophils themselves, may be the target of ROI produced by the neutrophil respiratory burst.[1] Endothelial cells, a crucial cell type in the interface phenomena between blood and tissues, are among those targets. This chapter first reviews evidence, from both in vitro and in vivo studies, of the involvement of the neutrophil respiratory burst in endothelial cell damage, followed by a discussion of the possible role of such a damage in human pathology.

2. EFFECT OF ACTIVATED NEUTROPHILS ON ENDOTHELIAL CELLS

2.1. In Vitro Studies

The experimental model generally employed is based on the use of monolayers of cultured endothelial cells exposed to neutrophils stimulated by different agents. Endothelial cell injury is then evaluated by several methods, including release of ^{51}Cr[2–13] or ^{111}In[14,15] or other radioactive tracers,[8,10] from prelabeled cells, as well as detachment of endothelial cells from the culture dish.[8,12–14]

ALDO DOBRINA and PIERLUIGI PATRIARCA • Institute of General Pathology, University of Trieste, 34127 Trieste, Italy.

Sacks et al.[2] first reported that neutrophilic leukocytes (PMN) stimulated by either zymosan-activated serum or purified C5a induced a small but statistically significant release of ^{51}Cr from subconfluent cultures of endothelial cells derived from human umbilical vein. ^{51}Cr release was abolished by pretreatment of PMN with cytochalasin B, a fungal metabolite that prevents granulocyte spreading and adherence to endothelial cells, suggesting that a close contact between target and effector cells was necessary for injury to take place. Neutrophils from a patient with chronic granulomatous disease (CGD), which are unable to respond with a respiratory burst to stimulation, were 50% less effective in producing endothelial cell injury than normal granulocytes, suggesting that at least two mechanisms underlie injury: one dependent on ROI generated by the respiratory burst and another that is oxygen independent. The role of oxygen metabolites as injurious agents was further supported by inhibition of ^{51}Cr release by addition of catalase to the incubation medium, whereas addition of superoxide dismutase (SOD) alone afforded variable and inconstant protection. These data also suggest that H_2O_2, but not O_2^-, was the most destructive species in that particular study.

The occurrence of catalase-inhibitable injury to endothelial cells by stimulated PMN has been subsequently confirmed using other stimuli such as opsonized zymosan, endotoxin, or its lipid A moiety[3] and phorbol myristate acetate (PMA).[4–7] It was also shown that the severity of endothelial injury correlated with the amount of H_2O_2 produced by stimulated PMN.[3,7,16] H_2O_2 seems to act as the toxic agent per se rather than in combination with myeloperoxidase (MPO) and an oxidizable substrate, as it occurs in the well-known microbicidal and cytotoxic MPO–H_2O_2–halide system. This conclusion is based on several observations: (1) there was no correlation between release of MPO by PMN and extent of injury[3]; (2) MPO-deficient neutrophils were as active as normal neutrophils in mediating cell damage; (3) scavengers of hypochlorous acid, the final mediator of the cytotoxic effects produced by the MPO–H_2O_2–halide system, did not inhibit PMN-induced damage to endothelial cells, and (4) MPO inhibitors were likewise ineffective in protecting endothelial cells from PMN-mediated injury.[4]

What emerges from these studies is that lytic injury of endothelial cells caused by the metabolic burst of neutrophils is attributable mostly to H_2O_2. The capacity of H_2O_2 to lyse endothelial cells has been also confirmed using either reagent-grade H_2O_2[6,8,17–19] or enzymatically generated H_2O_2.[2,4,6,9,16–18] Some workers have also attributed a role in endothelial cell injury to the hydroxyl radical (OH·), on the basis of protection against damage by PMA-stimulated PMN afforded by hydroxyl radical scavengers such as mannitol and dimethylsulfoxide, and the iron chelator deferoxamine, which inhibits the (OH·)-generating reactions.[7]

In some studies, however, the ability of PMN to cause endothelial cell lysis could not be demonstrated using several stimuli[8,14,15] including PMA.[9,11] It should be pointed out, however, that in four of those studies, no control is available on the degree of PMN activation, since the amount of H_2O_2 or $O_2^- \cdot$ produced by neutrophils was not indicated, while in the fifth study,[11] inhibition of $O_2^- \cdot$ production in PMA-stimulated neutrophils by endothelial cells has been demonstrated. Furthermore, when the ability of stimulated neutrophils to damage endothelial cells is evaluated, the possibility must be considered that the stimulating agent may compete with endothelial cells as the ROI target. This might be the case with opsonized zymosan,[8] since it has been shown to inhibit endothelial cells lysis caused by PMA-stimulated neutrophils, and neutrophils stimulated by opsonized zymosan were unable to lyse endothelial cells.[4] Finally, the priming effect[20] of some components on neutrophils should be taken into account. In fact, as shown by Smedly et al.,[14] although no damage to endothelial cells was obtained with C5a-, chemotactic formyl peptide (FMLP)-, or endotoxin (LPS)-stimulated PMN, lysis of endothelial cells occurred when endothelial cells were exposed to PMN primed with nonstimulating concentrations of LPS and then stimulated with C5a, FMLP, or LPS. Since priming of PMN is known to increase their metabolic response to stimuli,[21] the lack of lysis of endothelial cells by unprimed stimulated PMN may be attributable to generation of subthreshold amounts of H_2O_2. It is fair to conclude that there is enough experimental evidence in favor of a role of the products of neutrophil respiratory burst, particularly H_2O_2, and possibly hydroxyl radical to cause endothelial cell damage. The severity of injury seems to be dependent on the rate of generation of products of oxygen reduction by PMN (which, in turn, is dependent on the nature of the neutrophil stimulus, its concentration, neutrophil priming, and the ratio of neutrophil number to endothelial cell number), the close proximity of the activated neutrophil to endothelial cell,[2,13] the source of endothelial cells,[14] and the ability of endothelial cells to detoxify the oxygen products.

2.2. Isolated Perfused Lung

The role played by the neutrophil respiratory burst in causing endothelial damage has also been studied in perfused isolated rabbit lung.[22] The pulmonary artery and the left ventricle were cannulated in situ; after the blood was washed away by the perfusing fluid, the lungs were removed and placed in a humidified hood. The isolated lungs were ventilated and perfused in a closed circuit with balanced salt solution (BSS) containing albumin. PMN were then added to the perfusion fluid (2×10^8 PMN/300 ml perfusate) and were activated by addition of an appropriate stimulus to the venous terminal end of the perfusion circuit. Vascular permeability changes were then studied by monitoring lung weight and albumin concentration in the bronchial lavage fluid. This model has the advantage of minimizing contributions of humoral factors and other blood cells than neutrophils to lung edema formation. Significant increases in lung weight and albumin concentration in the bronchial lavage were obtained when rabbit granulocytes or human granulocytes from normal subjects stimulated with PMA were used, as compared with the infusion of PMA alone or unstimulated granulocytes, indicating that granulocyte stimulation is crucial in edema formation. When PMA-stimulated granulocytes from a patient with chronic granulomatous disease (CGD) were used, no significant increase of lung weight and albumin concentration in the lung lavage was obtained. Since the only recognized defect in CGD granulocytes is their inability to undergo a respiratory burst, failure of CGD granulocytes to cause edema suggests that the products of the respiratory burst are an essential element in causing edema. Examination of hystological sections from lung perfused with normal or CGD granulocytes showed similar numbers of granulocytes in and adjacent to small blood vessels, indicating that CGD neutrophils had been equally stimulated by PMA to adhere as normal neutrophils. The conclusion that the products of the respiratory burst in this experimental model are the main effectors of the vascular damage is further supported by studies showing that infusion of purine and xanthine–oxidase, an enzymatic source of oxyradicals, also produced an edema in the isolated lung that could be prevented by preinjection of oxygen radical scavengers.[23]

2.3. In Vivo Studies

The experimental models used are based, in general, on the evaluation of the microvascular permeability following neutrophil intravascular activation. Neutrophil activation is generally achieved by intravenous injection of either complement-activated plasma[24–33] or purified C5a[28,29] or complement activators such as cobra venom factor (CVF)[26,28,29,34–38] and endotoxin.[33,39–42]

Since the intravenously activated neutrophils tend to sequester in the pulmonary microvasculature, most studies have been focused on the lung. Changes in lung vascular permeability has been evaluated in different ways, including lung weight measurement,[22,42] determination of protein concentration in bronchial lavage fluid,[22] assay of radioactivity in the bronchial lavage fluid after intravascular injection of isotopically labeled albumin,[42] and measurement of lymph flow and limph protein concentration from a cannulated mediastinic lymphatic vessel draining the lung lymph.[24,30–33,39–41,43,44] In some studies, either light microscopic[24,26,32,34,35,38,39,41,44,45] or electron microscopic[27,32,34–36,38,41,42] examination of lung fragments has also been carried out to obtain more direct evidence of microvascular damage. Finally, in other studies, the effects of neutrophil activation on lung functions have been explored by measuring pulmonary function parameters such as arterial oxygen tension (PaO_2)[24,26,32,34,39,41,42,44] and pulmonary artery pressure.[24,32–34,39–41,43,44] Rabbit,[22,24–27,29,34,42,45] sheep[24,30–33,39–41,43,44] (particularly in the experimental model with chronic lymphatic fistulas), and rat[28,35–38] have been the experimental animals most frequently used. The results of these studies can be summarized as follows:

1. Intravenously stimulated neutrophils sequester in the lung microvasculature, is accompanied by systemic neutropenia. This result has been obtained by several authors in sheep,[24,30–33,39–41,43,44] rabbit,[22,24–27,29,34,42,45] and rat,[35–38] and using different stimuli such as zymosan-[24,25,27,29,30–33] or cellophane-[24–26] activated plasma, CVF,[26,29,34–36,38] endotoxin,[33,39–42] the neutrophil chemoattractant FNLP,[42] and thermal injury of the skin.[37] Evidence of pulmonary sequestration of neutrophils is based either on hystological evidence of engorgement of small vessels of the lung with

neutrophils,[25,27–29,32,34–39,41,42,44,45] or on measurement of lung radioactivity after injection of labeled PMN.[42,45] A detailed study of neutrophil body distribution after activation has been carried out by Shaw and Henson.[45] These investigators have shown, in rats, that more than 50% of the injected neutrophil radioactivity sequestered in the lungs, while only 12–15%, 2–3%, and 2–3% was associated with the liver, spleen, and kidney, respectively. It was also shown that neutrophil sequestration in the lung is a transient phenomenon.[42] In fact, after the initial, rapid accumulation in the lung, soon after the injection of 10 μg LPS, lung radioactivity rapidly declined in the subsequent few hours. Similar results have been reported by Meyrick and Brigham,[32] who determined neutrophil accumulation in the lung by enumerating interstitial neutrophils on lung biopsy tissue taken sequentially from various lobes of the lungs of open-chest sheep after infusion of zymosan-activated plasma. Hence, it would be more appropriate to speak of prolonged neutrophil transit through the lung rather than sequestration.[42]

2. Intravenous activation of neutrophils leads to increased pulmonary vascular permeability. This has been demonstrated in the sheep,[24,32,33,39–41,43,44] rabbit,[22,42] and rat[35–38] after injection of a variety of stimuli such as complement-activated plasma,[24,32,33] trypsin,[43] gram-negative bacteria,[39] endotoxin,[33,40,41] phorbol myristate acetate (PMA),[22] CVF,[35,36] and combined formyl peptide (FNLP)-endotoxin injection,[42] using several techniques, as described in the first part of this paragraph. In all studies, the maximum increase of vascular permeability is reached, after neutrophil accumulation in the lung has taken place, with a variable delay, depending on the experimental conditions used.
3. Impairment of the lung function has been also observed after intravenous activation of neutrophils. Thus, for example, a rapid fall in arterial oxygen tension (Pa_{O_2}) and a rapid increase in pulmonary artery pressure have been observed in sheep after intravenous injection of gram-negative bacteria[39] or cellophane- or zymosan-activated plasma.[24] However, only little impairment of gas exchange was observed in rabbit after IV injection of both LPS and formyl peptide, despite a demonstrated infiltration of neutrophils in the lung and an increased protein concentration in the air space.[42]
4. Morphological evidence of endothelial injury has been also obtained in some studies: (1) Meyrick and Brigham have seen increased signs of pinocytosis and focal disruption of the endothelial layer in the lung of sheeps that received infusion of zymosan-activated plasma[32] or endotoxin[41]; (2) Intensive blebbing of endothelial cells was demonstrated by Hohn et al.[27] to be associated with interstitial pulmonary edema and hemorrage in the lung of rabbit treated with repeated intravenous infusions of zymosan-activated plasma; (3) In electron microscopic studies of the lungs of rats after intravascular complement activation by purified CVF[35,36] or following thermal injury of the skin,[37] Till et al., found discontinuities in the endothelium lining the lung alveolar capillaries, with neutrophils in direct contact with

the alveolar basement membrane, fibrin deposits on the extraluminal surface of the capillary wall in these areas, and intra-alveolar hemorrhage; and (4) Worthen et al.[42] could demonstrate, in rabbit, swelling, blebbing, and vesciculation of the endothelium of alveolar capillaries after simultaneous injection of FNLP and LPS. However, the same group of investigators failed to obtain evidence, either by light microscopy[29,34,45] or electron microscopy,[34,45] of endothelial damage in rabbit treated with weekly infusions (8 weeks) of in vitro preactivated neutrophils[45] or by injection of the complement activator CVF.[29,34]

5. Given the general picture of lung microvascular injury described above, it should be stressed, however, that appreciable differences in response can be obtained using different animals and different procedures for intravascular neutrophil activation. A few examples have been already cited. Additional examples will be reported here. Thus, Craddock et al.[24] stated that more severe neutropenia and increase in lymph flux from the lungs was produced in the sheep after infusion of zymosan-activated plasma than after infusion of plasma incubated with hemodialysis cellophane membranes; in a recent comparison between the effects produced in sheep by a single injection of endotoxin or zymosan-activated plasma, Meyrick[33] reported important quantitative differences obtained with the two stimuli. In fact, the fall in the peripheral arterial leukocyte counts produced by zymosan-activated plasma resolved faster (within 1 hr) as compared with that produced by endotoxin (more than 4 hr), and the increase in pulmonary vascular permeability was minimal after zymosan-activated plasma as compared with that following endotoxin infusion. Moreover, morphological studies carried out on lung biopsy tissue taken from the lungs of open chest sheep showed that maximal granulocyte sequestration in the peripheral lung tissue occurred earlier after complement-activated plasma than after endotoxin infusion but tended to resolve within 1–2 hr and was accompanied by modest and transient signs of endothelial damage. Instead, after endotoxin injection, granulocyte accumulation in the peripheral lung tissue continued to increase for at least 4 hr and was accompanied by severe endothelial damage. Studies in rabbit[42] also showed important differences in lung permeability to [^{125}I]albumin after treatment of the animals with the chemotactic peptide FNLP alone, or LPS alone, or both stimuli. Either compound alone caused only a slight increase in lung accumulation of [^{125}I]albumin, which was more marked after injection of both compounds together. Accordingly, morphological evidence of endothelial injury was obtained only in rabbits injected with both compounds.
6. The causal role of neutrophils in the microvascular changes described above has been assessed in studies with animals depleted of neutrophils by means of chemical agents, such as hydroxyurea,[40,43] colchicine,[24] or nitrogen mustard[22,24,42] or antineutrophil antibodies.[35–37] In all studies, neutrophil-depleted animals did not show appreciable changes in lung per-

meability after treatment with several stimuli, including endotoxin,[40] endotoxin plus FNLP,[42] CVF,[35,36] PMA,[22] trypsin,[43] and activated plasma,[24] that were effective instead in normal animals. These results indicate that the neutrophil is an essential mediator of the endothelial change leading to the increased permeability of pulmonary microcirculation.

7. Direct evidence that neutrophils are metabolically activated after intravenous infusion of agents that are known to activate neutrophil metabolism in vitro has not been provided in any of the above-mentioned studies, due to the obvious technical difficulties of this approach. However, morphological observations of degranulated neutrophils in contact with endothelial cells in the lung microvessels[27,32,41] indirectly suggest that neutrophils that accumulated in the lung were actually activated. Another indirect, but more stringent indication, that neutrophil activation had really taken place, stems from the observation that partial protection of the animals from microvascular damage induced by intravenous activation of neutrophils was afforded by treatment of the animals with enzymes that detoxify the products of the respiratory burst $O_2^- \cdot$ and H_2O_2, i.e., SOD[31,35,37] and catalase,[35,37,38] respectively, or with scavengers of hydroxyl radical (OH·).[36,38]

A corollary of these data is that microvascular injury is, at least in part, attributable to the release from neutrophils of products of the respiratory burst. However, since protection by catalase, SOD, and OH· scavengers was never complete,[31,35–38] other mechanisms of injury should also be considered.

3. MECHANISM(S) OF ENDOTHELIAL DAMAGE BY NEUTROPHIL ACTIVATION

3.1. Role of the Respiratory Burst

The large body of evidence points to an important role of ROI in endothelial cell damage. The precise mechanism whereby ROI released by neutrophils may injure endothelial cells is unclear. A series of biochemical effects of ROI on tissue components have been demonstrated as extensively reviewed by Fridovich,[46] Weiss and Lo Buglio,[47] Fantone and Ward,[48] and Freeman and Crapo.[49] A few examples follow.

The reaction of unsaturated or sulfur-containing amino acids with ROI will lead to polymerization and aggregation of cytoplasmic and membrane proteins containing amino acids such as tyrosine, cysteine, histidine, phenylalanine, and tryptophan. Furthermore, enzymes such as glyceraldehyde 3-phosphate dehydrogenase (G3PD), which depend on these amino acids for activity, will be inhibited by exposure to ROI.

The reaction of ROI with the metals of hemeproteins and metalloproteins may account for inactivation of catalase and CuZn superoxide dismutase, which are essential in cell defense against peroxidation. In fact, catalase is converted by $O_2^- \cdot$ to the ferroxy and ferryl states, which are inactive forms of the enzyme, and H_2O_2 inactivates CuZn superoxide dismutase.

The unsaturated bonds of membrane cholesterol and fatty acids will readily react with free radicals and undergo peroxidation. The process can become autocatalytic and yield lipid peroxides, lipid alcohols, lipid epoxides, and aldehydic byproducts, which may affect intrinsic membrane properties, such as permeability, deformability, enzymatic activity, and the aggregation state of surface determinants.

ROI, peculiarly OH·, will also react with and modify deoxyribose and DNA bases, leading to chromosomal abberations and possibly, cell death.

Finally, degradation of matrix components, such as collagen,[50] proteoglycans,[50] and hyaluronic acid,[51] by O_2^- may lead to alterations of the cell basement membrane.

3.2. Role of Neutrophil Proteases

Superoxide dismutase, catalase, and hydroxyl radical scavengers were effective in protecting endothelial cells from neutrophil-induced damage but protection was never complete, and stimulated CGD neutrophils were able to damage endothelial cells to some extent. These two observations suggest that neutrophils possess an additional mechanism of endothelial injury, besides the ROI-dependent mechanism. Neutrophil proteases are probably responsible for this second mechanism. Elastase in particular has been studied in this respect. Neutrophil elastase is a neutral protease active on a variety of substrates,[52–54] including fibronectin[55] and collagen.[56,57] In principle, then, granulocyte elastase could affect endothelium by acting both on extracellular structures, such as subendothelial basement membrane[58] and on the endothelial cell membrane, which contains fibronectin. For example, Harlan et al.[8] demonstrated that zymosan-activated neutrophils caused detachment of human endothelial cell monolayers from the substrate accompanied by digestion of endothelial cell fibronectin and that both effects could be prevented by inhibitors of elastase and cathepsin G. Moreover, Le Roy et al.[59] presented evidence for a series of reversible metabolic changes by cultured endothelial cells exposed to nonlytic doses of elastase and other proteases of neutrophils; finally, a clear demonstration of endothelial lysis by neutrophil elastase was recently achieved by Smedly et al.[14] In fact, the release of ^{111}In from cultured human microvascular endothelial cells, induced by LPS-primed neutrophils stimulated with FNLP or C5a, was inhibitable by 60–70% by the elastase inhibitor AAPVCK.

However, when the role of leukocyte proteases in endothelial cell damage in vivo is considered, it should be taken into account that the high concentration of protease inhibitors in the extracellular space may prevent tissue damage by proteases.[54,60–62] Therefore, in order for leukocytes proteases to be effective on their substrates, either their activity must exceed the antiprotease activity or other mechanisms should be envisaged. Two possibilities have been suggested: (1) leukocyte proteinases may inactivate extracellular antiproteinases,[63] or (2) close contact between neutrophils and substrate may result in a microenvironment inaccessible to extracellular antiproteinases, thus allowing unrestrained proteolysis to occur.[64,65]

3.3. Cooperation between ROI-Dependent and Protease-Dependent Mechanisms

At least two examples of such a cooperation arc known. First, it has been shown that oxidants released by activated neutrophils can inactivate α_1-proteinase inhibitor (α_1PI), the primary regulator of neutrophil elastase activity.[60] In fact, stimulated neutrophils effectively decreased the ability of α_1PI to bind and inhibit the activity of pancreatic[62,67] or neutrophil elastase.[58,68] The results of studies with catalase, superoxide dismutase, or OH· scavengers[58,66,67] and with CGD cells[67] have led to the conclusion that ROI are involved in α_1PI inactivation. Also a myeloperoxidase-dependent mechanism seems to be involved in such inactivation, since inhibition of α_1PI activity by stimulated neutrophils was suppressed by the myeloperoxidase inhibitor sodium azide[58,66,67]; the in vitro-reconstituted system myeloperoxidase–H_2O_2–halide was able to suppress the antiprotease activity of α_1PI.[67] It seems that ROI may favor the neutrophil protease activity, particularly elastase, by suppressing antiprotease activity. Second, cooperation is shown by the increased substrate susceptibility to neutrophil proteases. It has been demonstrated, in fact, that pretreatment with hydrogen peroxide of proteins such as hemoglobin and fibronectin, as well as partially purified glomerular basement membranes, increases their rate of degradation by proteolytic enzymes, including elastase, plasmin, and the neutral proteases of leukocyte extracts.[69]

4. PROTECTIVE MECHANISMS OF ENDOTHELIAL CELLS AGAINST THE NEUTROPHIL RESPIRATORY BURST

Having established that activated neutrophils are able to injure endothelial cells, the question comes to mind whether endothelial cells behave as an unarmed bystander when activated neutrophils adhere to them and release their toxic products or whether they are endowed with protective mechanisms against the neutrophil-dependent damage. Increasing evidence from in vitro studies actually indicates that endothelial cells can protect themselves to some extent from the oxidative burst of the neutrophils in at least two ways, i.e., by inhibiting neutrophil activation and by degrading $O_2^{-}\cdot$ and H_2O_2.

4.1. Inhibition of the Neutrophil Respiratory Burst by Endothelial Cells

Hoover et al.[11] first reported that superoxide production by PMA-stimulated bovine granulocytes was inhibited by contact with bovine endothelial cell monolayers; more recently, using human neutrophils suspended in autologous plasma and stimulated with either the chemoattractant FMLP or complement-activated plasma or endotoxin, Fehr et al.[15] confirmed that the metabolic activation of the PMN was inhibited by contact with optimally adherent closed monolayers of cultured endothelial cells and fibroblasts. Inhibition was exerted in Fehr's experiments not only on the metabolic activation of the PMN, as measured by hexose monophosphate shunt stimulation, but also on their adhesion to the endothelial cell on fibroblast cultures and on degranulation. Furthermore, scanning electron microscopy (SEM) studies showed that PMN in contact with endothelial cell monolayers maintained a nonspread round morphology as compared with a highly flattened appearance of activated neutrophils adhering to a plastic surface.

Adenosine may be one possible mediator of the inhibition of neutrophil activation by endothelial cells. This hypothesis is based on the following observations: (1) cultured endothelial cells exposed to various stimuli such as collagenase, thrombin, trypsin,[70] and H_2O_2[18] can release large amounts of adenosine; and (2) exogenous adenosine (0.1–100 nM) added to PMN, cause a dose-dependent inhibition of superoxide anion release induced by a series of stimuli such as FMLP, A23187, Con A, and zymosan-treated serum.[71,72] It should be expected then, that endothelial cell damage is also inhibited by adenosine. In fact, Cronstein et al.[12] showed that the enzyme adenosine deaminase increased six times the release of ^{51}Cr from endothelial cells caused by FMLP stimulated PMN, while the compound 2-chloroadenosine, which also inhibits O_2^- production by PMN and is even more potent than adenosine,[71,73] was protective. Since 2-chloroadenosine did not protect against endothelial damage caused by enzymatically generated H_2O_2, and pretreatment of endothelial cells with 2-chloroadenosine followed by washing and then exposure of the cells to stimulated neutrophils was not protective either, it was concluded that 2-chloroadenosine exerted its action on PMN rather than on endothelial cells.[12] This is in agreement with the presence of an adenosine receptor on the neutrophil surface.[73] Finally, it is worth mentioning that adenosine is present in plasma at relatively high concentrations, i.e., 0.3–6.0 μM.[74] This might be of physiological importance in the regulation of intravascular activation of neutrophils and their interactions with the microvasculature.

4.2. Endothelial Cell Mechanisms for $O_2^- \cdot$ and H_2O_2 Degradation

Endothelial cells contain enzymes involved in $O_2^- \cdot$ and H_2O_2 metabolism. Superoxide dismutase activity has been demonstrated in cultured endothelial cells from pigs[75] and humans.[76,77] Moreover, Marklund[78] showed that human extracellular superoxide dismutase binds to the membrane of endothelial cells, thus suggesting that this extracellular superoxide dismutase may also take part in protecting endothelial cells against $O_2^- \cdot$. Glutathione peroxidase activity has been shown in endothelial cells from pig,[79] calf,[16] and humans[76,77] and catalase activity has been shown in endothelial cells isolated from pig[79] and calf.[16] Instead, in cultured endothelial cells isolated from the human umbilical vein[76,77] catalase activity has been found to be low.

There is evidence that these enzymes are actually effective in protecting endothelial cells from exogenously generated oxidants. It has been shown, for example, that either pretreatment of cultures of endothelial cells obtained from human umbilical vein, bovine aorta, and pulmonary artery with specific inhibitors of the glutathione cycle or depletion of endothelial cell glutathione caused lysis of endothelial cells by concentrations of H_2O_2 generated either enzymatically or by PMA-stimulated PMN, which were ineffective in the absence of the inhibitors or when intracellular glutathione concentration was normal.[9] Further evidence for a role of the glutathione redox cycle in protecting cultured endothelial cells isolated from the calf pulmonary artery against environmental H_2O_2 was recently revealed by Tsan et al.,[80] who showed an inverse linear correlation between cell glutathione levels and H_2O_2-mediated cytolysis. Dobrina and Patriarca[16] examined the capacity of bovine microvascular endothelial cells to degrade the H_2O_2 generated either enzymatically or by PMA-stimulated bovine neutrophils. The H_2O_2 degradative capacity of a single endothelial cell was shown to correspond to the H_2O_2 released by about three activated neutrophils, and it was noted that endothelial cells exposed to H_2O_2 concentrations within their degradative capacity did not undergo lytic damage or impairment of the mitochondrial function. However, when the cells were exposed to H_2O_2 exceeding their degradative ability, cell damage soon occurred, indicating that H_2O_2 degradation is an important means of self-protection for endothelial cells against the oxidative stress. In bovine endothelial cells, the glutathione redox cycle coupled with activation of the hexose monophosphate shunt and catalase are involved in H_2O_2 degradation. In human umbilical vein endothelial cells instead, despite an H_2O_2 degradative capacity comparable to that of bovine endothelial cells,[81] H_2O_2 degradation seems to be accounted for almost exclusively by the glutathione redox cycle.[81]

5. INTERACTIONS OF NEUTROPHILS WITH ENDOTHELIUM

Evidence has been provided that experimental activation of neutrophils, either in vivo or in vitro, leads to their adhesive interaction with endothelial cells, which, in turn, results in endothelium injury. There is no doubt, however, that most of the procedures used in the experimental studies discussed above to activate neutrophils in vivo (e.g., infusion of zymosan activated plasma, C5a, cobra venom factor, large doses of endotoxin) or in vitro (e.g., opsonized zymosan, C5a, endotoxin, PMA), are far different from what actually occurs in spontaneous pathology. One wonders then whether natural examples of endothelial cell–neutrophil interactions are accompanied by neutrophil activation and/or endothelial damage. The following possibilities come to mind.

5.1. Physiological Margination

About 50% of total blood neutrophils belong to the so-called marginated pool,[82] i.e., those neutrophils that adhere reversibly to the microvascular bed. Whether neutrophils are activated during the physiological margination is unknown. All in vitro studies performed so far have shown that contact of unstimulated neutrophils with normal endothelial cells is not accompanied by stimulation of the neutrophil respiratory burst[16] or by any detectable damage of endothelial cells.[2–9,14,16] Rather, contact with endothelial cells seems to prevent stimulation of neutrophils.[11,12,15]

5.2. Neutrophil–Endothelium Interaction during Diapedesis

A limited number of studies have been carried out on endothelial injury during neutrophil diapedesis in response to a chemotactic gradient. Meyrick et al.[83] approached the problem by studying migration of granulocytes through intimal explants of bovine pulmonary artery. The explants were mounted in chemotaxis chambers containing the chemoattractant zymosan-activated plasma in the lower part and bovine granulocytes in the upper part. It was shown electron microscopically that granulocytes and endothelial cells always maintained a close apposition during migration and, after completion of migration of granulocytes beyond the basal lamina, there were no signs of structural damage to the interendothelial junctions. Intactness of the endothelial layer was also confirmed by studies with radioisotopes, which showed that neither ^{3}H-labeled water, [^{14}C]sucrose, nor [^{125}I]albumin leaked from the upper compartment to the lower compartment. Similar conclusions were reached by Huang et al.,[84] who studied the changes of the electrical resistance produced by neutrophils traversing a monolayer of human umbilical vein endothelial cells cultured on the stromal surface of a collagenous matrix derived from a sheet of human amnion.

By contrast, Shaw et al.[85] showed that migration of neutrophils into the alveolar space following endotracheal injection of the chemoattractant C5 fragment was associated with alveolar edema and hemorrage, indicating disruption of the endothelium. In this case, however, it was not clear whether endothelial disruption occurred during the process of diapedesis or followed neutrophil activation once they had reached the extravascular space with subsequent release of toxic products, which could have damaged the vessel wall from the outside.

5.3. Interaction of Neutrophils with Altered Endothelial Cells

Adhesion of neutrophils to endothelial cells may be triggered not only by neutrophil stimulation but also by changes of endothelial cells themselves. It has been shown, for example, that pretreatment of endothelial cells with several agents such as endotoxin,[86] FMLP,[87] C5a,[88] interleukin-1 (IL-1),[86,89,90] tumor necrosis factor,[91] leukotrienes LTC_4 and LTD_4,[92] and thrombin,[92–94] markedly enhanced adhesion of unstimulated neutrophils to endothelial cells. The enhanced PMN adherence caused by thrombin[92–94] and by leukotrienes[92] has been shown to rely, in part, on the production of platelet activating factor (PAF), a stimulus for neutrophil activation,[95] by stimulated endothelial cells. Another possible mechanism of such an increased adhesion has recently been proposed. It has been shown, in fact, that viral infection of endothelial cells in the presence of endotoxin or complement leads to expression of Fc and C3b receptors on their surface.[96,97] Under these conditions, then, endothelial cells may bind immune complexes, with or without subsequent complement activation, which may favor PMN binding and their activation.[97] A similar mechanism for endothelial cell injury has recently been proposed for the acute vasculitis occurring in Kawasaki syndrome. In these patients, an increased generation of oxygen radicals by neutrophils has been demonstrated,[98] and treatment of the patients with SOD-containing liposomes proved beneficial.[99] It has also been demonstrated that sera of these patients contain IgM that react with cultured human endothelial cells treated with γ-interferon (IFN_{γ}).[100] The possibility then exists that immune complexes between IFN_{γ}-induced endothelial cells antigens and the relevant IgM antibody may lead to neutrophil activation and vascular injury. In addition, endothelial cell injury may result from complement activation at the endothelial cell surface by the IFN_{γ}-induced endothelial cell antigen–IgM complex. Actually, most vasculitises, including Kawasaki syndrome, are characterized by a marked immune activation and secretion of lymphokines.[100–102]

The postischemic tissue injury[103] may be another naturally occurring condition wherein circulating neutrophils are stimulated locally by altered endothelial cells. Reperfusion following ischemia has been demonstrated to alter and accelerate the process of cell death in such tissues as the heart[104] and to lead to hemorrage and other forms of severe tissue injury. This apparent paradox has been explained by the generation of toxic oxygen radicals in reperfused tissues. The sequence of reactions leading to oxyradicals formation would be as follows: (1) during the ischemic phase, ATP is converted to hypoxanthine, and xanthine dehydrogenase is converted into xanthine oxidase, (2) upon reoxygenation, hypoxanthine oxidation by xanthine oxidase leads to $O_2^- \cdot$ generation, and (3) $O_2^- \cdot$ and its derivatives (H_2O_2 and $OH\cdot$) would be responsible for additional tissue damage, hence extension of ischemic injury.[103] This injury may well involve endothelial cells, since (1) these cells contain large amounts of xanthine oxidase,[105] and (2) direct microvascular injury by injection of hypoxanthine and xanthine oxidase in vivo has been demonstrated.[106] Thus, endothelial cell injury may account, at least in part, for neutrophil accumulation in the reperfused area, which sometimes leads to microvascular obstruction.[107] That neutrophils may contribute to extension of the injured area after reperfusion is indicated by the protective effect of depletion of circulating neutrophils by antineutrophil serum.[108,109]

5.4. Intravascular Activation of Neutrophils

Activated complement is a possible neutrophil stimulator in spontaneous pathological conditions. Besides the already mentioned experimental procedures for complement activation in vivo, in several pathological events in humans, intravascular complement activation has been documented, as reviewed by Hammerschmidt[110]: (1) hemorragic shock, in which complement activation is attributed to endotoxinemia, presumably of gut origin; (2) acute pancreatitis, which leads to release in the blood of proteases which are responsible for complement activation; (3) severe trauma, wherein complement may be indirectly activated after activation of the clotting system or directly activated by products of crushed muscle; and (4) sepsis, wherein complement may be activated through the alternative pathway by bacterial toxins or bacterial cell wall fragments, or through the classical pathway by antibody-coated microbes. Alternatively, endotoxin may directly stimulate neutrophils or prime them to respond to other stimulating agents.

5.5. Extravascular Activation of Neutrophils

Subendothelial activation of neutrophils may occur in immune complex diseases, in which either immune complexes are deposited in the vascular intima[111–114] or antibodies react with planted antigens on basement membranes.[111,115] Following complement activation by the antigen–antibody complexes,[116] chemotactic attraction of PMN occurs. This is followed, in some experimental models, by damage to basement membrane.[111] The necessary requirement for neutrophils in the immune complex injury of tissues has been firmly established by a variety of studies.[115,117–119] However, the role of neutrophil-generated ROI in such damage has only recently been appreciated. In fact, in a model of immune complex disease, i.e., the lung[120] and skin[114] reverse passive Arthus reaction, tissue injury could be prevented by administration of catalase, hydroxyl radical scavengers, and iron chelators. Involvement of endothelium in this damage is indicated by the high-molecular-weight plasma protein and red blood cell (RBC) extravasation in the tissue.[114,120]

6. RELATIONSHIPS TO HUMAN PATHOLOGY

Some human diseases in which endothelial damage by activated neutrophils might be a relevant pathogenetic mechanism are considered below:

1. *Adult respiratory distress syndrome (ARDS):* ARDS, or shock lung syndrome,[110,121] is characterized by pulmonary edema due to increased vascular permeability with normal or low pulmonary microvascular hydrostatic pressure and respiratory failure and is associated with a case fatality rate of 65%.[122] The results of studies on ARDS patients can be summarized as follows:
   a. Some clinical situations, when combined, are known to predispose to the syndrome, so that they are thought of as risk factors, although none of them alone invariably leads to the syndrome[123]; such risk factors include (1) direct lung injury, caused, for example, by aspiration of gastric contents, inhalation of noxious gases, pulmonary infection, or

contusion, and (2) pathological processes not directly involving the lung, such as gram-negative sepsis, shock following nonthoracic trauma, multiple blood transfusions, burns, and pancreatitis.

b. Gram-negative sepsis is considered the crux of ARDS,[123] since in several large series of ARDS patients, gram-negative aerobic bacteria were the predominant infectious cause of the syndrome[122,124,125] or occurred as a complication of the syndrome.[122] Since endotoxin is detectable in the blood of patients with gram-negative sepsis,[126] endotoxinemia may be considered an important factor in the development of ARDS. However, the effects of neutrophil stimulation on endothelium damage. documented in in vitro and in vivo experimental studies described earlier in this chapter, cannot be extrapolated *sic et simpliciter* to the endotoxemia occurring in human beings, due to the wide difference in endotoxin concentrations in the two conditions.[14] In fact, during endotoxemia, concentrations of circulating endotoxin have been measured at 0.5–5 ng/ml,[126] while in experimental studies microgram concentrations of LPS as neutrophil stimulator have been used.[3,40] In some experimental studies, nanogram concentrations of LPS have also been used, but in this case LPS did not act as the stimulator, but as the primer, i.e., as a potentiator of a subsequent stimulus.[14]

c. Complement activation is also commonly present in ARDS patients. In fact, several conditions leading to ARDS, such as trauma, sepsis, and burns, have all been shown to be capable of activating complement, both in experimental animals and in humans.[127,128] Furthermore, in a prospective blind study of a large number of trauma and sepsis patients, Hammerschmidt et al.[129] were able to show a striking ($p < 10^{-5}$) correlation between activation of complement and the incidence of ARDS. Nevertheless, as pointed out by Smedly et al.[14] on the basis of reports on the absence of relationship between complement activation and ARDS,[130,131] generalized intravascular complement activation is, by itself, insufficient to produce ARDS.

d. In ARDS patients, neutrophils sequester in the lung microvasculature. In fact, lung biopsies of patients with the full-blown picture of ARDS, showed accumulation of granulocytes in the lung microvessels, and this was considered the histological hallmark of the ARDS lung (reviewed by Hammerschmidt[110]). Moreover, PMN were present also in the bronchoalveolar lavage fluid collected from ARDS patients, independent of ARDS etiology.[122,132]

e. There is only indirect evidence for activation of neutrophils sequestered in the lung microvessels of ARDS patients. For example, electron microscopic studies of lung biopsies showed that granulocytes adherent to the endothelium were partially degranulated,[110] which may represent a morphological indication of neutrophil activation. Moreover, peripheral granulocytes from patients with ARDS were shown to be in an excited state, as demonstrated by an enhanced spontaneous respiratory burst[133] and by an elevation of basal lysosomal enzyme release[134] relative to neutrophils from healthy donors. In addition, elastolytic activity[135] and oxidized (and inactivated) α_1-antiprotease[132] were found in pulmonary lavage fluid from patients with ARDS. It is conceivable that the elastolytic activity was contributed by the neutrophil elastase, and the oxidation of α_1-antiprotease was due to neutrophil generated ROI. Two clinical observations point to an important role of PMN in ARDS: (1) the syndrome has not been reported in any neutropenic patients, except those with overwhelming pneumonia,[136] and (2) in a study on the respiratory function of 15 patients during hemodialysis with cellophane membranes (in this situation, complement is activated and neutrophils sequester in the lung vasculature), signs of respiratory dysfunction were detected in all patients studied but one who had severe agranulocytosis.[24]

f. Finally, endothelial cell injury in the lung microvessels has been directly demonstrated by electron microscopic examination of lung biopsies from patients with the full-blown picture of ARDS. Vacuolization and swelling of endothelial cells of lung capillaries and, occasionally, loss of continuity between endothelial cells was shown, followed by prominent signs of fibrin deposition when the patients had clinically manifest respiratory dysfunction.[110] The concentration of high-molecular-weight plasma proteins in the bronchoalveolar lavage fluid is further indirect evidence of extensive lung microvascular injury in ARDS patients.

2. *Sudden blindness in acute pancreatitis:* Activation of circulating neutrophils has been proposed to play a pathogenetic role as well in the sudden blindness that may accompany a minority of cases of acute pancreatitis. In fact, in one such patient, blindness has been reported to be associated with a peculiar funduscopic appearance, known as Purtscher's retinopathy, consisting of retinal edema (cotton-wool spots) and hemorrages confined to an area limited by the optic disc and the macula. This funduscopic pattern can be induced, in the experimental animal, when small glass emboli are infused in the carotid artheries. In the serum of two thirds of patients with acute pancreatitis, C5-derived granulocyte aggregant activity is found during the acute stage of the disease. It has been then hypothesized that complement-induced intravascular activation and aggregation of PMN and lodgement of leukoemboli in the small arterioles of the posterior fundus vessels, which are of the terminal type, could produce the ischemic changes of Purtscher's syndrome.[137]
3. *Kawasaki syndrome:* This is an acute mucocutaneous lymphnode syndrome of early childhood[138] characterized histologically by a panvasculitis with endothelial necrosis, immunoglobulin deposition, and mononuclear cell infiltration of small and medium-sized blood vessels in several organs

(heart, kidney, spleen, lung).[139,140] The syndrome is associated with sudden death in approximately 15% of patients, due to coronary artheritis accompanied by coronary aneurism and thrombotic occlusion.[141,142] Peripheral blood neutrophils of these patients show increased basal generation of ROI as compared with normal neutrophils, suggesting that they might have been activated in the circulation.[98] Furthermore, the therapeutic effects of administration of superoxide dismutase entrapped in liposomes in the early stage of the disease suggest a pathogenetic role of neutrophil-generated ROI in vascular lesions.[99] The additional finding of IgM antibodies in the patients' serum that react in vitro with cultured umbilical vein endothelial cells, pretreated with IFN_{γ},[100] suggests that antibodies to endothelial cells could play a role in the pathogenesis of this syndrome. For example, endothelial damage can result from complement-dependent lysis of the IgM-coated endothelial cells.[100] It is then possible that neutrophils are stimulated in this disease by the altered endothelial cells, thereby causing further vascular damage.

4. *Ischemia–reperfusion:* A large body of evidence suggests that organ reperfusion following ischemia is a source of ROI-mediated injury. Important clinical ischemia–reperfusion examples include reperfusion of myocardial infaraction area with balloon dilation or streptokinase or tissue plasminogen activator, restoration of cardiac function after cardiac arrest or after elective hypothermic cardioplegia during cardiopulmonary bypass, brain reperfusion during cardiopulmonary resuscitation, treatment of hemorragic shock (which may be viewed as whole-body ischemia), and reperfusion of transplanted organs.[103,143] Studies on the pathogenesis of the ischemia–reperfusion syndrome have been done so far only in animals. Involvement of free radicals in the reperfusion injury has been documented on the basis of amelioration of the injury after treatment of animals with free radical scavengers or with the xanthine oxidase inhibitor, allopurinol.[103] It can be anticipated that such a treatment may prove useful as well in the human ischemia–reperfusion syndrome.[143]

REFERENCES

1. Clark RA: Extracellular effects of the myeloperoxidase–hydrogen peroxide–halide system. *Adv Inflammation Res* **5:**107–146, 1983.
2. Sacks T, Moldow CF, Craddock PR, et al: Oxygen radicals mediate endothelial cell damage by complement-stimulated granulocytes. An in vitro model of immune vascular damage. *J Clin Invest* **61:**1161–1167, 1978.
3. Yamada O, Moldow CF, Sacks T, et al: Deleterious effects of endotoxin on cultured endothelial cells. An in vitro model of vascular injury. *Inflammation* **5:**115–126, 1981.
4. Weiss SJ, Young J, Lo Buglio AF, et al: Role of hydrogen peroxide in neutrophil-mediated destruction of cultured endothelial cells. *J Clin Invest* **68:**714–721, 1981.
5. Suttorp N, Simon L: Lung cell oxidant injury. Enhancement of polymorphonuclear leukocyte-

mediated cytotoxicity in lung cells exposed to sustained in vitro hyporoxia. *J Clin Invest* **70:**342–350, 1982.

6. Martin WJ II: Neutrophils kill pulmonary endothelial cells by a hydrogen peroxide-dependent pathway. An in vitro model of neutrophil-mediated lung injury. *Ann Rev Respir Dis* **130:**209–213, 1984.
7. Varani G, Fligiel SEG, Till OG, et al: Pulmonary endothelial cell killing by human neutrophils. Possible involvement of hydroxyl radical. *Lab Invest* **53:**656–663, 1985.
8. Harlan JM, Killen PD, Harker LA, et al: Neutrophil-mediated endothelial injury in vitro. Mechanisms of cell detachment. *J Clin Invest* **68:**1394–1403, 1981.
9. Harlan JM, Levine JD, Callahan KS, et al: Glutathione redox cycle protects cultured endothelial cells against lysis by extracellularly generated hydrogen peroxide. *J Clin Invest* **73:**706–713, 1984.
10. Andreoli SP, Baehner RL, Bergstein JM: In vitro detection of endothelial cell damage using 2-deoxy-D-^{3}H-glucose: Comparison with chromium-51, ^{3}H-leucine, ^{3}H-adenine, and lactate dehydrogenase. *J Lab Clin Med* **106:**253–261, 1985.
11. Hoover RL, Robinson JM, Karnovsky MJ: Superoxide production by polymorphonuclear leukocytes is inhibited by contact with endothelial cells. *J Cell Biol* **95:**37A, 1982.
12. Cronstein BN, Levin RI, Belanoff J, et al: Adenosine: An endogenous inhibitor of neutrophil-mediated injury to endothelial cells. *J Clin Invest* **78:**760–770, 1986.
13. Diener AM, Beatty PG, Ochs HD, et al: The role of neutrophil membrane glycoprotein 150 (GP-150) in neutrophil-mediated endothelial cell injury in vitro. *J Immunol* **135:**537–543, 1985.
14. Smedly LA, Tonnesen MG, Sandhaus RA, et al: Neutrophil-mediated injury to endothelial cells. Enhancement by endotoxin and essential role of neutrophil elastase. *J Clin Invest* **77:**1233–1243, 1986.
15. Fehr J, Moser R, Leppert D, et al: Antiadhesive properties of biological surfaces are protective against stimulated granulocytes. *J Clin Invest* **76:**535–542, 1985.
16. Dobrina A, Patriarca P: Neutrophil–endothelial cell interaction. Evidence for and mechanism of the self protection of bovine microvascular endothelial cells from hydrogen peroxide-induced oxidative stress. *J Clin Invest* **78:**462–471, 1986.
17. Whorton AR, Montgomery ME, Kent RS: Effect of hydrogen peroxide on prostaglandin production and cellular integrity in cultured porcine aortic endothelial cells. *J Cell Invest* **76:**295–302, 1985.
18. Ager A, Gordon JL: Differential effects of hydrogen peroxide on indices of endothelial cell function. *J Exp Med* **159:**592–603, 1984.
19. Spragg RG, Hinshaw DB, Hyslop PA, et al: Alterations in adenosine triphosphate and energy charge in cultured endothelial and P388D$_1$ cells after oxidant injury. *J Clin Invest* **76:**1471–1476, 1985.
20. Guthrie LA, Mc Phail LC, Henson PM, et al: The priming of neutrophils for enhanced release of oxygen metabolites by bacterial lypopolysaccharide: evidence for increased activity of the superoxide producing enzyme. *J Exp Med* **160:**1656–1671, 1984.
21. Rossi F: The O_2^- forming NADPH oxidase of the phagocytes. Nature, mechanisms of activation and function. *Biochim Biophys Acta* **853:**65–89, 1986.
22. Shasby DM, Vanbenthuysen KM, Tate RM, et al: Granulocytes mediate acute edematous lung injury in rabbits and in isolated rabbit lungs perfused with phorbol myristate acetate: Role of oxygen radicals. *Am Rev Respir Dis* **125:**443–447, 1982.
23. Tate RM, Van Benthuysen KM, Shasby DM, et al: Dimethylthiourea, a hydroxyl radical scavenger, blocks oxygen radical-induced acute edematous lung injury in an isolated perfused lung. *Am Rev Respir Dis* **123:**243, 1981 (abst).
24. Craddock PR, Fehr J, Brigham KL, et al: Complement and leukocyte-mediated pulmonary dysfunction in hemodialysis. *N Engl J Med* **296:**769–774, 1977.
25. Craddock PR, Fehr J, Dalmasso AP, et al: Hemodialysis leukopenia. Pulmonary vascular leukostasis resulting from complement activation by dialyzer cellophane membranes. *J Clin Invest* **59:**879–888, 1977.

26. O'Faherty JT, Craddock PR, Jacob HS: Effect of intravascular complement activation on granulocyte adhesiveness and distribution. *Blood* **51:**731–739, 1978.
27. Hohn DC, Meyers AJ, Gherini ST, et al: Production of acute pulmonary injury by leukocytes and activated complement. *Surgery* **88:**48–58, 1980.
28. Hammerschmidt DE, Harris PD, Wayland JH, et al: Complement-induced granulocyte aggregation in vivo. *Am J Pathol* **102:**146–150, 1981.
29. Henson PM, Larsen GL, Webster RO, et al: Pulmonary microvascular alterations and injury induced by complement fragments: Synergistic effect of complement activation, neutrophil sequestration and prostaglandins. *Ann NY Acad Sci* **384:**287–300, 1982.
30. Gee MH, Perkowski SZ, Havill AM, et al: Role of prostaglandins and leukotrienes in complement-initiated lung vascular injury. *Chest* **83:**82S–85S, 1983.
31. Perkowski SZ, Havill AM, Flynn J, et al: Role of intrapulmonary release of eicosanoids and superoxide anion as mediators of pulmonary dysfunction and endothelial injury in sheep with intermittent complement activation. *Circ Res* **53:**574–583, 1983.
32. Meyrick BO, Brigham KL: The effect of a single infusion of zymosan-activated plasma on the pulmonary microcirculation of sheep. Structure–function relationships. *Am J Pathol* **114:**32–45, 1984.
33. Meyrick BO: Endotoxin-mediated pulmonary endothelial cell injury. *Fed Proc* **45:**19–24, 1986.
34. Webster RO, Larsen GL, Henson PM: Lack of inflammatory effects on the rabbit lung of intravascular complement activation. *Fed Proc* **40:**767, 1981 (abst).
35. Till GO, Johnson KJ, Kunkel R, et al: Intravascular activation of complement and acute lung injury. Dependency on neutrophils and toxic oxygen metabolites. *J Clin Invest* **69:**1126–1135, 1982.
36. Ward PA, Till GO, Kunkel R, et al: Evidence for role of hydroxyl radical in complement and neutrophil-dependent tissue injury. *J Clin Invest* **72:**789–801, 1983.
37. Till GO, Beauchamp C, Menapace D, et al: Oxygen radical dependent lung damage following thermal injury to rat skin. *J Trauma* **28:**269–277, 1983.
38. Till GO, Ward PA: Systemic complement activation and acute lung injury. *Fed Proc* **45:**13–18, 1986.
39. Brigham KL, Woolverton WC, Blake LH, et al: Increased sheep lung vascular permeability caused by Pseudomonas bacteremia. *J Clin Invest* **54:**792–804, 1974.
40. Heflin AC, Brigham KL: Prevention by granulocyte depletion of increased vascular permeability of sheep lung following endotoxemia. *J Clin Invest* **68:**1253–1260, 1981.
41. Meyrick B, Brigham KL: Acute effects of Escherichia coli endotoxin on the pulmonary microcirculation of anesthetized sheep. Structure–function relationships. *Lab Invest* **48:**458–470, 1983.
42. Worthen GS, Haslett C, Smedly LA, et al: Lung vascular injury induced by chemotactic factors: enhancement by bacterial endotoxins. *Fed Proc* **45:**7–12, 1986.
43. Garcia-Szabo RR, Johnson A, Malik AB: Leukocytes are required for the trypsin-induced increase in lung vascular permeability. *Am J Pathol* **124:**377–383, 1986.
44. Judges D, Sherkey P, Cheung H, et al: Pulmonary microvascular fluid flux in a large animal model of sepsis: Evidence for increased pulmonary endothelial permeability accompanying surgically induced peritonitis in sheep. *Surgery* **99:**222–234, 1986.
45. Shaw JO, Henson PM: Pulmonary intravascular sequestration of activated neutrophils. Failure to induce light-microscopic evidence of lung injury in rabbits. *Am J Pathol* **108:**17–23, 1982.
46. Fridovich I: The biology of oxygen radicals. *Science* **201:**875–880, 1978.
47. Weiss SJ, Lo Buglio AF: Biology of disease. Phagocyte-generated oxygen metabolites and cellular injury. *Lab Invest* **47:**5–18, 1982.
48. Fantone JC, Ward PA: Role of oxygen-derived free radicals and metabolites in leukocyte-dependent inflammatory reactions. *J Pathol* **107:**397–418, 1982.
49. Freeman BA, Crapo JD: Biology of disease. Free radicals and tissue injury. *Lab Invest* **47:**412–426, 1982.

50. Greenwald RA, Moy WW, Lazarus D: Degradation of cartilage proteoglycans and collagen by superoxide radical. *Arthritis Rheum* **19:**799, 1976 (abst).
51. Greenwald RA, Moy WW: Effect of oxygen-derived free radicals on hyaluronic acid. *Arthritis Rheum* **23:**455–463, 1980.
52. Klebanoff SJ, Clark RA (eds): *The Neutrophil: Function and Clinical Disorders.* Amsterdam, Elsevier/North-Holland, 1978.
53. Senior RM, Campbell EJ: Neutral proteinases from human inflammatory cells. A critical review of their role in extracellular matrix degradation, in Ward PA (ed): *Clinics in Laboratory Medicine.* Philadelphia, WB Saunders, 1983, vol 3: *Symposium on Tissue Immunopathology,* p 645.
54. Fritz H, Jochum M, Duswald KH, et al: Granulocyte proteinases as mediators of unspecific proteolysis in inflammation: A review, in Goldberg DM, Werner M (eds): *Selected Topics in Clinical Enzymology,* Berlin, de Gruyter, 1984 vol 2, p 305.
55. Mc Donald JA, Kelley DG: Degradation of fibronectin by human leukocyte elastase. Release of biologically-active fragments. *J Biol Chem* **255:**8848–8858, 1980.
56. Gadek JE, Fells JA, Wright DG, et al: Human neutrophil elastase functions as a type III collagen "collagenase." *Biochem Biophys Res Commun* **95:**1815–1822, 1980.
57. Mainardi CL, Dixit SN, Kang AH: Degradation of type IV (basement membrane) collagen by a proteinase isolated from human polymorphonuclear leukocyte granules. *J Biol Chem* **255:**5435–5441, 1980.
58. Weiss SJ, Regiani S: Neutrophils degrade subendothelial matrices in the presence of alpha-1-proteinase inhibitor. Cooperative use of lysosomal proteinases and oxygen metabolites. *J Clin Invest* **73:**1297–1303, 1984.
59. Le Roy EC, Ager A, Gordon JL: Effects of neutrophil elastase and other proteases on porcine aortic endothelial prostaglandin I_2 production, adenine nucleotide release, and responses to vasoactive agents. *J Clin Invest* **74:**1003–1010, 1984.
60. Ohlsson K: Interaction of granulocyte neutral proteases with alpha$_1$-antitrypsin, alpha$_2$-macroglobulin and alpha$_1$-antichymotrypsin, in: Havemann K, Janoff A (eds): *Neutral Proteases of Human Polymorphonuclear leukocytes.* Baltimore, Urban & Schwarzenberg, 1978, p 167.
61. Travis J, Giles PJ, Porcelli L, et al: Human leukocyte elastase and cathepsin G: Structural and functional characteristics. *Ciba Found Symp* **75:**51–68, 1980.
62. Weiss SJ: Oxygen as a weapon in the phagocyte armamentarium, in Ward PA (ed): *Handbook of Inflammation: Immunology of Inflammation,* Amsterdam, Elsevier, 1983, vol 4, p 37.
63. Banda MJ, Clark EJ, Werb Z: Regulation of alpha$_1$-proteinase inhibitor function by rabbit alveolar macrophages. *J Clin Invest* **75:**1758–1762, 1985.
64. Johnson KJ, Varani J: Substrate hydrolysis by immune complex-activated neutrophils: Effect of physical presentation of complexes and protease inhibitors. *J Immunol* **127:**1875–1879, 1981.
65. Campbell EJ, Senior RM, McDonald JA, et al: Proteolysis by neutrophils. Relative importance of cell–substrate contact and oxidative inactivation of proteinase inhibitors in vitro. *J Clin Invest* **70:**845–852, 1982.
66. Carp H, Janoff A: In vitro suppression of serum elastase-inhibitory capacity by reactive oxygen species generated by phagocytosing polymorphonuclear leukocytes. *J Clin Invest* **63:**793–797, 1979.
67. Carp H, Janoff A: Phagocyte-derived oxidants suppress the elastase-inhibitory capacity of alpha$_1$-proteinase inhibitor in vitro. *J Clin Invest* **66:**987–995, 1980.
68. Zaslow MC, Clark RA, Stone PJ, et al: Human neutrophil elastase does not bind to alpha$_1$-protease inhibitor that has been exposed to activated human neutrophils. *Am Rev Respir Dis* **128:**434–439, 1983.
69. Fligiel SEG, Lee EC, McCoy, JP, et al: Protein degradation following treatment with hydrogen peroxide. *Am J Pathol* **115:**418–425, 1984.
70. Pearson JD, Gordon JL: Vascular endothelial and smooth muscle cells in culture selectively release adenine nucleotides. *Nature (Lond)* **281:**384–386, 1979.

71. Cronstein BN, Kramer SB, Weismann G, et al: Adenosine: A physiological modulator of superoxide anion generation by human neutrophils. *J Exp Med* **158:**1160–1177, 1983.
72. Roberts PA, Newby AC, Hallet MB, et al: Inhibition by adenosine of reactive oxigen metabolite production by human polymorphonuclear leukocytes. *Biochem J* **227:**669–674, 1985.
73. Cronstein BN, Rosenstein ED, Kramer SB, et al: Adenosine: A physiologic modulator of superoxide anion generation by human neutrophils. Adenosine acts via an A_2 receptor on human neutrophils. *J Immunol* **135:**1366–1371, 1985.
74. Hirschhorn RV, Roegner-Maniscalco V, Kuritsky L, et al: Bone marrow transplanation only partially restores purine metabolites to normal in adenosine deaminase deficient patients. *J Clin Invest* **68:**1387–1393, 1981.
75. Ody C, Bach-Dieterle Y, Wand I, et al: Effect of hyperoxia on superoxide dismutase content of pig pulmonary artery and aortic endothelial cells in culture. *Exp Lung Res* **1:**271–279, 1980.
76. Marklund SL, Westman NG, Lundgren E, et al: Copper- and zinc-containing superoxide dismutase, catalase, and glutathione peroxidase in normal and neoplastic human cell lines and normal human tissues. *Cancer Res* **42:**1955–1961, 1982.
77. Shingu M, Yoshioka K, Nobunaga M, et al: Human vascular smooth muscle cells and endothelial cells lack catalase activity and are susceptible to hydrogen peroxide. *Inflammation* **9:**309–320, 1985.
78. Marklund SL: Physiological aspects of extracellular superoxide dismutase, in Rotilio G (ed): *Superoxide and Superoxide Dismutase in Chemistry, Biology and Medicine.* Amsterdam, Elsevier, 1986, p 438.
79. Housset B, Junod AF: Effects of culture conditions and hyperoxia on antioxidant enzymes in pig pulmonary artery and aortic endothelium. *Biochim Biophys Acta* **716:**283–289, 1982.
80. Tsan MF, Danis EH, Del Vecchio PJ, et al: Enhancement of intracellular glutathione protects endothelial cells against peroxidative damage. *Biochem Biophys Res Commun* **127:**270–276, 1985.
81. Dobrina A, Bulli G, Cecovini G, et al: Patterns of H_2O_2 degradation in bovine and human vascular endothelial cell. in Mauri C, Rizzo SC, Ricevuti G (eds): The Biology of Phagocytes in Health and Disease. *Advances in Bioscience.* Oxford, Pergamon, 1987, p 155.
82. Fehr J, Jacob HS: In vitro granulocyte adherence and in vivo margination: Two associated complement-dependent functions. Studies based on the acute neutropenia of filtration leukophoresis. *J Exp Med* **146:**641–652, 1977.
83. Meyrick B, Hoffman LH, Brigham KL: Chemotaxis of granulocytes across bovine pulmonary artery intimal explants without endothelial cell injury. *Tissue Cell* **16:**1–16, 1984.
84. Huang AG, Furie MB, Silverstein SC: Human neutrophil migration does not alter electrical resistance across cultured human endothelial cell monolayers, in *Fourth International Symposium on the Biology of the Vascular Endothelial Cell, August 19–23, 1986, Noordwijkerhout, The Netherlands,* p 20 (abst).
85. Shaw JO, Henson PM, Henson J, et al: Lung inflammation induced by complement-derived chemotactic fragments in the alveolus. *Lab Invest* **42:**547–558, 1980.
86. Schleimer RP, Rutledge BK: Cultured human vascular endothelial cells acquire adhesiveness for neutrophils after stimulation with interleukin 1, endotoxin and tumor-promoting phorbol diesters. *J Immunol* **136:**649–654, 1986.
87. Kirkpatrick CJ, Melzner I: Alterations in the biophysical properties of the human endothelial cell plasma membrane induced by a chemotactic tripeptide: Correlation with enhanced adherence of granulocytes. *J Pathol* **144:**201–211, 1984.
88. Zimmermann GA, Hill HR: Inflammatory mediators stimulate granulocyte adherence to cultured human endothelial cells. *Thromb Res* **35:**203–217, 1984.
89. Bevilacqua MP, Pober JS, Wheeler ME, et al: Interleukin 1 acts on cultured human vascular endothelium to increase the adhesion of polymorphonuclear leukocytes, monocytes and related leukocyte cell lines. *J Clin Invest* **76:**2003–2011, 1985.
90. Bevilacqua MP, Pober JS, Wheeler ME, et al: Interleukin 1 activation of vascular endothelium. Effects of proagulant activity and leukocyte adhesion. *Am J Pathol* **121:**393–403, 1985.

91. Gamble JR, Harlan JM, Klebanoff SJ, et al: Stimulation of the adherence of neutrophils to umbilical vein endothelium by human recombinant tumor necrosis factor. *Proc Natl Acad Sci USA* **82:**8667–8671, 1985.
92. McIntyre TM, Zimmermann GA, Prescott SM: Leukotrienes C_4 and D_4 stimulate human endothelial cells to synthesize platelet-activating factor and bind neutrophils. *Proc Natl Acad Sci USA* **83:**2204–2208, 1986.
93. Prescott SM, Zimmermann GA, McIntyre TM: Human endothelial cells in culture produce platelet-activating factor (1-alkyl-2-acetyl-*sn*-glycero-3-phosphocholine) when stimulated with thrombin. *Proc Natl Acad Sci USA* **81:**3534–3538, 1984.
94. Zimmermann GA, McIntyre TM, Prescott SM: Thrombin stimulates the adherence of neutrophils to human endothelial cells in vitro. *J Clin Invest* **76:**2235–2246, 1985.
95. Shaw JO, Pinckard RN, Ferrigni KS, et al: Activation of human neutrophils with 1-*O*-hexadecyl/octadecyl-2-acetyl-*sn*-glyceryl-3-phosphorylcholine (PAF). *J Immunol* **127:**1250–1255, 1981.
96. Ryan US, Schultz DR, Ryan JW: Fc and C3b receptors on pulmonary endothelial cells; induction by injury. *Science* **214:**557–559, 1981.
97. Ryan US: The endothelial surface and response to injury. *Fed Proc* **45:**101–108, 1986.
98. Niva Y, Shomija K: Enhanced neutrophilic functions in mucocutaneous lymph node syndrome, with special reference to the possible role of increased oxygen intermediate generation in the pathogenesis of coronary thromboarthritis. *J Pediatr* **104:**56–60, 1984.
99. Somiya K, Niva Y, Shimoda K, et al: Treatment with liposomal superoxide dismutase of patients with Kawasaki disease, in Rotilio G (ed): *Superoxide and Superoxide Dismutase in Chemistry, Biology and Medicine*. Amsterdam, Elsevier, 1986, p 513.
100. Leung DYM, Collins T, Lapierre LA, et al: Immunoglobulin M antibodies present in the acute phase of Kawasaki syndrome lyse cultured vascular endothelial cells stimulated by gamma interferon. *J Clin Invest* **77:**1428–1435, 1986.
101. Cines DB, Lyss AP, Reeher M, et al: Presence of complement fixing anti-endothelial cell antibodies in systemic lupus erythematosus. *J Clin Invest* **73:**611–625, 1984.
102. Paul LC, van Es LA, Balwinn WM III: Antigens in human renal allografts. *Clin Immunol Immunopathol* **19:**206–223, 1981.
103. McCord JM: Oxygen-derived free radicals postischemic tissue injury. *N Engl J Med* **312:**159–163, 1985.
104. Hearse DJ: Reperfusion of the ischemic myocardium. *Clin Res Rev* **4:**58–61, 1984.
105. Jarasch ED, Grund C, Bruder G, et al: Localization of xanthine oxidase in mammary gland epithelium and capillary endothelium. *Cell* **25:**67–82, 1981.
106. Del Maestro RF, Thaw IIII, Björk J, et al: Free radicals as mediators of tissue injury. *Acta Physiol Scand* **492**(suppl):43–57, 1980.
107. Engler RL, Schmid-Schönbein GW, Pavelec RS: Leukocyte capillary plugging in myocardial ischemia and reperfusion in the dog. *Am J Pathol* **111:**98–111, 1983.
108. Romson J, Hook B, Kunkel S, et al: Reduction in the extent of myocardial ischemic injury by neutrophil depletion in the dog. *Circulation* **67:**1016–1023, 1983.
109. Groegaard B, Schuerer L, Gerdin B, et al: Involvement of neutrophils in the cortical blood flow impairment after cerebral ischemia in rat; effects of antineutrophil serum and superoxide dismutase, in Rotilio G (ed): *Superoxide and Superoxide Dismutase in Chemistry, Biology and Medicine*. Amsterdam, Elsevier, 1986, p 608.
110. Hammerschmidt DE: Activation of the complement system and of granulocytes in lung injury: The adult respiratory distress syndrome. *Adv Inflammation Res* **5:**147–172, 1983.
111. Cochrane CG, Aikin BS: Polymorphonuclear leukocytes in immunologic reactions. The destruction of vascular basement membrane in vivo and in vitro. *J Exp Med* **124:**733–752, 1966.
112. DeShazo CV, Henson PM, Cochrane CG: Acute immunologic arthritis in rabbits. *J Clin Invest* **51:**50–57, 1972.

113. Johnson KJ, Ward PA: Acute immunologic pulmonary alveolitis. *J Clin Invest* **54:**349–357, 1974.
114. Fligiel SEG, Ward PA, Johnson KJ, et al: Evidence for a role of hydroxyl radical in immune-complex-induced vasculitis. *Am J Pathol* **115:**375–382, 1984.
115. Hawkins D, Cochrane CG: Glomerular basement membrane damage in immunological glomerulonephritis. *Immunology* **14:**665–681, 1968.
116. Ward PA, Hill JH: Biologic role of complement products. Complement-derived leukotactic activity extractable from lesions of immunologic vasculitis. *J Immunol* **108:**1137–1145, 1972.
117. Humphrey JH: The mechanisms of Arthus reactions. I. The role of polymorphonuclear leukocytes and other factors in reversed passive Arthus reactions in rabbits. *Br J Exp Pathol* **36:**268–282, 1955.
118. Cochrane CG, Weigle WO, Dixon FJ: The role of polymorphonuclear leukocytes in the initiation and cessation of the Arthus vasculitis. *J Exp Med* **110:**481–494, 1959.
119. Ward PA, Cochrane CG: Bound complement and immunologic injury of blood vessels. *J Exp Med* **121:**215–234, 1965.
120. Johnson KJ, Ward PA: Role of oxygen metabolites in immune complex injury of lung. *J Immunol* **126:**2365–2369, 1981.
121. Rinaldo EJ, Rogers RM: Adult respiratory distress syndrome: Changing concepts of lung injury and repair. *N Engl J Med* **306:**900–909, 1982.
122. Fowler AA, Hamman RF, Good JT, et al: The respiratory distress syndrome: Risk with common predisposition. *Ann Intern Med* **98:**593–597, 1983.
123. Hyers TM, Fowler AA: Adult respiratory distress syndrome: Causes, morbidity and mortality. *Fed Proc* **45:**25–29, 1986.
124. Pepe PE, Potkin RT, Reus DH, et al: Clinical indicators of the adult respiratory distress syndrome. *Am J Surg* **144:**124–130, 1982.
125. Weigelt JA, Snyder WH III, Mitchell RA: Early identification of patients prone to develop adult respiratory distress syndrome. *Am J Surg* **142:**687–691, 1981.
126. Levin J, Poore TE, Zauber NP, et al: Detection of endotoxin in the blood of patients with sepsis due to gram negative bacteria. *N Engl J Med* **283:**1313–1316, 1970.
127. Goldstein IM, Cala D, Radin A, et al: Evidence of complement catabolism in acute pancreatitis. *Am J Med Sci* **275:**257–264, 1978.
128. Morrison DC, Kline LF: Activation of the classical and properdin pathways of complement by bacterial lipopolysaccharides (LPS). *J Immunol* **118:**362–368, 1977.
129. Hammerschmidt DE, Weaver LJ, Hudson LD, et al: Association of complement activation and elevated plasma-C5a with adult respiratory distress syndrome: Pathophysiological relevance and possible prognostic value. *Lancet* **1:**947–949, 1980.
130. Weinberg PF, Matthay MA, Webster RO, et al: Lack of relationship between complement activation and acute lung injury. *Am Rev Respir Dis* **127:**95, 1983 (abst).
131. Spitzer RE, Vallota EH, Forrestal J, et al: Serum C3 lytic system in patients with glomerulonephritis. *Science* **164:**436–437, 1969.
132. McGuire WW, Spragg RG, Cohen AB, et al: Studies on the pathogenesis of the adult respiratory distress syndrome. *J Clin Invest* **69:**543–553, 1982.
133. Zimmerman GA, Renzetti AD, Hill HR: Functional and metabolic activity of granulocytes from patients with adult respiratory distress syndrome: Evidence for activated neutrophils in the pulmonary circulation. *Am Rev Respir Dis* **127:**290–300, 1983.
134. Fowler AA, Fisher BJ, Centor RM, et al: Development of the adult respiratory distress syndrome: progressive alteration of neutrophil chemotactic and secretory processes. *Am J Pathol* **116:**427–435, 1984.
135. Lee CT, Fein AM, Lippman M, et al: Elastolytic activity in pulmonary lavage fluid from patients with adult respiratory distress syndrome. *N Engl J Med* **304:**192–196, 1981.
136. Worthen GS, Henson PM: Mechanisms of acute lung injury, in Ward PA (ed): *Clinics in Laborato-*

ry Medicine, Philadelphia, WB Saunders, 1983, vol 3: *Symposium on Tissue Immunopathology,* p 601.

137. Jacob HS, Goldstein IM, Shapiro I, et al: Sudden blindness in acute pancreatitis: Possible manifestation of complement-induced retinal leukostasis. *Arch Intern Med* **141:**134–136, 1981.
138. Kawasaki T: Acute febrile mucocutaneous syndrome with lymphoid involvement with specific desquamation of the fingers and toes in children: clinical observations of 50 cases. *Jpn J Allerg* **16:**178–222, 1967.
139. Hirose S, Hamashima Y: Morphological observations on the vasculitis in the mucocutaneous lymph node syndrome. *Eur J Pediatr* **129:**17–27, 1978.
140. Landing BH, Larson JE: Are infantile periarteritis nodosa and fatal mucocutaneous lymph node syndrome the same? *Pediatrics* **59:**651–662, 1977.
141. Kawasaki T: Acute febrile mucocutaneous lymph node syndrome and sudden death. *Acta Pediatr Jpn* **75:**433–434, 1971.
142. Kato H, Koike S, Yamamoto M, et al: Coronary aneurisms in infants and young children with acute febrile mucocutaneous lymph node syndrome. *J Pediatr* **86:**892–898, 1975.
143. Bulkley GB, Morris JB: Role of oxygen derived free radicals as mediators of post-ischemic injury: A clinically oriented overview, in Rotilio G (ed): *Superoxide and Superoxide Dismutase in Chemistry, Biology and Medicine.* Amsterdam, Elsevier, 1986, p 565.

7

The Respiratory Burst of Eosinophils

GIULIANO ZABUCCHI, RENZO MENEGAZZI,
and PIERLUIGI PATRIARCA

1. INTRODUCTION

The first report on the ability of the eosinophils to perform the respiratory burst appeared in 1968.[1] At that time, the respiratory burst of neutrophils was already known in several details, and reviews on the subject were available.[2–5] This tells clearly enough what the rank of the eosinophil is, relative to the neutrophil, in the theme of respiratory burst—a situation similar to that of the second born son in a family whose performances are inevitably evaluated looking at the oldest son. In other words, the question underlying the research on the eosinophil respiratory burst has often, if not always, been whether the eosinophil could do what the neutrophil was already known to be able to do. It is therefore inevitable, in such a presentation, to refer frequently to neutrophils and to slip often into comparisons between eosinophils and neutrophils.

2. STUDIES WITH EOSINOPHILS FROM EOSINOPHILIC SUBJECTS

Because of the paucity of eosinophils in peripheral blood, most studies until 1980[6] have been carried out with eosinophils obtained from eosinophilic subjects, although the legitimacy of an extrapolation of the results obtained in this way to normal eosinophils was a matter of concern.[7] The results of these studies can be summarized as follows.

Stimulated eosinophils have an increased oxygen uptake,[8] superoxide anion (O_2^-)[9–11] and H_2O_2[7,8] generation, hexose monophosphate shunt (HMS) activity,[1,6,7,10] and iodination.[10,12] Therefore, the metabolic burst of eosinophils is

GIULIANO ZABUCCHI, RENZO MENEGAZZI, and PIERLUIGI PATRIARCA • Institute of General Pathology, University of Trieste, 34127 Trieste, Italy.

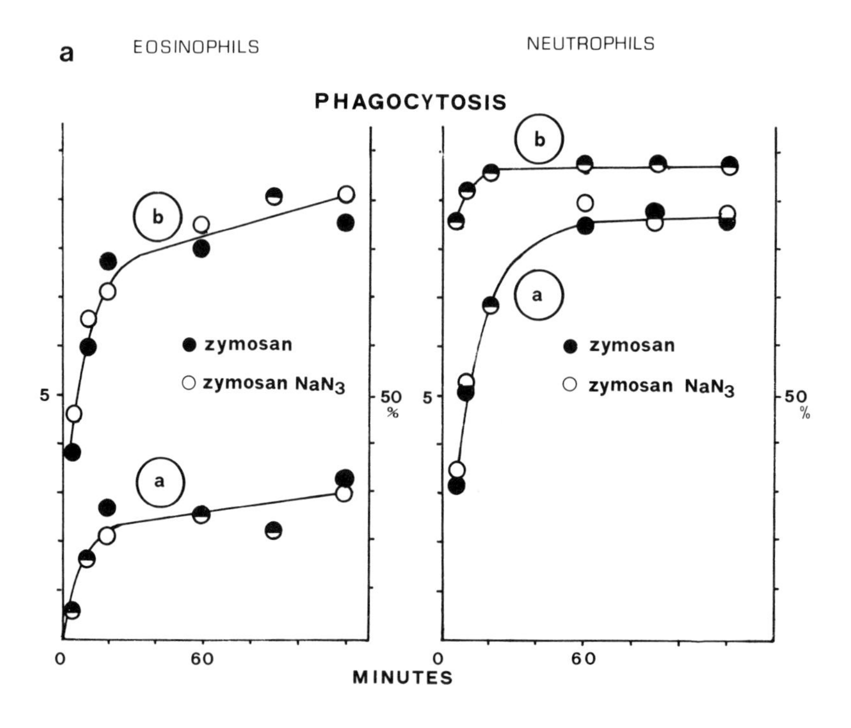

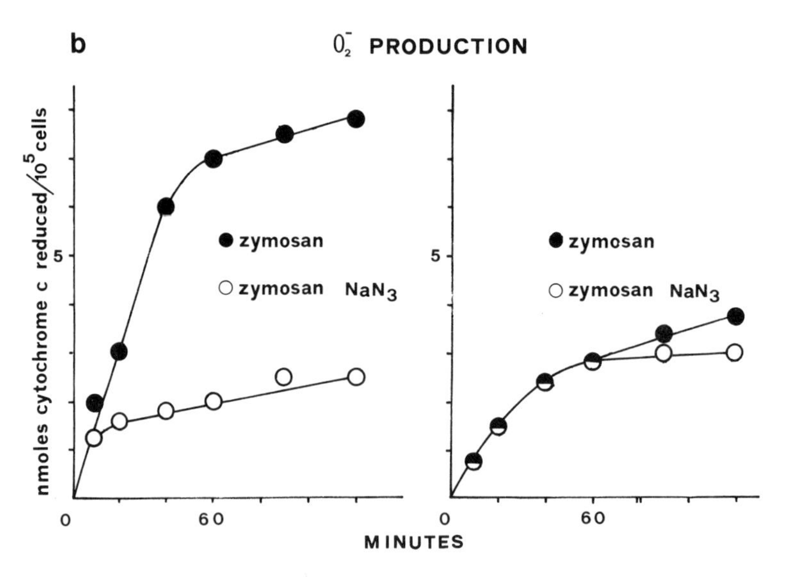

FIGURE 1. Phagocytosis and superoxide production by human eosinophils and neutrophils from normal blood. Neutrophils (>95% pure) were isolated by dextran sedimentation and centrifugation through Ficollpaque gradient. Eosinophils (90% pure) were purified on metrizamide discontinuous gradient. (a) Time course of the phagocytic process. Phagocytosis was evaluated

qualitatively similar to that of neutrophils. However, one major difference has been observed. While the respiratory burst of neutrophils is notoriously insensitive to NaN_3 and other mitochondrial inhibitors, Tauber et al.[11] showed that azide inhibits both oxygen uptake and O_2^- generation by eosinophils from eosinophilic subjects exposed to serum-opsonized zymosan. In that paper, the effect of NaN_3 on particle uptake was not reported, and therefore it could not be established whether the inhibitory effect of azide on the respiratory burst could be accounted for, at least in part, by inhibition of phagocytosis. In our laboratory, the inhibitory effect of NaN_3 on O_2^- generation has been confirmed with eosinophils from normal subjects stimulated by serum-opsonized zymosan (Fig. 1b), and it has been shown, in addition, that NaN_3 did not affect particle uptake (Fig. 1a) (P. Dri and R. Cramer, unpublished). The mechanism of the inhibitory effect of NaN_3 seems to be dependent on the stimulus that has elicited the burst, since inhibition of H_2O_2 generation is seen with phagocytosing eosinophils but not with phorbol 12-myristate 13-acetate (PMA)-stimulated eosinophils.[13] This result further suggests that there might be different mechanisms of elicitation of the respiratory burst in eosinophils, in analogy with what is well established in neutrophils.[14]

A quantitative comparison between eosinophils and neutrophils reveals that the metabolic activity of unstimulated eosinophils is in general higher than that of unstimulated neutrophils, whether oxygen consumption,[8] O_2^-[9–11] and H_2O_2[7,8] generation, HMS activity,[1,7,8,10] or iodinating capacity[10,12] are measured. The results of quantitative comparisons between stimulated eosinophils and stimulated neutrophils are more heterogeneous. In fact, several investigators have shown a higher metabolic burst, in terms of oxygen consumption,[8] O_2^- generation,[9,11] H_2O_2,[8] HMS activity,[8,10] and a higher iodinating capacity[10,12] in eosinophils than in neutrophils. Other investigators failed to show such a difference when HMS activity,[1] O_2^-,[10] and H_2O_2 generation[7] by stimulated neutrophils and eosinophils were compared. At this point, it is worth remembering that a careful choice of the parameter to be assayed is essential in order to obtain meaningful quantitative data on the magnitude of the respiratory burst in different cell types or in a single cell type under different experimental conditions. Thus, for example, assay of O_2^- and H_2O_2 merely reflects fluxes of these molecules from the cell to the extracellular medium but does not necessarily give quantitative information on the metabolic burst. Oxygen consumption is, in our opinion, the most objective index of the overall metabolic activation. and in those studies that have employed this technique eosinophils from eosinophilic subjects seem to be more active than neutrophils, both at rest and upon

on Wright- and Giemsa-stained smears. At least 150 neutrophils and 50 eosinophils were counted at any selected time. Values represent mean of two separate experiments. Assay conditions: serum opsonized zymosan 1 mg/ml, phagocytes 10^6/ml, final volume 1 ml Krebs-Ringer phosphate containing 5 mM glucose and 0.5 mM $CaCl_2$ (KRP); NaN_3 1 mM $t = 37°C$. a, particles per cell; b, percentage of phagocytosing cells. (b) Time course of superoxide production. Nanomoles of O_2^- were calculated from the SOD inhibitable reduction of cytochrome *c*. Assay conditions: cytochrome *c* 200 μM, opsonized zymosan 1 mg/ml, NaN_3 1 mM, 10^5 cells/ml, final volume 1 ml KRP; $t = 37°C$. Values are mean of two separate experiments.

stimulation. Two additional aspects deserve attention when the metabolic burst of eosinophils and neutrophils is compared: kinetics and phagocytic capacity.

Most investigators agree that the eosinophil respiratory burst, measured in different ways, is linear for a longer period of time as compared with that of neutrophils.[9,11,15,16] It follows that differences in the magnitude of the burst between the two cell types are better appreciated late after stimulation. Actually, some results that speak against a difference in metabolic burst between eosinophils and neutrophils should be viewed in this line,[11,16] since measurements were made at relatively early post-stimulation times.[10]

Eosinophils are less phagocytic than neutrophils with respect to several kinds of particles, such as Latex beads,[1] zymosan,[8] bacteria,[1,7,8,12,17] and antibody-coated erythrocytes.[1] The lower phagocytic activity of eosinophils can be appreciated both in terms of percentage of phagocytosing cells in the whole population and number of particles ingested by a single cell.[7,8] Thus, when the phagocytosis-associated respiratory burst of eosinophils is referred to a single ingested particle, its magnitude is unequivocally greater than in neutrophils, even in those instances in which no clear differences on a cell basis are observed between the two cell types.[1,7,10]

3. EOSINOPHILS FROM NORMAL SUBJECTS VERSUS EOSINOPHILS FROM EOSINOPHILIC SUBJECTS

In 1980 an important paper was published in which a careful comparison between the metabolic activity of eosinophils obtained from normal subjects (referred to as normal eosinophils from this point onward) and that of eosinophils from eosinophilic subjects (referred to as eosinophilic eosinophils) was carried out.[6] Profound differences were found between the cells from the two sources. Eosinophilic eosinophils had higher glucose transport, lower phagocytic activity, and a lower surface charge than did normal eosinophils. Furthermore, the surface charge of normal eosinophils decreased during stimulation, as it is known to occur in stimulated neutrophils, while eosinophilic eosinophils did not show any changes in their surface charge upon stimulation. Difference between normal and eosinophilic eosinophils were subsequently shown by Pincus et al.[18] in terms of respiratory burst. Another aspect emerging from those comparative studies is that the degree of variability of the results was much greater within eosinophilic eosinophils rather than within their normal counterpart. This is not at all surprising if one considers that eosinophilic eosinophils are obtained from subjects with a wide spectrum of diseases ranging, for example, from hypereosinophilia of unknown origin to hypereosinophilia associated with atopy, helminth infestations or neoplastic processes.[6,18] This heterogeneity among eosinophilic eosinophils may also explain some apparently conflicting results that have been discussed in the previous paragraph. Even in a single disease associated with hypereosinophilia, heterogeneity in the eosinophil population has been found. Thus, the eosinophils from hypereosinophilic syndrome have been resolved into two populations, one with normal density and the other with a density lower than that of normal eosinophils.[19–21] Differences in the ability to perform the respiratory burst were also found between normodense and hypodense eosinophils.[19,20] It is now clear that eosinophilic eosinophils are not a suitable material for studies aimed at an understanding of the physiology of normal eosinophils. Hence, more recently, normal eosinophils have been in generally used for comparative studies between eosinophils and neutrophils. By and large, these studies have confirmed that the eosinophil respiratory burst is greater than that of neutrophils[13,17,22–24] but suggest that the metabolic activity of resting eosinophils is similar to that of neutrophils,[13,17,22,25,26] in contrast with those conclusions drawn from studies with eosinophilic eosinophils.

4. FEATURES OF THE RESPIRATORY BURST OF EOSINOPHILS

There is considerable information on the respiratory burst of neutrophils that can be dissected in several well-identified steps, each of which can be studied with appropriate approaches. For details on the subject, the reader is referred to a recent review by F. Rossi.[14] Basically the following aspects of the respiratory burst of the neutrophils are discussed in that review: induction, products, enzymatic basis (NADPH-oxidase), mechanisms of activation of the NADPH-oxidase, functional level of the oxidase, and functions of the respiratory burst. In addition, within each of those aspects, several bodies of knowledge are further identified and discussed. For example, several steps can be recognized in the process, leading to activation of the NADPH-oxidase, such as recognition of the stimulus, transduction of the signal into the cell (which involves turnover of phosphoinositides, changes in calcium fluxes, protein kinase C activity, guanosine triphosphate-binding proteins), assembly of the oxidase components, and phosphorylation of one or more of such components. This brief introductory comment gives an idea of the great complexity of the neutrophil respiratory burst and on the large amount of knowledge available on it at present. With the eosinophil, the situation is different, since our knowledge is much less developed and some of the aspects of their respiratory burst are virtually unexplored, such as the transduction mechanism, the nature of the oxidase system, and the mechanism of its activation. Therefore, the review that follows is far more limited than a comparable review on neutrophils.

4.1. Induction of the Burst

The stimuli known to be able to induce a respiratory burst in eosinophils are listed in Table I. Their extreme heterogeneity is immediately apparent, as is the case with neutrophils. Most of the stimuli that induce a respiratory burst in neutrophils are effective on eosinophils as well. A few stimuli seem to be specific for eosinophils, as they are unable to induce a respiratory burst in neutrophils, e.g., histamine,[27–29] the 20-COOH derivative of leukotriene B_4 (LTB_4),[31] the monokine

TABLE I. Inducers of Eosinophils Respiratory Burst: Comparative Effects on Eosinophils and Neutrophils

	Eosinophils		Neutrophils	
Inducers	Effect[a]	Ref.	Effect[a]	Ref.
Phagocytosis	+	1,7,8	+	1,7,8
Concanavalin A	+	29,39	+	29
Phorbol-12-myristate-13-acetate (PMA)	+	13,22,29	+	13,22,29
Leukotriene B_4 (LTB_4)	+	31	+	31
Calcium ionophore A23187	+	29,40	+	29
Cytochalasin E	+	29	+	29
N-Formyl-methionyl-leucyl-phenylalanine (FMLP)	±	29,31,32	+	29,31
Eosinophil-cytotoxicity enhancing factor (M-ECEF, ECE-A) (TNF)[b]	−	37	+	33,34
Eosinophil-activating factor (EAF)	+ ▲	26	−	26
Val-gly-ser-glu (ECF-A)	− ▲	9,15,29,38	− ▼	15,29
Eosinophil-stimulation promoter (ESP)	− ▲	41		
Histamine	±	27–29	− ▼	27,28,29
20-COOH-LTB_4	+	31	−	31
Colony-stimulating factor (CSF-α)	+ ▲	42	+	35

[a]+, activating effect; −, no effect; ▲, potentiates phagocytosis induced respiratory burst; ▼, inhibits phagocytosis induced respiratory burst.
[b]A similarity between TNF and M-ECEF has been recently shown by Silberstein and David.[36]

eosinophil activating factor (EAF).[26] By contrast, the chemotactic peptide *N*-formyl-methionyl-leucyl-phenylalanine (FMLP), a known stimulator of neutrophils, does not stimulate normal eosinophils but seems to induce a burst in some populations of eosinophils isolated from eosinophilic subjects.[15,31,32] Also, the tumor necrosis factor (TNF) is a stimulator of neutrophils[33,34] but not of eosinophils.[37] The eosinophil chemotactic peptide Val-Gly-Ser-Glu is unable to induce a respiratory burst in eosinophils,[9,15,29,38] but it potentiates the burst induced in these cells by phagocytosable particles.[15,38] Rather, Val-Gly-Ser-Glu decreases the neutrophil metabolic burst induced by phagocytosable particles (zymosan).[15] A similar activity is attributed to histamine.[28] These data indicate that induction of the eosinophil respiratory burst shares many features in common with neutrophils, but there are situations that clearly distinguish the two cell types with regard to both the triggering stimulus and modulation of the burst. The differential effect of some stimuli on eosinophils and neutrophils may be best explained, at least in selective instances, by the different surface receptor profile and receptor density of the two cell types. Examples are shown in Table II.

4.2. Products of the Respiratory Burst

These are the same as for neutrophil respiratory burst, i.e., superoxide anion and hydrogen peroxide, and have been mentioned repeatedly. Instead, there is no direct evidence that stimulated eosinophils may generate hydroxyl radical and sin-

TABLE II. Receptors Identified on Eosinophils: Comparison with the Corresponding Receptors of Neutrophils

Receptor for:	Eosinophils	Ref.	Neutrophils	Ref.
Histamine	Present (H_1 type)	27	Present (H_2 type)	28,30
Fcγ	Present (different from neutrophil receptor)	13,43	Present	13
Fcϵ	Present	21,44	Absent	21,44
C3b, C4	Present } lower density than in neutrophils	13,45,reviewed in 46	Present	13
C3d	Present } lower density than in neutrophils	45	Present	45
C3bi	Present } lower density than in neutrophils	13	Present	13
FMLP	Absent from normodense cells	32	Present	32
Glucocorticoids	Present	51	Present	51
Estrogens	Present	48–50	Estrogen binding capacity is present only during phagocytosis	reviewed in 47
EO1 monoclonal antibody	Present	52	Absent	52

glet oxygen as it takes place in neutrophils.[47] Stimulated eosinophils have an enhanced chemiluminescence in both the presence[11,17,23,24] and absence of luminol.[10] Since generation of chemiluminescence involves, at least in neutrophils, a variety of oxidative products, such as O_2^-, H_2O_2, singlet oxygen, and HClO, the increased chemiluminescence of stimulated eosinophils suggests that even these cells may generate singlet oxygen and HClO.

4.3. Enzymatic Basis

Resting eosinophils exhibit particulate NADPH-oxidase activity,[8,29] which is several times stimulated in cells activated with phagocytosable particles,[10,11] PMA,[29] or the calcium ionophore A23187.[29] The oxidase generates superoxide anion[9,11,29] and has an apparent K_m of 25–38 μM for NADPH[9,11,29] and 0.77 mM for NADH.[11] These properties are reminiscent of those of the neutrophil NADPH-oxidase system, which is unanimously considered the key enzyme of the neutrophil respiratory burst. The incontrovertible proof of the truthfulness of such a statement is the absence of an activated NADPH-oxidase in stimulated neutrophils from patients with chronic granulomatous disease (CGD),[53] which are known to be unable to perform the respiratory burst.[54] Such proof is not available with eosinophils. In fact, stimulated eosinophils from CGD patients do not reduce nitroblue tetrazolium, indicating that neither they nor CGD neutrophils are able to mount a respiratory burst,[32,55] but their oxidase activity has never been assayed directly, as far as we are aware of. It is reasonable to admit, however, that a NADPH-oxidase activity is the key enzymatic mechanism of the respiratory burst in eosinophils as well. Several aspects of the eosinophil NADPH-oxidase remain virtually unexplored, such as subcellular localization, pH optimum, and the nature and mechanism of activation. Most of the NADPH-oxidase activity of both neutrophil and eosinophil homogenate sediments at $27{,}000 \times g$,[10,56] but a considerable part (25–40%) of the particulate oxidase activity of eosinophil homogenate and only a small amount of that of neutrophil homogenate (2–12%) sediments at $500 \times g$.[10,56] This suggests that the NADPH-oxidase may be associated, in eosinophils, to subcellular structures of higher density than in neutrophils. The nature of such structures is unknown. Eosinophil NADPH-oxidase has been assayed by different workers at either acid[10,56] or physiological pH,[8,9,11,29] but a complete pH curve has not been reported so far, so that pH optimum of the enzyme is unknown. Eosinophils contain a low potential cytochrome *b*[57] with properties in part similar to those of the low potential cytochrome *b*, which is part of the neutrophil oxidase system.[58] The amount of cytochrome *b* in eosinophils is about three times as much as in neutrophils. There is no evidence so far that this cytochrome is involved in the eosinophil respiratory burst.

4.4. Enzymes Indirectly Related to the Respiratory Burst

The primary product of the respiratory burst, i.e., the superoxide anion, is converted to hydrogen peroxide by the enzyme superoxide dismutase (SOD). In

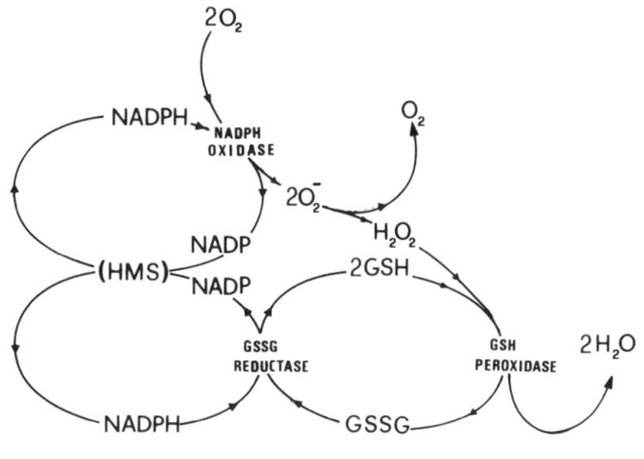

FIGURE 2. Schematic representation of glutathione redox cycle and its coupling with hexose monophosphate shunt (HMS) and NADPH-oxidase activity, GSH, reduced glutathione; GSSG, oxidized glutathione.

human neutrophils, two forms of this enzyme, one KCN-sensitive and the other KCN-insensitive, have been described. Most of the neutrophil SOD activity is found in the cytosol.[11,47] A superoxide dismutase activity has been reported to be present in eosinophils as well.[11,56] It is associated with a cytosol fraction,[11] and its activity per cell is comparable to that of neutrophils.[11,56] Data on its sensitivity to KCN and presence of isoenzymes are not available. Eosinophils are also provided with enzymes that degrade the hydrogen peroxide derived by O_2^- dismutation, i.e., eosinophil peroxidase (EPO),[11,59,60] catalase, and glutathione peroxidase.[56] Eosinophils also have glutathione reductase activity.[56] It is known, from the biochemistry of neutrophils, that glutathione peroxidase and glutathione reductase work in combination in the so-called glutathione redox cycle[47] (Fig. 2). According to the scheme depicted in Fig. 2, for each molecule of H_2O_2 degraded, one molecule of NADP is formed. This NADP, together with the NADP directly generated by the NADPH-oxidase, accounts for the stimulation of the HMS in neutrophils during the respiratory burst. Although there is no formal evidence for the glutathione cycle operativity and its coupling with HMS in eosinophils, several observations suggest that this may indeed be the case, as it is in neutrophils: (1) presence in the eosinophils of glutathione peroxidase and glutathione reductase, (2) presence in the eosinophils of glucose 6-phosphate dehydrogenase and 6-phospho-gluconate dehydrogenase, the first two enzymes of the HMS,[56] and (3) stimulation of the HMS during the respiratory burst in eosinophils.

5. FUNCTIONS OF THE RESPIRATORY BURST IN EOSINOPHILS

5.1. Microbicidal and Fungicidal Capacity

The microbicidal power of eosinophils against a variety of bacterial species, e.g., *Staphylococcus aureus,*[7,8,12,61] *Escherichia coli,*[7,61] and *Lysteria monocytogenes,*[62] has been reported to be lower than that of neutrophils. However, at high bacteria-to-cell ratios, eosinophils seem to be able to kill as many microorganisms as neutrophils.[17] The poor bactericidal activity of eosinophils, at least at a low bacteria-to-cell ratio, probably reflects the lower phagocytic capacity of these cells in comparison with neutrophils. The low microbicidal activity of eosinophils, together with the clinical observation that neutropenic children with high number of circulating eosinophils are nevertheless susceptible to bacterial infections,[63] has led to the suggestion that killing of micro-organisms may not be the major role of eosinophils. Whatever the importance of the eosinophils in host–bacteria interactions, a relevant question is whether there is any role for the respiratory burst in the microbicidal activity of these cells. There is no doubt that eosinophil peroxidase (EPO), in combination with H_2O_2 and an oxidizable factor (e.g., halides), is bactericidal in vitro.[64,65] The general mechanism of this bactericidal system is similar to that catalyzed by myeloperoxidase (MPO), the neutrophil peroxidase, although the optimal conditions for the two enzymes in terms of H_2O_2 concentration, affinity for halides, and pH are different.[64] The bactericidal activity of both systems is inhibited by the heme–protein inhibitor sodium azide (NaN_3).[65,66] Inhibition of bacterial killing by NaN_3 is also observed when whole neutrophils are used,[12,66] and this is considered as one of the points in favor of a physiological role of the MPO–H_2O_2–halide system in intact neutrophils. Instead, NaN_3, despite its inhibitory effect on the eosinophil respiratory burst and on the in vitro bactericidal capacity of the EPO–H_2O_2–halide system, has been reported not to inhibit, but rather to stimulate, killing of *Staphylococcus aureus* by eosinophils.[12] Since the NaN_3 concentrations used in that experiment were sufficient to inhibit the cell EPO drastically, the results would argue against a prominent role of the oxygen-dependent and EPO-mediated bactericidal mechanism in eosinophils. Studies with genetically EPO-deficient eosinophils would probably be definitive in this respect but, to our knowledge, no such experiment has been carried out so far, despite the availability of EPO-deficient cells.[67,68] There is another indication in the literature that seems to speak against the respiratory burst as a major determinant of the eosinophil bactericidal activity. It has been shown that the products of the respiratory burst caused perforation of *E. coli* envelope, which closely correlates with bacterial viability.[69] Perforation was dependent on the products of the burst, as clearly indicated by its absence in *E. coli* ingested by CGD neutrophils, but independent of MPO. When the same phenomenon was investigated in eosinophils, bacterial perforation was 75% lower than that observed with neutrophils, despite the presence of a respiratory burst,[17] suggesting that either the oxygen metabolite(s) required for perforation could not reach the bacteria ingested by the eosinophils or eosinophils did not produce the required metabolite(s). Instead, the respiratory burst of eosinophils seems to be involved in killing of fungi as suggested the decreased *Candida albicans* killing by CGD eosinophils as compared to normal eosinophils.[70]

5.2. Helminthotoxic Activity

A role for eosinophil in host defense against parasites was inferred in the late nineteenth century from the well-known association between helminth infestations and eosinophilia.[71] Direct evidence for such a role has been obtained, for the first time, by studying the effect of antieosinophil serum (AES) on the course of helminth infestation. In fact, treatment of a mixed neutrophil–eosinophil population with AES decreased the killing of larvae of *Schistosoma mansoni* by the cells.[72] Injection of AES into mice infected with cercariae of *S. mansoni* increased animal mortality, the number of schistosomula and adult worms in tissues,[73] and the number of eggs retained in tissues.[74] These effects were not observed after injection of an antineutrophil serum. The role of eosinophils in the damage to the invading stages of parasites has been confirmed using pure eosinophil suspensions that proved efficient in mediating ADCC to larvae of *S. mansoni*.[75,76] At first sight, the studies with AES seem to exclude a role for neutrophils in helminthotoxic activity. However, a cross-reaction of AES with neutrophils cannot be excluded in those studies. Actually, more recent studies in vitro with purified neutrophils have shown that even these cells are able to damage larvae of *S. mansoni* [77–79] and *T. spiralis*,[80,81] with an efficiency comparable to, and sometimes higher than, that of eosinophils. The effector mechanisms of eosinophil damage to parasite eggs and larvae are not completely understood. In vitro studies have provided evidence for the involvement of eosinophil cationic protein (ECP),[82] major basic protein (MBP),[82] and products of the respiratory burst.[76,83,84] Evidence for a role of the respiratory burst is based on the following observations: (1) eggs of *S. mansoni*[84] and larvae of *S. mansoni*[76] and *T. spiralis*[83] are sensitive to small doses of H_2O_2; (2) antibody-coated parasites and eggs stimulate O_2^- and H_2O_2 generation by eosinophils,[79,84] and the extent of damage was related to the amount of these products;[79] (3) exogenously added catalase prevents the damaging effect of eosinophils;[79,80] (4) SOD prevents damage to schistosoma eggs,[84] but it had no effect on damage to larvae of *T. spiralis*; [83] (5) a mixed neutrophil–eosinophil population from a CGD patient damaged larvae of *T. spiralis* less than normal cells;[80] and (6) purified neutrophils from CGD patients were unable to damage larvae of *S. mansoni,* while purified eosinophils from CGD patients had a 50% decreased larvicidal activity,[79] suggesting that eosinophils have at least two helminthotoxic mechanisms, one of which is dependent on the respiratory burst. However, there is one paper by Pincus et al.[76] that does not recognize any role of the respiratory burst in damaging schistosoma larvae. Under anaerobic conditions, eosinophils were still capable of efficiently killing antibody-coated schistosomula. The mechanism of action of the products of the respiratory burst in the helminthotoxic activity of the eosinophils has been studied, particularly with reference to the EPO–H_2O_2–halide cytotoxic system. It has been shown that the in vitro reconstituted system, with purified components, is toxic to larvae of *S. mansoni*[85] and *T. spiralis*.[86] Whether this applies as well to intact cells is a matter of controversy, since the studies are based on the use of the peroxidase inhibitors NaN_3 and KCN, and studies with EPO-deficient eosinophils are unavailable. NaN_3 has been shown either to reduce eosinophil-mediated damage to larvae of *S. mansoni*[79] or to amplify damage to eggs of *S. mansoni*[84] and larvae of *T. spiralis*,[80] while KCN has proved ineffective.[87] It should be recalled, however, that NaN_3 may exert other effects in addition to peroxidase inhibition, e.g., inhibition of the eosinophil respiratory burst and of helminth catalase. The relative importance of the various effects of NaN_3 in different experimental models might explain the apparently conflicting results.

5.3. Regulation of the Activity of the Inflammatory Mediators

Stimulated eosinophils have been shown to catalyze the conversion of leukotriene C_4 (LTC_4) to LTB_4.[88] The conversion is inhibited by NaN_3 and catalase, suggesting that the reaction is dependent on the H_2O_2 produced during the burst and on EPO. This was confirmed by demonstrating that the EPO–H_2O_2–halide in vitro-reconstituted system was able to catalyze the same reaction. Accordingly, catalase strongly decreased the formation of LTB_4 by stimulated eosinophils and enhanced formation of LTC_4 and LTD_4.[89] The EPO–H_2O_2–halide system has also been shown to inactivate LTC_4 and LTD_4 and, to a lesser extent, LTB_4[40] and to stimulate histamine release from mast cells.[90] These data suggest that eosinophils may play an important modulating role in the inflammatory reaction by regulating the concentration of some chemical mediators.

5.4. Cytotoxicity

The products of neutrophil respiratory burst are known to cause damage to the neutrophil itself and to a large spectrum of nucleated and non-nucleated cells.[91] Although studies of this type with eosinophils are much less numerous, it appears that the eosinophil respiratory burst may contribute to lyse or damage other cells. Again, the EPO–H_2O_2–halide system seems important in this respect, since it has been reported to damage, under appropriate conditions, tumor cells,[92] *Trypanosoma cruzi,*[93] pneumocytes,[94] *Toxoplasma,*[95] and mast cells.[90]

6. REGULATION OF EPO CONCENTRATION IN THE EXTRACELLULAR MEDIUM

What emerges is that EPO, in combination with H_2O_2 and halides, is a powerful toxic system to several targets, such as helminth larvae, bacteria, protozoa, and mammalian cells. It follows that release of EPO from stimulated eosinophils is a double-edged sword; on one side it can kill bacteria and parasites, but, on the other, it may damage tissue cells as well. The question arises: Is there any mechanism that regulates EPO concentration in interstitial fluids? We have found that neutrophils avidly and specifically bind EPO, but not MPO, and internalize it in small vesicles that then fuse and ultimately form multivesicular bodies.[96] The binding results in inhibition of some EPO catalytic activities. These findings suggest that neutrophils may serve to sequester EPO, a potentially toxic agent, thus limiting damage to the tissues in eosinophil-rich inflammatory lesions.

7. CONCLUSIONS

The metabolic performances of the eosinophil seem to be at least as good as those of his more renowned brother, the neutrophil. Actually, the eosinophil overdoes sometimes, for example, by responding with a more vigorous respiratory burst than the neutrophil to some stimuli. However, in the same way as even very alike brothers may differ in their personalities, the eosinophil has several distinctive metabolic features which have been underlined in the various paragraphs of the chapter.

The fine mechanisms of the eosinophil metabolic functions have been scarcely investigated as compared with neutrophils. This may depend on the surmised assumption that those mechanisms are similar to those of neutrophils. The assumption may not be completely correct, however. For example, the inhibitory effect of azide on the respiratory burst of eosinophils suggests that enzymatic mechanisms, different from those of neutrophils, may underlie the metabolic activation of eosinophils. Thus, an in-depth investigation of the molecular basis of eosinophil functions might well yield some surprises, rather than simply déjà vue results.

ACKNOWLEDGMENTS. This work was supported by grants from Ministero della Pubblica Istruzione (40% and 60%) and by grant 85-00872-52 from CNR (Progetto Finalizzato Controllo delle Malattie da Infezione).

REFERENCES

1. Cline MJ, Hanifin J, Lehrer RI: Phagocytosis by human eosinophils. *Blood* **32:**922–934, 1968.
2. Karnovsky ML: Metabolic basis of phagocytic activity. *Physiol Rev* **42:**143–168, 1962.
3. Cline MJ: Metabolism of the circulating leukocyte. *Physiol Rev* **45:**674–720, 1965.
4. Karnovsky ML: The metabolism of leukocytes. *Ser Haematol* **5:**156–165, 1968.
5. Beck WS: Leukocyte metabolism. *Semin Hematol* **1:**69–93, 1968.
6. Bass DA, Grover WH, Lewis JC, et al: Comparison of human eosinophils from normals and patients with eosinophilia. *J Clin Invest* **66:**1265–1273, 1980.
7. Mickenberg D, Root RK, Wolff SM: Bactericidal and metabolic properties of human eosinophils. *Blood* **39:**67–80, 1972.
8. Baehner RL, Johnston RB Jr: Metabolic and bactericidal activity of human eosinophils. *Br J Haematol* **20:**277–285, 1971.
9. Tauber AI, Goetzl EJ, Babior BM: The production of superoxide by human eosinophils. *Blood* **48:**968a, 1976 (abst).
10. DeChatelet LR, Shirley PS, McPhail LC, et al: Oxidative metabolism of the human eosinophil. *Blood* **50:**525–535, 1977.
11. Tauber AI, Goetzl EJ, Babior BM: Unique characteristics of superoxide production by human eosinophils in eosinophilic states. *Inflammation* **3:**261–272, 1979.
12. Bujak JS, Root RK: The role of peroxidase in the bactericidal activity of human blood eosinophils. *Blood* **43:**727–736, 1974.
13. Yazdanbakhsh M, Eckmann CM, Roos D: Characterization of the interaction of human eosinophils and neutrophils with opsonized particles. *J Immunol* **135:**1378–1384, 1985.
14. Rossi F: The O_2^- forming NADPH-oxidase of the phagocytes. Nature, mechanisms of activation and functions. *Biochim Biophys Acta Rev Bioenerg* **853:**65–89, 1986.

15. Beswick PH, Kay AB: The effect of an ECF-A and formyl methionyl chemotactic peptides on oxidative metabolism of human eosinophils and neutrophils. *Clin Exp Immunol* **43:**399–407, 1981.
16. Learn DB, Brestel EP: A comparison of superoxide production by human eosinophils and neutrophils. *Agents Actions* **12:**485–488, 1982.
17. Yazdanbakhsh M, Eckermann CM, Bot AAM, et al: Bactericidal action of eosinophils from normal human blood. *Infect Immun* **53:**192–198, 1986.
18. Pincus SH, Schooley WR, DiNapoli AM, et al: Metabolic heterogeneity of eosinophils from normal and hypereosinophilic patients. *Blood* **58:**1175–1181, 1981.
19. Winqvist I, Olofsson T, Olsson I, et al: Altered density, metabolism and surface receptors of eosinophils in eosinophilia. *Immunology* **47:**531–538, 1982.
20. Prin L, Charon J, Capron M, et al: Heterogeneity of human eosinophils. II. Variability of respiratory burst activity related to cell density. *Clin Exp Immunol* **57:**735–742, 1984.
21. Capron M, Spiegelberg HL, Prin L, et al: Role of IgE receptors in effector function of human eosinophils. *J Immunol* **132:**462–468, 1984.
22. Portmann P, Koppel P, Jörg A: An improved method for the isolation of eosinophilic leukocytes from normal human blood. *Experientia* **36:**139–141, 1980.
23. Shutt PA, Graziano FM, Wallow IH, et al: Comparison of superoxide generation and luminol dependent chemiluminescence with eosinophils and neutrophils from normal individuals. *J Lab Clin Med* **106:**638–645, 1985.
24. Shutt PA, Graziano FM, Busse WW: Enhanced eosinophil luminol-dependent chemiluminescence in allergic rhinitis. *J Allergy Clin Immunol* **77:**702–708, 1986.
25. Pincus SH, Butterworth AE, David JR, et al: Antibody-dependent eosinophil-mediated damage to schistosomula of *Schistosoma mansoni:* Lack of requirement for oxidative metabolism. *J Immunol* **126:**1794–1799, 1981.
26. Thorne JIK, Richardson BA, Veith MC, et al: Partial purification and biological properties of an eosinophil-activating factor. *Eur J Immunol* **15:**1083–1091, 1985.
27. Pincus SH, DiNapoli AM, Schooley WK: Superoxide production by eosinophils: Activation by histamine. *J Invest Dermatol* **79:**53–57, 1982.
28. Seligman BE, Fletcher MP, Gallin JI: Histamine modulation of human neutrophil oxidative metabolism, locomotion, degranulation, and membrane potential changes. *J Immunol* **130:**1902–1909, 1983.
29. Yamashita T, Someya A, Hara E: Response of superoxide anion production by guinea-pig eosinophils to various soluble stimuli: Comparison to neutrophils. *Arch Biochem Biophys* **241:**447–452, 1985.
30. Petty HR, Francis JW: Polymorphonuclear leukocyte histamine receptors: Occurrence in cell surface clusters and their redistribution during locomotion. *Proc Natl Acad Sci USA* **83:**4332–4335, 1986.
31. Palmblad J, Gyllenhammar H, Lindgren JA, et al: Effects of leukotrienes and f-Met-Leu-Phe on oxidative metabolism of neutrophils and eosinophils. *J Immunol* **132:**3041–3045, 1984.
32. Roberts RL, Gallin JI: Rapid method for isolation of normal human peripheral blood eosinophils on discontinuous Percoll gradients and comparison with neutrophils. *Blood* **65:**433–440, 1985.
33. Tsujimoto M, Yokota S, Vilček J, et al: Tumor necrosis factor provokes superoxide anion generation from neutrophils. *Biochem Biophys Res Commun* **137:**1094–1100, 1986.
34. Klebanoff SJ, Vadas MA, Harlan JM, et al: Stimulation of neutrophils by tumor necrosis factor. *J Immunol* **136:**4220–4225, 1986.
35. Lopez AF, Williamson DJ, Gamble JR, et al: Recombinant human granulocyte–macrophage colony-stimulating factor stimulates in vitro mature human neutrophil and eosinophil function, surface receptor expression and survival. *J Clin Invest* **78:**1220–1228, 1986.
36. Silberstein DS, David JR: Tumor necrosis factor enhances eosinophil toxicity to *Schistosoma mansoni* larvae. *Proc Natl Acad Sci USA* **83:**1055–1059, 1986.
37. Pincus SH, Dessein A, Lenzi H, et al: Eosinophil-mediated killing of schistosomula of *Schistosoma mansoni:* Oxidative requirement for enhancement by eosinophil colony stimulating factor (CSF-α) and supernatants with eosinophil cytotoxicity enhancing activity (E-CEA). *Cell Immunol* **87:**424–433, 1984.

38. Winqvist I, Oloffson T, Olsson I: Mechanisms for eosinophil degranulation; release of the eosinophil cationic protein. *Immunology* **51:**1–8, 1984.
39. Pincus SH: Eosinophil–particle interactions: A model system for study of cellular adherence and activation. *Cell Immunol* **86:**460–471, 1984.
40. Henderson WR, Jörg A, Klebanoff SJ: Eosinophil peroxidase-mediated inactivation of leukotrienes B_4, C_4 and D_4. *J Immunol* **128:**2609–2613, 1982.
41. Rand TH, Colley DG: Influence of a lymphokine fraction containing eosinophil stimulation promoter (ESP) on oxidative and degranulation responses of murine eosinophils. *Cell Immunol* **71:**334–345, 1982.
42. Vadas MA, Varigos G, Nicola N, et al: Eosinophil activation by colony-stimulating factor in man: Metabolic effects and analysis by flow cytometry. *Blood* **61:**1232–1241, 1983.
43. Kulczycki A Jr: Human neutrophils and eosinophils have structurally distinct Fc receptors. *J Immunol* **133:**849–853, 1984.
44. Capron M, Capron A, Dessaint JP, et al: Fc receptors for IgE on human and rat eosinophils. *J. Immunol* **126:**2087–2092, 1981.
45. Anwar ARE, Kay AB: Membrane receptors for IgG and complement (C4, C3b and C3d) on human eosinophils and neutrophils and their relation to eosinophilia. *J Immunol* **119:**976–982, 1977.
46. Weller PF, Goetzl EJ: The regulatory and effector roles of eosinophils. *Adv Immunol* **27:**339–371, 1979.
47. Klebanoff SJ, Clark RA: The metabolic burst, in Klebanoff SJ, Clark RA (eds): *The Neutrophil: Function and Clinical Disorders.* Amsterdam, North-Holland, 1978, p 283.
48. Klebanoff SJ, Durack DT, Rosen H, et al: Functional studies on human peritoneal eosinophils. *Infect Immun* **17:**167–173, 1977.
49. Tchernitchin A, Tchernitchin X: Characterization of the estrogen receptors in the uterine and blood eosinophil leukocytes. *Experientia* **32:**1240–1242, 1976.
50. Tchernitchin A, Tchernitchin X: The binding of estradiol to human polynuclear eosinophilic leukocytes. *CR Acad Sci (Paris)* **280:**1477–1480, 1975.
51. Peterson AP, Altman LC, Hill JS, et al: Glucocorticoid receptors in normal human eosinophils: comparison with neutrophils. *J Allergy Clin Immunol* **68:**212–217, 1981.
52. Saito H, Yamada K, Bréard J, et al: A monoclonal antibody reactive with human eosinophils. *Blood* **67:**50–56, 1986.
53. Babior BM: The nature of the NADPH-oxidase, in Gallin JI, Fauci AS (eds): *Advances in Host Defense Mechanisms.* New York, Raven, 1983, vol 3: *Chronic Granulomatous Disease,* p 91.
54. Mills EL, Quie PG: Inheritance of chronic granulomatous disease, in Gallin JI, Fauci AS (eds): *Advances in Host Defense Mechanisms, Chronic Granulomatous Disease,* vol 3. New York, Raven, 1983, p 25.
55. Nathan DG, Baehner RL, Weaver DK: Failure of nitro blue tetrazolium reduction in the phagocytic vacuoles of leukocytes in chronic granulomatous disease. *J Clin Invest* **48:**1895–1901, 1969.
56. DeChatelet LR, Migler RA, Shirley PS, et al: Enzymes of oxidative metabolism in the human eosinophil. *Proc Soc Exp Biol Med* **158:**537–541, 1978.
57. Segal AW, Garcia R, Goldstone AH, et al: Cytochrome b_{-245} of neutrophils is also present in human monocytes, macrophages and eosinophils. *Biochem J* **196:**363–367, 1981.
58. Segal AW: Chronic granulomatous disease: A model for studying the role of cytochrome b_{-245} in health and disease, in Gallin JI, Fauci AS (eds): *Advances in Host Defense Mechanisms.* New York, Raven, 1983, vol 3: *Chronic Granulomatous Disease,* p 121.
59. West BC, Gelb NA, Rosenthal AS: Isolation and partial characterization of human eosinophil granules. *Am J Pathol* **81:**575–588, 1975.
60. Dri P, Cramer R, Soranzo MR, et al: New approaches to the detection of myeloperoxidase deficiency. *Blood* **60:**323–327, 1982.
61. DeChatelet LR, Migler RA, Shirley PS, et al: Comparison of intracellular bactericidal activities of human neutrophils and eosinophils. *Blood* **52:**609–617, 1978.
62. Cline MJ: Microbicidal activity of human eosinophils. *J Reticuloendothel Soc* **12:**332–339, 1972.

63. Gilman PA, Jackson DP, Guild HG: Congenital agranulocytosis: Prolonged survival and terminal acute leukemia. *Blood* **36:**576–585, 1970.
64. Migler R, DeChatelet LR, Bass DA: Human eosinophilic peroxidase: Role in bactericidal activity. *Blood* **51:**445–456, 1978.
65. Jong EC, Henderson WR, Klebanoff SJ: Bactericidal activity of eosinophil peroxidase. *J Immunol* **124:**1378–1382, 1980.
66. Klebanoff SJ, Clark RA: Anti-microbial systems, in Klebanoff SJ, Clark RA (eds): *The Neutrophil: Function and Clinical Disorders.* Amsterdam, North-Holland, 1978, p 409.
67. Presentey B, Szapiro L: Hereditary deficiency of peroxidase and phospholipids in eosinophilic granulocytes. *Acta Haematol* **41:**359–362, 1969.
68. Presentey B, Joshua H: Peroxidase and phospholipid deficiency in human eosinophilic granulocytes. A marker in populations genetics. *Experientia* **38:**628–629, 1982.
69. Hamers MN, Bot AAM, Weening RS, et al: Kinetics and mechanism of bactericidal action of human neutrophils against *Escherichia coli. Blood* **64:**635–641, 1984.
70. Lehrer RI: Measurement of candidacidal activity of specific leukocyte types in mixed cell population. II. Normal and chronic granulomatous disease eosinophils. *Infect Immun* **3:**800–802, 1971.
71. Dessein AJ, David JR: The eosinophil in parasitic disease, in Gallin JI, Fauci AS (eds): *Advances in Host Defense Mechanisms.* New York, Raven, 1982, vol 1: *Phagocytic Cells,* p 243.
72. Butterworth AE, Sturrock RF, Houba V, et al: Eosinophils as mediators of antibody-dependent damage to schistosomula. *Nature (Lond)* **256:**727–729, 1975.
73. Mahmoud AAF, Warren KS, Peters PA: A role for the eosinophil in acquired resistence to *Schistosoma mansoni* infection as determined by antieosinophil serum. *J Exp Med* **142:**805–813, 1975.
74. Olds GR, Mahmoud AAF: Role of host granulomatous response in murine schistosomiasis mansoni. Eosinophil mediated destruction of eggs. *J Clin Invest* **66:**1191–1199, 1980.
75. Butterworth AE, David JR, Mahmoud AAF, et al: Antibody-dependent eosinophil-mediated damage to ^{51}Cr-labeled schistosomula of *Schistosoma mansoni:* Damage by purified eosinophils. *J Exp Med* **145:**136–150, 1977.
76. Pincus SH, Butterworth AE, David JR, et al: Antibody-dependent eosinophil-mediated damage to schistosomula of *Schistosoma mansoni:* Lack of requirement for oxidative metabolism. *J Immunol* **126:**1794–1799, 1981.
77. Anwar AR, Smithers SR, Kay AB: Killing of schistosomula of *Schistosoma mansoni* coated with antibody and/or complement by human leukocytes *in vitro:* Requirement for complement in preferential killing by eosinophils. *J Immunol* **122:**628–637, 1979.
78. Vadas MA, David JR, Butterworth A, et al: A new method for the purification of human eosinophils and neutrophils, and a comparison of the ability of these cells to damage schistosomula of *Schistosoma mansoni. J Immunol* **122:**1228–1236, 1979.
79. Kazura JW, Fanning MM, Blumer JL, et al: Role of cell-generated hydrogen peroxide in granulocyte-mediated killing of schistosomula of *Schistosoma mansoni in vitro. J Clin Invest* **67:**93–102, 1981.
80. Bass DA, Szejda P: Eosinophils versus neutrophils in host defense. Killing of newborn larvae of *Trichinella spiralis* by human granulocytes *in vitro. J Clin Invest* **64:**1415–1422, 1979.
81. Kazura JW, Aikawa M: Host defense mechanisms against *Trichinella spiralis* infection in the mouse: Eosinophil-mediated destruction of newborn larvae *in vitro. J Immunol* **124:**355–361, 1980.
82. Gleich GJ, Loegering DA: Immunobiology of eosinophils. *Annu Rev Immunol* **2:**429–459, 1984.
83. Bass DA, Szejda P: Mechanisms of killing of newborn larvae of *Trichinella spiralis* by neutrophils and eosinophils. *J Clin Invest* **64:**1558–1564, 1979.
84. Kazura JW, deBrito P, Rabbege J, et al: Role of granulocyte oxygen products in damage of *Schistosoma mansoni* eggs *in vitro. J Clin Invest* **75:**1297–1307, 1985.
85. Jong EC, Mahmoud AAF, Klebanoff SJ: Peroxidase-mediated toxicity to schistosomula of *Schistosoma mansoni. J Immunol* **126:**468–471, 1981.
86. Buys J, Wever R, Ruitenberg EJ: Myeloperoxidase is more efficient than eosinophil peroxidase in the *in vitro* killing of newborn larvae of *Trichinella spiralis. Immunology* **51:**601–607, 1984.

87. David JR, Butterworth AE, Remold HG, et al: Antibody-dependent, eosinophil-mediated damage to ^{51}Cr labeled schistosomula of *Schistosoma mansoni:* Effect of metabolic inhibitors and other agents which alter cell function. *J Immunol* **118:**2221–2229, 1977.
88. Goetzl EJ: The conversion of leukotriene C4 to isomers of leukotriene B4 by human eosinophil peroxidase. *Biochem Biophys Res Commun* **106:**270–275, 1982.
89. Ziltener HJ, Chavaillaz PA, Jörg A: Leukotriene formation by eosinophil leukocytes. Analysis with ion-pair high pressure liquid chromatography and effect of the respiratory burst. *Hoppe-Seylers Z Physiol Chem* **364:**1029–1037, 1983.
90. Henderson WR, Chi EY, Klebanoff SJ: Eosinophil peroxidase-induced mast cell secretion. *J Exp Med* **152:**265–279, 1980.
91. Clark RA: Extracellular effects of the myeloperoxidase–hydrogen peroxide-halide system, in Weissmann G (ed): *Advances in Inflammation Research.* New York, Raven, 1983, vol 5, p 107.
92. Jong EC, Klebanoff SJ: Eosinophil-mediated mammalian tumor cell cytotoxicity: Role of the peroxidase system. *J Immunol* **124:**1949–1953, 1980.
93. Nogueira NM, Klebanoff SJ, Cohn ZA: *T. cruzi:* Sensitization to macrophage killing by eosinophil peroxidase. *J Immunol* **128:**1705–1708, 1982.
94. Agosti JM, Ayars GH, Altman LC, et al: Eosinophil peroxidase–hydrogen peroxide–halide mediated pneumocyte injury. *J Allergy Clin Immunol* **77:**235, 458a, 1986.
95. Locksley RM, Wilson CB, Klebanoff SJ: Role for endogenous and acquired peroxidase in the toxoplasmacidal activity of murine and human mononuclear phagocytes. *J Clin Invest* **69:**1099–1111, 1982.
96. Zabucchi G, Menegazzi R, Soranzo MR, et al: Uptake of human eosinophil peroxidase by human neutrophils. *Am J Pathol* **124:**510–518, 1986.

8

The Respiratory Burst and Lymphocyte Function

ARTHUR L. SAGONE, Jr.

1. INTRODUCTION

Recent studies have established that phagocytic cells generate several reactive oxygen species (ROS).[1–24] Normally, under resting or baseline conditions, unstimulated granulocytes and monocytes appear to release only minimal, and probably nontoxic, amounts of ROS. However, during an immunological response, phagocytic cells may have a burst in oxidative metabolism (a respiratory burst), resulting in an enhanced release of ROS at an inflammatory site.[25–27] Since lymphocytes are an essential part of the inflammatory response, the potent compounds released by phagocytic cells might modify the functional capacity of these cells. This chapter discusses the possible effects of the ROS produced by phagocytic cells during the respiratory burst on the functional capacity of human lymphocytes at a site of inflammation.

Two potential effects might be anticipated. First, the overall effect might result in enhanced lymphocyte activity. Under these circumstances, the stimulated phagocytic cell would provide a helper cell function. Alternatively, the compounds released could impair lymphocytes function. Under these conditions, the stimulated phagocytic cell would exhibit suppressor cell activity.

The potential interaction between the ROS released by phagocytic cells and lymphocytes at an inflammatory site appears to be markedly complex and the effects of ROS on the functional capacity of lymphocytes could be modified by a number of factors generated during an immune response. Under most circumstances, the stimulus that induces a metabolic burst in phagocytic cells also stimulates simultaneously other independent cellular events. These include chemotax-

ARTHUR L. SAGONE, Jr. • Department of Medicine and Pharmacology, Division of Hematology and Oncology, Ohio State University, College of Medicine, Columbus, Ohio 43210.

is,[28,29] the release of lymphokines[30,31] and enzymes with diverse activity,[32,33] and the generation of prostaglandins and leukotrienes from arachidonic acid.[34–37]

In these complex systems, altered lymphocyte function could occur as a consequence of direct interaction of the ROS with the lymphocyte or indirectly due to the interaction of the ROS with another compound produced simultaneously by the phagocytic cell. For example, the ROS might react with a lymphokine, enzyme, or a prostaglandin, which then interacts with the lymphocyte resulting in an altered function. Furthermore, in some cases, the ROS could react with factors released from lymphocytes and required for optimal lymphocyte function.

Of particular importance in considering the effects of the ROS produced by phagocytic cells on the function of lymphoid cells is the relative sensitivity of subpopulations of these cells to oxidant damage.

2. CHARACTERISTICS OF THE METABOLIC BURST

2.1. Nature of the Metabolic Burst in Phagocytic Cells

We have discussed in detail the nature of the biochemical reactions associated with the metabolic burst occurring in the granulocyte following its activation by an appropriate stimulus (see Chapter 12, this volume). The metabolic burst is associated with an immediate release of several reactive oxygen species by these cells, including superoxide (O_2^-),[4] hydrogen peroxide (H_2O_2),[18] hydroxyl free radical ($OH\cdot$),[38–40,42] hypochlorous acid,[19–21] and the hypochlorous-derived stable oxidants-chloramines.[21–23] The stimulation of monocytes is associated with a similar rapid activation of oxidative metabolism.[15,41] The effect of the tumor promoter phorbol myristate acetate (PMA) on the HMPS activity and H_2O_2 production of blood mononuclear cell suspensions containing approximately 20% monocytes is given in Fig. 1. As seen, there is prompt activation of the HMPS pathway, as indicated by a marked increase in [1-^{14}C]glucose oxidation, and H_2O_2 production as indicated by enhanced [^{14}C]formate oxidation.

Overall, most studies indicate that stimulated monocytes release lower amounts of ROS than granulocytes.[6,15,42–48] These include the production of H_2O_2, HOCl, and $OH\cdot$.

In addition to granulocytes and monocytes, lymphocytes may also be exposed to macrophages at an inflammatory site, particularly in chronic infections. In this regard, there appear to be important differences in the oxidative metabolism of the macrophages compared to the other phagocytic cells. The functional and biochemical characteristics of animal macrophage from a variety of tissues has been the subject of considerable investigation.[49–54] These studies indicate an importance of O_2^-, H_2O_2, and possibly $OH\cdot$ radical in the functional capacity of these cells. However, macrophages appear to lack the capacity to produce HOCl even after activation because they have relatively little myeloperoxidase activity.[44,55] In contrast to animal macrophages, the oxidative metabolism of human macrophages is

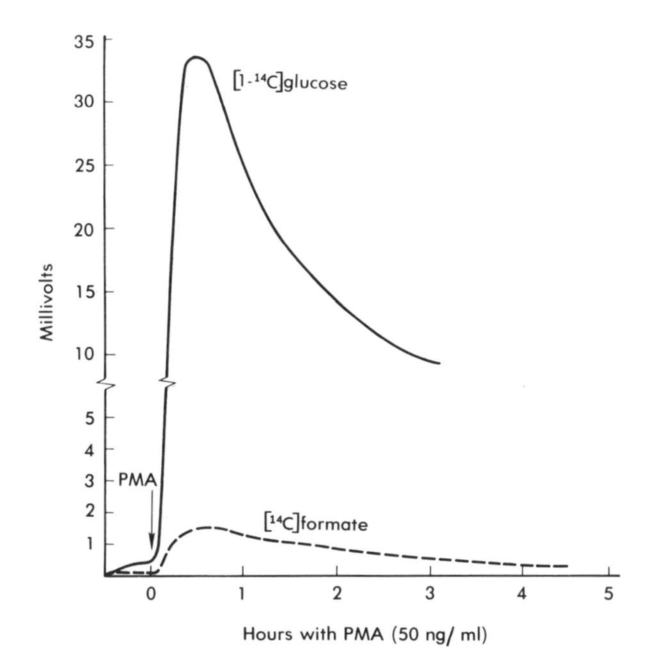

FIGURE 1. Oxidation of [1-^{14}C]glucose (——) and [^{14}C]formate (- - -) by suspensions of mononuclear cells (20 × 10^6 in 4.5 ml buffer). The curves represent a continuous measurement of $^{14}CO_2$ production from each substrate, using an ionization chamber electrometer system. The *y* axis indicates the electrical signal (in mV) generated by the $^{14}CO_2$ passing through the chamber. After steady state conditions were established, PMA was added as indicated by the arrow in a final concentration of 0.05 μg/ml. (From Sagone et al.[41])

less well delineated. Our knowledge of the biochemistry of this cell is derived primarily from the study of human alveolar macrophages.[42,44,56] Of importance is that these cells appear to release substantial amounts of O_2^- and H_2O_2, even under resting conditions. The alveolar macrophages derived by bronchopulmonary lavage from a normal person produces significant amounts of H_2O_2.[44,56] Similar to other phagocytic cells, the stimulation of human macrophages is associated with a burst in hexose monophosphate shunt pathway activity, increased oxygen consumption, and enhanced O_2^- and H_2O_2 production. Human macrophages appear to produce significantly less OH· than do other human phagocytic cells[42,44] and little HOCl. However, the O_2^- and H_2O_2 released by unstimulated macrophages might directly alter lymphocyte function while a similar alteration by PMN or monocytes may require stimulation of these cells.

Of particular interest is the capacity of phagocytic cells to become activated in vivo.[57–68] This activation is associated with an enhanced capacity to generate ROS. Phagocytic cells may become activated by a variety of mechanisms, such as exposure to interferon, lipopolysaccharide, and immune complexes.[69–74] While it may be somewhat confusing, the difference between activation and stimulation for the purposes of this discussion should be noted. Stimulation indicates the enhanced immediate release of ROS, enzymes, and so on, following the interaction of a phagocytic cell with an appropriate stimulus. Within this context, an activated cell, when stimulated, would release increased amounts of ROS, enzymes, lymphokines, and possibly new compounds compared with the nonactivated cells. Under some circumstances, however, the differences between stimulation and activation become somewhat hazy or arbitrary. For example, the resident human alveolar macrophage in healthy persons may already be activated as a consequence of chronic exposure to the noxious agents in our atmosphere.

2.2. Generation of ROS by Lymphocytes

Thus far, I have discussed the metabolic burst occurring in phagocytic cells following interaction with an appropriate stimulus. However, a fundamental physiological question arises: Does the lymphocyte or a specific subpopulation, such as T cells, B cells, or NK cells have the capacity to generate ROS in a manner similar to that of PMN, monocytes, and macrophages? Several groups have studied the oxidative metabolism of purified lymphocyte preparations.[15,45,46,48] Agents that stimulate the oxidative metabolism of phagocytic cells do not stimulate the oxidative metabolism of lymphocytes. This includes agents such as PMA and merezein, which do not require specific membrane receptors.[48] Therefore, most evidence indicates that human lymphocytes lack NADPH oxidase activity and do not have the capacity to undergo an immediate burst in oxidative metabolism with the generation and a release of ROS characteristic of phagocytic cells.

Whereas lymphocytes do not appear to have the immediate burst in oxidative metabolism following interaction with a variety of stimuli the stimulation of oxidative metabolism does occur in lymphocytes under some circumstances. For example, it has been known for some time that lymphocyte cultures undergoing lymphoblastic transformation to nonspecific mitogens have a marked augmentation of HMPS activity, which appears to start late the first day of culture and is greatest during the second day.[75–78] However, it is not clear whether the enhanced HMPS response is associated with the increased cellular production of ROS or relates to a specific requirement of the HMPS for cellular division.

There is some evidence that lymphocytes may produce OH· or another ROS under some conditions.[79–81] The addition of hydroxyl scavengers to mitogen-stimulated lymphocyte cultures impairs their proliferation, suggesting the generation of OH· under these conditions.[79] It has been suggested that OH· may be related to the NK activity of lymphocytes.[80] This conclusion is based on the capacity of hydroxyl scavengers to inhibit the NK activity of lymphocyte cultures. As discussed in Chapter 12 (this volume), OH· might be produced in lymphocytes during arachidonic metabolism or by their microsomal enzyme systems. Thus far, lymphocytes have not been shown to produce OH· directly using specific assays that measure the production of OH· in biological systems.[80] This may relate to the sensitivity of the assays available to measure this ROS.

2.3. Stimuli That Induce a Metabolic Burst in Phagocytic Cells

A number of diverse stimuli are known to induce a respiratory burst in phagocytic cells in vitro. In this regard, specific membrane receptors are of particular importance in the stimulation of the oxidative metabolism of phagocytic cells. The most notable of these are the Fc and complement receptors present on the membrane of all phagocytic cells.[82–90] Stimulation of these receptors would appear to represent an important final common pathway for the stimulation of the oxidative metabolism of phagocytic cells in vitro by a variety of stimuli. These include soluble immune complexes, insoluble immune complexes,[3,32,33] opzonized infectious organism such as bacteria[5,90] and viruses,[91,92] and antibody-sensitized target cells such as RBC and tumor cells.[93–99] A similar activation of phagocytic cells may occur due to immune-complex absorption on lymphocytes.[100] These stimuli trigger a metabolic burst by stimulation of the Fc receptor. In most cases, the simultaneous absorption of complement components with the antigen–antibody complex appears to enhance the overall response apparently by simultaneous stimulation of the complement and Fc receptors.[82,90,94] However, activated complement alone may trigger a metabolic burst in phagocytic cells in vivo. The complement stimulation of granulocyte metabolism may be related to the pulmonary distress syndrome reported in patients following hemodialysis and WBC cell transfusions.[101] Furthermore, the activation of serum by snake venom is known to induce ARDS in animals.[102,103] In this regard, a well-studied model for the stimulation of a respiratory burst in phagocytic cells by complement is the incubation of these cells with serum opsonized zymosan particles.[15]

The respiratory burst can be also induced by other mechanisms. The chemotactic peptide FMLP appears to stimulate a metabolic burst and enzyme release in phagocytes by interaction with a specific membrane receptor,[29] while compounds such as PMA and merezein appear to stimulate oxidative metabolism by direct acfivation of the enzyme C-kinase, thereby bypassing the need for the membrane stimulation.[104]

It is reasonable to conclude that a number of these stimuli will induce release of ROS by phagocytes at an inflammatory site in vivo. Such reactions have been postulated in patients with autoimmune disorders, bacteria and viral infections, as well, as patients with neoplasia.

A few additional comments seem appropriate when considering the potential role of the ROS released by specific phagocytes in modifying lymphocytic function in vivo. The first question is whether there is selective activation of specific phagocytic cells by immune complexes or sensitized target cells in vivo. For example, monocytes appear to be able to interact with target cells sensitized by low concentrations of antibody.[82] Monocytes bind and lyse RBC sensitized by the serum antibodies present in patients with autoimmune hemolytic anemia (IGG subtypes 1 and 3),[105] and RBC sensitized by Rhogam antibody (ADCC).[99] The interaction is associated with a metabolic burst in the monocyte, suggesting that ADCC is mediated in part by the ROS generated by these cells.[99] By contrast, granulocytes do not mediate ADCC in this system and are not stimulated. However, when granulocytes are incubated with target cells sensitized with high concentrations of antibody (i.e., human RBC sensitized with anti-A or anti-B antibody or rabbit RBC sensitized with sheep antirabbit antibodies), they mediate ADCC and have a metabolic burst similar to monocytes.[95,106]

Second, the capacity of unopsonized target cells to trigger an oxidative burst in phagocytic cells may vary considerably. The phagocytosis of unopsonized bacteria may be associated with some stimulation of the oxidative metabolism of the phagocyte cell although the degree response is much lower than that induced by opsonized organisms.[57,107] However, this is not always the case.[90,108] For example, histoplasma capsulation fails to trigger a metabolic burst in macrophages.[108] The failure of phagocytic cells to generate ROS may relate to the resistance of this organism to intracellular killing. Currently, there is little information concerning the capacity of viruses to induce the release of ROS in phagocytes. Abramson et al.[109] demonstrated that influenza virus stimulates a metabolic burst in both granulocytes and monocytes. We have demonstrated that the unopsonized mumps particles used for the mumps skin testing induce a oxidative metabolic burst in human monocyte suspensions but not granulocyte suspensions.[92] By contrast, we were unable to demonstrate a similar stimulation with inactivated feline leukemia virus[110] or inactivated poliovirus (unpublished observations). These observations suggest considerable variability in the capacity of some infectious agents and antigens to stimulate a oxidative burst in phagocytic cells. This seems to be a significant point, since the release of ROS by the phagocytic cell in addition to lymphokines might alter the primary immune response as a consequence of the capacity of the ROS to modify the functional capacity of lymphocytes.

3. EVIDENCE INDICATING THAT THE ROS PRODUCED BY PHAGOCYTIC CELLS MAY ALTER THE FUNCTIONAL CAPACITY OF HUMAN LYMPHOCYTES

Several in vitro model systems have addressed the effects of oxidant stress on the functional capacity of human lymphocytes. These systems include the effects of (1) periodate on lymphocyte proliferation, (2) hyperoxia on the proliferation of lymphocyte cultures, (3) the reactive oxygen species generated by enzyme systems on the T- and B-cell function of lymphocytes and finally, (4) oxidant injury in the radiation damage to lymphocytes. These systems have provided useful observations concerning the possible effects of the reactive oxygen species produced by phagocytic cells on a variety of lymphocyte function in vivo. For the most part, the results of these studies support the conclusion that oxidant stress to the lymphocyte in vivo may be associated with impaired T- and B-cell functions and further that these cells may be particularly sensitive to oxidant damage. The information generated in these model systems is discussed briefly in the following sections.

3.1. Effect of Hyperoxia on the Proliferation of Human Lymphocyte Cultures

Human blood lymphocytes that normally will not grow in culture can be induced to proliferate after stimulation with nonspecific mitogens such as phytohemogglutinin (PHA), concanavalin A (Con A), and pokeweed mitogen. The proliferation of such cultures appears optimal under experimental conditions in which the oxygen tensions in the atmosphere used to grow the cultures approximate those occurring in vivo.[111–116] In 1968, Anderson demonstrated that PHA stimulated human lymphocyte cultures incubated under an atmosphere containing only a small increase in oxygen content (30%, rather than the usual 20%) had significantly impaired lymphoblastic transformation, and the cultures incubated under 100% oxygen had virtually no proliferation at all.[111,112] This impaired proliferation could not be explained by a decreased viability of the cells, and no obvious morphological abnormality could be demonstrated in the lymphocytes, by electron microscopic examination. Studies in our laboratory have similarly demonstrated that hyperoxia impairs the LBT of human lymphocytes.[115] We have also characterized the carbohydrate metabolism of lymphocytes cultures under these conditions. Under hyperoxia, unstimulated cultures have an increased rate of glycolysis, a slight decrease in [1-^{14}C]glucose oxidation, and a significant decrease in [6-^{14}C]glucose oxidation (Table I). When the rate of [1-^{14}C]glucose oxidation is corrected for the amount oxidized by Krebs, or tricarboxylic acid (TCA) cycle activity, there is no significant

TABLE I. Mean Rates of Glucose Consumptions and $^{14}CO_2$ Production of 3-day Unstimulated and Mitogen-Stimulated Lymphocyte Cultures under Air and Oxygen[a]

	Unstimulated			PHA-stimulated		
	Glucose consumption	[1-^{14}C]glucose	[6-^{14}C]glucose	Glucose consumption	[1-^{14}C]glucose	[6-^{14}C]glucose
	Cultures incubated under air, 5% CO_2					
Mean	83	5.7	0.90	644	17.0	1.5
SD	35	0.28	0.15	70	3.3	0.17
N	6	3	3	7	4	3
Range	49–101	5.5–6.0	0.7–1.0	552–746	14.4–21.8	1.4–1.7
	Cultures incubated under 95% oxygen, CO_2					
Mean	136	5.0	0.33	417	7.9	0.33
SD	51	0.25	0.06	55	1.4	0.06
N	6	3	3	7	4	3
Range	105–225	4.7–5.2	0.3–0.4	350–503	6.3–9.7	0.3–0.4
p[b]	0.1	0.05	0.01	0.01	0.01	0.01

[a]Values are given in nanomoles/10^7 cell per hr.
[b]The data were analyzed according to tests for dependent or independent samples. Results are expressed as mean ± standard deviation (SD). (From Sagone.[115])

difference between the HMPS pathway activity of hyperoxic cultures and those grown under air. These data suggest that hyperoxia may cause subtle damage to the Krebs cycle of these cells without altering glycolysis or the activity of the HMPS pathway. Mitogen-stimulated cultures incubated under hyperoxia failed to demonstrate the augmentation in carbohydrate metabolism characteristic of cultures incubated under normal oxygen tensions.[75–77] This observation indicates that the impaired DNA synthesis of the cultures is not the only metabolic defect induced by hyperoxia and that the impaired carbohydrate metabolism of the hyperoxic cultures is an additional measure of the oxidant damage.

It is not clear whether the impaired carbohydrate metabolism induced by hyperoxia is the major cause for the impaired lymphocyte function. The effects of hyperoxia on the Krebs cycle activity of the unstimulated lymphocyte cultures suggest this possibility. However, the impaired DNA synthesis and carbohydrate metabolism may reflect other cellular defects induced by hyperoxia. These might include damage to one or more of the biochemical pathways required for normal cellular proliferation.

Presumably, the damage to lymphocytes caused by hyperoxia is mediated by the enhanced intracellular and extracellular production of toxic oxygen metabolites from oxygen. These would include an enhanced production of superoxide, hydrogen peroxide and hydroxyl free radical.[117] These are the same reactive oxygen species known to be generated by phagocytic cells during a metabolic burst. The relatively small increase in PO_2 (150 to 225 mm Hg) required to impair the proliferation of mitogen-stimulated human lymphocytes suggests that these cells are particularly sensitive to oxygen damage and indicates that lymphocytes may have suboptimal concentrations of cellular antioxidants and/or the protective enzyme system required to degrade ROS rapidly, particularly in sensitive cellular sites. The mammalian cell is known to have several mechanisms that normally protect it against oxidant damage by ROS. These defense mechanisms include the cellular enzymes superoxide dismutase, catalase, and glutathione peroxidase.[117–127] This latter enzyme requires glutathione as a cofactor, and is linked to the hexose monophosphate shunt pathway by the enzyme glutathione reductase. In addition, the mammalian cell has several antioxidants such as vitamins E and C. Lymphocytes have similar concentrations of superoxide dismutase, glutathione peroxidase, and glutathione reductase, compared with other blood cells.[120,121] However, lymphocytes have low concentrations of catalase which may relate directly to their sensitivity to oxidant damage.[122] It is clear that glutathione and glutathione peroxidase are important in cellular protection against oxidant damage to ROS. Cells deficient in glutathione or its related enzymes have increased sensitivity to oxidant damage even in cases in which there is an adequate cellular concentration of catalase.[118,119,124,125] However, the glutathione system may not be sufficient to protect the cell against H_2O_2 injury in the absence of cellular catalase, particularly in situations in which there is significant compartmentalization of these enzymes in the cell.

A second reason for the increased sensitivity of lymphocytes to oxidant damage may be related to a low cellular concentration of vitamin C. Vitamin C has the capacity to degrade a number of ROS rapidly, including OH·.[120] Lymphocytes have been reported to take up less vitamin C from the extracellular medium and to have decreased capacity to reduce dehydroascorbate compared with other blood cells.[120]

In summary, the effects of hyperoxia on the proliferation of human lymphocytes demonstrate three points: (1) lymphocytes appear sensitive to oxidant damage, (2) oxidant damage under these conditions does not induce cell death but rather nonlethal cellular damage, and (3) these observations suggest that a relatively minor increase in the release of ROS by phagocytic cells at an inflammatory site might suffice to impair lymphocyte function.

3.2. Role of ROS in Radiation Damage to Lymphocytes

Human lymphoid cells appear to be relatively sensitive to damage by radiation compared with other blood cells and have been demonstrated to have impaired function following low-dose irradiation in vitro.[128–132] A similar impairment in function has been reported in the blood lymphocytes of patients who have received radiation therapy.[133–135] The mechanism of radiation damage to mammalian cells is complex and appears to involve multiple mechanisms of damage to the cell that can be modified by several repair mechanisms.[136–148] It has been established for some time that oxygen enhances the radiation damage occurring in most cells, including those of mammals—the oxygen-enhancing ratio (OER).[136–140,147,148] This OER is variable from tissue to tissue. The mechanism for the OER is unclear. This OER may be mediated by oxygen itself due to the formation of peroxy radicals[136] (see Chapter 12, this volume, concerning the Howard Flanders model of oxygen mediated radiation damage). Alternatively, it has been suggested that this oxygen-related damage may be mediated by O_2^-, H_2O_2, or additional OH· generated by a Haber–Weiss mechanism.[137–140,144–148] This point is currently controversial and the subject of considerable study.

The importance of oxygen in the sensitivity of blood lymphocytes to radiation damage is illustrated in Fig. 2. Following low-dose irradiation in vitro, human blood mononuclear cells display impaired lymphoblastic transformation to nonspecific mitogens. As shown, the sensitivity of human lymphocytes to damage by low-dose irradiation was enhanced markedly by oxygen. This observation indicates that either oxygen itself or a ROS produced from oxygen during radiation has the capacity to cause major damage to the lymphocyte.

The sensitivity of the lymphocytes to this oxygen-mediated radiation damage, similar to their sensitivity to oxidant damage under other conditions, may also relate to a decreased capacity of these cells to rapidly degrade ROS.[120]

There is evidence that GSH may protect complex cells against radiation damage.[147–149] This possibility has been raised by a number of observations. One is the study of the effects of radiation on bacteria and human fibroblasts deficient in GSH.[123,149] These mutant cells are equally sensitive to radiation under anaerobic and aerobic conditions and the degree to damage under both conditions is similar to

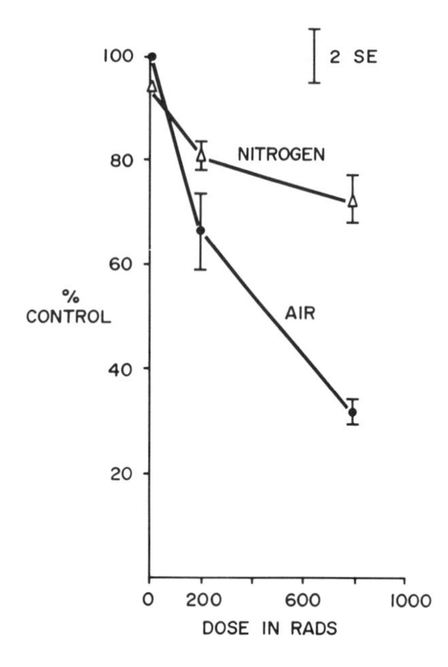

FIGURE 2. Comparison of lymphoblastic transformation to PHA after the irradiation of human lymphocyte cultures under an air and nitrogen atmosphere. DNA synthesis was quantitated by the [^{3}H]thymidine method. Results are expressed as uptake of unirradiated air control. (From Roberts et al.[130])

that occurring to the normal cells under oxygen. In addition, a number of studies indicate that the sensitivity of mammalian cells to radiation injury can be changed by modifying their sulfhydryl content.[123,147,148] In general, increasing the cellular concentrations of sulfhydryls is associated with resistence, while agents that deplete cellular sulfhydryl compounds appear to enhance sensitivity. In some cases, depletion of cellular GSH causes a substantial reduction in the OER. In other cases, GSH depletion is associated with sensitization of both aerated and hypoxic cultures resulting in no net change in the OER. Overall, the effect of sulfhydryl depletion may depend on the cell type used for study.

In this regard, GSH may also protect the lymphocyte against radiation damage. Diamide, an agent that rapidly oxidizes GSH, enhances the sensitivity of lymphocytes (Fig. 3) to radiation damage under hypoxic conditions but not aerobic conditions.[150] Similar to other tissues, the mechanism by which GSH protects the lymphocyte against radiation damage is unclear but may relate to its capacity to reduce organic radicals formed by OH·.[130,150]

3.3. Effect of Chemical Oxidation on Lymphocyte Function

About the same time that Anderson demonstrated that mild hyperoxia impairs the proliferative response of lymphocytes to nonspecific mitogens, several groups reported that human lymphocytes could be induced to proliferate by periodate, a potent oxidizing agent.[151–154] This activation may be secondary to oxidation of the terminal sialic acid of membrane glycoproteins or glycolipids to an aldehyde, result-

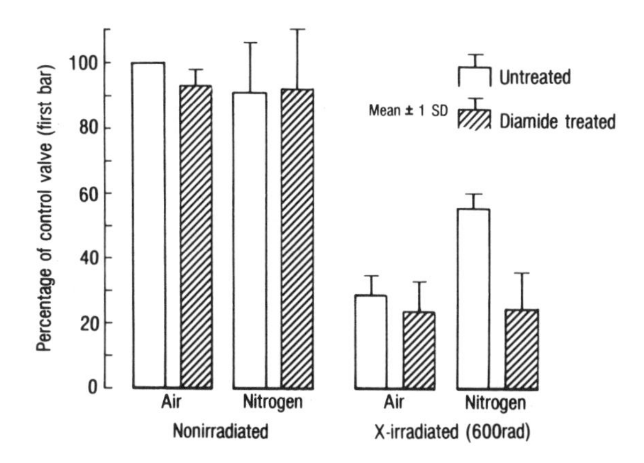

FIGURE 3. Effect of diamide and irradiation on the lymphoblastic transformation (LBT) of human lymphocyte cultures to PHA. The [^{3}H]Thymidine uptake during the final 4 hr of the 72-hr culture was used to quantitate DNA synthesis. Bars indicate mean ±SD of three experiments. Experimental values are expressed as percentage uptake of the unirradiated control under air (bar 1). The LBT of unirradiated lymphocyte cultures preincubated under hypoxic conditions (bar 3) were similar to the controls preincubated under air (bar 1). These control values range from 4–5 $\times$ 10^4 cpm/0.2 $\times$ 10^6 lymphocytes. Preincubation of lymphocyte suspensions with diamide under air (bar 2) or hypoxic conditions (bar 4) did not impair LBT. Following X-irradiation (600 rad) in an air atmosphere, diamide-treated suspension (bar 6) had an impairment in LBT similar to that in the irradiated suspensions without diamide (bar 5). Suspensions irradiated under hypoxic conditions were protected (bar 7). However, suspensions incubated with diamide under nitrogen (bar 8) had an impairment in LBT similar to suspensions incubated under air (bars 5 and 6). (From Sagone.[150])

ing a cellular signal for mitogenesis.[151] Periodate-induced proliferation requires careful control of the experimental conditions and is favored by low temperature, low concentrations of periodate, a short incubation time, and low pH. The excessive oxidation of the membrane by periodate results in a decreased response.[151] A similar induction of proliferation occurs following the incubation of lymphocytes with neuraminidase and galactose oxidase.[151] This treatment also results in the production of membrane aldehydes from galactose. Agents that react with aldehydes impair the response.[152] Therefore, it seems well established that membrane aldehydes are involved in chemical mitogenesis. Chemical mitogenesis is associated with the production of interleukin-2 (IL-2) and the induction of IL-2 receptors on sensitive T cells.[155] Therefore, this process appears to be analogous to the proliferation induced by nonspecific mitogens. Other lymphocyte functions might be similarly activated. It has also been reported that oxidation of cells with periodate or neuramindase–galactose oxidase induces enhanced lymphocyte cytotoxicity.[151]

Overall, the results of these experiments suggest that under some circumstances mild oxidation of the glycoconjungates on the membrane of lymphocytes might induce enhanced lymphocyte function in vivo. Chemical mitogenesis appears to represent altered lymphocyte function as a consequence of an extracellular oxidant reacting primary with the cell membrane. The type of oxidant stress in system therefore is different from that occurring under hyperoxic conditions or as a consequence of radiation in which enhanced amounts of oxidants may be generated both intracellularly or extracellularly.

3.4. Effects of the ROS Generated by Enzyme Systems on the Functional Capacity of Lymphocytes in Vitro

Lymphocytes may be exposed to high fluxes of several ROS during an immunological reaction in vivo. Under these conditions, the lymphocyte would be exposed to oxidants generated primarily outside the cell. Theoretically, this situation would favor the interaction of the ROS with the membrane but, in some cases, particularly under conditions in which O_2^-, H_2O_2, or chloramines might be present in a high extracellular concentrations, significant amounts of the ROS could reach sensitive intracellular sites, causing oxidant damage. Under these conditions, damage could occur to the membrane and/or sensitive intracellular sites. Several methods have been used to induce oxidant damage to lymphocytes in vitro. These include xanthine oxidase enzyme system, the glucose–glucose oxidase enzyme system, the H_2O_2–myeloperoxidase enzyme system, and H_2O_2. This approach has permitted direct evaluation of the capacity of ROS to damage lymphocytes in the absence of other factors that may be released by phagocytic cells during cell–cell interaction. The results of these studies indicate that ROS impair several of the functions of lymphocytes and are discussed below.

3.4.1. Effect of Oxidant Damage on the Proliferation of Lymphocytes to Nonspecific Mitogens

Shortly after the reports that hyperoxia and chemical oxidation altered lymphocyte proliferation, we became interested in the possible effects of the ROS generated by phagocytic cells on the functional capacity of normal blood human lymphocytes.[156] Initially, we chose to study the effect of the ROS generated by the xanthine–xanthine oxidase system (X–XO) on the lymphoblastic transformation of human lymphocyte to nonspecific mitogens (PHA). This enzyme system seemed ideal, since it had already been reported to generate a sustained release of three of the ROS generated by phagocytic cells (O_2^-, H_2O_2, and OH·) and appeared analogous to the physiology of the metabolic burst known to occur in phagocytic cells.

The lymphoblastic transformation of mononuclear cell cultures was significantly impaired and delayed compared with controls following incubation with the X–XO enzyme system (Fig. 4). To identify the ROS causing the damage, we studied the protective effects of several agents that are scavengers of reactive oxygen species. Cultures supplemented with catalase were almost completely protected against the effect of the enzyme system. By contrast, cultures supplemented with superoxide dismutase and mannitol scavengers of O_2^- and OH·, respectively, were not. Therefore, the oxidant damage to the lymphocytes was mediated pri-

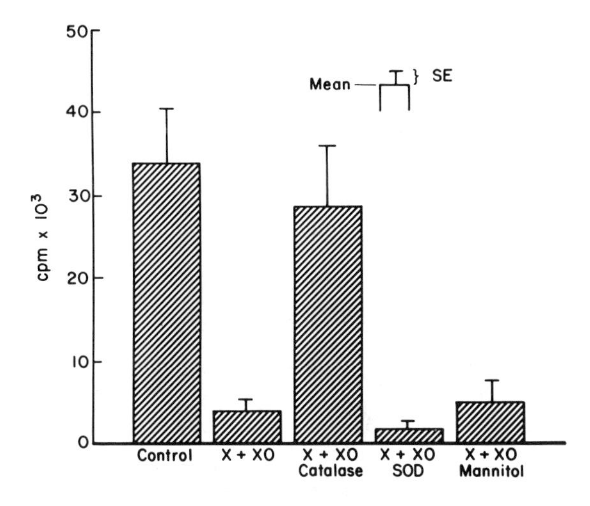

FIGURE 4. Effect of the ROS scavengers catalase, SOD, and mannitol on the transformation of lymphocyte cultures exposed to xanthine (X) and xanthine oxidase (XO). Bar 1 indicates the transformation of cultures incubated with PHA alone (controls). Bar 2 indicates cultures incubated with the PHA and the enzyme system. The remaining cultures were supplemented with catalase (bar 3), SOD (bar 4), or mannitol (bar 5). The scavengers alone did not alter the proliferation of the cultures in the concentrations used. (From Sagone et al.[156])

marily by H_2O_2 and not by the O_2^- or the OH· generated by the enzyme system. Recently, Zoschke and Messner[46] reported a similar impairment of LBT to concanavalin A following the incubation of human blood mononuclear cell with the X–XO enzyme system. They also concluded that H_2O_2 was the ROS mediating most of the damage. However, they also believed that OH· might be related to some of the damage, since ethanol as well as catalase was protective.

The impaired LBT cannot be explained by a decreased cell viability since the carbohydrate metabolism of unstimulated cultures exposed to the X–XO enzyme system was similar to controls[156] (Fig. 5). However, PHA-stimulated cultures did not have the augmented carbohydrate metabolism following H_2O_2 injury characteristics of normal cultures. This abnormality is similar to the impaired metabolism noted in PHA-stimulated lymphocyte cultures under hyperoxia (see Section 3.1).

We also studied the capacity of the H_2O_2 generated by the X–XO oxidase enzyme system to permeate the lymphocyte. We demonstrated that there was a marked stimulation of the HMPS shunt activity of lymphocytes during incubation with the enzyme system indicating the active degradation of H_2O_2 by glutathione peroxidase (Fig. 6). The addition of catalase to the incubation prevented the augmented HMPS activity. This experiment confirmed that H_2O_2, and not another ROS produced by the enzyme system, was related to the stimulation. This observation also provides solid evidence that a significant intracellular influx of H_2O_2 and presumably O_2^- was occurring under these experimental conditions. By contrast, purified lymphocyte cultures incubated with X–XO enzyme system oxidized only

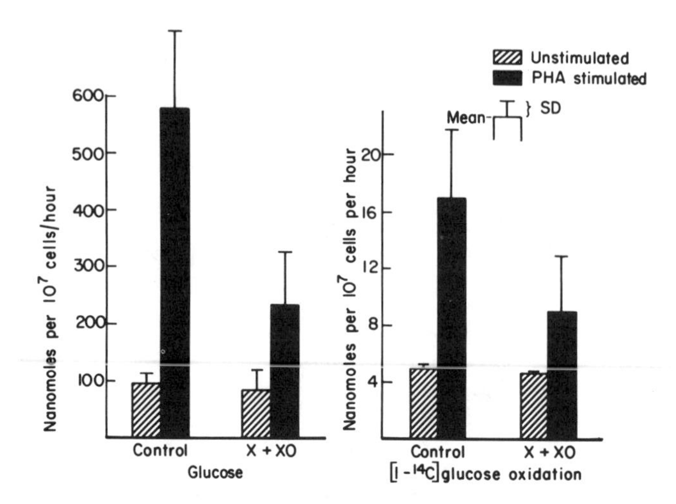

FIGURE 5. Effect of oxidant injury on the glucose metabolism of unstimulated and PHA-stimulated lymphocyte cultures. The mean glucose consumptions and [1-^{14}C]glucose oxidations of the cultures over the 3 days of incubation are given at the left and right, respectively. The values for the unstimulated cultures (three paired experiments) are indicated by the hatched bars and the values for mitogen-stimulated cultures (five paired experiments) by the solid bars. There is no significant difference in the rate of glucose consumption ($p > 0.2$) or [1-^{14}C]glucose oxidation ($p > 0.5$) in the unstimulated cultures incubated with the xanthine–xanthine oxidase (X–XO) enzyme system compared with the controls. The rate of glucose consumption ($p < 0.01$) and [1-^{14}C]glucose oxidation ($p < 0.05$) of the PHA-stimulated culture, incubated with the X–XO enzyme system, are significantly lower than the corresponding values of the PHA-stimulated controls. Values for the PHA-stimulated cultures are significantly higher than those of the corresponding unstimulated cultures. (From Sagone et al.[156])

small amounts of formate, suggesting little H_2O_2 degradation by intracellular catalase. Furthermore, the addition of azide to the cultures enhanced oxidant damage, suggesting an importance of intracellular catalase in the protection of the blood mononuclear cells against oxidant damage.

Other methods have been used to induce oxidant damage to lymphocytes in addition to the X–XO enzyme system. These include the addition of hydrogen peroxide to lymphocyte cultures,[41] the production of H_2O_2 by the glucose–glucose oxidase system,[46] H_2O_2 produced by the interaction of penicillamine and ceruloplasmin or copper,[157] and the H_2O_2–myeloperoxidase enzyme system.[158] In the latter system, both H_2O_2 and H_2O_2 generated by the glucose–glucose oxidase system have been used as a source of substrate. In all cases, H_2O_2 impairs the proliferation of lymphocytes to nonspecific mitogens.

In summary, H_2O_2 impairs the capacity of human lymphocytes to proliferate in response to nonspecific mitogens in a similar manner to hyperoxia (see Section 3.1) and by a mechanism that does not involve the significant death of the cell. It is of interest that this damage to lymphocytes occurs in spite of active intracellular

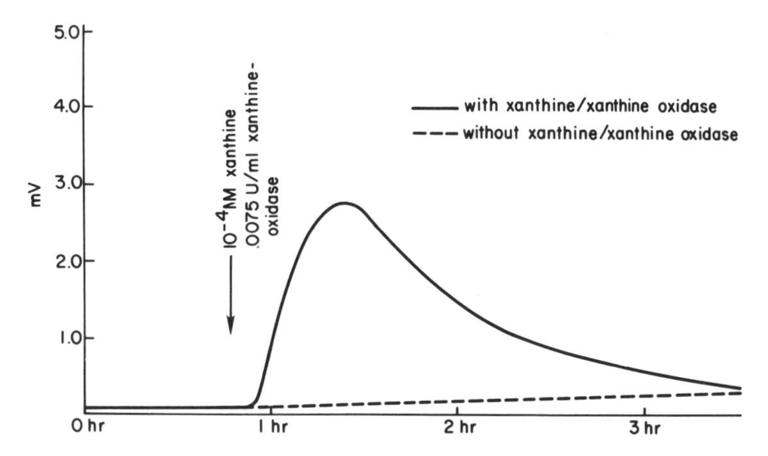

FIGURE 6. Effect of the xanthine and xanthine oxidase on the oxidation of [1-^{14}C]glucose by lymphocytes in short-term culture. The curves represent a continuous measurement of $^{14}CO_2$ production by lymphocyte suspensions in buffer using the ionization chamber electrometer system. The curves were drawn using the data points from a single experiment and are representative of four experiments performed. The *y* axis indicates the electric signal (in mV) generated by $^{14}CO_2$ in the ionization chamber. Signal generated by: untreated suspensions (- - -); suspensions supplemented with xanthine and xanthine oxidase (—). As indicated by the arrow, the addition of the enzyme system resulted in a prompt but transient increase in the oxidation of [1-^{14}C]glucose. The addition of substrate or the enzyme alone did not effect the oxidation of [1-^{14}C]glucose. (From Sagone et al.[156])

degradation of the H_2O_2 by the HMPS pathway of these cells and may relate to a low concentration of catalase in these cells.

3.4.2. Effect of ROS on the NK Activity and ADCC of Lymphocytes

Other studies have demonstrated that H_2O_2 similarly impairs other lymphocyte functions. These include the NK activity of lymphocytes against tumor target cells as well as their capacity to kill antibody-sensitized tumor cells (ADCC).[45,159–160] Also, H_2O_2 impairs the Staphylococcus protein A induction of human B-cell colony formation.[161] The concentrations of H_2O_2 required to impair these latter functions are in the range of 0.1 mM.[45,159–161] It is of interest that similar concentrations of H_2O_2 do not impair the capacity of monocytes to mediate ADCC suggesting that lymphocytes are more sensitive to oxidant damage than monocytes.[41] Furthermore, it has been suggested that the capacity of monocytes to rapidly degrade H_2O_2 may protect lymphocytes against oxidant damage.[157,158] This seems consistent with the observation that monocytes have a much higher concentration of catalase than lymphocytes.[122]

There is also evidence that H_2O_2 damage to T and B lymphocytes is associated with membrane alteration and/or impair microtubular function. These cells lose their capacity to bind sheep RBC (rosette assay), have a decreased percentage of cells with surface IGG and have impaired Con A capping following H_2O_2 injury.[162]

3.4.3. Possible Role of the Myeloperoxidase Enzyme System in Oxidant Damage to Lymphocyte

The H_2O_2–myeloperoxidase system has been used to induce oxidant injury to lymphocytes. This physiological enzyme system serves an important role in the capacity of phagocytic cells, particularly granulocytes, to kill microorganisms.[159,163] Recent studies indicate that the ROS produced by this enzyme system may also cause oxidant damage to a variety of target tissues.[159] These include endothelial cells, tumor cells, platelets, and RBC.[159,164,165] In addition, HOCl and chloramine have the capacity to inactivate or activate extracellular enzymes.[166,167] Under some circumstances, myeloperoxidase may be released from phagocytic cells in vivo. Therefore, significant concentrations of HOCl or the stable oxidants derived from HOCl (chloramine) might develop at an inflammatory site.[21–23] Either of these reactive species potentially could induce altered lymphocyte function.

Recently, El-Hag and Clark and co-workers studied the capacity of the H_2O_2 myeloperoxidase enzyme system to impair lymphocyte functions.[158,159] They found that the addition of myeloperoxidase significantly reduces the concentration of H_2O_2 required to induce oxidant damage to both T and B lymphocytes. These observations suggest the possible importance of this enzyme system in oxidant damage to lymphocytes during cell–cell interaction.

3.5. Possible Differences in the Sensitivity of T and B Cells to Oxidant Damage

A number of studies have suggested that T and B cells may have a different sensitivity to oxidant damage.[158,168] Furthermore, some subsets of T cells, such as suppressor cells, may be particularly sensitive.[132] Recent studies by El-Hag et al.[158] have compared the sensitivity of a variety of T- and B-cell function to oxidant damage. These workers studied the effect of H_2O_2 injury with and without myeloperoxidase on the NK activity of lymphocytes, the proliferation of lymphocytes to the nonspecific mitogens and the generation of immunoglobulins secreting cells by pokeweed. All these lymphocyte functions could be impaired by oxidant injury. In all cases, the injury was greatest when the cultures were incubated with both H_2O_2 and MPO. However, similar results were found with H_2O_2 alone when the concentration used in the cultures was high enough. The formation of antibody-secreting cells to pokeweed proved most sensitive, while the proliferative response to pokeweed mitogen was found most resistant. The sensitivity of the lymphocyte proliferative responses to PHA and Con A as well as NK appeared to be intermediate. El-Hag et al. suggested the following rank order of sensitivity: B (or T helper) $>$ NK $\geq$ T. These results suggest that there may be a range of sensitivities of T- and B-lymphocyte subpopulations to oxidant damage. The complex nature of these systems, however, makes it difficult in some cases to identify the precise mechanism involved in the altered response.

4. EVIDENCE THAT THE PRODUCTION OF ROS BY PHAGOCYTIC CELLS CAN ALTER LYMPHOCYTE FUNCTION

The results generated in the model systems discussed in Section 3 provide relatively solid evidence that oxidants may modify lymphocyte function. However, the obvious question is whether similar impairment of lymphocyte function occurs during cell–cell interaction in vivo in patients with a variety of diseases. The studies discussed in this section suggest that this is a likely possibility.

4.1. Evidence That the Release of ROS by Phagocytic Cells May Impair the NK Activity of Lymphocytes

A subset of T lymphocytes have the capacity to selectively recognize and kill target cells (NK activity)[169] and may therefore play an important role in the primary host defense, particularly against tumors, and infections. Similarly, a subset of lymphocytes is able to mediate cytotoxicity against the antibody-sensitized target cells (ADCC).[170] In 1981, Seaman et al.[169] reported that the NK activity of human blood mononuclear cells could be suppressed by the phorbol diester tumor promoters. The suppression of NK activity by these agents correlated with their potency as tumor promoters. The suppressive effect could be reduced if peripheral blood mononuclear cells (PBMC) were depleted of adherent cells indicating the importance of the monocyte in the suppressor activity.[169] In a subsequent study,[45] the same group evaluated the role of the ROS generated by monocytes as a mechanism for this suppressor activity. The NK activity of PMBC against the K 562 cells was decreased following incubation with (12-*o*-tetradecanoylphorbol-13-acetate) (TPA) or opsonized zymosan. Both agents stimulated the production of O_2^- and H_2O_2 by mononuclear cells. When PBMC were depleted of adherent cells, both the production of ROS and suppression of NK activity were virtually ablated. The activity could be restored by adding back either monocytes or granulocytes to the incubation. By contrast, the addition of granulocytes of patients with chronic granulomatous disease (CGG) that lack the capacity to generate ROS did not restore the suppressive effects. Scavenger studies were then done to identify the ROS primarily related to the damage. Catalase provided the most protection, while scavengers of OH· (mannitol, ethanol), were ineffective. Some protective effect of SOD was also observed. Azide, cyanide, and aminotriazole were also studied to evaluate the possible role of myeloperoxidase in the reaction. None of these agents provided a protective effect, and in fact the suppresion was enhanced in the experiments with zymosan. Therefore, MPO activity did not appear to be essential for the suppressor activity. It remains unclear what role the MPO pathway may play in the ROS induced injury to lymphocytes, even though injury by the H_2O_2 of MPO enzyme has been shown to impair several lymphocyte function (see Section 3.4). Furthermore, the possibility that the stable oxidants produced from this system (chloramines) may have the capacity to induce oxidant damage has, as yet, not been the subject of extensive study. This seems an important point since the half life of these ROS at an inflammatory site may be much longer than the other ROS generated by phagocytic cells.[21–23]

There are other reports confirming that stimulated phagocytic cells may mediate suppressor cell activity against lymphocyte NK.[171] In some studies, nonoxidative mechanisms have also been suggested for the phagocytic cell-mediated suppression of NK or ADCC activity.[172,173] This latter observation is consistent with other reports indicating that monocytes and granulocytes have two mechanisms for mediating cytotoxicity.[93,95,98,99,174]

The capacity of TPA to suppress the NK activity of PBMC has led to one additional area of investigation, the capacity of TPA to stimulate the NK activity of purified lymphocyte cultures.[171] Under these conditions, lymphocytes incubated with PMA have increased NK activity after approximately 24 hr in culture. The capacity of PMA to increase an activity is synergistic with other agents known to induce the NK activity of lymphocytes such as IFN or IL-2.

4.2. Evidence That ROS Released by Phagocytic Cells Can Alter Lymphocyte Proliferation

4.2.1. Experiments in Which Phagocytic Cells Have Been Stimulated to Release ROS during Coculture with Lymphocytes

Several recent studies have established that the ROS released by monocytes and granulocytes suppress human lymphocyte mitogenesis.[41,46,175] Zoschke and Messner studied the capacity of monocyte and granulocytes stimulated with a variety of agents to suppress lymphocyte mitogenesis to Con A. The agents studied included PMA, the chemotactic peptide *n*-formyl L-methionyl-leucyl,phenylalanine (FMLP), opsonized zymosan particles, and heat-aggregated γ-globulin.

The results with PMA are of particular interest, since this compound is known to be a potent stimulator of the oxidation metabolism of phagocytic cells (see Fig. 1), as well as a mitogen for human lymphocytes. This is not true of other mitogens such as PHA, Con A, and pokeweed mitogen. The PMA induced transformation of human blood mononuclear cells containing 20% monocytes and 80% lymphocytes could be augmented by the addition of catalase to the cultures, indicating that the H_2O_2 released by the monocytes could suppress mitogenesis. By contrast, the addition of scavengers of O_2^- and hydroxyl radical had no effect. As expected, the protective effect of catalase was not observed in PMA-stimulated lymphocyte cultures depleted of monocytes ($<5\%$) confirming that the monocyte were related to the suppressor activity. In contrast to PMA cultures, there was no augmentation of LBT in Con A-stimulated cultures supplemented with catalase. This observation confirms that Con A, similar to PHA, does not stimulate a significant metabolic burst or release of ROS in phagocytic cells. When PMA was added to Con A-stimulated cultures as a comitogen, the LBT was not greater than that observed with either mitogen alone. However, the addition of catalase to the cultures augmented their proliferation twofold, again indicating suppressor activity due to the release of H_2O_2 from monocytes. In this regard, the catalase had to be added early in the cultures to provide a protective effect. Similar results were found in experiments in which lymphocytes were cocultured with granulocytes rather than monocytes.

Suppression of LBT to Con A was also observed when lymphocyte–granulocyte cultures were stimulated with opsonized zymosan particles, FMLP, and aggregated immunoglobulin. Varying degrees of suppression were observed, depending on the stimulant used. Catalase added to the cultures again protected against this suppressive effect, although in some cases this protection was not complete. Finally, similar experiments were done using the blood cells of patients with CGD. As opposed to normals, the LBT of the lymphocyte cultures of CGD patients was not inhibited by PMA stimulation of their phagocytic cells. However, catalase-reversible suppression could be shown when normal phagocytic cells were cocultured with the lymphocytes of these patients.

We have also studied the capacity of the H_2O_2 released by PMA-stimulated blood mononuclear cell cultures to alter PMA-induced mitogenesis.[41] We found that the addition of catalase to PMA-stimulated cultures augmented their LBT (see Fig. 7). As expected, catalase did not alter the proliferation of mononuclear cell cultures stimulated by PHA, since the mitogen does not stimulate an immediate release of ROS.

Our measurements indicate the concentration of H_2O_2 reached values as high as 0.008 mM under our culture conditions. Since factor(s) released by monocytes also appears to augment lymphocytic transformation,[176] we evaluated the possibility that the monocyte might be autooxidized by its own ROS, thereby impairing its own function. However, we were unable to demonstrate impaired monocyte ADCC following incubation with 0.1 mM H_2O_2, a concentration known to ablate the NK and ADCC of lymphocytes.[160] Of interest is the observation that monocytes pretreated with azide and then exposed to H_2O_2 had a marked suppression of the ADCC activity ($<90\%$). This observation suggests that catalase protects these cells against oxidant damage; it also supports the idea that the greater sensitivity of lymphocytes to oxidant damage is related to a low concentration of catalase in these cells.

Overall, the results of the experiments discussed above indicate that the H_2O_2 released by phagocytic cells directly injures one or more subpopulation(s) of lymphocytes required for optimal T-cell proliferation. However, the possibility that the effect of low fluxes of H_2O_2 can augment the function of some cells such as suppressor cells cannot be excluded. Also, the effects could be secondary to an altered function of the lymphokines required for LBT. Furthermore, it is possible that oxidant damage may be associated with multiple sites of injury.

Other studies have documented that oxidant damage impairs B-cell function.

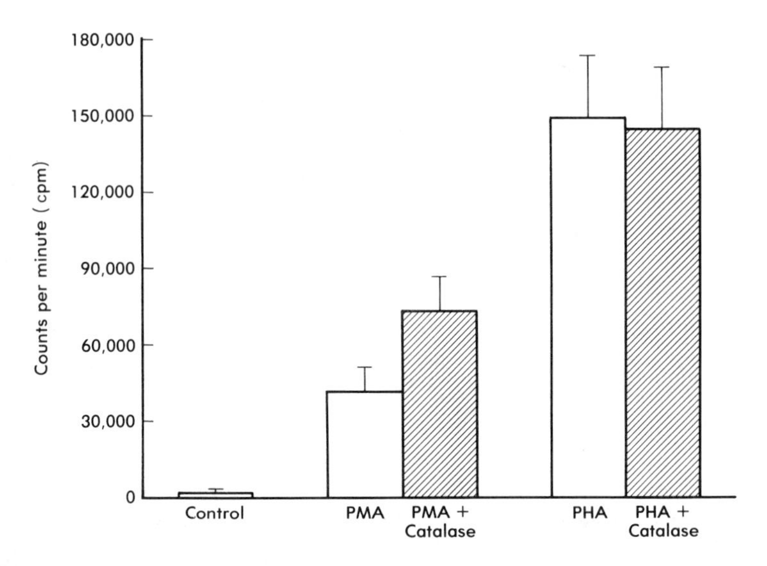

FIGURE 7. Effect of catalase on the LBT of mononuclear cells to PMA and PHA. The *y* axis indicates the thymidine uptake of the cultures (in cpm). The results indicate the mean ±1 SD for four paired experiments. The mean value for cultures with PMA is significantly lower than that for PMA-treated cultures supplemented with catalase $p < 0.02$). (From Sagone et al.[41])

Whisler and Newhouse[161] reported that the endogenous release of H_2O_2 as well as prostaglandins by mononuclear cells inhibited the formation of human B-cell colonies stimulated by protein Staph A. Also, the inhibition of colony formation following the addition of lipopolysaccharide (LPS) to the cultures could be reversed by the addition of catalase or indomethacin. These observations indicate that the H_2O_2 and prostaglandins released by monocytes were related to the LPS-induced suppressive effect. These results provide evidence that the amount of H_2O_2 released by phagocytic cells is also sufficient to alter B-cell function. In both cases, the stimulus (protein Staph A and LPS) probably induce the direct release of H_2O_2 by the monocytes.

Archibald et al.[100] recently demonstrated that an antigen–antibody complex—bovine serum albumin (BSA) with anti-BSA—or aggregated γ-globulin bound to B lymphocytes triggers a respiratory burst in neutrophils. This is indicated by the generation of chemiluminescence by the PMN during coculture with the lymphocytes. These investigators also demonstrated that the lymphocytes had an impaired proliferation to pokeweed mitogen but not to PHA mitogen after periods of coculture with the granulocytes. The impaired response could not be explained by decreased viability of the cells. This is an interesting system in which the lymphocytes trigger the PMN to release ROS, which then impair the functional capacity of the lymphocytes. The authors suggest that the stimulation of PMNs by lymphocyte surface-bound immune complexes might lead to similar lymphocyte damage in vivo.

4.2.2. Studies That Have Evaluated the Effect of the Spontaneous Release of ROS by Phagocytes on Lymphocyte Proliferation

Numerous recent reports have established the importance of ROS in the functional capacity of activated macrophages, particularly in their bacterial and tumorcidal activities. These cells are also known to have the capacity to suppress lymphocyte function.[177–180] Similarly, the blood monocytes and granulocytes of patients with inflammatory and malignant disorders have been reported to have activated oxidative metabolism and function.[175,181–186] Presumably, this activation is associated with the release of ROS in vivo and suggests that the release of ROS by activated phagocytes might be one mechanism for the suppression of lymphocyte function found in several groups of patients. In this regard, several recent studies have evaluated the possible role of ROS in the suppressor cell activity of activated phagocytic cells against lymphocytes.

Metzger et al.[179] in 1980 reported that peritoneal exudate cells (PEC) induced by *Corynebacter parvum* or thioglycolate could suppress the proliferation of murine splenic lymphocytes to Con A in vitro.[179] At 2 : 10 ratio, the macrophages totally suppressed lymphocyte transformation. At a lower macrophage-to-lymphocyte ratio, partial suppression occurred (60–68%). Partial suppression of approximately 50% could be demonstrated in cultures incubated with prostaglandins E_1 and E_2 (PGE_1 and PGE_2) in high concentrations (10^{-8}–10^{-5} M). The suppression did not occur with similar concentrations of $PGF_{2\alpha}$. These concentrations of prostaglandins were higher than those generated by the PEC during cocultures with the lymphocytes. Therefore, the release of prostaglandins from the PEC did not entirely explain the suppressive effect on lymphocyte function. The possible role of H_2O_2 and of prostaglandins in this suppressor cell activity was then evaluated by adding catalase and indomethacin to the cultures. The addition of either indomethacin or catalase could protect the partially suppressed cultures. These agents were relatively ineffective, however, in protecting the totally suppressed cultures. The simultaneous addition of catalase and indomethacin to these cultures provided complete protection against macrophage-mediated suppression. These experiments suggest that the suppressive effects of PEC against lymphocytes was mediated by both prostaglandins and H_2O_2. While the precise mechanism involved in the protective effects of catalase and indomethacin in this system is not entirely clear, the experiments provide evidence that H_2O_2 may be involved in the suppressor activity of macrophages in vivo and seem to approximate conditions in vivo in which lymphocytes would be exposed chronically to H_2O_2.

It has been established for some time that patients with Hodgkin's disease have impaired cellular immunity.[180,187,188] These patients have impaired skin reactivity, and their blood lymphocytes have impaired proliferation following stimulation by nonspecific mitogens in vitro. Recent studies evaluated the possibility that this impaired lymphocyte function might be due to the release of H_2O_2 by the monocytes in the cultures.[180,187] However, while the results of these studies are of interest, they proved to be complex.[180,187] As expected, the blood mononuclear cells (PMBC) of most Hodgkin's disease patients had an impaired proliferation to PHA. When the PBMC were depleted of monocytes, the proliferation was enhanced indicating that monocytes were suppressing the response by some mechanism. When catalase was added to the cultures, no protective effect was observed, apparently excluding H_2O_2 as a major mediator of the suppressor activity. By contrast, the addition of indomethacin augmented the proliferation of the cultures in a similar manner to monocyte depletion, suggesting that release of prostaglandin by the monocytes was one mechanism for the suppression. Unexpectedly, the simultaneous addition of catalase and indomethacin further augmented the LBT of cultures, including those of normals as well as those from the patients. Overall, the response still remained significantly impaired in the patient cultures compared with those of normals. The failure of catalase and indomethacin to provide complete protection indicates that additional factor(s) besides prostaglandins and H_2O_2 relate to the impaired LBT of these cells. Fisher and Bostick-Bruton[180] concluded that the T cells of patients with Hodgkin's disease probably have an intrinsic defect as a major etiology for their impaired function. The mechanism for the enhanced response in cultures with catalase and indomethacin is unclear. One explanation for the effect may be in the capacity of prostaglandins to inhibit the production of H_2O_2 by monocytes.[189] While these studies do not establish a causal relationship between the release of ROS by monocytes and the impaired T-cell function of Hodgkin patients, they do provide additional evidence that the release of ROS by phagocytes may alter lymphocyte proliferation.

Thus far, the effects of the ROS released by activated granulocytes on the functional capacity of lymphocytes does not appear to have been the subject of extensive evaluation. Niwa et al.[175] recently reported that the stimulated neutrophils from the synovial fluid with rheumatoid arthritis can alter the function of lymphocytes. The release of ROS by the unstimulated and zymosan-stimulated granulocytes from the joints of these patients was enhanced compared with blood granulocytes of normal persons. The coculture of these activated granulocytes with patient and normal lymphocytes for 17 hr resulted in a decreased number of $OKT_4{}^+$ and $OKT_8{}^+$ cells and an impaired response to nonspecific mitogens. The lymphocytes could be protected by the addition of catalase and SOD, indicating that the suppressor effect was mediated by the release of ROS from these cells. These results demonstrate that the ROS released by activated granulocytes have the capacity to impair lymphocyte function.

4.3. Role of the Lymphocyte in the Metabolic Burst

Numerous reports indicate that lymphocytes, in contrast to phagocytic cells, lack the capacity to undergo a metabolic burst. There is some evidence, however, that OH· may be involved in the functional capacity of these cells. Hydroxyl scavengers have been reported to inhibit the proliferation of human lymphocyte cultures to nonspecific mitogens.[79] The mitogenesis induced by PMA is much more sensitive to inhibition than that induced by PHA or Con A. Hydroxyl scavengers also impair the proliferation of mouse thymocyte cultures to IL-2.

These observations suggest that OH· is involved in some way in mitogenesis. The mechanism is not clear and needs further study. These studies are of interest, however, since they represent one of the few observations in the literature that indicate that oxidants may enhance lymphocyte function during cell–cell interaction.

Also, some reports indicated that OH· may be related to the NK activity of lymphocytes.[80] If so, these cells may have a burst in OH· production after conjugate formation with the target cell. This evidence is based primarily on scavenger studies and therefore represents indirect evidence for the production of OH· in this cell. However, recent reports have also suggested a role of the lipoxygenase pathway in NK activity.[190] This pathway has been shown to produce OH· in other cells,[191] so it is possible that OH· production could occur in the lymphocyte by this pathway as well. This explanation could link the role of the lipoxygenase pathway and OH· in the NK activity of lymphocytes.[190] Another possible source of OH· production in lymphocytes could be microsomes, which have been established to generate OH· (see Chapter 12). If lymphocytes do produce OH· under some circumstances, this ROS might alter the function of other cells, particularly some subpopulations of lymphocytes.[79]

Finally, activated lymphocyte subsets are known to produce a number of factors (lymphokines) such as interleukin, interferon, and peroxide stimulating factor that may activate both lymphocytes and phagocytic cells.[69–72] Such activated phagocytic cells, once stimulated, should release increased amounts of ROS and have increased suppressor cell activity. Whether such a feedback mechanism plays a physiological role in the immune response would seem to be of interest.

5. SUMMARY

The studies that have been discussed indicate that the ROS released by phagocytic cells can impair the functional capacity of blood T and B lymphocytes. This sensitivity of lymphocytes to H_2O_2 may relate to a low concentration of catalase and/or other antioxidants in these cells. The major ROS shown to cause injury in these in vitro systems has been H_2O_2, but it seems likely that the other ROS released by phagocytes might also impair lymphocyte function under some conditions in vivo. While chemical mitogenesis suggests that oxidants might augment lymphocyte function under some conditions, thus far, only a few reports indicate that this occurs during cell–cell interaction.

Presumably, impaired lymphocytes function might occur in patients with a variety of diseases. However, establishing a cause–effect relationship between the release of ROS by phagocytic cells in vivo and an impaired lymphocyte function is more complicated. Phagocytic cells are known to have nonoxidative mechanisms as well as oxidative ones for injurying target cells. Therefore, the differentiation of oxidative and nonoxidative damage to lymphocytes by phagocytic cells may prove difficult. Also, other factors may alter lymphocyte function in vivo. These include prostaglandins, lymphokines, and serum factors, to mention a few. Furthermore, the situation may be even more complex in some groups of patients, such as those with cancer. Recently, neoplastic cells have been shown to release factors that impair the functional capacity of both lymphocytes and phagocytic cells.[192–194] In spite of the complex nature of the immune system, it seems likely that future studies will establish that the ROS generated by phagocytes have an important role in regulating the immune response in vivo.

REFERENCES

1. Klebanoff SJ: Antimicrobial mechanism in neutrophilic polymorphonuclear leukocytes. *Semin Hematol* **12:**117–142, 1975.
2. Babior BM: The respiratory burst of phagocytes. *J Clin Invest* **73:**559–601, 1984.
3. Johnston RB, Lehmeyer JE: Elaboration of toxic oxygen by-products by neutrophils in a model of immune complex disease. *J Clin Invest* **51:**836–841, 1976.
4. Babior BM, Kipnes R, Curnette JT: Biological defense mechanisms. The production of leukocytes of superoxide, a potential bactericidal agent. *J Clin Invest* **52:**741–744, 1973.
5. Johnston RB Jr, Keele BB, Misra HP, et al: The role of superoxide anion generation in phagocytic bactericidal activity—Studies with normal and chronic granulomatous disease leukocytes. *J Clin Invest* **55:**1357–1372, 1975.
6. Lehmeyer JE, Johnston RB Jr: Effect of anti-inflammatory drugs and agents that elevate intracellular cyclic AMP on the release of toxic oxygen metabolites by phagocytes: Studies in a model of tissue-bound IgG. *Clin Immunol Immunopathol* **9:**482–490, 1978.
7. Iyer GY, Islam MF, Quastel JH: Biochemical aspect of phagocytosis. *Nature (Lond.)* **192:**535–541, 1961.
8. Sbarra AJ, Karnovsky ML: The biochemical basis of phagocytosis I, Metabolic changes during the ingestion of particles by polymorphonuclear leukocytes. *J Biol Chem* **234:**1355–1362, 1959.
9. Karnovsky ML: Metabolic basis of phagocytic activity. *Physiol Rev* **42:**143–168, 1962.
10. Curnette JT, Babior B: Biological defense mechanisms, the effect of bacteria and serum on superoxide production by granulocyte. *J Clin Invest* **53:**1662–1672, 1974.
11. Homan-Muller J, Weening RS, Roos D: Production of hydrogen peroxide by phagocytosing human granulocytes *J Lab Clin Med* **85:**198–207, 1975.
12. Allen RC, Stjernholm R, Steele RH: Evidence for the generation of an electronic excitation state(s) in human polymorphonuclear leukocytes and its participation in bactericidal activity. *Biochem Biophys Res Commun* **47:**679–684, 1972.
13. Webb LS, Keele BB Jr, Johnston RB Jr: Inhibition of phagocytosis-associated chemiluminescence by superoxide dismutase. *Infect Immun* **9:**1051–1056, 1974.
14. Baehner RL, Murrman SK, Davis J, et al: The role of superoxide anion and hydrogen peroxide in phagocytosis-associated oxidative metabolic reactions. *J Clin Invest* **56:**571–576, 1975.
15. Sagone AL Jr, King GW, Metz EN: A comparison of the metabolic responses to phagocytosis in human granulocytes and monocytes. *J Clin Invest* **57:**1352–1358, 1976.
16. Klebanoff SJ, Pincus SH: Hydrogen peroxide utilization in myeloperoxidase-deficient leukocytes: A possible microbicidal control mechanism. *J Clin Invest* **50:**2226–2229, 1971.

17. Weening RS, Roos P, Loos JA: Oxygen consumption of phagocytizing cells in human leukocyte and granulocyte preparations: A comparative study. *J Lab Clin Med* **83:**570–576, 1974.
18. Root RK, Metcalf J, Oshino N: H_2O_2 release from human granulocytes during phagocytosis. I. Documentation, quantitation, and some regulating factors. *J Clin Invest* **55:**945–955, 1975.
19. Tsan Min Fu, Chen JW: Oxidation of methionine by human polymorphonuclear leukocyte. *J Clin Invest* **65:**1041–1050, 1980.
20. Weiss SJ, Klein R, Slivka A, et al: Chlorination of taurine by human neutrophils. Evidence for hypochlorous acid generation. *J Clin Invest* **70:**598–607, 1982.
21. Thomas EL, Grisham MB, Jefferson MM: Myeloperoxidase-dependent effect of amines on function of isolated neutrophils. *J Clin Invest* **72:**441–454, 1983.
22. Weiss SJ, Lambert MD, Test ST: Long-lived oxidants generated by human neutrophils: Characteristics and bioactivity. *Science* **222:**625–628, 1983.
23. Sagone AL Jr, Husney RM, O'Dorisio MS, et al: Mechanisms for the oxidation of reduced glutathione by zymosan stimulated granulocytes. *Blood* **63:**96–104, 1984.
24. Pincus S, Klebanoff SJ: Quantitative leukocyte iodination. *N Engl J Med* **284:**744–750, 1984.
25. McCord J: Free Radicals and inflammation: Protection of synovial fluid by superoxide dismutase. *Science* **185:**529–531, 1974.
26. Oxygen free radicals and tissue damage, in Fitzsimons DW (ed): *Ciba Foundation Symposium 65*, Excerpta Medica Amsterdam, New York, 1979, 1–381.
27. Weissman G, Smolen JE, Korchak EM: Release of inflammatory mediators from stimulated neutrophils. *N Engl J Med* **303:**27–34, 1980.
28. Weissman G, Serhan C, Korchak HM, et al: Neutrophils generate phosphatidic acid, an "endogenous calcium ionophore: Before releasing mediators of inflammation. *Trans Assoc Am Phys* **94:**357–365, 1981.
29. Boxer LA, Yoder M, Bonsib S, et al: Effect of a chemotactic factor, N-formylmethionyl peptide on adherence, superoxide anion generation, phagocytosis and microtubule assembly of human polymorphonuclear leukocytes. *J Lab Clin Med* **93:**506–514, 1979.
30. Koretzky GA, Daniele RP, Nowell PC: A phorbol ester (TPA) can replace macrophages in human lymphocyte cultures stimulated with a mitogen but not with an antigen. *J Immunol* **128:**1776–1780, 1982.
31. Simon PL, Willoughby WF: The role of subcellular factors in pulmonary immune function: Physicochemical characterization of two distinct species of lymphocyte-activating factor produced by rabbit alveolar macrophages. *J Immunol* **126:**1534–1541, 1981.
32. Goldstein IM, Roos D, Kaplan HB, Weissman G: Complement and immunoglobulin stimulate superoxide production by human leukocytes independently of phagocytosis. *J Clin Invest* **56:**1155–1163, 1975.
33. Henson PM, Oades ZG: Stimulation of human neutrophils by soluble and immunoglobin aggregates—Secretion of granule constituents and increased oxidation of glucose. *J Clin Invest* **56:**1053–1061, 1975.
34. Weksler BB, Goldstein IM: Prostaglandin: Interactions with platelets and polymorphonuclear leukocytes in hemostasis and inflammation. *Am J Med* **68:**419–428, 1980.
35. Segal ML, Fertel RH, Kraut EH, Sagone AL: The role of reactive oxygen species in thromboxane B_2 generation by polymorphonuclear leukocytes. *J Lab Clin Med* **102:**788–794, 1983.
36. Goldstein IM, Malmsten CL, Kindahl H, et al: Thromboxane generation by human neutrophils polymorphonuclear leukocytes. *J Exp Med* **148:**787–792, 1978.
37. Mallery SR, Zeligs BJ, Ramwell PW, Bellanti JA: Gender-related variations of human neutrophil cyclo-oxygenase and oxidative burst metabolites. *J Leukocyte Biol* **40:**133–148, 1986.
38. Sagone AL Jr, Decker MA, Wells RM, Democko C: A new method for the detection of hydroxyl radical production. *Biochim Biophys Acta* **628:**90–97, 1980.
39. Alexander MS, Husney RM, Sagone AL Jr: Metabolism of benzoic acid by stimulated polymorphonuclear cells. *Biochem Pharmacol* **35:**3649–3651, 1986.
40. Sagone AL, Husney RM: Oxidation of salicylates by stimulated granulocytes—Evidence that these drugs act as free radical scavengers in biological systems. *J Immunol* **138:**1–7, 1987.

41. Sagone AL, Husney R, Guter H, Clark L: The effect of catalase on the proliferation of human lymphocytes to phorbol myristate acetate. *J Immunol* **133:**1488–1494, 1984.
42. Repine JE, Eaton JW, Anders MW, et al: Generation of hydroxyl radicals by enzymes, chemicals, and human phagocytes in vitro. Detection with the anti-inflammatory agent, dimethyl sulfoxide. *J Clin Invest* **64:**1642–1651, 1979.
43. Klassen D, Conkling P, Sagone AL: Activation of monocyte and granulocyte antibody-dependent cytotoxicity by phorbol myristate acetate. *Infect Immun* **35:**818–825, 1982.
44. Papermaster-Bender G, Whitcomb M, Sagone AL Jr: Characterization of the metabolic responses of human pulmonary alveolar macrophage. *J Reticulendothel Soc* **28:**129–139, 1980.
45. Seaman WE, Gindhart TD, Blackman MA, et al: Suppression of natural killing in vitro by monocytes and polymorphonuclear leukocytes. *J Clin Invest* **69:**876–888, 1982.
46. Zoschke DC, Messner RP: Suppression of human lymphocyte mitogenesis mediated by phagocyte-released reactive oxygen species: comparative activities in normals and in chronic granulomatous disease. *Clin Immunol Immunopathol* **32:**29–40, 1984.
47. Pryzwansky KB, Martin LE, Spitznagel JK: Immunocytochemical localization of myeloperoxidase, lactoferrin, lysozyme and neutral proteases in human monocytes and neutrophilic granulocytes. *J Reticuloendothel Soc* **24:**295–309, 1978.
48. Zabos P, Kyner D, Mendelsohn N, et al: Catabolism of 2-deoxyglucose by phagocytic leukocytes in the presence of 12-*O*-tetradecanoyl phorbol-13-acetate. *Proc Natl Acad Sci USA* **75:**5422–5426, 1978.
49. Nathan CF, Brukner LH, Silverstein SC, et al: Extracellular cytolysis by activated macrophages and granulocytes. I. Pharmacologic triggering of effector cells and the release of hydrogen peroxide. *J Exp Med* **49:**84–99, 1979.
50. Murray HW, Juangbhanich CW, Nathan CF, et al: Macrophage oxygen-dependent antimicrobial activity. *J Exp Med* **150:**950–964, 1979.
51. Cohen MS, Taffet ST, Adams DO: The relationship between competence for secretion of H_2O_2 and completion of tumor cytotoxicity by BCG-elicited murine marcophages. *J Immunol* **128:**1781–1785, 1982.
52. Saito H, Tomioka H, Watanabe T: H_2O_2 releasing function of macrophages activated with various mycobacteria based on wheat germ agglutinin and phorbol myristate acetate triggering. *J Reticuloendothel Soc* **29:**193–204, 1981.
53. Pennline KJ, Herscowitz HB: Dual role for alveolar macrophages in humoral and cell-mediated immune responses: evidence for suppressor and enhancing functions. *J Reticuloendothel Soc* **30:**205–217, 1981.
54. Drath DB, Karnovsky ML, Huber GL: Hydroxyl radical formation in phagocytic cells of the rat. *J Appl Physiol* **46:**136–140, 1979.
55. Heifets L, Imai K, Goren MB: Expression of peroxidase-dependent iodination by macrophages ingesting neutrophil debris. *J Reticuloendothel Soc* **28:**391–404, 1980.
56. Hoidal J, Repine J, Beall G et al: The effect of phorbol myristate acetate on the metabolism and ultrastructure of alveolar macrophages. *Am J Pathol* **91:**469–476, 1978.
57. Murray H, Cohn Z: Macrophage oxygen-dependent antimicrobial activity III Enhanced oxidative metabolism as an expression of macrophage activation. *J Exp Med* **152:**1596–1609, 1980.
58. North RJ: Opinions: The concept of thc activatcd macrophage. *J Immunol* **121:**806–816, 1978.
59. Turpin J, Hersh EM, Lopez-Berestein G: Characterization of small and large human peripheral blood monocytes: effects of in vitro maturation on hydrogen peroxide release and on the response to macrophage activators. *J Immunol* **136:**4194–4198, 1986.
60. Montarroso AM, Myrvik QN: Oxidative metabolism of BCG-activated alveolar macrophages. *J Reticuloendothel Soc* **25:**559–574, 1979.
61. Douglas SD: Macrophage nomenclature: Where are we going? *J Reticuloendothel Soc* **27:**241–245, 1980.
62. Breton-Gorius J, Guichard J, Vainchenker W, et al: Ultrastructural and cytochemical changes induced by short and prolonged culture of human monocytes. *J Reticuloendothel Soc* **27:**289–301, 1980.

63. Anderson SE Jr, Remington JS: Effect of normal and activated human macrophages on toxoplasma gondii. *J Exp Med* **139:**1154–1173, 1974.
64. Karnovsky ML, Simmons S, Glass EA: Metabolism of macrophages, in Van Furth R (ed): *Mononuclear Phagocytes* Philadelphia, F. A. Davis, 1970, p 103–120.
65. Musson RA, Shafran H, Henson PM: Intracellular levels and stimulated release of lysosomal enzymes from human peripheral blood monocytes and monocyte-derived macrophages. *J Reticuloendothel Soc* **28:**249–264, 1980.
66. Rosenthal AS: Medical intelligence. Current concepts: Regulation of the immune response—Role of the macrophage. *N Engl J Med* **303:**1153–1156, 1980.
67. Speert DP, Silverstein SC: Phagocytosis of unopsonized zymosan by human monocyte-derived macrophages: Maturation and inhibition by mannan. *J Leukocyte Biol* **38:**655–658, 1985.
68. Soler P, Basset F, Mazin F, et al: Peroxidatic activities of human alveolar macrophages in some pulmonary granulomatous disorders. *J Reticuloendothel Soc* **31:**511–521, 1982.
69. Gately CL, Wahl SM, Oppenheim JJ: Characterization of hydrogen peroxide-potentiating factor. A lymphokine that increases the capacity of human monocytes and monocyte-like cell lines to produce hydrogen peroxide. *J Immunol* **131:**2853–2858, 1983.
70. Chapes SK, Haskill S: Synergistic effect between neutrophils and corynebacterium parvum in the process of macrophage activation. *Cancer Res* **44:**31–34, 1984.
71. Weisbart RH, Kwan L, Golde DW: Human GM-CSF primes neutrophils for enhanced oxidative metabolism in response to the major physiological chemoattractants. *Blood* **69:**18–21, 1987.
72. Nathan CF, Murray HW, Wiebe ME, et al: Identification of interferon-γ as the lymphokine that activates human macrophage oxidative metabolism and antimicrobial activity. *J Exp Med* **158:**670–689, 1983.
73. Wilson ME, Jones DP, Munkenbeck P, et al: Serum-dependent and independent effects of bacterial lipopolysaccharides on human neutrophil oxidative capacity in vitro. *J Reticuloendothel Soc* **31:**43–57, 1982.
74. Nakagawara, A, Desantis NM, Nogueira N, et al: Lymphokines enhance the capacity of human monocytes to secrete reactive oxygen intermediates. *J Clin Invest* **70:**1042–1048, 1982.
75. Sagone AL Jr, LoBuglio AT, Balcerzak SP: Alteration in hexose monophosphate shunt during lymphoblastic transformation. *Cell Immunol* **14:**443–452, 1974.
76. Hadden JW, Hadden EM, Good RA: Adrenergic mechanisms in human lymphocyte metabolism. *Biochim Biophysic Acta* **237:**330–347, 1971.
77. Roos D, Loos JA: Changes in the carbohydrate metabolism of mitogenically stimulated human peripheral lymphocytes. *Biochim Biophys Acta* **222:**565–582, 1970.
78. Stjernholm RL: Early biochemical changes in phytohemagglutinin-stimulated human lymphocytes of blood and lymph. *J Reticuloendothel Soc* **7:**471–483, 1970.
79. Novogrodsky A, Ravid A, Rubin A, et al: Hydroxyl radical scavengers inhibit lymphocyte mitogenesis. *Proc Natl Acad Sci USA* **79:**1171–1174, 1982.
80. Duwe AK, Werkmeister J, Roder JC: Natural killer cell-mediated lysis involves an hydroxyl radical-dependent step. *J Immunol* **134:**2637–2644, 1985.
81. Mookerjee BK, Wakerle H, Sharon N, et al: Chemiluminescence and lymphocyte proliferation: parallelism in collaboration between subpopulations of thymus cells for both types of responses. *J Leukocyte Biol* **35:**427–438, 1984.
82. Kurlander RJ, Rosse WF, Logue GL: Quantitative influence of antibody and complement coating of red cells on monocyte-mediated cell lysis. *J Clin Invest* **61:**1309–1319, 1978.
83. Adler R, Glorioso JC, Cossman J, et al: Possible role of Fc receptors on cells infected and transformed by herpesvirus: Escape from immune cytolysis. *Infect Immun* **21:**442–447, 1978.
84. Ehlenberger AG, Nussenzweig V: The role of membrane receptors for C3b and C3d in phagocytosis. *J Exp Med* **145:**357–371, 1977.
85. Wuest D, Crane R, Rinehart JJ: Enhancement of Fc receptor function during human monocyte differentiation in vitro. *J Reticuloendothel Soc* **30:**147–155, 1981.
86. Newman SL, Becker S, Halme J: Phagocytosis by receptors for C3b (CR_1), iC3b, (CR_3), and IgG (Fc) on human peritoneal macrophages. *J Leukocyte Biol* **38:**267–278, 1985.

87. Leijh PC, Van Den Barselaar M, van Zwet TL, et al: Requirement of extracellular complement and immunoglobulin for intracellular killing of micro-organisms by human monocytes. *J Clin Invest* **63:**772–784, 1979.
88. Lawrence DA, Weigle WO, Spiegelberg HL: Immunoglobulins cytophilic for human lymphocytes, monocytes, and neutrophils. *J Clin Invest* **55:**368–376, 1975.
89. Whaley K: Biosynthesis of the complement components and the regulatory proteins of the alternative complement pathway by human peripheral blood monocytes. *J Exp Med* **151:**501–516, 1980.
90. Shigeoka AO, Hall RT, Hemming VG, et al: Role of antibody and complement in opsonization of group B streptococci. *Infect Immun* **21:**34–40, 1978.
91. Wardley RC, Rouse BT, Babiuk LA: Antibody dependent cytotoxicity mediated by neutrophils: A possible mechanism of antiviral defense. *J Reticuloendothel Soc* **19:**323–332, 1976.
92. Brogen M, Sagone AL Jr: The metabolic response of human phagocytic cells to killed mumps particles. *J Reticulothel Soc* **27:**13–22, 1980.
93. Katz P, Simone CB, Henkart PA, et al: Mechanisms of antibody-dependent cellular cytotoxicity. *J Clin Invest* **65:**55–63, 1980.
94. Clark RA, Klebanoff SJ: Studies on the mechanism of antibody-dependent polymorphonuclear leukocyte-mediated cytotoxicity. *J Immunol* **119:**1413–1418, 1977.
95. Borregaard N, Kragballe K: Role of oxygen in antibody-dependent cytotoxicity mediated by monocytes and macrophages. *J Clin Invest* **66:**676–683, 1980.
96. Seim S, Espevik T: Toxic oxygen species in monocyte-mediated antibody-dependent cytotoxicity. *J Reticuloendothel Soc* **33:**417–428, 1983.
97. Hafeman DG, Lucas ZJ: Polymorphonuclear leukocyte-mediated, antibody-dependent, cellular cytotoxicity against tumor cells: Dependence on oxygen and the respiratory burst. *J Immunol* **123:**55–62, 1979.
98. Storkus WJ, Dawson JR: Oxygen-reactive metabolites are not detected at the effector–target interface during natural killing. *J Leukocyte Biol* **39:**547–557, 1986.
99. Klassen DK, Sagone AL Jr: Evidence for both oxygen and non-oxygen dependent mechanisms of antibody sensitized target cell lysis by human monocytes. *Blood* **56:**985–992, 1980.
100. Archibald AC, Cheung K, Robinson MF: The interaction of lymphocyte surface-bound immune complexes and neutrophils. *J Immunol* **131:**207–211, 1983.
101. Sacks T, Moldow CF, Craddock PR, et al: Oxygen radicals mediate endothelial cell damage by complement-stimulated granulocytes an in vitro model of immune vascular damage. *J Clin Invest* **61:**1161–1167, 1978.
102. Till GO, Johnson RJ, Kunkel R, Ward PA: Intravascular activation of complement and acute lung injury. *J Clin Invest* **69:**1126–1135, 1982.
103. Ward PA, Till, Kunkel R, et al: Evidence of role of hydroxyl radical in complement and neutrophil-dependent tissue injury. *J Clin Invest* **72:**789–801, 1983.
104. Cooke E, Hallett MB: The role of C-kinase in the physiological activation of the neutrophil oxidase. *Biochem J* **232:**323–327, 1985.
105. van der Meulen FW, van der Hart M, Fleer A, et al: The role of adherence to human mononuclear phagocytes in the destruction of red cells sensitized with non-complement binding IgG antibodies. *Br J Haematol* **38:**541–549, 1978.
106. Conkling PR, Klassen DK, Sagone AL: Comparison of antibody-dependent cytotoxicity mediated by human polymorphonuclear cells, monocytes, alveolar macrophages. *Blood* **60:**1290–1297, 1982.
107. Lee D, Hoidal J, Garlich D et al: Opsonin-independent phagocytosis of surface-adherent bacteria by human alveolar macrophages. *J Leukocyte Biol* **36:**689–701, 1984.
108. Eissenberg LG, Goldman WE: Histoplasma capsulatum fails to trigger release of superoxide from macrophages. *Infect Immun* **55:**29–34, 1987.
109. Abramson JS, Mills EL, Giebink GS, et al: Depression of monocyte and polymorphonuclear leukocyte oxidative metabolism and bactericidal capacity by influenza virus. *Infect Immun* **35:**350–355, 1982.

110. Copeland E, Rinehart J, Lewis M, et al: The mechanism of retrovirus suppression of T cell proliferation. *J Immunol* **131:**2017–2020, 1983.
111. Anderson V, Hellung-Larsen FP, Shrenson SF: Optimal oxygen tension for human lymphocytes in culture. *J Cell Physiol* **72:**149–152, 1968.
112. Anderson V, Anderson AB, Skovbjerg H, et al: Structure of mitochondria in lymphocytes culture at different oxygen tensions. *Acta Pathol Microbiol Scand [A]* **78:**537–544, 1970.
113. Mizrahi A, Vosseller GV, Yagi Y, et al: The effect of dissolved oxygen partial pressure on growth, metabolism and immunoglobulin production in a permanent human lymphocyte cell line culture. *Proc Soc Exp Biol Mod* **139:**118–122, 1972.
114. Kilburn DG, Morley M, Yensen J: The influence of dissolved oxygen on the mitogen responses of mouse lymphocytes. *J Cell Physiol* **87:**307–311, 1975.
115. Sagone AL Jr: Effect of hyperoxia on the carbohydrate metabolism of human lymphocytes. *Am J Hematol* **18:**269–274, 1985.
116. Rabinowitz Y, Lubrano T, Wilhite BA, et al: Lactic dehydrogenase of cultured lymphocytes: Response to environmental conditions. *Exp Cell Res* **48:**675–678, 1967.
117. Fridovich I: The biology of oxygen radicals. *Science* **201:**875–880, 1978.
118. Harlan JM, Levine JD, Callahan KS, et al: Glutathione Redox cycle protects cultured endothelial cells against lysis by extracellularly generated hydrogen peroxide. *J Clin Invest* **73:**706–713, 1984.
119. Arrick BA, Nathan CF, Cohn ZA: Inhibition of glutathione synthesis augments lysis of murine tumor cells by sulfhydryl-reactive antineoplastics. *J Clin Invest* **71:**258–267, 1983.
120. Stankova L, Rigas DA, Keown P, et al: Leukocyte ascorbate and glutathione: Potential capacity for inactivating oxidants and free radicals. *J Reticuloendothel Soc* **21:**97–102, 1977.
121. Marklund S, Nordensson I, Back O: Normal CuZn superoxide dismutase, Mn superoxide dismutase, catalase and glutathione peroxidase in Werner's syndrome. *J Gerontol* **36:**405–409, 1981.
122. Meerhof LJ, Roos D: An easy, specific and sensitive assay for the determination of the catalase activity of human blood cells. *J Reticuloendothel Soc* **28:**419–425, 1980.
123. Arrick B, Nathan C: Glutathione metabolism as a determinant of therapeutic efficacy. A review. *Cancer Res.* **44:**4224–4232, 1984.
124. Roos D, Weening R, Voetman A, et al: Protection of phagocytic leukocytes by exogenous glutathione: studies in a family with glutathione deficiency reductase deficiency. *Blood* **53:**851–866, 1979.
125. Spielberg SP, Boxer LA, Oliver JM, et al: Oxidative damage to neutrophils in glutathione synthetase deficiency. *Br J Haematol* **42:**215–223, 1979.
126. Freeman BA, Crapo JD: Biology of disease-free radicals and tissue injury. *Lab Invest* **47:**412–426, 1982.
127. Chance B, Sies H, Boveris A: Hydroperoxide metabolism in mammalian organs. *Physiol Rev* **59:**527–605, 1979.
128. Baral E, Blomgen H: Response of human lymphocytes to mitogenic stimuli after irradiation in vitro. *Acta Radiol Ther Phys Biol* **15:**149–161, 1976.
129. Braeman J, Moore J: The lymphocyte response to phytohemagglutinin after in vitro radiation. *Br J Radiol* **47:**297, 1974.
130. Roberts W, Kartha M, Sagone AL Jr: Effect of irradiation on the hexose monophosphate shunt pathway of human lymphocytes. *Radiat Res* **79:**601–610, 1979.
131. Wasserman J, Petrini B, Blomgren H: Radiosensitivity of T-lymphocyte subpopulations. *J Clin Lab Immunol* **7–8:**139–140, 1982.
132. Kleinerman ES, Decker JM, Muchmore AV: In vitro cellular regulation of monocyte function: Evidence for a radiosensitive suppressor. *J Reticuloendothel Soc* **30:**373–380, 1981.
133. Stefani S, Kerman R: Lymphocyte response to phytohemagglutinin before and after radiation therapy in patients with carcinomas of the head and neck. *J Laryngol Otol* **91:**605–610, 1977.
134. Anderson RE, Standefer JC, Scaletti JV: Radiosensitivity of defined populations of lymphocytes. *Cell Immun* **33:**45–61, 1977.
135. Merz T, Hazra T, Ross M, et al: Transformation delay of lymphocytes in patients undergoing radiation therapy. *AJR* **127:**337–339, 1976.

136. Howard-Flanders P, Levin J, Theriot L: Reactions of DNA radicals with sulfhydryl in x-irradiated bacteriophage systems. *Radiat Res* **18:**593–606, 1963.
137. Goscin S, Fridovich I: Superoxide dismutase and the oxygen effect. *Radiat Res* **56:**565–569, 1973.
138. Scholes G, Weiss J: Oxygen effects and formation of peroxides in aqueous solution. *Radiat Res* **1** (suppl)**:**177–189, 1959.
139. Oberley L, Lindgren A, Baker et al: Superoxide ion and the cause of the oxygen effect. *Radiat Res* **68:**320–328, 1976.
140. Lavelle F, Michelson M, Mitrijene: Biological protection by superoxide dismutase. *Biochem Biophys Res Commun* **55:**350–357, 1973.
141. Konings AWT, Oosterloo SK: Radiation effects on membranes. *Radiat Res* **81:**200–207, 1980.
142. Stankova L, Rigas D, Head C, et al: Determinants of resistance to radiation injury in blood granulocytes from normal donors and from patients with myeloproliferative disorders. *Radiat Res* **80:**49–60, 1979.
143. Edsmyr F, Huber W, Menander KB: Orgotein efficacy in ameliorating side effects due to radiation therapy. *Current Ther Res* **19:**198–211, 1976.
144. Westman G, Marklund SL: Diethyldithiocarbamate, a superoxide dismutase inhibitor, decreases the radioresistance of chinese hamster cells. *Radiat Res* **83:**303–311, 1980.
145. McLennan G, Oberley LW, Autor AP: The role of oxygen-derived free radicals in radiation-induced damage and death of nondividing eucaryotic cells. *Radiat Res* **84:**122–132, 1980.
146. Kiefer J: Does singlet oxygen contribute to the oxygen effect. *Int J Radiat Biol* **34:**587–588, 1978.
147. Helf KD: Interactions of radioprotectors and oxygen in cultured mammalian cells. *Radiat Res* **101:**424–433, 1985.
148. Russo A, Mitchell JB, Finkelstein E, et al: The effects of cellular glutathione elevation on the oxygen enhancement ratio. *Radiat Res* **103:**232–239, 1985.
149. Morse M, Dahl R: Cellular glutathione is a key to the oxygen effect in radiation damage. *Nature (Lond)* **271:**660–662, 1978.
150. Sagone AL Jr: Role of the hexose monophosphate shunt in cellular protection against radiation damage, in Rodgers M, Powers E (ed): *Oxygen and Oxy-radicals in Chemistry and Biology*. New York, Academic, 1981, pp 725–729.
151. O'Brien RL, Parker JW: Oxidation-Induced lymphocyte transformation. *Cell* **7:**13–20, 1976.
152. Novogrodsky A, Katchalski E: Membrane site modified on induction of the transformation of lymphocytes by periodate. *Proc Natl Acad Sci USA* **69:**3207–3210, 1972.
153. Biniaminov M, Ramot B, Novogrodsky A: Effect of macrophages on periodate-induced transformation of normal and chronic lymphatic leukaemia lymphocytes. *Clin Exp Immunol* **16:**235–242, 1974.
154. Monahan TM, Fritz RR, Abell CW: Sodium periodate stimulation of normal and chronic lymphocytic leukemia lymphocytes. *Exp Cell Res* **103:**263–269, 1976.
155. Roffman E, Wilchek M: The extent of oxidative mitogenesis does not correlate with the degree of aldehyde formation of the T lymphocyte membrane. *J Immunol* **137:**40–44, 1986.
156. Sagone AL Jr, Kamps S, Campbell R: The effect of oxidant injury on the lymphoblastic transformation of human lymphocytes. *Photochem Photobiol* **28:**909–915, 1978.
157. Lipsky PE: Immunosuppression by D-penicillamine in vitro. *J Clin Invest* **73:**53–65, 1984.
158. El-Hag A, Lipsky PE, Bennett M, et al: Immunomodulation by neutrophil myeloperoxidase and hydrogen peroxide: Differential susceptibility of human lymphocyte functions. *J Immunol* **136:**3420–3426, 1986.
159. El-Hag A, Clark RA: Down-regulation of human natural killer activity against tumors by the neutrophil myeloperoxidase system and hydrogen peroxide. *J Immunol* **133:**3291–3297, 1984.
160. Grever MR, Thompson VN, Balcerzak SP, Sagone AL Jr: The effect of oxidant stress on human lymphocyte cytotoxicity. *Blood* **56:**284–288, 1980.
161. Whisler RL, Newhouse YG: Inhibition of human B lymphocyte colony responses by endogenous synthesized hydrogen peroxide and prostaglandins. *Cell Immunol* **69:**34–45, 1982.
162. Kraut EH, Sagone AL Jr: The effect of oxidant injury on the lymphocyte membrane and functions. *J Lab Clin Med* **98:**697–703, 1981.

163. Sugar AM, Chahal RS, Brummer E, et al: The iron-hydrogen peroxide-iodide system is fungicidal: Activity against the yeast phase of blastomyces dermatitidis. *J Leukocyte Biol* **36:**545–548, 1984.
164. Clark RA, Klebanoff SJ, Einstein AB, et al: Peroxidase–H_2O_2–halide system: Cytotoxic effect on mammalian tumor cells. *Blood* **45:**161–170, 1975.
165. Weiss SJ: The role of superoxide in the destruction of erythrocyte targets by human neutrophils. *J Biol Chem* **255:**9912–9917, 1980.
166. Weiss SJ: Oxidative autoactivation of latent collangenase by human neutrophils. *Science* **227:**747–749, 1985.
167. Carp H, Janoff A: Potential mediator of inflammation phagocytic-derived oxidants suppress the elastase-inhibitory capacity of alpha$_1$-proteinase inhibitor in vitro. *J Clin Invest* **66:**987–995, 1980.
168. Farber CM, Liebes LF, Kanganis DN, et al: Human B lymphocytes show greater susceptibility to H_2O_2 toxicity than T lymphocytes. *J Immunol* **132:**2543–2546, 1984.
169. Seaman WE, Gindhart TD, Blackman MA, et al: Natural killing of tumor cells by human peripheral blood cells. *J Clin Invest* **67:**1324–1333, 1981.
170. Urbaniak SJ: ADCC (K-cell) lysis of human erythrocytes sensitized with rhesus alloantibodies. *Br J Haematol* **42:**315–328, 1979.
171. Ramos OF, Mascucci M, Klein E: Modulation of human blood lymphocyte cytotoxicity for the phorbol ester tumor promoter $P(Bu)_2$: Increase of target binding, impairment of effector recycling and activation of lytic potential which is independent of IL-2. *Cell Immunol* **91:**178–192, 1985.
172. Kay HD, Smith DL: Regulation of human lymphocyte-mediated natural killer (NK) cell activity. *J Immunol* **130:**475–483, 1983.
173. Dallegri F, Patrone F, Frumento G: Down-regulation of K cell activity by neutrophils. *Blood* **65:**571–577, 1985.
174. Nissen-Meyer J, Kildahl-Andersen O, Austgulen R: Human monocyte-released cytotoxic factor: Effect on various cellular functions, and dependency of cytolysis on various metabolic processes. *J Leukocyte Biol* **40:**121–132, 1986.
175. Niwa Y, Sakane T, Shingu, M, et al: Effect of stimulated neutrophils from the synovial fluid of patients with rheumatoid arthritis on lymphocytes—A possible role of increased oxygen radicals generated by the neutrophils. *J Clin Immunol* **3:**228–240, 1983.
176. de Vries JE, Caviles AP Jr, Bont WS, et al: The role of monocytes in human lymphocyte activation by mitogens. *J Immunol* **122:**1099–1107, 1979.
177. Lause DB: Suppressed lymphocyte reactivity following lymphocyte interaction with macrophages. *J Reticuloendothel Soc* **26:**775–786, 1979.
178. Novogrodsky A, Patya M, Rubin A, et al: Inhibition of B-adrenergic stimulation of lymphocyte adenylate cyclase by phorbol myristate acetate is mediated by activated macrophages. *Biochem Biophys Res Commun* **104:**389–393, 1982.
179. Metzger Z, Hoffeld JT, Oppenheim JJ: Macrophage-mediated suppression. *J Immunol* **124:**983–988, 1980.
180. Fisher RI, Bostick-Bruton F: Depressed T cell proliferative responses in Hodgkin's disease: Role of monocyte-mediated suppression via prostaglandins and hydrogen peroxide. *J Immunol* **129:**1770–1774, 1982.
181. Bomalaski JS, Clark MA, Douglas SD: Enhanced Phospholipase A_2 and C activities of peripheral blood polymorphonuclear leukocytes from patients with rheumatoid arthritis. *J Leukocyte Biol* **38:**649–654, 1985.
182. Randazzo B, Hirschberg T, Hirschberg H: Cytotoxic effects of activated human monocytes and lymphocytes to anti-D-treated human erythrocytes in vitro. *Scand J Immunol* **9:**351–358, 1979.
183. Nyholm RE, Currie GA: Monocytes and macrophages in malignant melanoma in Lysis of antibody-coated human erythrocytes as an assay of monocyte function. *Br J Cancer* **37:**337–344, 1978.
184. Bass DA, Olbrantz P, Szejda P, et al: Subpopulations of neutrophils with increased oxidative product formation in blood of patients with infection. *J Immunol* **136:**860–866, 1984.
185. Bass DA, Grover WH, Lewis JC, et al: Comparison of human eosinophils from normals and patients with eosinophilia. *J Clin Invest* **66:**1265–1273, 1980.

186. Kitahara M, Eyre HJ, Hill HR: Monocyte functional and metabolic activity in malignant and inflammatory diseases. *J Lab Clin Med* **93:**472–479, 1979.
187. Deshazo RD, Ewel C, Londono S, et al: Evidence for the involvement of monocyte-derived toxic oxygen metabolites in the lymphocyte dysfunction of Hodgkin's disease. *Clin Exp Immunol* **46:**313–320, 1981.
188. Nagai H, Fisher RI, Cossman J, et al: Decreased expression of Class II major histocompatibility antigens on monocytes from patients with Hodgkin's disease. *J Leukocyte Biol* **39:**313–321, 1986.
189. Metzger Z, Hoffeld JT, Oppenheim JJ: Regulation by PGE_2 of the production of oxygen intermediates by LPS-activated macrophages. *J Immunol* **127:**1109–1113, 1981.
190. Bray RA, Brahmi Z: Role of lipoxygenation in human natural killer cell activation. *J Immunol* **136:**1783–1798, 1986.
191. Singh D, Greenwald J, Bianchine J, et al: Evidence for the generation of hydroxyj radical during arachidonic acid metabolism by human platelets. *Am J Hematol* **11:**233–240, 1981.
192. Harrel RA, Cianciolo CJ, Copeland TD: Suppression of the respiratory burst of human monocytes by a synthetic peptide homologous to envelope proteins of human and animal retroviruses. *J Immunol* **186:**3517–3528, 1986.
193. Cianciolo GJ, Copeland TD, Oroszlan S, et al: Inhibition of lymphocyte proliferation by a synthetic peptide homologous to retroviral envelope proteins. *Science* **230:**453–455, 1985.
194. Rhodes J, Bishop M, Benfield J: Tumor Surveillance: How tumors may resist macrophage-mediated host defense. *Science* **203:**179–182, 1979.

9

Modulation of the Respiratory Burst by Naturally Occurring Substances

STANLEY D. SOMERFIELD and EMIL SKAMENE

1. INTRODUCTION

The respiratory burst and subsequent production of oxygen radicals and metabolites by the leukocyte are considered important contributing processes in nonspecific microbicidal systems.[1] In addition, in disease states in which inflammation, tissue destruction, and the apparent absence of bacterial invasion are predominant features, excessive and inappropriate production of oxygen radicals and metabolites is thought to play a substantial role in continuing tissue destruction.[2] Despite a lack of substantial direct evidence for a primary role, there is nonetheless much indirect evidence incriminating oxygen metabolites as playing a major contributing role in inflammation.[2] Oxygen radicals have been strongly implicated as pathogenic in a wide variety of inflammatory disease states involving most physiological organ systems.[3]

These disease states include pulmonary damage and fibrosis,[3] cardiac disease,[4] central nervous system (CNS) disease,[5] and rheumatic disease.[6–8] Evidence for a pathological role of oxygen metabolites in joint inflammation and destruction is the demonstration of free radical oxygenation products in rheumatic joint fluid[6] and oxygen radical-mediated damage of connective tissue constituents, including hyaluronic acid[7] and collagen. Hydrogen peroxide production by inflammatory cells has been found to suppress proteoglycan synthesis of intact articular cartilage, as measured by $^{35}SO_4$ uptake.[9] This oxygen metabolite production, particularly H_2O_2 production, may contribute to the cartilage destruction, as seen in rheumatoid inflammation. The H_2O_2 production is prolonged and not controllable. Furthermore, the generation of long-acting oxidants similar to α-chloroamines, with long half-lives of the order of eighteen hours, and with sufficient oxidizing potential to

STANLEY D. SOMERFIELD and EMIL SKAMENE • Division of Clinical Immunology and Allergy, The Montreal General Hospital, Montreal, Québec H3G 1A4, Canada.

attack sulfhydryl- or thioester-containing compounds by neutrophils has been described by Weiss et al.[10] The same group has described the activation of collagenase by H_2O_2.[11] These results highlight the central role of H_2O_2 in activating mechanisms, some quite stable, capable of further tissue breakdown and destruction, in addition to oxygen metabolites themselves being directly tissue destructive.

Much previous elegant work has described mechanisms and processes whereby inflammatory cells are activated.[12] This work has led to advances in unravelling the biochemistry of inflammatory cell activation, including not only the respiratory burst, but more classical leukocyte functions of phagocytosis, chemotaxis, and chemokinesis.

By contrast, despite an accumulation of knowledge concerning mechanisms of leukocyte respiratory burst activation, scant attention has been paid to mechanisms of inhibition of this important leukocyte function. Control of function in all biological systems involves not only activation but also suppression, and suppression of the respiratory burst by naturally occuring substances forms the subject of discussion in this chapter. Suppression of oxidative products production by the inflammatory cell is of obvious fundamental importance to the regulation of inflammation in vivo, and there may be potential for new measures for control by exogenous biological agents having similar actions.

2. CONTROL OF NEUTROPHIL RESPIRATORY BURST BY DEFINED BIOLOGICAL SUBSTANCES

The fact that inflammatory systems and disease models can be controlled by biological substances such as peptides and proteins is now well established.[13–17] The role of steroid-induced proteins and inhibition of inflammation by antiphospholipase A_2 (PLA_2)[18] and the elucidation of a protective role of superoxide dismutase on inflammation support the concept that during inflammatory events endogenous proteins are capable of exerting an inhibitory effect on inflammation in vivo. Antiphospholipase A_2 substances require steroid–receptor interactions and new protein synthesis for their production in vivo before their effects can be achieved; thus, antiinflammatory effects will be delayed and any effect of a similar substance on the respiratory burst may not be physiologically relevant in vivo, as the burst will be over within minutes. This release of anti-PLA_2 proteins may be a stress-induced phenomenon with effects later during inflammatory events and over longer periods of time. Whether anti-PLA_2 substances or fractions thereof are involved in modulating the respiratory burst awaits further research.

3. ANTIPROTEINASES

α_1-Antiprotease inhibitor has been shown to react with active oxygen species. The capacity of α_1-antiprotease to inhibit elastase has been shown to be lost in the presence of oxygen metabolite generation from inflammatory cells.[21] Experiments have shown that methionine residues in α_1-protease inhibitor are sensitive to oxidation by oxygen species.[20] It has been suggested that this antiprotease acts as an oxidant trap by reacting in this way. However, inactivation of α_1-antiprotease has been shown to increase radical production from neutrophils and macrophages.[21]

Endogenous serine proteases have been shown to stimulate superoxide production from cells stimulated with either cytochalasin B or conconavalin A (Con A).[22–24] Furthermore, superoxide generation has been shown to be inhibited by treating cells with serine protease inhibitor in experiments in which human basophils were stimulated with anti-IgE.[25]

It has been postulated that these proteins act not only as a means by which oxygen radicals may be inactivated, but also α_1-protease inhibitor may be synthesized by monocytes, and they may also act as regulatory molecules on the membrane of the cell, modulating superoxide production.[13]

It is conceivable that, in circumstances wherein there is an accumulation of inflammatory cells with production of large amounts of oxygen metabolites and a local deficiency of antiproteinase, further augmentation of oxygen radical production could occur in accordance with this mechanism.

4. DIALYZABLE SUBSTANCES IN EXUDATES OF SPONGE-INDUCED INFLAMMATION: POLYAMINES

Dialyzable substances released from the exudates of sponge-induced inflammation have been shown to be anti-inflammatory. Lewis et al.[26] found that polyamines contained in these exudates are anti-inflammatory in vivo.[26] They have failed to demonstrate any peptides with anti-inflammatory activity in exudate dialyzable concentrates.[26] Trypsin treatment accelerated the breakdown of the exudate but did not affect the anti-inflammatory activity of the dialysate. Polyamines have been shown to inhibit protein kinase C,[27] and this could represent a mechanism for their inhibition of oxygen metabolite production in these circumstances. Thus, polyamines such as putrescine and spermidine could both be formed endogenously in inflammatory exudates and have an effect in vivo in modulating oxygen metabolite production.

5. SUBSTANCE P

Substance P is a neuropeptide consisting of 11 amino acids with widespread distribution throughout a variety of systems, particularly the CNS.[28] The prime activity of substance P is nosiception; however, it has recently been identified as being involved in immunological and inflammatory responses, particularly as a potential agonist. Substance P has been reported to induce degranulation of mast cells, promote neutrophil chemotaxis, provoke lysosome enzyme release from PMNs, and enhance phagocytosis by neutrophils and macrophages. Furthermore, substance P has been shown to mediate activation of macrophages as reflected by initiation of the respiratory burst with release of oxygen metabolites.[29] Substance P was capable of causing direct release of O_2 and H_2O_2 after 1-hr exposure to substance P; this effect was dose dependent. A rapid release of both mediators was found, most of which occurred during the first 30 min of exposure. Subsequently, Hartung et al.[30] identified a homogeneous class of specific binding sites with high affinity for substance P on macrophages; subsequent activation was found to have all the characteristics of a receptor-mediated mechanism. Substance P may be an endogenously produced proinflammatory substance released from nervous tissue that can stimulate inflammatory cells via a specific receptor to release oxygen metabolites. This may be important in neurogenic inflammation.

6. PNEUMOLYSIN

Pneumolysin is a sulfhydryl-activated toxin produced by *Streptococcus pneumoniae*. Paton and Ferrante[31] showed that incubation of human PMNs with highly purified pneumolysin inhibited the respiratory burst in response to stimulation. The inhibition was observed at very low doses of <2 ng/10^6 PMNs. Pneumolysin apparently binds cholesterol in the plasma membrane of host cells, as described by Johnson et al.[32]; these workers attribute the mechanism of action to this property of pneumolysin.

Paton and Ferrante[31] examined the effect of pneumolysin on PMLs and showed a progressive decrease in HMP shunt activity in the presence of zymosan to stimulate the cells, with increasing dose, inhibition becoming statistically significant at 0.25 hemolytic units/10^6 PMN. H_2O_2 production was similarly reduced following preincubation of PMNLs with pneumolysin. With 2 HU, a 50% reduction of H_2O_2 production was observed, compared with untreated controls, following stimulation with PMA. Inhibitory effects of pneumolysin on cells were not observed when PMNLs were treated with cholesterol-inactivated toxin. It is of interest that this toxin is able to inhibit the HMP shunt activity and H_2O_2 production. This inhibition correlated with the liability of human PMNLs to kill *S. pneumoniae* in vitro. This toxin thus affords the microbe a measure of protection and may influence its survival within the phagocytosing cell due to inhibition of oxygen-dependent killing mechanisms.

There is obviously much scope for exploring this type of relationship between microbes and inflammatory cells, as low-molecular-weight substances with potent inhibitory effects on inflammatory cells following membrane interaction may further understanding of membrane events involved during activation of the cytotoxic events.

7. STREPTOLYSIN O

It is of interest to contrast the findings regarding pneumolysin with those reported by Anderson and Duncan for streptolysin O.[33] These investigators reported that the related streptococcal toxin streptolysin O stimulated metabolite activity of resting human PMN as measured by chemiluminescence and O_2 consumption. This stimulation was only slight compared to that induced by zymosan. Also, Cavalieri and Snyder[34] found that the *Escherichia coli* α-toxin stimulated chemiluminescence of PMNLs but inhibited chemotaxis, phagocytosis, and bactericidal activity.

It would appear that a number of bacterial products are capable of modulating aspects of neutrophil oxidative metabolism; however, the range of parameters that need to be used to describe this modulation is incomplete, and conclusions as to precise mechanisms of action cannot be reached. It is tempting to conclude that pneumolysin acts by binding cholesterol, but it may act by inhibiting other events not yet demonstrated, e.g., protein kinase C inhibition or binding calmodulin.

8. ACID PHOSPHATASE

Acid phosphatase produced by *Leishmania donovani* promastigotes was shown by Remaley et al.[36] to inhibit superoxide production. The acid phosphatase is produced from the outer coat of the microbe.

This group of investigators has also shown that *Legionella micdadei* produces at least two acid phosphatases. One, designated ACP_2, has been shown to inhibit superoxide anion production by FMLP-stimulated human neutrophils. These workers showed that 200 units of ACP_2 reduced the rate of superoxide production by 50%. In addition, an inhibitor of ACP_2 activity, complex G, prevented the phosphatase from inhibiting FMLP-induced superoxide anion production. The molecular weight of ACP_2 was found to be 86,000 and that of *L. donovani* acid phosphatase 128,000. Since ACP_2 catalyzes the dephosphorylation of phosphoproteins, sugar phosphates, and phosphorylated amino acids, it is an important tool to explore the relevant pathophysiological substrates as outlined by Saba et al.[37]

9. CALMODULIN-BINDING PEPTIDES

It has been proposed[38] that superoxide and other oxygen metabolite production in vivo must be regulated by biochemical systems. Obviously, as outlined, the potential for invading microorganisms to exploit that regulatory system to their advantage is conceivable. Studies of factors produced by microorganisms that inhibit leukocyte oxidative metabolism will not only demonstrate mechanisms for evasion of host defenses but will facilitate understanding of normal NADPH oxidase activation, regulation, and inhibition in vivo as well.

The possibility has been suggested[39] that toxic molecules produced by other species, working through inhibition of basic biochemical mechanisms, could also serve as probes of the regulatory systems of leukocyte oxygen metabolite production. Melittin, the main constituent of honey bee venom, has been found to inhibit neutrophil superoxide and hydrogen peroxide production. This effect was dose dependent at 1–3 μg/ml and occurred at levels below the toxic range of 5 μg/ml.

Melittin inhibited production whether it was preincubated or added after the burst of superoxide production was in progress. In this respect, it resembles the effects of trifluoperazine as described by others[40] and suggests that the mechanism of inhibition of superoxide production by melittin may involve inhibition of a calmodulin-dependent step in NADPH activation. Melittin is indeed a calmodulin-binding peptide.[41] However, it also inhibits phospholipid and calcium-sensitive protein kinases and may be inhibiting through such a mechanism.[42] Whether other calmodulin-binding peptides similar to mellitin are involved in in vivo regulation of neutrophil superoxide production remains to be demonstrated.

It was previously postulated that if superoxide production is calmodulin dependent and can be inhibited by calmodulin-binding drugs and peptides,[38] this could be a mechanism for regulation of neutrophil oxidation metabolism. A number of calmodulin-binding peptides have been described (Table I).[43,44] Of these, melittin has been found to be inhibitory but, despite having a higher K_d for calmodulin, vasoactive intestinal peptide (VIP) is without inhibitory effect (S. D. Somerfield, unpublished data). It would appear, at this preliminary stage, that other properties of a peptide such as melittin, in addition to its calmodulin-binding functions, are important in determining its capacity to inhibit superoxide production. These may include its amphipathic nature or its ability to inhibit protein kinase C activity. All peptides that bind calmodulin appear to share structural features thought to be important for interactions with calmodulin: (1) clusters of basic residues, (2) hydrophobic residues adjacent to the basic residues, and (3) predicted or observed α-helical structure. These features may thus serve to project a useful basic model for a peptide that could inhibit neutrophil superoxide production.

Recently, Blumenthal et al.[45] investigated the calmodulin–protein interaction of myosin light-chain kinase, one of the better-characterized calmodulin-regulated enzymes. The amino acid sequence of the calmodulin-binding portion of this mole-

TABLE I. Amino Acid Sequences of Peptides Interacting with Calmodulin[a]

Peptide	Sequence																K_d (nM)
Mastoparan	I	N	L	K	A	L	A	A	L	A	K	K	I	L			0.3
Mastoparan X	I	N	W	K	G	I	A	A	M	A	K	K	L	L			0.9
P. mastoparan	V	D	W	K	K	I	G	Q	H	I	L	S	V	L			3.5
LK2			L	K	K	L	L	K	L	L	K	K	L	L	K	L	3.0
Melittin	G	A	N	L	K	V	L	T	T	G	L	P	A	L			3.0
VIP	M	A	V	K	K	V	L	N	S	I	L	N					50
GIP	I	R	Q	Q	D	F	V	N	W	L	L	A	Q	Q			70
δ-Hemolysin	S	T	I	G	D	L	V	K	W	I	I	D	T	V	K	N	
M 13	A	N	R	F	K	K	I	S	S	S	G	A	L	M			10
M 12	R	R	L	K	S	Q	I	L	L	K	K	Y	L	M			

[a]Diagram composed of data from Malencik and Anderson,[43] Cox et al.,[44] and Blumenthal et al.[45]

cule has been determined and is listed in Table I. It also bears a striking similarity to other calmodulin-binding peptides. It is thus conceivable that inhibitory peptides could be cleaved from precursor molecules, which could then act on the neutrophil-inhibiting superoxide production. Alternatively, degranulation of a suppressor cell could induce release of a precursor or native calmodulin-binding peptide, which could inhibit the neutrophil when it is cleaved by proteolytic enzymes released by the neutrophil. While these are conceptual models and lack direct proof, there is nonetheless tangible evidence of the existence of these types of substances in vivo; further work directed toward elucidating any role or involvement in the regulation of superoxide metabolism is clearly needed.

10. PLATELET ACTIVATING FACTOR

Platelet activating factor (PAF), a product of membrane activation, has been implicated as a phlogistic agent in a number of inflammatory diseases.[46,47] Direct stimulation of neutrophil oxidative metabolism was observed to occur at concentrations of the order of 10^{-7} M, leading to only a brief burst of superoxide production.[48] However, marked synergy was noted when PAF was supplemented with either PMN or FMLP, compared with control cells incubated with either stimuli alone. The enhancement occurred with PAF concentrations as low as 10^{-9} M/liter and occurred with equal magnitude whether cells were exposed to PAF 60 min before stimulation or immediately prior to stimulation. The response was independent of calcium and was due to an increase in the maximal rate of superoxide production. Structure–function studies with similar substances showed the necessity of a saturated fatty acid at position 1 in the molecule to be critical. Thus, PAF may serve to amplify oxygen metabolite responses in inflammatory reactions.

11. FRUCTOSE 1,6-DIPHOSPHATE

Fructose 1,6-diphosphate (FDP) is a product of intermediary metabolism that has been shown to interact with cell membranes, despite an inability to cross them, and to regulate the glycolytic pathway with consequent adenosine triphosphate (ATP) production.[49] Schinetti et al.[50] were thus prompted to investigate the effect of FDP on neutrophil superoxide production. They found that inhibition was dose dependent and that 80% inhibition occurs at 5 mM FDP concentration when cells were stimulated with PMA. FDP does not act as a scavenger or dismute O_2^-. Previously, Badwey et al.[51] had shown that FDP did not inhibit the solubilized NADPH oxidase. In Schinetti's study, they were able to show that PMA stimulation induced a drop in ATP content that was countered by FDP.[50] This continued stabilization of ATP may inhibit NADPH oxidase, as ATP has been shown to be inhibiting to the solubilized NADPH oxidase.[51,52]

12. ADENOSINE

Adenosine has been shown[53] to inhibit the ability to neutrophils to generate oxygen metabolites. This inhibition is mediated by a neutrophil adenosine A_2 receptor.[53] Adenosine has been shown to be a neurotransmitter and to act through the A_2 receptor to modulate intracellular cyclic adenosine monophosphate (cAMP).[54,55] Stimulation of the lower affinity A_2 receptors, as present on the neutrophil, causes a rise in cAMP. It was shown that superoxide production was inhibited by adenosine in response to FMLP, Con A, A23187, PMA, and C_5A.[56] Furthermore, the effect was specific for superoxide production with no effects noted on enzyme release or FMLP-induced aggregation. Pike and Snyderman[57] showed that adenosine inhibits superoxide anion production in macrophages by inhibiting transmethylation reactions; however, this mechanism was not found to be operating in polymorphonuclear leukocytes, nor was metabolism via adenosine deaminase necessary for inhibition to occur.[56] It was thus concluded[56] that (1) the adenosine-induced inhibition of superoxide generation was reversible, (2) neutrophils release significant inhibitory concentrations of adenosine into the medium in vitro, (3) exogenous adenosine deaminase reversed the effects of exogenous adenosine, and (4) the site of action is on the cell surface. It is postulated that adenosine is a retaliatory metabolite[55] that is released from normal and damaged tissue, that interacts with a specific receptor on the polymorphonuclear cell, and that prevents further neutrophil-mediated tissue damage, without interfering with lysosomal enzyme release.

13. MODULATION OF RESPIRATORY BURST BY VIRAL PRODUCTS

Immunosuppression due to viral infection may be due to the release of immunosuppressive factors from viruses or the generation of such factors from infected cells. Workers from Snyderman's group[58] previously noted inhibitory effects of retroviral transmembrane envelope protein p15E on several immune functions; these workers synthesized a 17-amino acid peptide portion termed CKS-17 and tested it for effects on monocyte function. They found that CKS-17 attached to bovine serum albumin (BSA) inhibited the respiratory burst. The peptide alone was inactive. Oxygen consumption was suppressed, as was H_2O_2 production. The peptide alone was not a scavenger. The ID (50) for O_2^- production was 7.5 mM and was about 30 times higher than for the parent protein. Other peptides, homologous to other regions of the retroviral envelope protein, were inactive. CKS-17 caused no inhibition of monocyte chemotaxis but did inhibit lymphocyte proliferative responses. The mechanism of action remains unknown but may involve effects on oxygen utilization.

14. SUMMARY AND CONCLUSIONS

A number of naturally occurring substances have been shown to inhibit and enhance oxygen metabolite production by the stimulated neutrophil. In many instances, these substances are products of invasive microorganisms obviously designed to inhibit this oxygen metabolite production in order to gain a survival advantage. No doubt more such substances will be discovered in the future. Furthermore, a more diverse range of natural products is being described from other natural sources, e.g., bee venom. It is possible that toxic molecules such as mellitin, which can be found to inhibit neutrophil superoxide production in vitro, will do so via a basic biochemical mechanism and thus provide further insight not only into the basic physiology of this system but into its potential for endogenous and exogenous regulation as well.

REFERENCES

1. Babior BM, Kipnés RS, Curnutte JT: Biological defense mechanisms: The production by leukocytes of superoxide, a potential bactericidal agent. *J Clin Invest* **52:**741–744, 1973.
2. McCord JM, Wong K: Phagocyte-produced free radicals: Roles in cytotoxicity and inflammation, in *Oxygen-Free Radicals and Tissue Damage*. CIBA Foundation Symposium 65, Amsterdam, Elsevier, 1979, pp. 343–369.
3. Fantone JC, Ward PA: Role of oxygen-derived free radicals and metabolites in leukocyte-dependent inflammatory reactions. *Am J Pathol* **107:**395–418, 1982.
4. Singal PK, Kapur N, Dhillon KS: Role of free radicals in catecholamine-induced cardiomyopathy. *Can J Physiol Pharmacol* **60:**1390–1397, 1982.
5. Donaldson J, McGregor D, La Bella F: Manganese neurotoxicity: A model for free radical-mediated neurodegeneration. *Can J Physiol Pharmacol* **9:**443–445, 1982.
6. Lunec J, Halloran SP, White RA, et al: Free radical oxidation peroxidation) products in serum and synovial fluid in rheumatoid arthritis. *J Rheumatol* **8:**233–245, 1981.
7. McCord JM: Free radicals and inflammation: Protection of synovial fluid by superoxide dismutase. *Science* **185:**529–531, 1974.
8. Hyrst NP, Bessac B, Nuki G: Monocyte superoxide anion production in rheumatoid arthritis: Preliminary evidence for enhanced rates of superoxide anion production by monocytes from patients receiving penicillamine, sodium aminothiomalate and corticosteroids. *Ann Rheum Dis* **43:**28–33.
9. Schalkwijk J, van den Berg WBl, van de Putte LBA, et al: Hydrogen peroxide suppresses the proteoglycan synthesis of intact articular cartilage. *J Rheumatol* **12:**205–210, 1985.
10. Weiss SJ, Lampert MB, Test ST: Long-lived oxidants generated by human neutrophils: Characterization and bioactivity. *Science* **222:**625–628, 1985.
11. Weiss SJ, Pepping G, Ortiz X, et al: Oxidative autoactivation of latent collagenase by human neutrophils. *Science* **227:**747–749, 1985.
12. Babior BM: The respiratory burst of phagocytes. *J Clin Invest* **73:**599–601, 1984.
13. Lewis DA: Endogenous anti-inflammatory factors. *Biochem Pharmacol* **33:**1705–1714, 1984.
14. Menander-Huber KB: Double blind controlled clinical trials in man with bovine copper zinc superoxide dismutase (orgotein), in Bannister WH, Bannister JV (eds): *Biological and Clinical Aspects of Superoxide and Superoxide Dismutase*. Amsterdam, Elsevier/North-Holland, 1980, pp. 408–423.
15. Koyrounakis LH, Yiangou M: Bee venom and adjuvant-induced disease. (Letter.) *J Rheumatol* **10:**522, 1983.

16. Chang YH, Bliven ML: Antiarthritic effect of bee venom. *Agents Actions* **9:**205–211, 1979.
17. McCord JM, Wong K, Stokes SH, et al: Superoxide and inflammation: A mechanism for the anti-inflammatory activity of superoxide dismutase. *Acta Physiol Scand* **494**(suppl)**:**25–30, 1980.
18. Flower R, Blackwell AJ: Anti-inflammatory steroids induce biosynthesis of a phospholipase A_2 inhibitor which prevents prostaglandin generation. *Nature (Lond)* **278:**456–459, 1979.
19. Hirata F, Schiffmann E, Venkatasubramanian K, et al: A phospholipase A_2 inhibitory protein in rabbit neutrophils induced by glucocorticoids. *Proc Natl Acad Sci USA* **77:**2533–2536, 1980.
20. Travis J, Johnson D: Human $alpha_1$ protease inhibitor. *Methods Enzymol* **80:**754–765, 1981.
21. Clark RA, Stone PJ, Haga GD, et al: Myeloperoxidase-catalysed inactivation of $alpha_1$-protease inhibitor by human neutrophils. *J Biol Chem* **256:**3348–3353, 1981.
22. Cochrane CG: Plasma proteins and inflammatory disease. *Pharmacol Rev* **340:**39–42, 1982.
23. Kitagawa S, Takaki F, Sakamota S: Serine protease inhibitors inhibit superoxide production by human basophils stimulated by anti-IgE. *Biochem Biophys Res Commun* **95:**801–806, 1980.
24. Kitagawa S, Takai F, Sakamoto S: Evidence that proteases are involved in superoxide production by human polymorphonuclear leukocytes and monocytes. *J Clin Invest* **65:**74–81, 1980.
25. Kitagawa S, Takai F, Sakamoto S: Possible involvement of proteases in superoxide production by human polymorphonuclear leukocytes. (Letter.) *FEBS Lett* **99:**275–282, 1979.
26. Bird J, Mohd-Hidir S, Lewis DA: Putrescine—A potent endogenous anti-inflammatory substance in inflammatory exudates. *Agents Actions* **13:**348–353, 1983.
27. Qi DF, Schatzman RC, Mazzei GJ, et al: Polyamines inhibit phospholipid-sensitive and calmodulin-sensitive Ca^{2+}-dependent protein kinases. *Biochem J* **281:**281–288, 1983.
28. Porter R, O'Connor (eds): *Substance P in the Nervous System.* CIBA Foundation Symposium 91. London, 1982.
29. Hartung HP, Toyka KV: Activation of macrophages by substance P. Induction of oxidative burst and thromboxane release. *Eur J Pharmacol* **89:**301–305, 1983.
30. Hartung HP, Walters K, Toyka KV: Substance P binding properties and studies on cellular responses in guinea pig macrophages. *J Immunol* **136:**3856–3863, 1986.
31. Paton JC, Ferrante A: Inhibition of human polymorphonuclear leukocytes respiratory burst, bactericidal activity and migration by pneumolysin. *Infect Immun* **41:**1212–1216, 1983.
32. Johnson MK, Geoffroy C, Alouf JE: Binding of cholesterol by sulfhydryl-activated cytolysins. *Infect Immun* **27:**97–101, 1980.
33. Anderson BR, Duncan JL: Activation of human neutrophil metabolism by streptolysin O. *J Infect Dis* **141:**680–685.
34. Cavalieri SJ, Snyder IS: Effect of E. Coli alpha-hemolysin on human peripheral leucocyte function in vitro. *Infect Immun* **37:**966–974, 1982.
35. Gemmell CJ, Peterson PM, Schmelling DG, et al: Effect of staphylococcal alpha-toxin on phagocytosis of staphylococci by human polymorphonuclear leucocytes. *Infect Immun* **38:**975–980, 1982.
36. Remaley AT, Kuhns BD, Basford RE, et al: Leishmania phosphatase blocks neutrophil O_2 production. *J Biol Chem* **259:**11173–11175, 1984.
37. Saha AK, Dowling JD, La Marco K, et al: Properties of an acid phosphatase from Legionella micdadei which blocks superoxide anion production by human neutrophils. *Arch Biochem Biophys* **243:**150–160, 1985.
38. Somerfield SD: Bee venom and arthritis. *NZ Med J* **99:**281–283, 1986.
39. Somerfield SD, Stach JL. Mraz C, et al: Bee venom melittin blocks neutrophil O_2 production. *Inflammation* **10:**175–182, 1986.
40. Takeshige K, Minakami S: Involvement of calmodulin in phagocytic respiratory burst of leukocytes. *Biochem Biophys Res Commun* **99:**484–490, 1981.
41. Comte M, Maulet Y, Cox JA: Ca^{2+}-dependent high-affinity complex formation between calmodulin and melittin. *Biochem J* **209:**269–272, 1983.
42. Kato N, Raynor RL, Wise BC, et al: Inhibition by melittin of phospholipid-sensitive and calmodulin-sensitive Ca^{2+}-dependent protein kinases. *Biochem J* **202:**217–224, 1982.
43. Malencik DA, Anderson SG: Binding of hormones and neuropeptides by calmodulin. *Biochemistry* **22:**1995–2001, 1983.

44. Cox JA, Comte M, Fitton JE, et al: The interaction of calmodulin with amphiphilic peptides. *J Biol Chem* **260:**2527–2534, 1985.
45. Blumenthal DK, Takio K, Eldman A, et al: Identification of the calmodulin-binding domain of skeletal muscle light chain kinase. *Proc Natl Acad Sci USA* **82:**3187–3191, 1985.
46. Archer CB, Page CP, Paul W, et al: Inflammatory characteristics of platelet activating factor (PAF) in human skin. *Br J Dermatol* **110:**45–50, 1984.
47. Camussi G, Palowski I, Teita C, et al: Acute lung inflammation induced in the rabbit by local instillation of 1-*O*-octadectyl-2-acetyl-*sn*-glyceryl-3 phosphorylcholine or of native platelet-activating factor. *Am J Pathol* **112:**78–88, 1983.
48. Gay JC, Beckman JK, Zaboy KA: Modulation of neutrophil oxidative responses to soluble stimuli by PAF. *Blood* **67:**931–936, 1986.
49. Kirtley ME, McKay M: Fructose-1,6-biphosphate, a regulator of metabolism. *Mol Cell Biochem* **18:**141–149, 1977.
50. Schinetti ML, Lazzarino G: Inhibition of phorbol ester stimulated chemiluminescence and superoxide production in human neutrophils by fructose 1-6-diphosphate. *Biochem Pharmacol* **35:**1762–1764, 1986.
51. Badwey JA, Karnovsky ML: Production of superoxide and hydrogen peroxide by an NADH-oxidase in guinea pig polymorphonuclear leukocytes. *J Biol Chem* **254:**11530–11537, 1979.
52. Babior BM, Peters WA: The O_2-producing enzyme of human neutrophils. *J Biol Chem* **256:**2321–2323, 1981.
53. Cronstein BN, Kramer SB, Weissmann G, et al: Adenosine: A physiological modulator of superoxide anion generation by human neutrophils. *J Exp Med* **158:**1160–1177, 1983.
54. Sattin BN, Rall TW: The effect of adenosine and adenine nucleotides on the cyclic adenosine 3′,5′-phosphate content of guinea pig cerebral cortex slices. *Mol Pharmacol* **6:**13–23, 1970.
55. Van Calker D, Muller M, Hamprecht B: Adenosine regulates via two different types of receptors, the accumulation of cyclic AMP in cultured brain cells. *J Neurochem* **33:**999–1005, 1979.
56. Cronstein BN, Kramer SB, Rosenstein ED, et al: Adenosine modulates the generation of superoxide anion by stimulated human neutrophils via interaction with a specific cell surface receptor. *Ann NY Acad Sci* **451:**291–301, 1985.
57. Pike MC, Snyderman R: Transmethylation reactions regulate affinity and functional activity of chemostatic factor receptors on macrophages. *Cell* **28:**107–114, 1982.
58. Harrell RA, Cianciolo GJ, Copeland TD, et al: Suppression of the respiratory burst of human monocytes by a synthetic peptide homologous to envelope proteins of human and animal retroviruses. *J Immunol* **136:**3517–3520, 1986.

10

Oxidants Formed by the Respiratory Burst

Their Physiological Role and Their Involvement in the Oxidative Metabolism and Activation of Drugs, Carcinogens, and Xenobiotics

PETER J. O'BRIEN

1. INTRODUCTION

Phagocytic neutrophils exhibit a burst in oxygen consumption that is cyanide resistant—the respiratory burst—when the cells are exposed to various soluble stimuli (e.g., chemotactic oligopeptides, complement fragments) or particulate stimuli (e.g., opsonized bacteria).[1] The respiratory response involves a recognition site on the plasma membrane surface that leads to the release of an intracellular signal.[1] Transduction of this signal causes the activation of a plasma membrane NADPH oxidase that we suggested was responsible for the respiratory burst,[2] and that was confirmed by others.[1] The superoxide anion $O_2^{-}\cdot$ formed is released outside the cell, where it dismutates into H_2O_2. H_2O_2 may also diffuse back into the neutrophil, where it is decomposed by intracellular myeloperoxidase, GSH/GSH–peroxidase and catalase. When particulate stimuli are completely engulfed within a phagocytic vacuole, superoxide and the dismutation product H_2O_2 accumulate in the vacuole and diffuse into the cytoplasm.

We have shown that protons are released by activated neutrophils, probably as a result of activation of the plasma membrane NADPH oxidase.[3] Vacuolar membrane NADPH oxidase could also be responsible for the acidification of the vacuoles following phagocytosis. At pH 5, ferric iron is released from plasma transferrin and possibly neutrophil lactoferrin. Superoxide radicals reduce the ferric iron, and the ferrous iron formed can then form hydroxyl radicals from hydrogen perox-

PETER J. O'BRIEN • Faculty of Pharmacy, University of Toronto, Toronto, Ontario M5S 1A1, Canada.

ide.[4] The sequence of $O_2^{-}\cdot$–H_2O_2–$OH\cdot$ formation has been demonstrated for neutrophils following activation[5]:

$$\begin{aligned} NADPH + H^+ + 2O_2 &\rightarrow NADP^+ + 2O_2^{-}\cdot + 2H^+ \\ 2O_2^{-}\cdot + 2H^+ &\rightarrow H_2O_2 + O_2 \\ O_2^{-}\cdot + Fe^{3+} &\rightarrow O_2 + Fe^{2+} \\ H_2O_2 + Fe^{2+} &\rightarrow OH\cdot + Fe^{3+} \end{aligned}$$

Following the respiratory burst, specific (secondary) granules and then azurophil granules merge with the forming phagosome, and their contents are discharged into the phagocytic vacuole.[6] Specific granules release lactoferrin, lysozyme, and collagenase, whereas azurophil granules release myeloperoxidase, various lysosomal enzymes, elastase, and cationic proteins. Even in the absence of phagocytosis, some enzyme secretion from the neutrophil occurs with these stimuli, particularly when neutrophils are treated with cytochalasin B to impair normal actin function. The secretion is then markedly enhanced and the leukocytes become a secretory organ first for the specific granules and then for the azurophil granules.[6] The released myeloperoxidase decomposes the extracellular H_2O_2 and forms HOCl with plasma chloride.

Myeloperoxidase uses chloride as a substrate and forms hypochlorite with H_2O_2. The optimal pH for the myeloperoxidase-mediated antimicrobial system is ~pH 5.0; after phagocytosis, the pH value in the vacuolar space may fall to this level. Other oxidizing agents include singlet oxygen, which forms hypochlorite with H_2O_2. Plasma membrane arachidonate, under some circumstances, may also be transformed by phospholipases and lipoxygenase into hydroperoxides, which then transform into leukotrienes. The reductive activation of oxygen results in the formation of oxidants (e.g., H_2O_2, $OH\cdot$, OCl^-, lipid hydroperoxides, and higher oxidation states of peroxidase), which enables neutrophils recruited to the inflammatory sites to (1) engulf and kill infective agents and tumor cells, and (2) destroy damaged tissues and produce signals for the development and evolution of the inflammatory process, as well as other activities.

The following discussion compares the role played by these oxidants in the function of the neutrophil, e.g., bactericidal activity, inactivation of various chemotactic peptides, leukotrienes, α_1-antitrypsin, and prostaglandins, and activation of proteases associated with the inflammatory process. The role of these oxidants in tumor cell lysis and the turnover of plasma thyroxine (T_3) and oestradiol is also discussed. Antithyroid drugs, nonsteroidal anti-inflammatory drugs, and antidepressants are also oxidized by activated neutrophils, and neutrophils could therefore contribute to their in vivo metabolism.

Neutrophil-catalyzed activation of carcinogens could be responsible for the enhanced incidence of cancer associated with chronic inflammatory states. Recently we found that the arylamine carcinogens, [^{14}C]benzidine, [^{14}C]-*N*-methylaminoazobenzene, and [^{14}C]aminofluorene incubated with leukocytes are co-oxidized and bind irreversibly to leukocyte nuclear DNA if the leukocytes are activated by the tumor promoter phorbol myristate acetate (PMA).[7] The binding was dependent on oxygen and was decreased by sulfhydryl inhibitors and phenolic antioxidants that inhibit the respiratory burst. It is likely that the NADPH oxidase and protein kinase C are thiol-dependent enzymes. Both the binding and the respiratory burst were increased by low concentrations of azide, presumably as a result of enhanced H_2O_2 levels due to intracellular catalase inhibition. However, higher azide and cyanide concentrations prevented binding without affecting the respiratory burst, indicating that myeloperoxidase is a catalyst for the binding. A 20- or 40-fold higher level of DNA binding was formed when H_2O_2 and myeloperoxidase (or granules from the activated leukocytes) were used to catalyze the irreversible binding of these carcinogens to exogenous calf thymus DNA. This suggests that proteins and hydrogen donors in the neutrophils bind or reduce the oxidized products of the arylamine carcinogens. [^{14}C]phenol and [^{14}C]hydroquinone also caused considerable macromolecular binding in activated neutrophils. Phenol and hydroquinone are metabolites of benzene, a leukemogen; their oxidative activation by bone marrow via eosinophil peroxidase or myeloperoxidase could be an important event in the initiation of benzene induced leukemia. Benzo(α)pyrene or its 7,8-diol metabolite can also be oxidized to the ultimate carcinogen by activated leukocytes and thus could contribute to the processes involved in the induction of lung cancer by polycyclic aromatic hydrocarbons (PAH).

2. EXTRACELLULAR VERSUS INTRACELLULAR OXIDATION AND HALOGENATION

The role of the myeloperoxidase system in neutrophils is thought to operate both intravacuolarly as well as extracellularly. The former is believed to catalyze the killing of ingested microorganisms. However, the secretion of myeloperoxidase (up to 28% of the neutrophil content) and H_2O_2 into the extracellular fluid may serve such biological functions as (1) lysis of mammalian cells including tumor cells, (2) killing of hyphal forms of invasive fungi, (3) destruction of metazoan parasites, (4) oxidative inactivation of mediators of inflammation such as chemotactic factors and protease inhibitors (e.g., α-1-trypsin inhibitor) in the serum, and (5) oxidative inactivation of microbial toxins.

Activated leukocytes also release a neutral membrane-bound serine proteinase into the medium, which is synergistic with H_2O_2 in its cytolytic activity.[8] Latent collagenase released by stimulated leukocytes[9] is also activated by the released hypochlorite. These effects could result in significant damage to the surrounding connective tissue structures, associated with lung injury, such as adult respiratory distress syndrome (ARDS) or emphysema. The secreted myeloperoxidase reacts with superoxide to form a superoxoferric or oxyferrous form (compound III). However, although compound III is less reactive with most peroxidase substrates than peroxidase compound I or II formed with H_2O_2, it forms HOCl from H_2O_2 and Cl^- at the same rate as the ferric enzyme.[10] Many investigators believe that HOCl could be the reactive species for the above biological effects and chlorination of a variety of compounds has been demonstrated.[11] Halogenation can occur in the phagocytic vacuole as autoradiography shows that iodination is found in the intravacuolar space following phagocytosis.[12] Furthermore, serum-opsonized fluorescein covalently bound to zymosan becomes chlorinated and loses its fluorescence when ingested by activated leukocytes in a chloride medium.[13] However some investigators have suggested that chlorination is mostly extracellular.[11] The antimicrobial activity instead was attributed to oxygen-independent mechanisms, e.g., cationic granular proteins.[10,14] How these proteins kill bacteria is a mystery, but they may damage the bacterial membrane. A survey of antimicrobial activity of leukocytes under anaerobic versus aerobic conditions concluded that effective antimicrobial activity rest on the coexistence of O_2-independent bactericidal proteins specific for certain gram-negative bacteria and O_2-requiring systems that attack all cells.[14]

3. TRAPPING ACTIVE CHLORINE GENERATED BY THE RESPIRATORY BURST

Unlike chloroperoxidase that oxidizes acceptor molecules via the peroxidase compound I–chloride complex and a ferriperoxidase hypochlorite complex,[15] myeloperoxidase preferentially catalyzes the peroxidation of the chloride ion to free hypochlorous acid.[16] Taurine chloramine was formed in stoichiometric amounts when a large excess of taurine (5 mM) was used to trap the hypochlorite generated by a myeloperoxidase, H_2O_2, Cl^- reaction system. With phorbol myristate acetate (PMA) or zymosan-stimulated neutrophils,

$$\underset{\text{taurine}}{HOCl + H_2N\text{–}CH_2CH_2SO_3H} \rightarrow \underset{\text{taurine chloramine (250 nm)}}{ClNH\text{–}CH_2CH_2SO_3H} + H_2O$$

15 mM taurine also effectively trapped the hypochlorite formed.[17] Taurine chloramine formation was measured by the oxidation of iodide to iodine or *S*-thio-2-nitrobenzoic acid to disulfide. Neutrophil formation was prevented by the myeloperoxidase inhibitors azide, cyanide, and aminotriazole. Most of the chloramine formed was in the extracellular milieu, and its formation was inhibited by catalase, which suggests the chloramine formation was mainly extracellular. In the absence of Cl^- (0.1 M), only 8% chloramine formation by the stimulated neutrophils occurred. No further oxidation to dichloramines occurred in the leukocytes. Endogenous amines such as taurine, amino acids, peptides, and granule proteins could also form hydrophilic chloramines (RNHCl) that were released into the medium and were of low toxicity.[18] However, the hydrophilic chloramines or HOCl could react with endogenous NH_4^+ to yield the lipophilic oxidizing agent monochloramine that was bactericidal, cytotoxic, and cytolytic[18]:

$$RNHCl + NH_4^+ = RNH_2 + NH_2Cl + H^+$$
$$HOCl + RNH_2 = H_2O + RNHCl$$
$$HOCl + NH_4^+ = H_2O + NH_2Cl + H^+$$

Monochlorodimedon (0.8 mM) is a more reactive trap for hypochlorite formation by stimulated neutrophils.[10] Monochlorodimedon absorbs at 290 nm (E mM = 17.7), unlike the product dichlorodimedon. This method was used to show that PMA-stimulated neutrophils formed 60 μM hypochlorite in 12 min.[10] This corresponds to one half the H_2O_2 formed. Myeloperoxidase was involved as azide prevented hypochlorite formation. Added catalase also inhibited suggesting that the chlorination was extracellular:

At high H_2O_2 concentrations, less hypochlorite is formed in the myeloperoxidase system.[19] This was attributed to the inactivation of myeloperoxidase. However, it is more likely the result of oxidation of hypochlorite by H_2O_2 resulting in O_2 formation (apparent catalatic activity)

$$H_2O_2 + HOCl \rightarrow O_2 + HCl + H_2O$$

first reported for OBr^-.[20] Part of the oxygen may be released as singlet oxygen in the case of HOCl or HOBr[20] but not I_2.

Ascorbate stimulates hypochlorite formation by the myeloperoxidase system threefold, as measured with monochlorodimedone. The optimum pH was also increased from 5.2 to 6.0. This was attributed to reduction of inactive peroxidase compound II formed during turnover, to the native peroxidase.[20] Ascorbate is also present in leukocytes at about 0.7 mM and would protect the leukocyte by reacting with the hypochlorite formed. However, the concentration of ascorbate in the phagolysosome may be much lower and substantial hypochlorite could be formed in this location. Compound II could accumulate under low chloride concentrations. Chloride or electron donors oxidized by compound I would prevent formation of compound II.

Carbon-14-1,3,5-trimethoxybenzene (100 μM) has also been used to trap active chlorine generated by human leukocytes stimulated with polystyrene beads. The 2-chloro-1,3,5-trimethoxybenzene product was determined by high-performance liquid chromatography (HPLC). At least 28% of the oxygen consumed could be accounted for by active chlorine formation.[21] This value represents a lower limit, as the active chlorine formed will react with other molecules in the leukocyte, such as proteins or glutathione (GSH). Little active chlorine could be trapped with myeloperoxidase-deficient polymorphonuclear leukocytes, leukocytes from patients with chronic granulomatous disease (CGD), or azide-treated normal leukocytes.

Dyes have been used extensively to show that the surface membrane potential decreases and that the intravacuolar pH initially increases following leukocyte activation.[22] However, recently similar changes have been found with hypochlorite or the myeloperoxidase–H_2O_2–chloride system which suggests a reevaluation of the leukocyte results is needed. Fluorescein, the pH-sensitive dye used,[22] forms sequentially 4′-chlorofluorescein, 4′,5′-dichlorofluorescein and other products, causing decreased fluorescence and a shift of proton equilibrium constants to more acidic values. Fluorescein (1.5 μM), covalently bound to zymosan also formed these products when the unopsonized particles were incubated with phorbol myristate acetate or *N*-formyl-methionyl-leucyl-phenylalanine-stimulated human neutrophils. The decreased fluorescence was consistent with chlorination rather than with alkalinization of the zymosan-containing vacuoles.[21] This research also provides proof that active chlorine is formed in the phagocytic vacuole where it could chlorinate and oxidize bacteria. A variety of sulfonphthalein dyes, including bromothymol blue, bromphenol blue, bromcresol green, and bromcresol purple dyes, underwent oxidative bleaching to yellow solutions. The apparent second-order rate constants as determined by stop-flow methods[13,23] are shown in Table I.

The fluorescent dye 3,5-dipropylthiodicarboxylamine used to probe cellular membrane potential also undergoes a halide-dependent myeloperoxidase-catalyzed fluorescence quenching both in cell-free enzymatic systems and in the presence of

TABLE I. Comparison of the Rates of Oxidative Bleaching of Various Dyes and Biomolecules by Hypochlorite

Dye/biomolecule oxidized	Rate constants (apparent second order) (M^{-1} sec^{-1})	Reference
Fluorescein pH 7.4	4×10^3	13
pH 6.4	7×10^3	13
Bromcresol green	91	13
Bromphenol blue	15	13
Bromcresol purple	3.1×10^3	13
Bromthymol blue	6.9×10^3	13
Spinach ferrodoxin	$\geq 10^5$	23
carotene	2.3×10^4	23
Cytochrome *c*	2×10^3	23
Adenosine monophosphate	2×10^3	23

activated neutrophils.[22] In this case, thioether oxidation probably occurred. The consequence is that whereas this probe still fluoresces normally as a result of the membrane depolarization that correlates with the respiratory burst, there is a rapid decay of fluorescence due to active chlorine. The fluorescence was much more stable in myeloperoxidase-deficient leukocytes, as would be expected. The sulfur-free 3,3′-dipentyloxadicarbocyanine also used for determining membrane potential was not affected. The membrane potential dyes used are believed not to permeate the leukocyte, so that the decreased fluorescence of the former dye may serve as an indicator of the secretion of myeloperoxidase and H_2O_2. To assess membrane potential and proton changes with the above probes, it is therefore important to use a halide free medium or myeloperoxidase inhibitors or myeloperoxidase deficient leukocytes if oxidative bleaching is to be avoided.

4. PHYSIOLOGICAL ROLE OF HALOGENATION AND OXIDATION

4.1. Bactericidal Activity and Cell Death

Bacteria were chlorinated following phagocytosis by neutrophils with $^{36}Cl^-$ in the medium. No chlorination of the neutrophils occurred in the absence of bacteria. Chlorination was inhibited by the myeloperoxidase inhibitors azide or cyanide or the competitive substrate iodide. Lysosomal proteins were also chlorinated, which suggests that only the contents of the phagosome were chlorinated. The cell-free system myeloperoxidase–H_2O_2 also chlorinated added serum albumin.[24] No information is available as to what chemical changes have occurred. However, serum albumin or B lactoglobulin incubated with myeloperoxidase–H_2O_2–iodide resulted in the oxidation of sulfhydryls to sulfenyl iodide with some disulfide formation.[25] Iodide was required. Other investigators have shown that iodination of tyrosine and histidine occurs in insulin and other proteins when treated with iodine.[15] Tryptophan and methionine can also be oxidized.[15] Iodination of unsaturated membrane lipids occurs as well. Arachidonic acid is transformed into a variety of iodinated compounds, including at least three distinct iodolactones.[26] Halogenation may lead to irreversible damage of the bacteria presumably by decarboxylation of amino acids and by oxidative peptide cleavage and/or halogenations and bleaching of DNA, NADPH, and ATP.[27]

However, bactericidal activity seems to correlate with the oxidation of bacterial thiol groups rather than chlorination of the bacterial cell wall.[27] Oxidation of thiol groups is, however, mediated by halogenation. Another suggestion is that the microbicidal action of hypochlorite arises from the respiratory loss caused by bleaching of cytochrome *b*.[23] Oxidation of iron–sulfur proteins or cytochromes should be reversible and not toxic. The rate constants for the reaction of hypochlorite with the iron sulfur protein ferredoxin and cytochrome *c* are shown in Table I.

4.1.1. Hypochlorite- or Chloramine-Mediated Mammalian Cell Lysis

Stimulated neutrophils may play a physiological role in killing mammalian tumor cells. However, this cytotoxic activity in various inflammation acute disease states could kill other cells (e.g., endothelial and epithelial cells) and cause lung damage in ARDS. Tumor cell killing involves the respiratory burst and may be due to hypochlorite when leukocytes are activated by phagocytosis, lectin or PMA.[28–31] However, myeloperoxidase does not seem to be involved in antibody dependent cellular cytotoxicity in which effector cells are triggered by antibody-coated target cells. Oxygen-independent mechanisms may be important in this case, although H_2O_2 also plays a role.[32]

Cytotoxicity against murine lymphoma cells by lectin-activated neutrophils is rapid, with significant killing (about 20 tumor cells per neutrophil). The respiratory burst is required as no cytotoxicity occurred with leukocytes obtained from patients with chronic granulomatosis disease (CGD). Furthermore, addition of H_2O_2 restored the ability of these leukocytes to cause cytotoxicity. Halides are also required.[29] Cytotoxicity was prevented by catalase or myeloperoxidase inhibitors and did not occur with neutrophils deficient in myeloperoxidase. The activation was shown to involve the binding of lectin to the effector cell and the secretion of the lysosomal enzyme myeloperoxidase.[29] Tumor cell injury was confirmed by release of ^{51}Cr, loss of ability to exclude trypan blue, inhibition of glucose oxidation, and loss of ability to cause tumors in test mice.[30] With iodide as a cofactor, cytotoxicity is associated with iodination of tumor cell proteins.[30] A myeloperoxidase, H_2O_2 and halide system or hypochlorous acid is highly cytotoxic to tumor cells eg. a human-T-lymphoblast cell line,[28] mouse ascitic lymphoma cells.[29,31] PMA-activated human T lymphoblast cells are also cytotoxic to a wide range of different tumor cells as a target.

Erythrocytes treated with chloramine underwent a rapid, dose-dependent oxidation of hemoglobin and oxidation of GSH to GSSG and mixed protein disulfides. Glucose partially restored the hemoglobin and GSH levels of the erythrocyte. The partial nature of the restoration suggests that inactivation of glucose 6-phosphate dehydrogenase (G6PD) or glutathione reductase had occurred.[33] Prior alkylation of GSH with *N*-ethymaleimide prior to treatment with NH_2Cl increased methemoglobin formation.[34] GSH depletion by chlorite also led to H_2O_2 formation in the erythrocyte.[35]

4.1.2. H_2O_2-Mediated Mammalian Cell Lysis

The cytolytic activity of macrophages and peritoneal granulocytes activated by PMA, however, seems to be mediated by H_2O_2 because although catalase prevented cytotoxicity, azide or cyanide did not inhibit cytotoxicity.

Tumor cell targets differ widely in their sensitivity to killing by macrophages. There seems to be a strong inverse correlation between their sensitivity to lysis by H_2O_2 (2–100 μM) and their GSH content but not with levels of catalase, glutathione peroxidase, or glutathione reductase. Inactivation of glutathione reductase or depletion of glutathione peroxidase (with a selenium deficient diet) markedly increased the sensitivity of the more resistant tumor cells. Inhibition of catalase also increased their sensitivity.[36] This suggests that the glutathione redox system is an important defense against H_2O_2-induced oxidative stress. The antitumor effect of H_2O_2 has been recently demonstrated.[37]

Neutrophils stimulated with PMA destroyed suspensions or monolayers of cultured human endothelial cells. The cytotoxicity was attributed to H_2O_2 and not to hypochlorous acid.[38] Endothelial cells are 14-fold more sensitive to H_2O_2 than are alveolar epithelial cells as a result of their lower H_2O_2 metabolism capacity. Injury to the endothelial cells can lead to a loss of vascular integrity and is probably important in the pathogenesis of infection or immune-mediated atherosclerosis and vasculitides.

Bovine aortic endothelial cells were more resistant and were lysed 50% by 500 μM H_2O_2 or hypochlorite. However, after inactivation of glutathione reductase, 10 μM H_2O_2 caused 50% lysis, even though the cells were more resistant to hypochlorite.[39] Depletion of cellular GSH or inhibition of GSH synthesis also increased endothelial cell lysis by H_2O_2. The GSH redox cycle therefore plays an important role as an endogenous defense mechanism against H_2O_2 and could explain the cytotoxic sensitivity of different target cells. Some investigators believe that hydroxyl radicals generated in the target cell may be responsible for the H_2O_2-induced bovine pulmonary artery endothelial cell cytotoxicity, as protection occurred with hydroxy radical scavengers or iron chelators but not with myeloperoxidase inhibitors.[40] These results suggest that the prevention of oxidant-induced endothelial damage leading to pulmonary edema and then ARDS could involve enhancing endothelial cell GSH and the use of iron chelators. Stimulated neutrophils probably kill alveolar epithelial cells in an oxygen metabolite-independent manner, e.g., with cationic granule proteins and proteases, and was promoted by adherence of the neutrophils to epithelial cell monolayers.[41]

4.1.3. Bromination–Oxidation of Bacteria and Cells

The eosinophil plays a uniquely beneficial role in destroying parasite worms and tumor cells. They also play a detrimental role in inflammatory disease states by damaging host tissues. Eosinophils are less bactericidal than neutrophils, although they have a greater respiratory burst. Tumor cell cytotoxicity is similar and is abolished by peroxidase inhibitors.[42]

Eosinophil peroxidase is a different gene product from myeloperoxidase and differs in structure and properties. This peroxidase has the ability as a strongly positive cationic protein to bind firmly to negatively charged target surfaces. In vivo extracellular fluids contain 20–100 μM bromide. Eosinophil peroxidase is also much more effective than myeloperoxidase in oxidizing bromide. A comparison of eosinophils and leukocytes is presented in Table II.[43] Eosinophils can also release their peroxidase. Cells with low peroxidase activity (e.g., mast cells, alveolar

TABLE II. Comparison of Oxidant Formation by Activated Eosinophils and Neutrophils

Oxidant formation activity	nmoles per 10^6 cells	
	Eosinophils	Neutrophils
$O_2^{-}\cdot$ formation[a]	83.2	43.1
HOCl formation[a]	25.8	121.1
Cell bromination[b]	5.2	0.4

[a]Activated by FMLP.
[b]Activated by PMA.

macrophages, and monocytes), which have some oxygen-dependent cytotoxic mechanisms, could therefore increase their cytotoxic potential by using extracellular or phagocytosed eosinophil peroxidase bound to target surfaces. These cells could then oxidize bromide to cytotoxic levels of active bromine.[43] Eosinophil peroxidase levels are normal in patients with myeloperoxidase deficiency, so that phagocytosis by the PMN of eosinophil peroxidase-coated bacteria could compensate for the myeloperoxidase deficiency.

Although active chlorine has a higher redox potential and is a stronger oxidizing agent than active bromine, active bromine is more effective at aromatic and aliphatic halogenation and is a stronger bleaching agent. Iodine is weakest as an oxidizing agent but is strongest in halogenation or halogen-mediated oxidation. In vivo iodination of tumor cells has been demonstrated in mice given antitumor antibody conjugated to glucose oxidase, when lactoperoxidase and ^{125}I were also administered.[32]

4.1.4. Role and Toxicity of *N*-Chloroamines

Hypochlorous acid reacts with a wide variety of neutrophil nitrogen (RNH_2) (e.g., taurine) containing compounds to form *N*-chloroamines (see Section 3). This reaction may occur endogenously or extracellularly following secretion of these nitrogen-containing compounds. The *N*-chloroamines (RNHCl and $RNCl_2$) are still powerful oxidizing agents, even though they have a lower redox potential than HOCl but have a much longer half-life than HOCl. Thus, they still oxidize protein thiols and methionine, halogenate tyrosine, and histidine and are incorporated into proteins.[44] They are mutagenic, presumably as a result of halogenating DNA bases.[45] Chloramine induces DNA single-strand breaks and inactivates endonucleases in *Bacillus subtilis*.[46] They are generated by neutrophils, have a half-life of approximately 18 hr, are bactericidal, and inactivate methionine-dependent chemotactic peptides and a plasma α_1-protease inhibitor.[47] Monochloroamine NH_2Cl) formed from the reaction between hypochlorite and plasma ammonium ions differs from other *N*-chloroamines and hypochlorite in that it is lipophilic and therefore readily crosses membranes. These properties enable NH_2Cl to inactivate the protease inhibitor, even when the neutrophils and protease inhibitor are physically separated.

Activated leukocytes readily cause the oxidation of oxyhemoglobin and lysis in intact erythrocytes. It is likely that the former was caused by NH_2Cl, whereas the latter was caused by HOCl.[48] NH_2Cl was 50 times more effective than H_2O_2 or HOCl as an oxidant for hemoglobin in intact erythrocytes. whereas HOCl was 10 times more effective than NH_2Cl as a lytic agent.[48] The effectiveness of NH_2Cl in oxidizing hemoglobin may reflect its ease of penetrating the hydrophobic heme pocket.

4.1.5. Biomolecule Modification and Cytotoxicity Mechanisms

4.1.6a. H_2O_2 Mediated. Cytotoxicity in various different endothelial cells seems to correlate with the ability of H_2O_2 to deplete intracellular GSH. This reflects initial GSH levels and the reductive processes required to restore GSH following oxidation. This leads to the oxidation of NADPH and activation of the hexose monophosphate shunt.[49] Extensive studies were recently carried out into the sequence of distal events induced by H_2O_2 that precipitates irreversible injury to cultured bovine aortic endothelial cells culminating in lysis. GSH oxidation in the mitochondria leads to extensive NADPH oxidation and hydrolysis. In turn, this activates ADP-ribose polymerase,[50] causing ADP ribosylation of the mitochondrial membrane and the release of Ca^{2+} stored in the mitochondria. Increased cytosolic Ca^{2+} levels[51] may increase microtubule polymerization and thereby cause surface-membrane bleb formation. Activation of membrane phospholipases and proteases may then lead to irreversible changes and cell cytolysis.

Other processes contribute to the irreversibility of the above changes induced by H_2O_2, including a fall in intracellular ATP, which thereby prevents the operation of ionic pumps. DNA strand breaks also initiates the DNA repair process by activating polyadenosine diphosphate-ribose polymerase. The marked activation of the DNA repair process leads to depletion of cellular NAD^+.[50]

4.1.6b. HOCl or NH_2Cl Mediated. The above cytotoxicity mechanisms ascertained for H_2O_2 have not yet been investigated for active chlorine oxidizing agents. However, the key molecules initially modified, i.e., GSH, NADPH, ATP, and DNA, have been shown to be modified by active chlorine in the following ways. Glutathione oxidation by H_2O_2 to oxidized glutathione is poorly catalyzed by myeloperoxidase but is increased 5.5-fold by chloride.[52] Iodide was much more effective and bromide was intermediate in its effectiveness. An optimum pH of 5.5–6 was found. No evidence of halogenation was found, although it is likely that sulfenyl chloride is an intermediate.

$$GSH + HOCl \rightarrow GS\text{-}Cl + H_2O$$

$$GS\text{-}Cl + GS^- \rightarrow GSSG + Cl^-$$

The oxidized glutathione formed can be readily reduced back to GSH by NADPH and glutathione reductase.

On the other hand, NADPH is oxidized to chlorinated $NADP^+$, which, unlike $NADP^+$, is not reduced back by alcohol dehydrogenase and isocitric dehydrogenase. This irreversible oxidation should have significant toxicological consequences. Two atoms of ^{36}Cl were incorporated, and an increase at 260 mm was observed.[53] Chloroperoxidase and H_2O_2 incorporate 1 atom of ^{36}Cl into NADH to form a 264-nm absorbing product oxidized by a second mole of H_2O_2 to a chlorinated NAD^+.[54] A similar product was formed with 1 M equivalent of hypochlorite. The following reactions were postulated:

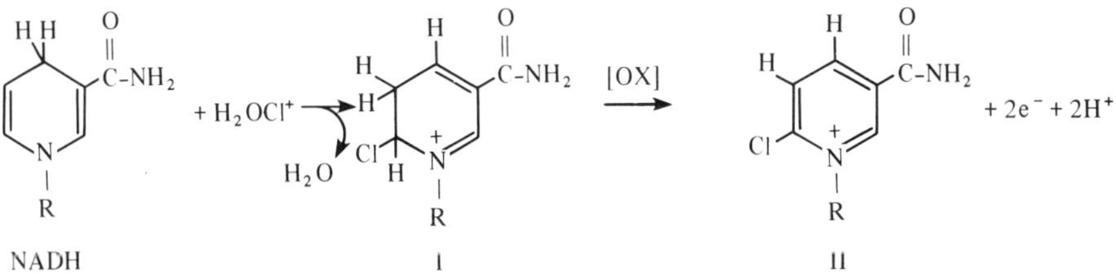

H H O || C-NH2 N R + H_2OCl^+ → H H O || C-NH2 H Cl H H2O + N R [OX] → H O || C-NH2 H Cl + N R + $2e^-$ + $2H^+$

NADH I II

Chlroinated NAD^+ was formed from NADH, with 2 MEq of hypochlorite. The chlorinated NAD^+ was not reduced by glucose 6-phosphate dehydrogenase (G6PD) and glucose 6-phosphate (G6P).[54]

Hypochlorite added to an equimolar mixture of GSH and NADH, oxidized GSH before NADH (O'Brien, P. J., unpublished data). Others have also shown that GSH and methionine were more than 100 times more reactive to hypochlorite generated by a myeloperoxidase–H_2O_2–Cl reaction system than NADH, other amino acids, and uric acid.[55] This finding suggests that GSH will protect the cell from hypochlorite-induced cytotoxicity until it is oxidized. However, chlorination of NADPH will lead to irreversible oxidation and prevent restoration of GSH levels.

Hypochlorous acid reacts with DNA bases. Results with cytosine[44] are as follows:

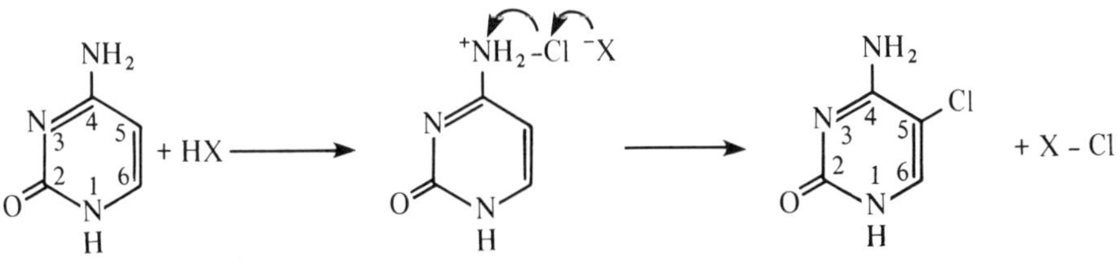

NH_2 N 3 4 5 2 1 6 O N H + HX → $^+NH_2$-Cl $^-$X N O N H → NH_2 Cl N 3 4 5 2 1 6 O N H + X - Cl

Similar reactions presumably occur with guanine and adenine of DNA or RNA and could explain the mutagenicity and genotoxic effects of hypochlorite.[56]

4.2. Oxidative Inactivation of Microbial Toxins

The myeloperoxidase–H_2O_2–chloride reaction system inactivates diphtheria toxin, tetanus toxin, and a wide variety of other lethal cytolytic toxins produced by gram-positive bacteria.[31] Some of these toxins are thiol activated and oxygen sensitive.

4.3. Oxidative Inactivation of Methionine-Dependent Inflammation Mediators

Methionine is oxidized by H_2O_2 or hypochlorite to the sulfoxide. Myeloperoxidase does not catalyze the oxidation by H_2O_2 but does so in the presence of chloride.[57] Bromide and iodide were more effective. A pH optimum of 5.0–6.0 was found. It is unlikely that singlet oxygen formed from the reaction was responsible, as H_2O_2 did not increase the effectiveness of hypochlorite and the myeloperoxidase system was not affected by D_2O, in which the lifetime of singlet oxygen ($'O_2$) is much longer. Phagocytosing human polymorphonuclear

$$OCl^- + H_2O_2 \rightarrow Cl^- + H_2O + {'O_2}$$

neutrophils, but not resting neutrophils, oxidized both intracellular and extracellular methionine to sulfoxide.[58] The oxidation was prevented by the myeloperoxidase inhibitors NaN_3 and KCN, with NaN_3 10 times more potent. The effect of halides was not investigated. However, oxidation catalyzed by the granular fraction and H_2O_2 was markedly reduced by removing Cl^- from the medium. Tsan and Chen[58] suggested that $'O_2$ may participate because of stimulation by D_2O and inhibition by $'O_2$ quenchers.

Myeloperoxidase in the presence of H_2O_2 and halide ion catalytically oxidizes methionine in various biologically active peptides.[30] Because of this, leukocytes could play an important role in the maintenance of tissue homeostasis in response to injury. Neutral proteases released at the same time as myeloperoxidase or located as an ectoenzyme in the plasma membrane may also play such a role. *N*-formylmethionyl peptides (e.g., FMLP) released by bacteria during protein synthesis are important chemotactic factors that attract leukocytes to sites of infection or inflammation. Chemotaxis involves a highly directional migration of leukocytes from the circulation to the tissues by passage through the endothelial cell barrier. The end result of this complex process is the generation and/or release of agents from the leukocytes that destroy micro-organisms. The factor activates the respiratory burst of leukocytes following binding to surface-membrane glycoprotein receptors and causes granule release and exocytosis of myeloperoxidase, proteases, and lysozyme involved in the bactericidal process. Potent vasoactive agents such as the leukotrienes are also formed and released. The activation differs from phorbol myristate acetate in that cytosolic calcium is increased both by influx through the plasma membrane and by release from intracellular stores.[59] Whether calcium and increased phosphoinositide metabolism contributes to the regulation of these processes is not clear.[60] The NADPH oxidase activity of the thyroid follicular cell that forms H_2O_2 required for thyroxine formation requires calcium for its activity.[61]

When antigens combine with circulating antibodies, cells are lysed, bacteria are opsonized, and leukocytes are attracted to the antigen. Histamine, released from granulocytes, mast cells, and platelets, then dilates blood vessels and increases capillary permeability. An oligopeptide C5a is released by a system of plasma enzymes called the complement system, from the antigen–immunoglobulin complex. This oligopeptide and others cause histamine release and act as chemotactic factors that attracts leukocytes and macrophages to the site of the antigen–antibody reaction. Other complement oligopeptides cause cell lysis and bacterial opsonization. The C5a oligopeptide contains methionine and a partially purified preparation was readily inactivated by the myeloperoxidase–H_2O_2–halide system.[62] The ability to bind to intact PMN was lost indicating that methionine oxidation results in a decreased affinity for membrane receptors. The chemical nature of this modification still needs to be confirmed.

Leukocytes attracted to the lungs by inflammatory stimuli release proteases, including elastase, which attacks the elastic tissue in the lungs. At the same time, α_1-antitrypsin, a plasma protein that normally inactivates elastase and other proteases, is itself inhibited. A genetic deficiency of α_1-antitrypsin can increase the risk of contracting emphysemia. Methionine is at the active center of this enzyme and seems to be required for binding to the elastin substrate. Myeloperoxidase, H_2O_2, and a halide result in a loss of human α_1-proteinase inhibitor activity that accompanies the oxidation of four methionine residues to sulfoxides.[63] Furthermore, α_1-antitrypsin inactivation by activated PMN is partially blocked by azide or catalase, suggesting the involvement of myeloperoxidase and H_2O_2.[64] The effect of halides on the activated PMN system remains to be investigated to determine whether the oxidizing species is Cl^+, OH·, or $'O_2$.

4.4. Oxidative Activation of Proteases

Activated neutrophils release and activate extracellularly the latent metalloenzyme, gelatinase, which directly attacks denatured collagens, solubilises some collagens and potentiates neutrophil interstiteal collagenase. Hypochlorite also activates gelatinase. The release and activation of gelatinase by neutrophils is partially inhibited by agents known to inhibit or scavenge hypochlorite.[65] Oxygen-independent processes may contribute as well.

4.5. Inactivation of Prostaglandins and Leukotrienes

Leukocytes stimulated with the calcium ionophore A23187 have increased phospholipid turnover, leading to the release of free arachidonic acid, which is oxidized by 5-lipoxygenase to form leukotriene A_4, a 5,6-oxide compound. This reacts with GSH to form the sulfidopeptide leukotrienes (C_4, D_4, and E_4 and the slow-reacting substance) that cause increased vascular permeability and constriction of peripheral lung airways. Activated leukocytes readily inactivate the leukotrienes. This inactivation was also related to activation by PMA and paralleled the dose response of the respiratory burst. This may be the result of modification by hydroxyl radicals to unknown products[66] or oxidation by hypochlorite to form sulfoxides[67] and chlorolactones.[26] Myeloperoxidase inhibitors prevent the inactivation and the same sulfoxide products were formed with myeloperoxidase, H_2O_2, and chloride.[67] It is likely that most of the inactivation occurs extracellularly. Myeloperoxidase deficient leukocytes inactivate leukotrienes presumably as a result of modification by hydroxyl radicals,[66] but the products formed are unknown.

Leukotriene B_4, a dihydroxy compound, is formed from leukotriene A_4 by enzymatic hydrolysis. It is the main leukotriene formed in A23187 activated leukocytes and has potent chemokinetic and chemotactic activity. Leukocytes activated by *N*-formylmethionine-leucine-phenylalanine peptide (FMLP) may terminate leukotriene-mediated inflammation by Ω-oxidation of the leukotrienes B_4 to C_{20}-hydroxyl and C_{20}-carboxyl metabolites or by sulfur oxidation of leukotriene C_4.[68] Leukotriene inactivation also occurs with eosinophil peroxidase, H_2O_2, and halide[69] or hydroxyl radicals.[66]

6-keto-Prostaglandin F_1, prostaglandin E_2, and prostaglandin F_2 (PGF_1, PGE_2, and PGF_2, respectively) are also inactivated by PMA-stimulated neutrophils.[68] Myeloperoxidase, H_2O_2. and chloride seem to be responsible, so that it is likely that inactivation occurs as a result of chlorination and chlorolactone formation.[46]

4.6. Thyroxine Turnover

Do activated leukocytes have the potential to act as a standby to synthesize T_3 in adverse physiological conditions related to loss of thyroid function? Myeloperoxidase, H_2O_2, and iodide readily converts tyrosine into monoidodotyroxine.[70] Diiodotyrosine coupling to T_3 also occurred with myeloperoxidase and H_2O_2.[71] Activated leukocytes also accumulate iodide fourfold against an electrochemical gradient presumably as a result of myeloperoxidase-catalyzed oxidation.[72] Fixed iodide is localized, in part, in the phagocytic vacuole in association with ingested bacteria. Iodination of extracellular proteins also occurs.[73] Thus, intravacuolar and secreted myeloperoxidase catalyze iodide oxidation. No iodination occurred in myeloperoxidase-deficient cells or with leukocytes isolated from CGD patients. It is not known whether activated leukocytes could convert added tyrosine to T_3 in the presence of iodide. Leukocytes also synthesize T_3 by iodinating endogenous tyrosine-rich proteins. Pronase hydrolysis of these leukocyte proteins can release T_3.[72]

Could leukocytes be another site of tri-iodothyronine (T_4) formation to the liver or kidney? The monodeiodination of the outer ring of L-T_3 in the phagocytosing leukocytes may involve myeloperoxidase or H_2O_2 and is inhibited by the antithyroid agent propylthiouracil.[74] However, a thiol-dependent plasma membrane enzyme may be responsible. The iodide released by deiodination could then be oxidized by myeloperoxidase and play an antibacterial role. Thyroxine has also been shown to act as a cofactor in the myeloperoxidase-mediated antimicrobial system.[75] Accelerated T_3 turnover in vivo also occurs during bacterial infection.[76]

4.7. Estradiol Inactivation

Can leukocytes play some role in the oxidative inactivation of estradiol in vivo? Estradiol is readily oxidized by leukocytes to macromolecule-binding species during the respiratory burst, resulting in the uptake of estradiol by leukocytes.[77] Autoradiographic studies show that estradiol is bound to the granules. The myeloperoxidase inhibitors azide and cyanide inhibited estradiol binding, whereas the catalase inhibitor aminotriazole stimulated.[77] The nature of the species involved is unknown. However, our studies on the nature of the oestradiol oxidation products formed by uterine peroxidase that binds to protein[78] show that most of the binding is noncovalent and involves polymeric products. Intracellular GSH and NADPH also prevent binding by reducing the estradiol phenoxy radicals to estradiol. Furthermore, catalase inhibited leukocyte binding. This suggests that the estradiol is oxidized and bound extracellularly. Whether hypochlorite is involved or whether chlorinated products are formed is also unknown. However, bromine readily brominates estradiol to the 2-bromo and 4-bromo derivatives.[79]

5. OXIDATIVE METABOLISM OF DRUGS

5.1. Antithyroid Drugs

Graves disease is probably an autoimmune disorder in which hyperthyroidism is the result of thyroid-stimulating antibodies binding to the thyroid cell membrane and stimulating thyroid hormone biosynthesis. The thionamide antithyroid drugs are the mainstays of treatment by inhibiting thyroid hormone production.[80] The organification of iodide and the coupling of iodothyronines are inhibited. In addition, propylthiouracil, but not methimazole, inhibits T_3 outer-ring monodeiodinase and thus decreases total body T_3 production. These drugs may also accelerate remission by their immunosuppressive effects in inhibiting the production of pathogenic antibodies. Toxic reactions to these drugs include hypersensitivity reactions that occur during the first 3 weeks of treatment and granulocytopenia (<1500 granulocytes/mm^3), which may appear after 4–8 weeks of treatment.[80] In this section, evidence is described suggesting that the therapeutic and toxicological effects of these drugs are related to their inhibition of various peroxidases in the thyroid, bone marrow, and granulocytes and their ability to remove peroxides, hydroxyl radicals, and activated iodine species.

Thioureylene antithyroid drugs (e.g., propylthiouracil or methimazole) are used in the treatment of Graves disease to inhibit excessive thyroid hormone biosynthesis. Propylthiouracil has recently been shown to inactivate thyroid peroxidase in vivo and was still marked 18 h after injection.[81] This would partly explain the inhibition of thyroid hormone biosynthesis. Propylthiouracil inactivates thyroid peroxidase irreversibly in the presence of H_2O_2 in vitro probably by a mechanism-based (suicide) mode of inhibition as a result of the covalent binding of the oxidized drug product to the heme moiety of the peroxidase to form a sulfheme.[82] S-oxygenation probably occurs when the drug interacts with the compound II oxidation state of thyroid peroxidase.[83] Iodide greatly enhances S-oxygenation and the peroxidase inactivation in vitro in the presence of high concentrations of iodide is reversible.[84] In this case the drug is oxidized via a sulfenyl iodide intermediate by an active iodine species formed by the peroxidase.[85] The oxidized products would be expected to be cytotoxic as a result of covalent binding to essential cellular proteins. The major metabolites identified in thyroid after propylthiouracil administration are the sulfinic and sulfonic acids. Methimazole on the other hand is metabolized in the thyroid to sulfate.[81] Propylthiouracil was also extensively bound to cellular proteins probably as a mixed disulfide or a thiosulfonic ester.[86] Antithyroid drugs accumulate in the thyroid follicular cells (e.g., methimazole accumulates 940-fold at 96 hr[80]) presumably as a result of the thyroid peroxidase catalyzed intracellular oxidation. Thyroid-stimulating hormone or iodide markedly increases the thyroidal accumulation and oxidation of antithyroid drugs.[87] A low iodide diet decreases the accumulation and oxidation of these drugs.

Part of the therapeutic effectiveness of antithyroid drugs against Graves disease has been attributed to their immunosuppressive action. This may be the result of their ability to inhibit antibody production by inhibiting oxygen radical production by monocytes[87] and thereby prevent their priming by autoantigen. The effectiveness of methimazole in reacting with hydroxyl and iodine radicals[88,89] and inactivating myeloperoxidase[87] could contribute to its immunosuppressive activity.

Adverse reactions such as bone marrow depression, i.e., agranulocytosis, occasionally follow treatment with propylthiouracil or methionazole[90] and have an inci-

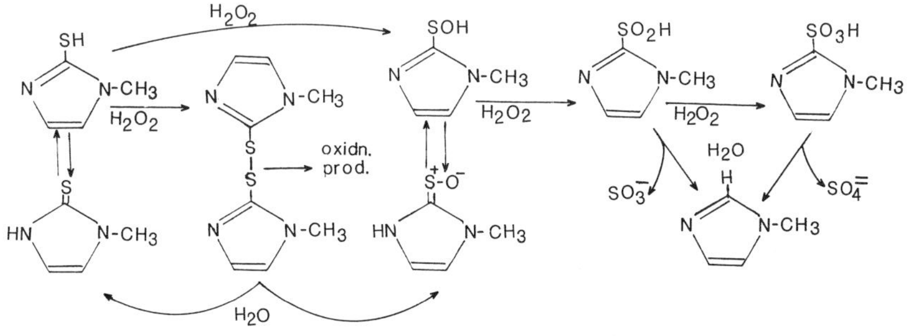

FIGURE 1. Suggested mechanism for peroxidase catalyzed methimazole oxidation.

dence of approximately 1% and a fatal outcome in 10% of these cases. The repeated administration of propylthiouracil in vivo leads to a decrease in leukocyte numbers and a decrease in bone marrow peroxidase activity. The myeloperoxidase activity of bone marrow homogenate is also inhibited by propylthiouracil in vitro.[91] Propylthiouracil was oxidized to its sulfonic acid and propyluracil by a myeloperoxidase-hydrogen peroxide-chloride system probably via hypochlorous acid. Propyluracil and elemental sulfur are acid dissociation products of propylthiouracil, sulfinic, and sulfonic acid formed by a further peroxidase catalyzed oxidation of the disulfide.[92] Phagocytosing polymorphonuclear leukocytes also accumulate propylthiouracil (8-to-28-fold)[93] presumably as a result of myeloperoxidase catalyzed oxidation.[94] Sulfinic acid, sulfonic acid and propyluracil were also identified as metabolites.[95] The covalent binding of these S-oxide metabolites to essential cellular proteins[95], e.g., by forming mixed protein disulfides, may be responsible for the cytotoxicity. The oxidative activity of polymorphonuclear leukocytes could conceivably also contribute to the liver necrosis associated with propylthiouracil treatment.[90] Methimazole is oxidized by myeloperoxidase-H_2O_2 to 1-methyl-2-imidazole (O'Brien, unpublished). A mechanism is outlined in Figure 1.

Serum samples from patients suffering agranulocytosis inhibit the respiratory burst of leukocytes from normal donors suggesting that an immune complex of drug and drug-specific antibody may have formed in the patients.[88] Subsequent exocytosis or cytotoxic release of the protein bound antithyroid drug could lead to antibody formation and subsequent extensive destruction of other bone marrow cells following adsorption of the antibody-drug complex. Inhibition of cell division in the bone marrow may also be an important mechanism.

5.2. Oxygenation of Drugs

The nonsteroidal anti-inflammatory drug phenylbutazone is metabolized by an extract of rat polymorphonuclear leukocytes, H_2O_2, and chloride to the 4-hydroxy, 4-hydroperoxy, and 4 chloro derivatives.[96] In the absence of chloride, 4-hydroxy and 4-hydroperoxy derivatives were formed to the same degree. Azide prevented the formation of the 4-hydroperoxy and 4-chloro derivatives. Hypochlorous acid formed only the 4-chloro derivative. These results suggest that the myeloperoxidase–H_2O_2 system forms 4-hydroperoxy whereas 4-chloro was formed by hypochlorite.[96] The peroxidase may oxidize phenylbutazone to a carbon-centered radical that reacts with oxygen to form a peroxy radical. The 4-hydroperoxy derivative is then reduced to the 4-hydroxy derivative. The anti-inflammatory therapeutic effects of phenylbutazone may be because of its ability to act as a donor for myeloperoxidase, thereby aiding in the removal of H_2O_2 or the trapping of hypochlorite released by activated leukocytes.

Tenoxicam is a new nonsteroidal anti-inflammatory drug. The major metabolite found in urine are thiophenecarboxylic acid as well as the minor metabolites *N*-methylthiophenesulfimide and pyridyloxamic acid. Recently, a leukocyte extract containing myeloperoxidase and eosinophil peroxidase were found to catalyze the formation of these metabolites.[97] The mechanism seems to involve an initial oxidation at the C-3 position. Presumably, the radical formed reacts with oxygen to form a peroxy radical, which causes cleavage.

Monochlorodimedone is also oxidized by chlorperoxidase and H_2O_2 to a radical that takes up oxygen and forms an oxidized unknown product. In the presence of chloride or bromide, halogenation to dichlorodimedone occurs, a process inhibited by oxygen. This suggests a radical mechanism of halogenation.[98]

5.3. Electronic Excitation

Leukocytes on activation produce weak light that can be detected by a scintillation counter. A variety of xenobiotics markedly increase this chemiluminescence, presumably as a result of a leukocyte-catalyzed epoxidation or oxidation via excited intermediates[99–103] (Table III).

Imipramine, a widely used tricyclic antidepressant, induces liver injury in vivo, which has been attributed to a cytochrome P-450-dependent epoxidation. Epoxidation catalyzed by a liver microsomal system is accompanied by chemiluminescence. Imipramine oxidation by activated leukocytes is also accompanied by chemiluminescence.[99] Adverse effects and agranulocytosis of imipramine could involve immunochemical toxicity due to imipramine oxidation. The carcinogenic PAH, benzo(α)pyrene is also oxidized by a liver microsomal system to a 7,8-dihydrodiol, which is further oxidized to a highly mutagenic bay region epoxide via an excited intermediate.[104] Chemiluminescence is also increased on addition of benzo(α)pyrene to activated leukocytes.[100] Such chemiluminescence would suggest that activated leukocytes could oxidize polycyclic aromatic hydrocarbons to carcinogenic metabolites.

The chemical processes resulting in chemiluminescence are still unknown. However, it is believed that when imipramine or benzo(α)pyrene-*trans*-7,8-dihydrodiol are oxidized to an epoxide some dioxetane is also formed. In the case of benzo(α)pyrene derivatives, a double bond at the 9,10-position was required. Similar chemiluminescence was observed during the oxidation of benzo(α)pyrene 7,8-

TABLE III. Comparison of the Effectiveness of Activated Neutrophils at Catalyzing Xenobiotic Oxidation to Excited Intermediates

	Chemiluminescence activated leukocytes	Reference
None	1.2×10^5 cpm (10^7 PMN)	99
Imipramine (10^{-4} M)	3×10^5 cpm (10^7 PMN)	99
Luminol (10^{-7} M)	3×10^5 cpm (10^5 PMN)	101
Lucigenin (2×10^{-4} M)	0.024 mV	102
Indole-3-acetic acid (1.5×10^{-3} M)	3.3×10^5 cpm (10^7 PMN)	103
Indole-3-propionic acid (1.5×10^{-3} M)	3.1×10^5 cpm (10^7 PMN)	103
Benzo(α)pyrene 7,8-dihydrodiol (3×10^{-6} M)	13.5×10^5 cpm (7×10^6 PMN)	100
Benzo(α)pyrene (3×10^{-6} M)	2.1×10^5 cpm (7×10^6 PMN)	100

dihydrodiol by singlet oxygen.[104] Imipramine inhibits superoxy radical formation by the activated neutrophil, which suggests that superoxy radicals are used for the oxidation of imipramine. Oxidation of other closely related compounds by activated leukocytes, leading to chemiluminescence has also been reported. These include desipramine, opipramol, iprindole,[105] and various benzo(α)pyrene derivatives.[100]

The cyclic hydrazide luminol is oxidized to an excited aminophthalate anion by various oxidant species, particularly at an alkaline pH.[106] However, the chemiluminescence induced with activated leukocytes is completely inhibited by inactivation of myeloperoxidase or by the use of myeloperoxidase-deficient leukocytes.[101] Hypochlorite or a reaction mixture containing myeloperoxidase and H_2O_2 and/or choride cause luminol oxidation to chemiluminescent products. Granules (isolated from activated leukocytes), H_2O_2, and chloride also form chemiluminescent products with luminol.[106] Chemiluminescence formed by activated leukocytes first reaches a maximum at 2–3 min, then decays, and reaches a second smaller maximum at 12 min.[101] The first maxima was attributed to myeloperoxidase released extracellularly, whereas the second maxima was attributed to intracellular oxidation.[107]

Superoxide or a H_2O_2–peroxidase system may oxidize imipramine via a blue free radical to an epoxide. Thus, superoxy dismutase, but not catalase or hydroxyl radical scavengers, inhibit the chemiluminescence that accompanies imipramine oxidation by xanthine oxidase-hypoxanthine.[99] However, alveolar macrophages that lack peroxidase but produce superoxide and H_2O_2, both at rest and following activation, cause only a small increase in chemiluminescence.[99] Inactivation of myeloperoxidase in leukocytes with azide prevents the formation of chemiluminescence with imipramine or benzo(α)pyrene-7,8-dihydrodiol, suggesting that myeloperoxidase. H_2O_2, and/or hypochlorite causes imipramine oxidation in leukocytes.

Lucigenin also amplifies the chemiluminescent response of leukocytes. Only one peak of emission was observed, corresponding to the initial peak with luminol.[108] Superoxide dismutase inhibited the chemiluminescence but azide stimulated the leukocyte chemiluminescence suggesting myeloperoxidase is not involved.[102] The oxidation reaction therefore seems to involve a direct oxidation by H_2O_2

TABLE IV. Comparison of the Effectiveness of Various Oxidant Species at Oxidizing Xenobiotics to Excited Products

	H_2O_2	H_2O_2[a]	OCl^-	$'O_2$	OH·	$O_2^-·$
None						
Imipramine	−	+			(−)	+
Luminol	(+)	+	+		+	
Lucigenin	+	−				
Indol-3-acetic acid		+				+
Benzo(α)pyrene-7,8-dihydrodiol				+		

[a]With peroxidase.

formed by dismutation of superoxide radicals. Lucigenin is fairly hydrophilic and probably does not penetrate the cell membrane. The H_2O_2 detected is therefore likely to be extracellular. The mechanism may involve oxidation to a dioxetane which cleaves to two molecules of *N*-methylacridone, one of which is electronically excited.

Indole analogues (e.g., indole-3-acetic acid or indole-3-propionic acid) are also excited by activated leukocytes. Superoxide radical formation was inhibited, suggesting that $O_2^{-}\cdot$ is used for oxidation of the indole analogues. The indole acetic acid was modified to products that were not those derived from singlet oxygen. It has been suggested that excitation is due to the transfer of energy from an excited carbonyl group to the indole ring rather than the result of the oxidative cleavage of the indole ring.[103] The effect of myeloperoxidase inhibitors was not investigated, although excitation can occur with peroxidase and H_2O_2.

The synthetic analogue of *Cypridina luciferin,* 2-methyl-6-phenyl-3,7-dihydroimidazo-[1,2-1]pyrazin-3-one was recently used as a chemiluminescent probe to detect superoxide anion generation in activated leukocytes. The chemiluminescence in this system was little affected by catalase or myeloperoxidase inhibition, suggesting that this probe may be highly specific for superoxide anions.[109]

6. OXIDATIVE ACTIVATION OF CARCINOGENS AND XENOBIOTICS

6.1. Arylamines and Phenol

6.1.1. *N*-Hydroxylation of Arylamines

The *N*-oxidation of aromatic amines to *N*-hydroxy derivatives is generally regarded as an important activation step in the formation of more toxic and/or carcinogenic products. Cytochrome P-450-dependent microsomal mixed-function oxidase is believed to be the catalyst usually involved. However, this type of reaction occurs for the oxidation of 4-chloroaniline to the hydroxylamine and nitroso derivative in a prostaglandin synthetase-arachidonate-catalyzed oxidation. Chloroperoxidase and H_2O_2 in the absence of chloride also oxidizes various substituted anilines to the nitroso derivative.[110] Myeloperoxidase and lactoperoxidase may also catalyse *N*-oxidation. It is not clear whether a one-electron oxidation or a two-electron oxidation is involved. However, horseradish peroxidase (HRP) and H_2O_2 catalyze the oxidation of anilines by a free radical mechanism to a complex mixture of dimers and polymers. Bromide or chloride ions markedly accelerate the *N*-oxidation of 4-chloroaniline catalyzed by chloroperoxidase[111]:

NH_2 (Cl) → NHOH (Cl) $\xrightarrow{2H^+}$ NO (Cl)

This was not due to HOBr or HOCl formation, as ring halogenation (probably via *N*-halogenation), and not *N*-oxidation, occurred. Presumably the bromide or chloride reacts with compound I to form a more reactive compound I halide complex that catalyzes *N*-oxidation. Ring halogenation also occurred in the peroxidase reaction resulting in 2-bromo-4-chloroaniline and 2,6-dibromo-4-chloroaniline formation with bromide ions and 2,4-dichloroaniline formation with chloride ions.[112] Halogenation was optimal at pH 4, a more acid pH than that optimal for *N*-hydroxylation.

6.1.2. Phenol

Chronic exposure to benzene results in numerous blood and bone marrow disorders including pancytopenia, aplastic anemia, and leukemia. Bone marrow contains high levels of myeloperoxidase that could be involved in the localized activation of the metabolites of benzene. Other peroxidase-containing organs such as Zymbal gland and Harderian gland are also target organs for benzene. We have shown that [^{14}C]phenol, the principal metabolite of benzene in vivo, is converted to DNA and protein-binding species during the oxidative burst of human neutrophils.[113] Inactivation of myeloperoxidase prevented the binding, whereas inactivation of catalase increased endogenous H_2O_2 formation and binding. Peroxidase catalyzed [^{14}C]phenol oxidation to *o,o*-biphenol, *p,p*-biphenol, *p,p*-biphenoquinone, and "melanin"-like polymers. DNA was extensively bound and was shown to be due to the noncovalent binding of polymers principally formed from *o,o*-biphenol.[114,115] Protein binding was also shown to involve noncovalent binding of polymers. The GSH conjugate formed with *p,p*-biphenoquinone was identified as 3-(glutathion-S-yl)-p,p′-biphenol.[116] A phenoxy radical was the first formed intermediate, as in the presence of GSH a thiyl radical was trapped and extensive GSH oxidation to GSSG occurred. No phenol oxidation occurred, so that the GSH functioned by reducing the phenoxy radical back to phenol. This suggests that macromolecular binding that could possibly lead to genotoxic damage can occur in the intact leukocyte only after GSH and NADPH have been oxidized.

The effect of halides on the myeloperoxidase-catalyzed oxidation of phenol has been investigated. Preliminary findings show that with an iodide–myeloperoxidase–H_2O_2 mixture, the phenol underwent iodination chiefly to 2-iodophenol. The rate of dimerization to *p,p*-biphenol was increased by iodide, but the amount of *p,p*-biphenol formed was lower, presumably as a result of competition by the iodination reaction. Iodine reacted with phenol to form 2-iodophenol, but no dimerization was detected. Chloride increased the rate of myeloperoxidase-catalyzed oxidation of phenol to *p,p*-biphenoquinone, and only a small amount of 2,2′-dichloro-4,4′-biphenoquinone was formed. Hypochlorite reacted with phenol to form 2-chlorophenol and some 4-chlorophenol, but no dimerization was detected. Chloroamine which is also formed by activated leukocytes reacted with phenol at a much slower rate.

Hydroquinone, the principal metabolite of phenol formed by a mixed-function oxidase, is rapidly oxidized to *p*-benzoquinone by myeloperoxidase and H_2O_2. In the presence of chloride, the rate of hydroquinone oxidation was increased. Both *p*-benzoquinone and chlorobenzoquinone are highly cytotoxic[117] and antibactericidal,[118] probably as a result of their ability to extensively deplete cellular GSH and proteins thiols by alkylation. They are also genotoxic and thus could be the active metabolites of benzene formed in bone marrow that initiate the processes resulting in leukemia.

N,N-dimethyl-para-anisidine is *N*-demethylated when oxidized by myeloperoxidase and hydrogen peroxide. A cation radical is also formed. In the presence of chloride, however, benzoquinone and the para-*N,N*-dimethyl-aminophenoxy radical are formed.[119]

6.2. Oxygen Radicals and Singlet Oxygen

The mechanism of H_2O_2 formation by the activated leukocyte is believed to involve the activation of an NADPH oxidase located in the plasma membrane, which catalyzes a one-electron reduction of oxygen to superoxide radicals. Dismutation of the superoxide radicals to H_2O_2 could be nonenzymatic but is catalyzed by superoxide dismutase. Superoxide radicals readily reduce ferric ions. Thus, activated neutrophils can enhance the effectiveness of the anticancer drug bleomycin at inducing DNA strand breaks by reducing the bleomycin–Fe^{3+} complex and thereby increasing oxygen activation.[120] Superoxide radicals formed from activated neutrophils can reduce and thereby mobilize iron from plasma ferritin.[121] Superoxide may also reduce lactoferrin released intravacuolarly or extracellularly from activated neutrophils.[122] The released ferrous iron could then react with H_2O_2 to form hydroxyl radicals and would explain the formation of hydroxyl radicals by activated neutrophils.[123] 5,5-dimethyl-1-pyrroline-*N*-oxide was used to trap the hydroxyl radical. However, other investigators have recently shown that the spin adduct formed was likely formed by the glutathione peroxidase catalyzed reduction of 5,5-dimethyl-2-hydroperoxypyrrolidine-*N*-oxide, the spin adduct formed with superoxide radicals.[124]

Hydroxyl radicals may be responsible for the bactericidal activity, hydroxylation of benzoate,[123] and extracellular hyaluronic acid breakdown catalysed by activated neutrophils. Activated neutrophils also catalyze leukotriene oxidation to sulfoxides,[66] methional[3] or 2-ketomethylthiobutyric acid[6] (KMBA) oxidation to ethylene, dimethylsulfoxide oxidation to methane; sulindac sulfide (the metabolite of the prodrug sulindac) oxidation to sulindac,[125] and luminol oxidation to excited aminophthalate ion.[106] These reactions, however, can also occur with HOCl:

$$OH\cdot + \underset{\text{methional}}{HCO\cdot CH_2\cdot CH_2\cdot SCH_3} \longrightarrow C_2H_4 + \tfrac{1}{2}(CH_3S)_2 + HCOOH$$

$$OH\cdot + \underset{\text{KMBA}}{CH_3S\cdot CH_2CH_2\cdot CO\cdot COOH} \rightarrow C_2H_4 + \tfrac{1}{2}(CH_3S)_2 + HCOOH + CO_2$$

$$OH\cdot + CH_3\cdot SO\cdot CH_3 \longrightarrow CH_3\cdot S(OH)O\cdot CH_3 \longrightarrow CH_3\cdot + CH_3SO_2H$$

$$CH_3\cdot + RH \longrightarrow CH_4 + R\cdot$$

Thus, it is difficult to ascertain whether their oxidation by leukocytes is due to OH· or HOCl, unless the effects of a halide free medium are studied. The myeloperoxidase H_2O_2 system may also catalyze some of these oxidations, but azide is a hydroxyl radical scavenger and other inhibitors need to be researched. However, peroxidase-deficient neutrophils or macrophages also catalyze these oxidations.

Electron-spin resonance (ESR) has been used to trap $O_2^- \cdot$ and OH· using the spin trap 5,5-dimethyl-pyrroline-*N*-oxide. The myeloperoxidase inhibitor azide stimulated $O_2^- \cdot$ formation, leading Bannister and Bannister[122] to suggest that HOCl reacted with $O_2^- \cdot$:

$$HOCl + O_2^- \cdot \rightarrow Cl^- + O_2 + OH \cdot$$

The cytotoxicity of cultured endothelial cell killing induced by activated leukocytes is prevented by hydroxyl radical scavengers or desferoxamine, an iron chelator.[126] This further suggests that hydroxyl radicals are formed by activated leukocytes and cause cytotoxicity. Hydrogen peroxide also probably kill bacteria by reacting with bacterial iron to form hydroxyl radicals.[3] Hydroxyl radical scavengers also protect phagocytosing leukocytes from self-inflicted damage (autotoxicity).[127]

Myeloperoxidase with H_2O_2 and bromide ion as cosubstrates has recently been shown to generate singlet oxygen efficiently.[128] This reaction as well as the reaction between $HOCl^- + H_2O_2$ has been known to emit red chemiluminescence characteristic of singlet oxygen for some time. However, recently a novel ultrasensitive infrared (IR) spectrophotometer has been used to detect the near-IR electronic emission of singlet oxygen at 1268 nm. In 50% D_2O (pH + pD = 4.5), the singlet oxygen generated was comparable to that of the standard chemical reaction $OCl^- + H_2O_2$ at identical peroxide concentrations. Other investigators have obtained similar findings with lactoperoxidase and chloroperoxidase. Iodide was much less effective than chloride or bromide, possibly because iodine is formed that quenches singlet oxygen.[129] The oxidation of diphenylfuran or 1,3-diphenylisobenzofuran is enhanced in a D_2O solvent and has therefore been partially attributed to singlet oxygen.[20] However, other investigators found that when cholesterol or linoleate was incubated with activated neutrophils, the products formed were characteristic of that formed by a free radical reaction and not of a singlet oxygen reaction.[125] We have found that cholesterol is a poor singlet oxygen trap and was ineffective with any dark singlet oxygen-forming reaction, e.g., $OCl^- + H_2O_2$, peroxidase–H_2O_2–Br^-. It has also been suggested that singlet oxygen formation is sharply decreased at pH <8.5,[3] whereas others have suggested that the oxidation of the singlet oxygen chemical traps is due to OCl^-.[129] However, clearly singlet oxygen is formed at pH <8.5, and the enhanced oxidation of chemical traps in a D_2O solvent suggests singlet oxygen is involved.[130] This skepticism about the formation of singlet oxygen has hindered serious consideration of the involvement of singlet oxygen in leukocyte-catalyzed oxidations.

REFERENCES

1. Rossi F: The $O_2^{\bar{}}\cdot$-forming NADPH oxidase of the phagocytes: Nature, mechanisms of activation and function. *Biochim Biophys Acta* **853:**65–89, 1986.
2. Takanaka K, O'Brien PJ: Mechanism of H_2O_2 formation by leukocytes. 1. Evidence for a plasma membrane location. *Arch Biochem Biophys* **169:**428–435, 1975.
3. Takanaka K, O'Brien PJ: Proton release by polymorphonuclear leukocytes during phagocytosis or activation by digitonin or fluoride. *Biochem Int* **11:**127–135, 1985.
4. Aruoma OI, Halliwell B: Superoxide-dependent and ascorbate-dependent formation of hydroxyl radicals from hydrogen peroxide in the presence of iron. *Biochem J.* **241:**273–278, 1987.
5. Takanaka K, O'Brien PJ: Generation of activated oxygen species by polymorphonuclear leukocytes. *FEBS Lett* **110:**283–287, 1980.
6. Bentwood BJ. Henson PM: The sequential release of granule constituents from human neutrophils. *J Immunol* **124:**855–862, 1980.
7. Tsuruta Y, Subrahmanyan VV, Marshall W et al: Peroxidase mediated irreversible binding of arylamine carcinogens to DNA in intact polymorphonuclear leukocytes activated by a tumor promoter. *Chem-Biol Interact* **53:**25–35, 1985.
8. Adams DO, Johnson WJ, Fiorito E, et al: Hydrogen peroxide and cytolytic factor can interact synergistically in effecting cytolysis of neoplastic targets. *J Immunol* **127:**1973–1977, 1981.
9. Weiss SJ, Pepin G, Ortig X, et al: Oxidative autoactivation of latent collagenase by human neutrophils. *Science* **227:**747–749, 1985.
10. Winterbourn CC, Garcia RC, Segal AW: Production of the superoxide adduct of myeloperoxide by stimulated human neutrophils and its reactivity with hydrogen peroxide and chloride. *Biochem J* **228:**583–592, 1985.
11. Nauseef WM, Metcalf JA, Root RK: Role of myeloperoxidase in the respiratory burst of human neutrophils. *Blood* **61:**483–492, 1983.
12. Root RK, Stossel TP: Myeloperoxidase-mediated iodination by granulocytes. Intracellular site of operation and some regulating factors. *J Clin Invest* **53:**1207–1215, 1974.
13. Hurst JK, Albrich JM, Green TR, et al: Myeloperoxidase-dependent fluorescein chlorination by stimulated neutrophils. *J Biol Chem* **259:**4812–4821, 1984.
14. Spitznagel JK, Shafer WM: Neutrophil killing of bacteria by oxygen-independent mechanisms: A historical survey. *Rev Infect Dis* **7:**398–401, 1985.
15. Morrison M, Schonbaum GR: Peroxidase catalyzed halogenation. *Annu Rev Biochem* **45:**861–888, 1976.
16. Harrison JE, Schultz J: Studies on the chlorination activity of myeloperoxidase. *J Biol Chem* **251:**1371–1374, 1976.
17. Weiss SJ, Klein R, Shirka A, et al: Chlorination of taurine by human neutrophils. *J Clin Invest* **70:**598–607, 1982.
18. Grisham MB, Jefferson MM, Melton DF, et al: Chlorination of endogenous amines by isolated neutrophils. *J Biol Chem* **259:**10404–10408, 1984.
19. Bakkenist ARJ, DeBoer JEG, Plat H, et al: The halide complexes of myeloperoxidase and the mechanism of the halogenation reactions. *Biochim Biophys Acta* **613:**337–348, 1976.
20. Piatt J, O'Brien PJ: Singlet oxygen formation by a peroxidase, H_2O_2 and halide system. *Eur J Biochem* **93:**323–332, 1979.
21. Foote CS, Goyne TE, Lehrer RI: Assessment of chlorination by human neutrophils. *Nature (Lond)* **301:**715–716, 1983.
22. Whitin JC, Clark RA, Simons ER, et al: Effects of the myeloperoxidase system on fluorescent probes of granulocyte membrane potential. *J Biol Chem* **256:**8904–8906, 1981.
23. Allrich JM, McCarthy CA, Hurst JK: Biological reactivity of hypochlorous acid: Implications for microbicidal mechanisms of leukocyte myeloperoxidase. *Proc Natl Acad Sci USA* **78:**210–214, 1981.

24. Zgliczyński JM, Stelmaszynska T: Chlorinating ability of human phagocytosing leukocytes. *Eur J Biochem* **56:**157–162, 1975.
25. Thomas EL, Anne TM: Peroxidase-catalyzed oxidation of protein sulfhydryls mediated by iodine. *Biochemistry* **16:**3581–3586, 1977.
26. Turk J, Henderson WR, Klebanoff SJ, et al: Iodination of arachidonic acid mediated by eosinophil peroxidase, myeloperoxidase and lactoperoxidase. *Biochim Biophys Acta* **751:**189–200, 1983,
27. Thomas EL: Myeloperoxidase, hydrogen peroxide, chloride antimicrobial system, nitrogen-chloride derivatives of becterial components in bactericidal action against *Escherichia coli. Infect Immun* **23:**522–531, 1979.
28. Shirka A, Lobuglio AF, Weiss SJ: A potential role for hypochlorous acid in granulocyte mediated tumor cell cytotoxicity. *Blood* **55:**347–350, 1980.
29. Clark RA, Klebanoff SJ: Role of the myeloperoxidase–H_2O_2–halide system in concanavalin A-induced tumor cell killing by neutrophils. *J Immunol* **122:**2605–2610, 1979.
30. Clark RA: Extracellular effects of the myeloperoxidase–H_2O_2–halide system. *Adv Inflammation Res* **5:**107–146, 1983.
31. Clark RA, Klebanoff SJ, Einstein AB, et al: Peroxidase–H_2O_2–halide system: Cytotoxic effect on mammalian tumor cells. *Blood* **45:**161–170, 1975.
32. Hafeman DG, Lucas ZJ: Polymorphonuclear leukocyte mediated, antibody dependent, cellular toxicity against tumor cells. Dependence on oxygen and the respiratory burst. *J Immunol* **123:**55–61, 1979.
33. Eaton J, Klopin C, Swofford H: Chlorinated urban water: A cause of dialysis-induced hemolytic anemia. *Science* **181:**463–464, 1973.
34. Buckley JA: Intoxication of trout erythrocytes from monochloroamine in vitro oxidative alteration and recovery of hemoglobin and GSH. *Comp Biochem Physiol* **69C:**337–344, 1981.
35. Heffeman WP, Guion C, Bull RJ: Oxidative damage to the erythrocyte induced by sodium chlorite, in vivo. *J Exp Pathol. Toxicol* **2:**1487–1499, 1982.
36. Nathan CF, Arrick BA, Murray HW, et al: Tumor cell antioxidant defenses. Inhibition of the glutathione redox cxcle enhances macrophage-mediated cytolysis. *J Exp Med* **153:**766–782, 1981.
37. Nathan CF, Cohn ZA: Antitumor effects of hydrogen peroxide *in vivo. J Exp Med* **154:**1539–1553, 1981.
38. Weiss SJ, Young J, LoBuglio AF. et al: Role of hydrogen peroxide in neutrophil-mediated destruction of cultured endothelial cells. *J Clin Invest* **68:**714–721, 1981.
39. Harlan JM, Levine JD, Callahan KS, et al: Glutathione redox cycle protects cultured endothelial cells against lysis by extracellularly generated hydrogen peroxide. *J Clin Invest* **73:**706–713, 1984.
40. Varani J, Fligiel SEG, Till GO, et al: Pulmonary endothelial cell killing by human neutrophils. *Lab Invest* **53:**656–663, 1985.
41. Simon RH, DeHart PD, Todd RF: Neutrophil-induced injury of rat pulmonary alveolar epithelial cells. *J Clin Invest* **78:**1375–1386, 1986.
42. Jong EC, Klebanoff SJ: Eosinophil-mediated mammalian tumor cell cytotoxicity: Role of the peroxidase system. *J Immunol* **124:**1949–1953, 1980.
43. Weiss SJ, Test ST, Eckmann CM, et al: Brominating oxidants generated by human eosinophils. *Science* **234:**200–202, 1986.
44. Kurokawa Y, Takayama S, Konishi Y, et al: Long term *in vivo* carcinogenicity test on K bromate and hypochlorite conducted in Japan. *Environ Health Persp.* **69:**221–230, 1986.
45. Palton W. Bacon V, Duffield AM, et al: The reaction of aqueous hypochlorous acid with cytosine. *Biochem Biophys Res Commun* **48:**880–884, 1972.
46. Shih KL, Lederberg J: Effects of chloramine on *Bacillus subtilis* deoxyribonucleic acid. *J Bacteriol* **125:**934–945, 1976.
47. Weiss SJ, Lampert MB, Test St: Long-lived oxidants generated by human neutrophils: Characterization and bioactivity. *Science* **222:**625–627, 1983.
48. Grisham MB, Jefferson MM, Thomas EL: Role of monochloroamine in the oxidation of erythrocyte hemoglobin by stimulated neutrophils. *J Biol Chem* **259:**6766–6772, 1984.

49. Schraufstatter IU, Hinshaw DB, Hyslop PA, et al: Glutathione cycle activity and pyridine nucleotide levels in oxidant-induced injury of cells. *J Clin Invest* **76:**1131–1139, 1985.
50. Schraufstatter IU, Hinshaw DB, Hyslop PA, et al: Oxidant injury of cells: DNA strand breaks lead to depletion of nicotinamide adenine dinucleotide. *J Clin Invest* **77:**1312–1320. 1986.
51. Hyslop PA, Hinshaw DB, Schraufstatter IU, et al: Intracellular calcium homeostasis during hydrogen peroxide injury to cultured P388D$_1$ cells. *J Cell Physiol* **129:**356–366, 1986.
52. Turkall RM, Tsan MF: Oxidation of glutathione by the myeloperoxidase system. *J Reticuloendothel Soc* **31:**353–360, 1982.
53. Selvaraj Rj, Zgliczyński JM, Paul BB, et al: Chlorination of reduced nicotinamide adenine nucleotides by myeloperoxidase: A novel bactericidal mechanism. *J Reticuloendothel Soc* **27:**31–38, 1980.
54. Griffin BW, Haddox R: Chlorination of NADH: Similarities of the HOCl-supported and chloroperoxidase-catalysed reactions. *Arch Biochem Biophys* **239:**305–309, 1985.
55. Winterbourn CC: Comparative reactivities of various biological compounds with myeloperoxidase–hydrogen peroxide–chloride, and similarity of the oxidant to hypochlorite. *Biochim Biophys Acta* **840:**204–210, 1985.
56. Rice RG, Gromez-Taylor M: Occurrence of by-products of strong oxidants reacting with drinking water contaminants. *Environ Health Persp* **69:**31–44, 221–230, 1986.
57. Tsan MF: Myeloperoxidase-mediated oxidation of methionine. *J Cell Physiol* **111:**49–54, 1982.
58. Tsan MF, Chen JW: Oxidation of methionine by human polymorphonuclear leukocytes. *J Clin Invest* **65:**1041–1050, 1980.
59. Nakagawara M, Takeshige K, Suminoto H, et al: Superoxide release and intracellular free calcium of calcium depleted human neutrophils stimulated by *N*-formyl-methionyl-leucyl-phenylalanine. *Biochim Biophys Acta* **805:**97–104, 1984.
60. Gryeskowiak M, Della Bianca V, Cassatella MA, et al: Complete dissociation between the activation of phosphoinositide turnover and of NADPH oxidase in human neutrophils depleted of Ca^{2+}. *Biochem Biophys Res Commun* **135:**785–794, 1986.
61. Dupuy C, Virion A, Hammori NA, et al: Solubilisation and characteristics of the thyroid NADPH-dependent H_2O_2 generating system. *Biochem Biophys Res Commun* **141:**839–846, 1986.
62. Clark RA, Szot S, Venkatasubramanian K, et al: Chemotactic factor inactivation by myeloperoxidase mediated oxidation of methionine. *J Immunol* **124:**2020–2026, 1980.
63. Matheson NR, Wong PS, Travis J: Enzymatic inactivation of human alpha-1-proteinase inhibitor by neutrophil myeloperoxidase. *Biochem Biophys Res Commun* **88:**402–409, 1979.
64. Carp H, Janoff A: *In vitro* suppression of serum elastase-inhibitory capacity by reactive oxygen species generated by phagocytosing polymorphonuclear leukocytes. *J Clin Invest* **63:**793–797, 1979.
65. Peppin GJ, Weiss SJ: Activation of the endogenous metalloproteinase, gelatinase, by triggered human neutrophils. *Proc Natl Acad Sci* **83:**4322–4326, 1986.
66. Henderson WR, Klebanoff SJ: Leukotrienes B_4, C_4, D_4 and E_4 inactivation by hydroxyl radicals. *Biochem Biophys Res Commun* **110:**266–272, 1983.
67. Lee CW, Lewis RA, Tauber AI. et al: Myeloperoxidase dependent metabolism of leukotrienes C_4, D_4 and E_4 to sulfoxides. *J Biol Chem* **258:**15004–15010, 1983.
68. Paredes JM, Weiss SJ: Human neutrophils transform prostaglandins by a myeloperoxidase-dependent mechanism. *J Biol Chem* **257:**2738–2740, 1982.
69. Henderson WR, Jorg A, Klebanoff SJ: Eosinophil peroxidase-mediated inactivation of leukotrienes B_4, C_4 and D_4. *J Immunol* **128:**2609–2613, 1982.
70. Klebanoff SJ, Yip C, Kessler D: The iodination of tyrosine by beef thyroid preparations. *Biochim Biophys Acta* **58:**563–574, 1962.
71. Yip C, Klebanoff SJ: Synthesis *in vitro* of thyroxine from diiodotyrosine by myeloperoxidase and by a cell-free preparation of beef thyroid glands. *Biochim Biophys Acta* **74:**747–755, 1963.
72. Stolc V: Stimulation of iodoproteins and thyroxine formation in human leukocytes by phagocytosis. *Biochem Biophys Res Commun* **45:**159–168, 1971.

73. Odeberg H, Olofsson T, Olsson I: Myeloperoxidase-mediated extracellular iodination during phagocytosis in granulocytes. *Scand J Hematol* **12:**155–160. 1974.
74. Woeber KA: L-Triiodothyronine and L-reverse-triiodothyronine generation in the human polymorphonuclear leukocyte. *J Clin Invest* **62:**577–584. 1978.
75. Klebanoff SJ, Green WC: Degradation of thyroid hormones in phagocytosing human leukocytes. *J Clin Invest* **52:**60–72, 1973.
76. Davidson B, Soodak M, Strout HC, et al: Thiourea and cyanamide as inhibitors of thyroid peroxidase: The role of iodide. *Endocrinology* **104:**919–924, 1979.
77. Klebanoff SJ: Oestrogen binding by leukocytes during phagocytosis. *J Exp Med* **145:**983–998, 1977.
78. Bennett S, Marshall W, O'Brien PJ: Metabolic activation of diethylstilbestrol by prostaglandin synthetase as a mechanism for its carcinogenicity, in Powles TJ, Bockman RC, Honn HV, Ramwell P (eds): *Prostaglandins and Cancer*. New York, Liss, 1982, pp 143–148.
79. Slaunwhite WR, Neely L: Bromination of phenolic steroids. *J Org Chem* **27:**1749–1752, 1962.
80. Ingbar SH, Werner SC: *The Thyroid—A Fundamental and Clinical Text*. New York, Harper & Row, 1986.
81. Shiroozu A, Taurog A, Engler H, et al: Mechanism of action of thioureylene antithyroid drugs in the rat. *Endocrinology* **113:**362–370, 1983.
82. Nakamura S, Nakamura M, Yamazaki I, et al: Reactions of ferryl lactoperoxidase (compound II) with sulfide and sulfhydryl compounds. *J Biol Chem* **259:**7080–7085, 1985.
83. Ohtaki S, Nakagawa M, Nakamura M, et al: Reactions of thyroid peroxidase with H_2O_2, tyrosine and methylmercaptoimidazole in comparison with bovine lactoperoxidase. *J Biol Chem* **257:**761–766, 1982.
84. Engler H, Taurog A, Luthy C, et al: Reversible and irreversible inhibition of thyroid peroxidase catalysed iodination by thioureylene drugs. *Endocrinology* **112:**190–196, 1983.
85. Engler H, Taurog A, Luthy C, et al: Reversible and irreversible inhibition of thyroid peroxidase-catalyzed iodination by thioureylene drugs. *Endocrinology* **112:**86–95, 1983.
86. Lindsay RH, Kelly K, Hill JB: Oxidative metabolites of [2-^{14}C]propylthiouracil in rat thyroid. *Endocrinology* **104:**1686–1697, 1979.
87. Mitchell ML, Whitehead WO, O'Rourke ME, et al: Metabolism of thiourea-^{35}S by the thyroid gland of the guinea pig. *Endocrinology* **70:**540–545, 1962.
88. Weetman AP, Holt MG, Campbell AK, et al: Methimazole and generation of oxygen radicals by monocytes: Potential role in immunosuppression. *Br Med J* **288:**518–520, 1984.
89. Taylor JJ, Wilson RL, Kendall-Taylor P: Evidence for direct interactions between methimazole and free radicals. *Febs Lett* **176:**337–340, 1984.
90. Neal RA, Halpert J: Toxicology of thiono-sulfur compounds. *Annu Rev Pharmacol Toxicol* **22:**321–339, 1982.
91. Kariya K, Lee E, Hirouchi M: Relationship between leukopenia and bone marrow myeloperoxidase in the rat treated with propylthiouracil. *Jpn J Pharmacol* **36:**217–222, 1984.
92. Morris DR, Hager LP: Mechanism of the inhibition of enzymatic halogenation by antithyroid agents. *J Biol Chem* **241:**3582–3589, 1966.
93. Lam DCC, Lindsay RH: Accumulation of 2-^{14}Cpropylthiouracil in human polymorphonuclear leukocytes. *Biochem Pharmacol* **28:**2289–2296, 1979.
94. Ferguson MM, Alexander WD, Connell JMC, et al: Peroxidase activity in relation to iodide, 17 β-oestradiol and propylthiouracil uptake in human polymorphoneutrophils. *Biochem Pharmacol* **33:**757–762, 1984.
95. Lam DCC, Lindsay RH: Oxidation and binding of [2-^{14}C] propylthiouracil in human polymorphonuclear leukocytes. *Drug Metab Dispos* **7:**285–289, 1979.
96. Ichihara S, Tomisawa H, Fukazawa H, et al: Involvement of leukocytes in the oxygenation and chlorination reaction of phenylbutazone. *Biochem Pharmacol* **35:**3935–03939, 1986.
97. Ichihara S, Tomisawa H, Fukazawa H, et al: Involvement of leukocyte peroxidase in the metabolism of tenoxicam. *Biochem Pharmacol* **34:**1337–1338, 1985.

98. Griffin BW, Ashley PL: Evidence for a radical mechanism of halogenation of monochlorodimedone catalysed by chloroperoxidase. *Arch Biochem Biophys* **233:**188–196, 1984.
99. Trush MA, Reason MJ, Wilson ME, et al: Oxidant-mediated electronic excitation of imipramine. *Biochem Pharmacol* **33:**1401–1410, 1984.
100. Trush MA, Seed JL, Kensler TW: Oxidant-dependent metabolic activation of polycyclic aromatic hydrocarbons by phorbol ester stimulated human polymorphonuclear leukocytes: Possible link between inflammation and cancer. *Proc Natl Acad Sci USA* **82:**5194–5198, 1985.
101. Dahlgren C, Briheim G: Comparison between the luminol-dependent chemiluminescence of polymorphonuclear leukocytes and of the myeloperoxidase–H_2O_2 system. *Photochem Photobiol* **41:**605–610, 1985.
102. Williams AJ, Cole PJ: The onset of polymorphonuclear leukocyte membrane-stimulated metabolic activity. *Immunology* **43:**733–739, 1981.
103. Ushijima Y, Nakano M: Excitation of indole analogs by phagocytosing leukocytes. *Biochem Biophys Res Commun* **82:**853–858. 1978.
104. Seliger HH, Thompson A, Hamman JP, et al: Chemiluminescence of benzo(a)pyrene 7,8-diol. *Photochem Photobiol* **36:**359–265, 1982.
105. Trush MA, Reason MJ, Wilson ME, et al: Comparison of the interaction of tricyclic antidepressants with human polymorphonuclear leukocytes as monitored by the generation of chemiluminescence. *Chem Biol Interact* **28:**71–81, 1979.
106. DeChatelet LR, Long GD, Shirley PS, et al: Mechanism of the luminol-dependent chemiluminescence of human neutrophils. *J Immunol* **129:**1589–1593, 1982.
107. Briheim G, Stendahl O, Dahlgren C: Intra- and extracellular events in luminol-dependent chemiluminescence of polymorphonuclear leukocytes. *Infect Immun* **45:**1–45, 1984.
108. Dahlgreen C, Aniansson H, Magnusson KE: Pattern of formylmethionyl-leucyl-phenylalanine-induced luminol- and lucigenin-dependent chemiluminescence in human neutrophils. *Infect Immun* **47:**326–328, 1985.
109. Nishida A, Sugioka K, Nagano M, et al: Use of 2-methyl-6 phenyl-3,7-dihydroimidazo (1,2-1)-pyrazin-3-one for assay of superoxide anions in phagocytosing granulocytes. *J Clin Biochem Nutr* **1:**5–10, 1986.
110. Corbett MD, Baden DG, Chipko BR: Arylamine oxidation by chloroperoxidase. *Bioorg Chem* **8:**91–95, 1979.
111. Corbett MD, Chipko BR, Baden DG: Chloroperoxidase-catalysed oxidation of 4-chloroaniline to 4-chloronitrosobenzene. *Biochem J* **175:**353–360, 1978.
112. Corbett MD, Chipko BR, Batchelor AO: The action of chloride peroxidase on 4-chloroaniline. *Biochem J* **187:**893–903, 1980.
113. O'Brien PJ, Gregory B, Tsuruta Y: One electron oxidation in cell toxicity, in Boobis AR, Caldwell J, et al (eds): *Microsomes and Drug Oxidations.* London, Ellis Horwood Publ, 1985, pp 284–295.
114. Subrahmanyan VV, O'Brien PJ: Peroxidase mediated ^{14}C-phenol binding to DNA. *Xenobiotica* **15:**859–872, 1985.
115. O'Brien PJ: Free-radical mediated DNA binding. *Environ Health Persp* **64:**219–232, 1985.
116. McGirr LG, Subrahmanyan VV, Moore GA, et al: Peroxidase-catalysed-3-glutathione-Syl-*p,p*-biphenol formation. *Chem Biol Interact* **60:**85–99, 1986.
117. Rossi L, Orrenius S, O'Brien PJ: Quinone toxicity in hepatocytes without oxidative stress. *Arch Biochem Biophys* **251:**25–35, 1986.
118. Beckman JS, Siedow JN: Bactericidal agents generated by the peroxidase-catalysed oxidation of hydroquinones. *J Biol Chem* **260:**14604–14609, 1985.
119. Sayo H, Hosokawa M, Lee E, et al: ESR studies on the oxidation of *N,N*-dimethyl-*p*-anisidine and its analogues catalysed by myeloperoxidase. *Biochim Biophys Acta* **874:**187–192, 1986.
120. Trush MA: Activation of bleomycin A_2 to a DNA-damaging intermediate by phorbol ester-stimulated human polymorphonuclear leukocytes. *Toxicol Lett* **20:**3649–3651, 1986.
121. Biemond P, van Eüjk HG, Koster JF, et al: Iron mobilisation from ferritin by superoxide derived from stimulated polymorphonuclear leukocytes. *J Clin Invest* **73:**1576–1579, 1984.

122. Bannister JC, Bannister WH: Production of oxygen-centered radicals by neutrophils and macrophages as studied by electron spin resonance. *Environ Health Persp* **64:**37–43, 1986.
123. Alexander MS, Husney RM, Sagone AL: Metabolism of benzoic acid by stimulated polymorphonuclear cells. *Biochem Pharm* **35:**3649–3651, 1986.
124. Britigan BE, Rosen GM, Chai Y, Cohen MS: Do human neutrophils make hydroxyl radicals? *J Biol Chem* **261:**4426–4431, 1986.
125. Egan RW, Gale PH, Huehl FA: Reduction of hydroperoxides in the prostaglandin biosynthetic pathways by a microsomal peroxidase. *J Biol Chem* **254:**3295–3302, 1979.
126. Varani J, Fligiel SEG, Ward PA, et al: Pulmonary endothelial cell killing by human neutrophils: Possible involvement of hydroxyl radical. *Lab Invest* **53:**656–663, 1985.
127. Klock JC, Stossel TP: Detection, pathogenesis and prevention of damage to human granulocytes. *J Clin Invest* **60:**1183–1191, 1977.
128. Khan AU: Myeloperoxidase catalysed singlet molecular oxygen generation detected by direct infrared electronic emission. *Biochem Biophys Res Commun* **122:**668–675, 1986.
129. Kanofsky JR: Singlet oxygen production by chloroperoxidase–hydrogen peroxide–halide systems. *J Biol Chem* **259:**5596–5600, 1984.
130. Foote CS, Abakerli RB, Clough RL, et al: On the question of singlet oxygen production in polymorphonuclear leukocytes. *Biolumin Chem* **2:**81–88, 1981.

11

Drug-Induced Agranulocytosis and Other Effects Mediated by Peroxidases during the Respiratory Burst

JACK P. UETRECHT

1. METABOLISM OF DRUGS TO REACTIVE INTERMEDIATES

There are numerous examples demonstrating strong evidence that toxic reactions to drugs or other chemicals are due to chemically reactive metabolites.[1] The form this toxicity takes can vary from cancer (presumably due to reaction of the metabolite with DNA) to anaphylactic reactions [presumably due to the metabolite acting as a hapten and reacting with a protein leading to the induction of immunoglobulin E (IgE) antibodies]. The greatest activity of enzymes capable of metabolizing xenobiotics is found in the liver; however, such enzymatic activity has been found in numerous other organs. One important aspect of this extrahepatic activity is that most reactive metabolites formed in the liver are too reactive to reach other target organs.

The combination of peroxide and peroxidase has strong oxidant properties. The combination of prostaglandin synthase (a peroxidase) and prostaglandin G_2 (a hydroperoxide) has been demonstrated to oxidize several drugs and carcinogens to reactive metabolites.[2–6] Other peroxidases found in the body include myeloperoxidase and thyroid peroxidase. Tsuruta et al.[7] demonstrated that several arylamine carcinogens are oxidized by polymorphonuclear leukocytes to metabolites which covalently bind to DNA. This metabolism appeared to be mediated by myeloperoxidase, since it required a respiratory burst and was inhibited by cyanide.

Myeloperoxidase is found in the granules of neutrophils and monocytes.[8,9] A similar peroxidase is found in eosinophils. In the resting cell, there is little opportunity for these peroxidases to oxidize substrates; however, during a respiratory burst, hydrogen peroxide is generated and myeloperoxidase-containing granules are

---

JACK P. UETRECHT • Faculties of Pharmacy and Medicine, University of Toronto and Sunnybrook Hospital, Toronto, Ontario M5S 1A1, Canada.

usually released. Hydrogen peroxide converts myeloperoxidase to compound I, which is capable of a one- or two-electron oxidation of a substrate.[10,11] The principal natural substrate oxidized by compound I is chloride. The result is an active chlorinating agent.[12,13]

Other substrates can also be oxidized by myeloperoxidase. Recently Ichihara et al.[14,15] demonstrated that myeloperoxidase–hydrogen peroxide can oxidize the drugs tenoxicam and phenylbutazone. The initial product of tenoxicam oxidation is unstable and breaks down to secondary products. The oxidation of phenylbutazone by leukocyte peroxidases leads to three products: an alcohol, a hydroperoxide, and a chloride. Apparently, the alcohol is formed by a different mechanism than the hydroperoxide because the hydroperoxide was not converted to the alcohol under the conditions of the reaction. The chloride metabolite could arise from oxidation by hypochlorite generated by myeloperoxidase–hydrogen peroxide–chloride, but there is evidence that some peroxidase-mediated chlorinations do not involve free hypochlorite ion.[16] Ichihara and co-workers suggest that the myeloperoxidase mediated oxidation of phenylbutazone may be responsible for both its anti-inflammatory and its toxic effects; however, some of the anti-inflammatory effects of phenylbutazone are presumably due to inhibition of prostaglandin synthesis. The mechanism of inhibition of prostaglandin synthesis and myeloperoxidase inhibition are probably similar, since the prostaglandin synthetase system has also been shown to oxidize phenylbutazone to the same alcohol and hydroperoxide metabolites.[17,18]

2. PHARMACOLOGICAL EFFECTS OF DAPSONE AND OTHER ARYLAMINES MEDIATED BY MYELOPEROXIDASE

Although the major use of dapsone is for the treatment of leprosy, it also has anti-inflammatory properties.[19] It seems to be most effective in diseases, such as dermatitis herpetiformis, in which the inflammation is mediated by neutrophils.[20,21] Stendahl et al.[22] found that dapsone did not inhibit the respiratory burst or neutrophil migration but does inhibit myeloperoxidase as measured by protein iodination. Dapsone also does not appear to inhibit complement activation.[23–25] We have found that dapsone is metabolized by human polymorphonuclear leukocytes (PMN) to its nitro derivative.[26,27] This only occurs during a respiratory burst, induced by either phorbol myristate acetate (PMA) or opsonized zymosan, when hydrogen peroxide is generated. The oxidation is a six-electron oxidation of dapsone; it is unlikely that this occurs in one step. The intermediate hydroxylamine is oxidized by hydrogen peroxide to the nitro derivative in the absence of enzyme, but dapsone itself is not oxidized by hydrogen peroxide. The nitroso metabolite is the other presumed two-electron oxidation intermediate, but it is very reactive and was not detected. The addition of ascorbic acid to the incubation stops the reaction at the hydroxylamine. These reactions are summarized in Fig. 1.

Human peripheral mononuclear cells also oxidize dapsone but, in this case, the oxidation stops at the hydroxylamine, even in the absence of ascorbic acid. This is

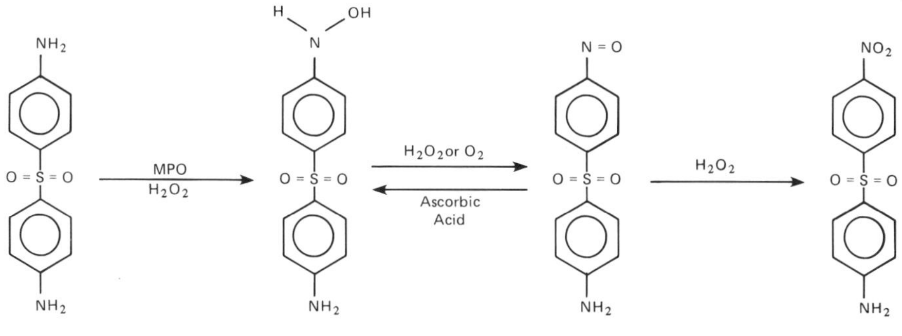

FIGURE 1. Metabolism of dapsone by leukocytes or myeloperoxidase (MPO) and hydrogen peroxide.

presumably because less hydrogen peroxide is generated by monocytes in the mononuclear cells than is generated by the PMN.[26,27] The oxidation by both polymorphonuclear and mononuclear leukocytes is inhibited by the myeloperoxidase inhibitor azide. Dapsone is also oxidized by purified myeloperoxidase–H_2O_2 to the hydroxylamine.[27]

Just as Ichihara suggested that the anti-inflammatory effects of phenylbutazone and tenoxicam are associated with its oxidation by myeloperoxidase–hydrogen peroxide, it is likely that the anti-inflammatory properties of dapsone are also associated with its oxidation by myeloperoxidase. Since the anti-inflammatory properties of dapsone are reported to be due to inhibition of myeloperoxidase, it is important to determine whether oxidation of dapsone by myeloperoxidase could lead to its inhibition. Although it has been proposed that dapsone could act by scavenging hydrogen peroxide, this is not a tenable hypothesis because dapsone does not react with hydrogen peroxide in the absence of enzyme.[28] It is possible that the hydroxylamine metabolite could scavenge hydrogen peroxide, since it does react with hydrogen peroxide; however, it would seem unlikely that this would be quantitatively important. Since dapsone is oxidized by myeloperoxidase, it could act as a competitive inhibitor and decrease the oxidation of other substances such as chloride. Another possibility is that the hydroxylamine product could bind to the heme iron of myeloperoxidase and inhibit its activity. Hydroxylamines are known to be good ligands for heme iron.[29] In addition, although it is reactive and we have not detected it directly, the nitroso derivative must be produced as an intermediate in the production of the nitro derivative, and it would also be expected to bind to the heme iron of myeloperoxidase.

In addition to its role in the anti-inflammatory properties of dapsone, it is likely that the oxidation of dapsone to the hydroxylamine and nitroso metabolites is responsible for the induction of agranulocytosis in some patients.[30–34] Agranulocytosis is a severe decrease in the peripheral polymorphonuclear cell count. At a count of less than 500 cells/μl, the patient is at high risk of developing a life-threatening infection. Aplastic anemia is similar to agranulocytosis except that all blood cell lines are decreased. The mortality rate varies but can be as high as 50%. Thus, drug-induced agranulocytosis is a serious side effect.

Weetman et al.[35] demonstrated that the hydroxylamine of dapsone is toxic to bone marrow. Thus a respiratory burst leading to the oxidation of dapsone could damage bone marrow. The drug classically associated with aplastic anemia or agranulocytosis is chloramphenicol.[36] Although choramphenicol is not an arylamine like dapsone, it is reduced in vivo to an arylamine.[37,38] The mechanism of chloramphenicol-induced aplastic anemia and agranulocytosis has been studied more extensively than that of dapsone. It was found that the arylamine metabolite of chloramphenicol is further oxidized by the liver to the hydroxylamine and presumably the nitroso metabolites.[39] The nitroso metabolite, unlike chloramphenicol and the amine metabolite, is toxic to bone marrow;[40,41] however, the reactivity of the nitroso metabolite is too great for it to reach the bone marrow if it were only formed in the liver or intestine.[39] Therefore, if the amine of chloramphenicol, as we have found with dapsone and other amines, is oxidized by activated leukocytes, formation of the reactive metabolite could occur in the bone marrow. These metabolic steps could represent the initial steps in chloramphenicol-induced aplastic anemia.

The other arylamine drugs that we have studied, sulfadiazine and procainamide, are also oxidized to hydroxylamine metabolites by activated neutrophils.[26,42] We have shown that the hydroxylamines of procainamide and sulfamethoxazole are toxic to leukocytes.[43,44] Procainamide and the sulfas are also associated with agranulocytosis.[45–48] Thus, oxidation by activated neutrophils may be a general metabolic pathway for aromatic amines and may mediate many of their pharmacological and toxic effects. Other drugs that are arylamines and that have been associated with agranulocytosis include aminosalicyclic acid, aprindine, and aminoglutethimide.[49–51]

3. PHARMACOLOGICAL AND TOXIC EFFECTS OF OTHER DRUGS METABOLIZED BY MYELOPEROXIDASE

It would seem that the ability of myeloperoxidase to oxidize a drug depends more on the presence of a functional group with a high election density than on the more selective binding seen with cytochrome P-450 isozyme-mediated oxidations. Amines represent such a functional group because of the heteroatom nitrogen. Sulfur is another heteroatom that is present in several functional groups that are easily oxidized. Propylthiouracil and methimazole contain a thionosulfur that is easily oxidized. Its oxidation by thyroid peroxidase is probably responsible for the therapeutic antithyroid activity of these drugs.[52–54] In the presence of iodine, oxidation of the drug to a sulfenyliodide is thought to occur and result in competitive inhibition. However, at low iodine concentrations, direct oxidation of the drug appears to occur, forming a reactive metabolite that binds covalently to thyroid peroxidase resulting in irreversible inhibition.[55–57] Propylthiouracil and methimazole also inhibit neutrophil and monocyte chemiluminescence evoked by PMA and other stimulants[58–59]; therefore, it is likely that these drugs are also metabolized by myeloperoxidase. Like the aromatic amines, the most common serious side effect of propylthiouracil and methimazole is agranulocytosis. Furthermore, in an animal model there is a correlation between the inhibition of myeloperoxidase and the degree of leukopenia.[60] Other thiol-containing drugs that are easily oxidized, such as captopril and penicillamine, are also associated with the induction of agranulocytosis.[61,62]

Returning to phenylbutazone, it appears as if compounds having a 1,3-diketo group are oxidized by peroxidases. Such a compound is monochlorodimedone, which is a standard substrate used to measure the chlorinating activity of peroxidases.[63] The analgesics antipyrine, aminopyrine, and dipyrone have structures similar to phenylbutazone and all have been associated with agranulocytosis.[64–66]

4. POSSIBLE MECHANISMS OF DRUG-INDUCED AGRANULOCYTOSIS

There are two plausible mechanisms by which oxidation of these compounds by myeloperoxidase during a respiratory burst could lead to agranulocytosis.

4.1. Direct Mechanism

The metabolite generated by myeloperoxidase is chemically reactive and could simply be toxic to bone marrow. This would seem to be the case for chloramphenicol because the nitroso metabolite has been demonstrated to be directly toxic to bone marrow.[40,41] In addition, the usual picture seen with chloramphenicol-induced aplastic anemia is destruction of bone marrow including pluripotential stem cells and a decrease in the production of new cells.[67] This mechanism is more likely to lead to depletion of all blood cell lines, i.e., aplastic anemia, rather than to the more selective agranulocytosis. Because of the reactivity of the nitroso metabolite, it would have to be formed in the bone marrow. For such metabolism to occur, a stimulus would be required to produce a respiratory burst and hydrogen peroxide. It is the presence or absence of such a stimulus that may determine who develops chloramphenicol-induced aplastic anemia.

4.2. Immune-Mediated Mechanism

In contrast to the usual picture seen with chloramphenicol, some types of agranulocytosis appear to involve immune-mediated destruction of granulocytes. The classic example of this is aminopyrine.[68] Sulfonamides and propylthiouracil also appear to be able to cause agranulocytosis by this mechanism.[69,70] If a reactive metabolite is formed on the surface of a neutrophil, it is likely to bind to the neutrophil cell membrane. This could lead to the induction of antineutrophil antibodies. This mechanism would require activation of neutrophils such that hydrogen peroxide was formed and myeloperoxidase released but, unlike the direct mechanism, this would not have to occur in the bone marrow.

5. PHARMACOLOGICAL EFFECTS MEDIATED BY THYROID PEROXIDASE

Thyroid peroxidase is similar to myeloperoxidase. Unlike some peroxidases, it readily catalyzes two-electron oxidation.[71] It is likely that myeloperoxidase and thyroid peroxidase oxidize many of the same substrates. Propylthiouracil and methimazole inhibit both thyroid peroxidase and myeloperoxidase, probably by a mechanism involving oxidation of the drugs. Likewise, arylamines metabolized by myeloperoxidase probably inhibit thyroid peroxidase. The sulfonamides inhibit thyroid peroxidase.[72] Aminosalicyclic acid, aminoglutethimide, and phenylbutazone can cause hypothyroidism.[73–75] Dapsone is not known to cause significant hypothyroidism in humans, but it has been reported to cause thyroid cancer in rats at a relatively low dose, and this could also be due to a reactive metabolite formed by thyroid peroxidase.[76]

6. IDIOSYNCRATIC REACTIONS POSSIBLY DUE TO METABOLISM BY MONOCYTES

Other side effects associated with many of these drugs (i.e., drugs containing an aromatic amine, thiono-sulfur or 1,3-diketo group) are drug-induced lupus and idiosyncratic reactions. Drug-induced lupus is an autoimmune disease in which autoantibodies can affect almost any system in the body. In general, acute life-threatening complications are uncommon, and the reaction abates when the drug is discontinued. By contrast, other idiosyncratic reactions can be very serious and life-threatening. Common manifestations of idiosyncratic reactions include fever, rash, lymphadenopathy, arthraligias, and damage to various organs, such as the liver. Such reactions are unpredictable and represent a large clinical problem.

These effects also seem to be mediated by the immune system. Monocytes contain myeloperoxidase and could therefore also metabolize drugs to reactive metabolites. Monocytes are an integral part of the immune system and are responsible for processing antigen for presentation to lymphocytes. Because of this, it has been suggested that binding of haptens to monocyte/macrophage membranes may be important to the generation of an allergic response.[77,78] This combination of functions may make monocytes critical for the initiation of idiosyncratic reactions.

For example, procainamide is an aromatic amine which is responsible for a high incidence of drug-induced lupus. We have shown that it is metabolized by the liver to reactive hydroxylamine and nitroso metabolites which bind to histone pro-

TABLE I. Summary of Pharmacological and Toxic Effects of Drugs Known, or Suspected, of Being Metabolized by Peroxidases

Drug	Agranulocytosis	Anti-inflammatory or inhibition of myeloperoxidase	Hypothyroidism or inhibition of thyroid peroxidase	Lupus	Idiosyncratic reaction
Arylamine					
Dapsone	+	+	?	±	++
Sulfonamides	+	+	+	+	++
Procainamide	+	–	?	++	+
Aminosalicylic acid	+	+	+	+	+
Aprindine	++	?	?	?	?
Aminoglutethimide	+	?	+	+	+
Chloramphenicol[a]	++	–	–	–	–
Thionosulfur or thiol					
Methimazole	++	+	++	+	+
Propylthiouracil	++	+	++	+	+
Penicillamine	+	++	+	+	+
Captopril	+	?	?	+	+
Pyrazolone Derivatives					
Phenylbutazone	++	++	+	+	+
Aminopyrine	++	+	?	?	+
Antipyrine	+	+	?	?	+

[a]Chloramphenicol has a nitro group rather than an arylamine, but it is reduced by colonic bacteria to an arylamine.

tein (a common antigen in procainamide-induced lupus) and are toxic to lymphocytes.[43,79,80] Although it does not appear as if these metabolites escape the liver, they are also formed by activated monocytes.[42] Sulfonamides are also metabolized to a hydroxylamine by myeloperoxidase and this hydroxylamine appears to be responsible for idiosyncratic reactions to sulfonamides.[44,81,82] The effects of the drugs mentioned are summarized in Table I.

7. SUMMARY

We have found that leukocytes containing myeloperoxidase can metabolize some drugs to reactive metabolites. It is likely that many other drugs with similar functional groups can also be metabolized to reactive metabolites. Other peroxidases, such as thyroid peroxidase, can also oxidize drugs to reactive metabolites. Many of the therapeutic and adverse effects of these drugs involve leukocytes and the thyroid, therefore, it is appealing to speculate that these effects are mediated by the reactive metabolites formed by peroxidases in these cells. This association could be a coincidence but that seems unlikely. Since a stimulus is required for activation of leukocytes and a respiratory burst to generate hydrogen peroxide, this hypothesis may explain a major factor that places some patients at increased risk for many of the idiosyncratic reactions associated with these drugs. Other types of drugs may also be metabolized by myeloperoxidase in the presence of hydrogen peroxide. Thus, the respiratory burst may have important implications for the toxicity of drugs.

ACKNOWLEDGMENTS. This work was supported by grants from the Medical Research Council of Canada, Sunnybrook Hospital and the Banting Research Foundation.

REFERENCES

1. Anders MW: *Bioactivation of Foreign Compounds*. Orlando, Academic, 1985.
2. Moldeus P, Andersson B, Rahimtula A, Berggren M: Prostaglandin synthetase catalyzed activation of paracetamol. *Biochem Pharmacol* **31:**1363–1368, 1982.
3. Kadlubar F, Frederick C, Weis C, Zenser T: Prostaglandin endoperoxide synthetase-mediated metabolism of carcinogenic aromatic amines and their binding to DNA and protein. *Biochem Biophys Res Commun* **108:**253–258, 1982.
4. Zenser T, Mattammal M, Armbrecht H, Davis B: Benzidine binding to nucleic acids mediated by peroxidative activity of prostaglandin endoperoxide synthetase. *Cancer Res* **40:**2839–2845, 1980.
5. Krauss R, Eling T: Formation of unique arylamine: DNA adducts from 2-aminofluorene activated by prostaglandin H synthase. *Cancer Res* **45:**1680–1686, 1985.
6. Krauss R, Eling T: Arachidonic acid-dependent cooxidation: A potential pathway for the activation of chemical carcinogens *in vivo*. *Biochem Pharmacol* **33:**3319–3324, 1984.
7. Tsuruta Y, Subrahmanyam V, Marshall W, O'Brien P: Peroxidase-mediated irreversible binding of arylamine carcinogens to DNA in intact polymorphonuclear leukocytes activated by a tumor promoter. *Chem Biol Interact* **53:**25–35, 1985.
8. Schultz J, Kaminker K: Myeloperoxidase of leukocyte of normal human blood. I. Content and localization. *Arch Biochem* **96:**465–467, 1962.
9. van Furth R, Raeburn J, van Zwet T: Characteristics of human mononuclear phagocytes. *Blood* **54:**485–500, 1979.
10. Harrison J, Araiso T, Palcic M, Dunford H: Compound I of myeloperoxidase. *Biochem Biophys Res Commun* **94:**34–40, 1980.
11. Bolscher B, Zoutberg G, Cuperus R, Wever R: Vitamin C stimulates the chlorinating activity of human myeloperoxidase. *Biochim Biophys Acta* **784:**189–191, 1984.
12. Zgliczynski J, Selvaraj R, Paul B, et al: Chlorination by the myeloperoxidase–H_2O_2–Cl^- antimicrobial system at acid and neutral pH. *Proc Soc Exp Biol NY* **154:**418–422, 1977.
13. Winterbourn C: Comparative reactivities of various biological compounds with myeloperoxidase–hydrogen peroxide–chloride, and similarity of the oxidant to hypochlorite. *Biochim Biophys Acta* **840:**204–210, 1985.
14. Ichihara S, Tomisawa H, Fukazawa H, Tateishi M: Involvement of leukocyte peroxidases in the metabolism of tenoxicam. *Biochem Pharmacol* **34:**1337–1338, 1985.
15. Ichihara S, Tomisawa H, Fukazowa H et al: Involvement of leukocytes in the oxidation and chlorination reaction of phenylbutazone. *Biochem Pharmacol* **35:**3935–3939, 1986.
16. Libby R, Thomas J, Kaiser L, Hager L: Chloroperoxidase halogenation reactions: Chemical versus enzymatic halogenating intermediates. *J Biol Chem* **257:**5030–5037, 1982.
17. Marnett L, Bieukowski M, Pagels W, Reed G: Mechanism of xenobiotic cooxidation coupled to prostaglandin H_2 biosynthesis, in Samuelson P, Ramewell W, Paoletti R (eds): *Advances in Prostaglandin and Thromboxane Research,* New York, Raven, 1980, vol 6, pp 149–151.

18. Reed G, Griffin I, Eling T: Inactivation of prostaglandin H synthase and prostacyclin synthase by phenylbutazone. *Mol Pharmacol* **27:**109–114, 1985.
19. Lang P: Sulfones and sulfonamides in dermatology today. *J Am Acad Dermatol* **1:**479–492, 1979.
20. Morgan J, Marsden C, Coburn J, et al: Dapsone in dermatitis herpetiformis. *Lancet* **1:**1197–1200, 1955.
21. Alexander J: Dapsone in the treatment of dermatitis herpetiformis. *Lancet* **1:**1201–1202, 1955.
22. Stendahl O, Molin L, Dahlgren C: The inhibition of polymorphonuclear leukocyte cytotoxicity by dapsone. *J Clin Invest* **62:**214–220, 1978.
23. Katz S, Hertz K, Crawford P, et al: Effect of sulfones on complement deposition in dermatitis herpetiformis and on complement-mediated guinea-pig reactions. *J Invest Dermatol* **67:**688–690, 1976.
24. Schifferli J, Jones R: Dapsone and complement. *Lancet* **2:**368–369, 1981.
25. Drummond L, Gemmell D: The effect of dapsone on complement activation. *Agents Actions* **13:**435–437, 1983.
26. Uetrecht J, Shear N, Biggar W: Dapsone is metabolized by human neutrophils to a hydroxylamine. *Pharmacologist* **28:**239, 1986.
27. Uetrecht J, Zahid N, Shear N, Biggar W: Metabolism of dapsone to a hydroxylamine by human neutrophils and mononuclear cells. *J Pharmacol Exp Ther* **245:**1–6, 1988.
28. Niwa Y, Sakane T, Miyachi Y: Dissociation of the inhibitory effect of dapsone on the generation of oxygen intermediates—In comparison with that of colchicine and various scavengers. *Biochem Pharmacol* **33:**2355–2360, 1984.
29. Ortiz de Montellano P, Reich N: Inhibition of Cytochrome P-450 enzymes, in Ortiz de Montellano P (ed): *Cytochrome P-450: Structure, Mechanism, and Biochemistry.* New York, Plenum, 1986, p 283.
30. McKenna W, Chalmers A: Agranulocytosis following dapsone therapy. *Br Med J* **1:**324–325, 1958.
31. Ognibene A: Agranulocytosis due to dapsone. *Ann Intern Med* **72:**521–524, 1970.
32. Firkin F, Mariani A: Agranulocytosis due to dapsone. *Med J Aust* **2:**247–251, 1977.
33. Wilson J, Harris J: Hematologic side-effects of dapsone. *Ohio State Med J* **73:**557–560, 1977.
34. Lahuerta-Palacios J, Gomez-Pedraja J, Montalban M, et al: Proliferation of Ig Dx plasma cells after agranulocytosis induced by dapsone. *Br Med J* **290:**282–283, 1985.
35. Weetman R, Boxer L, Brown M, et al: *In vitro* inhibition of granulopoiesis by 4-amino-4′-hydroxylaminodiphenyl sulfone. *Br J Haematol* **45:**361–370, 1980.
36. Wallerstein R, Condit P, Brown J, Morrison F: Statewide study of chloramphenicol—Therapy and fatal aplastic anemia. *JAMA* **208:**2045–2050, 1969.
37. Fouts J, Brodie B: The enzymatic reduction of chloramphenicol, *p*-nitrobenzoic acid and other aromatic nitro compounds in mammals. *J Pharmacol Exp Ther* **119:**197–207, 1957.
38. Scheline R: Metabolism of foreign compounds by gastrointestinal microorganisms. *Pharmacol Rev* **25:**451–523, 1973.
39. Ascherl M, Eyer P, Kampffmeyer H: Formation and disposition of nitrosochloramphenicol in rat liver. *Biochem Pharmacol* **34:**3755–3763, 1985.
40. Yunis A, Miller A, Salem Z, et al: Nitroso-chloramphenicol: Possible mediator in chloramphenicol-induced aplastic anemia. *J Lab Clin Med* **96:**36–46, 1980.
41. Gross B, Branchflower R, Burke T, et al: Bone marrow toxicity *in vitro* of chloramphenicol and its metabolites. *Toxicol Appl Pharmacol* **64:**557–565, 1982.
42. Uetrecht J, Zahid N, Rubin R: Metabolism of procainamide to a hydroxylamine by human neutrophils and mononuclear cells. *Chem Res Toxicol,* 1988 (in press).
43. Rubin R, Uetrecht J, Jones J: Cytotoxicity of oxidative metabolites of procainamide. *J Pharmacol Exp Ther* **242:**833–841, 1987.
44. Reider M, Uetrecht J, Shear N, Spielberg S: Synthesis and in vitro toxicity of hydroxylamine metabolites of sulfonamides. *J Pharmacol Exp Ther* **244:**724–728, 1988.
45. Kutscher A, Lane, Segall R: The clinical toxicity of antibiotics and sulfonamides: A comparative review of the literature based on 104,672 cases treated systematically. *J Allergy* **25:**135–150, 1954.

46. Berger B, Hauser D: Agranulocytosis due to new sustained-release procainamide. *Am Heart J* **105:**1035–1036, 1983.
47. Ellrodt A, Murata G, Riedinger M, et al: Severe neutropenia associated with sustained-release procainamide. *Ann Intern Med* **100:**197–201, 1984.
48. Nelson J, Lutton J, Fass A: Procainamide-induced agranulocytosis with reversible myeloid sensitivity. *Am J Hematol* **17:**427–432, 1984.
49. Rab S, Alam M: Severe agranulocytosis during para-aminosalicylic acid therapy. *Br J Dis Chest* **64:**164–168, 1970.
50. Bodenheimer H, Samarel A: Agranulocytosis associated with aprindine therapy. *Arch Intern Med* **139:**1181–1182, 1979.
51. Lawrence B, Sarter R, Lipton A, et al: Pancytopenia induced by aminoglutethimide in the treatment of breast cancer. *Cancer Treatm Rep* **62:**1581–1583, 1978.
52. Taurog A: The mechanism of action of the thioureylene antithyroid drugs. *Endocrinology* **98:**1031–1046, 1976.
53. Nagasaka A, Hidaka H: Effect of antithyroid agents 6-propyl-2-thiouracil and 1-methyl-2-mercaptoimidazole on human thyroid peroxidase. *J Clin Endocrinol Metab* **43:**152–158, 1976.
54. Nakashima T, Taurog A, Riesco G: Mechanism of action of thioureylene antithyroid drugs: Factors affecting intrathyroidal metabolism of propylthiouracil and methimazole in rats. *Endocrinology* **103:**2187–2197, 1978.
55. Davidson B, Soodak M, Neary J, et al: The irreversible inactivation of thyroid peroxidase by methylmercaptoimidazole, thiouracil, and propylthiouracil *in vitro* and its relationship to *in vivo* findings. *Endocrinology* **103:**871–882, 1978.
56. Engler H, Taurog A, Nakashima T: Mechanism of inactivation of thyroid peroxidase by thioureylene drugs. *Biochem Pharmacol* **31:**3801–3806, 1982.
57. Engler H, Taurog A, Luthy C, Dorris M: Reversible and irreversible inhibition of thyroid peroxidase-catalyzed iodination by thioureylene drugs. *Endocrinology* **112:**86–95, 1983.
58. Imamura M, Aoka N, Saito T, et al: Inhibitory effects of antithyroid drugs on oxygen radical formation in human neutrophils. *Acta Endocrinol (Copenh)* **112:**210–216, 1986.
59. Weetman A, Holt M, Campbell A, et al: Methimazole and generation of oxygen radicals by monocytes: Potential role in immunosuppression. *Br Med J* **288:**518–520, 1984.
60. Kariya K, Lee E, Hirouchi M: Relationship between leukopenia and bone marrow myeloperoxidase in the rat treated with propylthiouracil. *Jpn J Pharmacol* **36:**217–222, 1984.
61. Forsland T, Borgmastars F, Fyhrquist F: Captropril associated leukopenia confirmed by rechallenge in patient with renal failure. *Lancet* **1:**166, 1981.
62. Weiss A, Markenson J, Weiss M, Kammerer W: Toxicity of d-penicillamine in rheumatoid arthritis. *Am J Med* **64:**114–120, 1978.
63. Harrison J, Schultz J: Studies on the chlorinating activity of myeloperoxidase. *J Biol Chem* **251:**1371–1374, 1976.
64. Kadar D, Kalow W: Acute and latent leukopenic reaction to antipyrine. *Clin Pharmacol Ther* **28:**820–822, 1980.
65. Barrett A, Weller E, Rozengust N, et al: Amidopyrine agranulocytosis: Drug inhibition of granulocyte colonies in the presence of patient's serum. *Br Med J* **2:**850–851, 1976.
66. Huguley C: Agranulocytosis induced by dipyrone, a hazardous antipyretic and analgesic. *JAMA* **189:**938–941, 1964.
67. Vincent P: Drug-induced aplastic anaemia and agranulocytosis: Incidence and mechanisms. *Drugs* **31:**52–63, 1986.
68. Pisciotta V: Drug-induced agranulocytosis. *Drugs* **15:**132–143, 1978.
69. Weitzman S, Stossel T: Drug-induced immunological neutropenia. *Lancet* **1:**1068–1072, 1978.
70. Fibbe W, Claas F, Van der Star-Dijkstra W, et al: Agranulocytosis induced by propylthiouracil: evidence of a drug dependent antibody reacting with granulocytes, monocytes, and haematopoietic progenitor cells. *Br J Haematol* **64:**363–373, 1986.
71. Nakamura M, Yamazaki I, Kotoni T, Ohtaki S: Thyroid peroxidase selects the mechanism of either

1- or 2-electron oxidation of phenols, depending on their substituents. *J Biol Chem* **260:**13546–13552, 1985.
72. Takayama S, Aihara K, Onodera T, Akimoto T: Antithyroid effects of propylthiouracil and sulfamonomethoxine in rats and monkeys. *Toxicol Appl Pharmacol* **82:**191–199, 1986.
73. Haynes R, Murad F: Thyroid and antithyroid drugs, in Gilman A, Goodman L, Rall T, Murad F (eds): *The Pharmacological Basis of Therapeutics,* ed 7. New York, Macmillan, 1985, p 1402.
74. Hughes SW, Burley D: Aminoglutethimide: Side-effect turned to therapeutic advantage. *Postgrad Med J* **46:**409–416, 1970.
75. Vrhovac B: Anti-inflammatory analgesics and drugs used in gout, in Dukes M (ed): *Meyler's Side Effects of Drugs,* ed 10. Amsterdam, Elsevier, 1984, p 154.
76. Griciute L, Tomatis L: Carcinogenicity of dapsone in mice and rats. *Int J Cancer* **25:**123–129, 1980.
77. Park BK, Coleman J, Kitteringham N: Drug disposition and drug hypersensitivity. *Biochem Pharmacol* **36:**581–590, 1987.
78. Ahlstedt S, Kristofferson A: Immune mechanisms for induction of penicillin allergy. *Prog Allergy* **30:**67–134, 1982.
79. Uetrecht J, Sweetman B, Woosley, R, Oates J: Metabolism of procainamide to a hydroxylamine by rat and human hepatic microsomes. *Drug Metab Dispos* **12:**77–81, 1984.
80. Uetrecht J: Reactivity and possible significance of hydroxylamine and nitroso metabolites of procainamide. *J Pharmacol Exp Ther* **232:**420–425, 1985.
81. Shear N, Spielberg S: In vitro evaluation of a toxic metabolite of sulfadiazine. *Can J Physiol Pharmacol* **63:**1370–1372, 1985.
82. Reider M, Uetrecht J, Miller M, Spielberg S: Toxicity of a reactive intermediate of sulfadiazine in an in vitro system. *Pharmacologist* **28:**124, 1986.

12

The Respiratory Burst and the Metabolism of Drugs

ARTHUR L. SAGONE, Jr.

1. INTRODUCTION

It has only been about 20 years since it was first demonstrated that the granulocytes of patients with chronic granulomatous disease (CGD) lack a burst in oxidative metabolism.[1] This observation stimulated extensive study of this aspect of granulocyte physiology and has established the importance of reactive oxygen species (ROS) in the functional capacity of phagocytic cells.[2–6] The complex nature of this biochemical pathway as well as an ever-expanding role of ROS in diverse physiological processes is becoming more apparent with each new report. In addition, there is now evidence that these oxygen metabolites may be major mediators of the inflammatory process occurring in patients with several diverse clinical disorders.[7–25]

We became interested in the study of the oxidative metabolism of granulocytes several years ago shortly after Babior et al.[26] reported that these cells produce superoxide (O_2^-). We had just developed modifications of the method reported by Davidson[27] that used an ionization chamber-electrometer system for the measurement of the CO_2 produced by cell suspensions. This system seemed ideal for the study of the metabolic burst of phagocytic cells, since it allows for the continuous measurement of the hexose monophosphate shunt activity of cells suspensions as well as the production of several reactive oxygen species.

One goal of our studies has been to identify the reactive oxygen species generated in phagocytic cells during the metabolic burst and to evaluate their importance in the functional capacity of these cells. Of particular interest has been the possible role of the hydroxyl free radical in the inflammatory process.

Several observations indicate that granulocytes generate OH· during the metabolic burst.[9,15,16,22,28–39] Therefore, the possible role of OH· in the inflammatory

ARTHUR L. SAGONE, Jr. • Department of Medicine and Pharmacology, Division of Hematology and Oncology, Ohio State University, College of Medicine, Columbus, Ohio 43210.

response and the microbicidal capacity of phagocytic cells seems reasonable.[40] This consideration led us, similarly to others,[9,22,41–45] to evaluate the possibility that the anti-inflammatory nature of several clinically useful anti-inflammatory drugs was directly related to their capacity to scavenge or impair the production of ROS by granulocytic cells. While the results of these experiments seem to support this concept, a major role of OH· in the inflammatory response, and the microbicidal activity of the phagocyte is still considered circumstantial.[32,46,47]

However, these recent studies have raised another consideration that has significant implications; this is the capacity of activated granulocytes to metabolize drugs in a manner that appears analogous to the well-documented microsomal system of the liver.[48] The capacity of granulocytes to metabolize xenobiotics and its clinical implications is the subject of this discussion. However, before proceeding with this subject, I would like to review briefly our current understanding of the metabolic burst as it occurs in phagocytic cells. The literature in this area is extensive and represents a major effort of a large number of investigators.

2. CHARACTERISTICS OF THE METABOLIC BURST

The respiratory burst in the human granulocytes is characterized by a rapid, marked increase in the rate of the hexose monophosphate shunt pathway, a rapid increase in cellular oxygen consumption, and the generation of several reactive oxygen species.[9,15,16,28–39,49–68] These include superoxide (O_2^-), hydrogen peroxide (H_2O_2), hypochlorous acid (HOCl), stable oxidants (now identified as chloramines), and hydroxyl free radical (OH·). The burst also results in the oxidation of I^- to I_2, and the subsequent fixation of I_2 to protein.[69] The possibility that singlet oxygen is generated has been suggested,[70,71] but currently, there is no solid evidence to support the generation of this ROS.[32] The following scheme summarizes the nature of the biochemical reactions as currently accepted and involves several

$$\mathrm{NADPH} + O_2 \xrightarrow[\text{NADPH oxidase}]{} \mathrm{NADP}^+ + O_2^- \tag{1}$$

$$2H^+ + O_2^+ + O_2^+ \xrightarrow[\text{Superoxide dismutase}]{} H_2O_2 + O_2 \tag{2}$$

$$H_2O_2 + Cl^- + H^+ \xrightarrow[\text{Myeloperoxidase}]{} \mathrm{HOCl} + H_2O \tag{3}$$

$$\underset{\text{amino acid}}{\mathrm{HOCL} + \mathrm{RNH}_2} \longrightarrow \underset{\text{chloramine}}{\mathrm{RNHCl} + H_2O} \tag{4}$$

$$O_2^- + H_2O_2 \xrightarrow[\text{Haber–Weiss reaction}]{} \mathrm{OH}\cdot + \mathrm{OH}^- + \mathrm{O2} \tag{5}$$

cellular enzymes, including NADPH oxidase, myeloperoxidase, and superoxide dismutase.

The initial reaction appears to be a one-electron transfer from NADPH to molecular oxygen, resulting in the production of superoxide (O_2^-). The reaction requires the activation of the enzyme NADPH oxidase [Eq. (1)]. This enzyme(s) may be cytochrome *b* (either 245, 559, or both).[72,73] The NADPH required for the reaction is provided by the hexose monophosphate shunt pathway, explaining its rapid activation. The O_2^- is then rapidly converted to H_2O_2 by spontaneous dismutation or by the enzyme superoxide dismutase [Eq. (2)]. Once produced, H_2O_2 is metabolized by several routes. Some H_2O_2 is degraded by cellular catalase[55,62,74] and glutathione peroxidase.[20,75] H_2O_2 is also metabolized by myeloperoxidase with the production of hypochlorous acid[64–66] [Eq. (3)] and the subsequent production of stable oxidants by the interaction of HOCl with amino acids[66–68] [Eq. (4)]. In activated granulocytes, the degradation of H_2O_2 by glutathione peroxidase may be impaired by a fall in cellular GSH,[76–78] thereby facilitating the activity of the myeloperoxidase pathway. The production of hydroxyl free radical has been suggested to occur as the consequence of a Haber–Weiss or Fenton-type reaction [Eq. (5)]. The mechanism is still not entirely clear.[6,32,36,37] The oxidation of I^- may occur as a consequence of three mechanisms: the reaction of I^- with HOCl,[79] chloramines,[66–68] and hydroxyl free radical.[79]

Two aspects of the metabolic burst are important. First, it appears to occur relatively rapidly after the activation of the granulocytes by a variety of stimuli.[1–6,10–12,16,25,80] In some cases, the enhanced oxygen consumption and H_2O_2 production may start within 10–20 sec.[55,61,62,81] However, the lag varies with the stimulus.[6,25] Second, the reaction with some stimuli is sustained and appears to continue for an extended period of time after the phagocytic cell has been activated.

The kinetics of activation of the hexose monophosphate pathway as measured by the continuous production of $^{14}CO_2$ from [1-^{14}C]glucose is illustrated in Fig. 1. The method used in these experiments is the ionization chamber electrometer system as reported from our laboratory.[59] As indicated by the arrow, which marks the time of the addition of opsonized zymosan particles to the granulocyte suspension, there is a marked and rapid increase in the oxidation of [1-^{14}C]glucose. There is a built-in lag in the ionization chamber electrometer system for the detection of changes in $^{14}CO_2$ production that is dependent on the flow rate used in the experiment as well as the size of the ionization chamber. When a flow rate of 100 ml/min is used with a 275-ml ionization chamber, a lag of 2–3 min is expected. Therefore, the enhanced HMPS activity probably occurs less than 1 min after the addition of the zymosan particles. Once stimulation occurs, it is sustained for some time, peaking within 20–30 min after the addition of the particles. The kinetics of hydrogen peroxide production as measured by the oxidation of [^{14}C]formate is similar to that of [1-^{14}C]glucose (Fig. 1), indicating that these are closely related biochemical events. Our studies correlate well with those of others who have addressed the kinetics of oxygen consumption, H_2O_2 and hypochlorous acid production.[48–68]

The same stimuli that induce an oxidative burst in the granulocyte also initiates

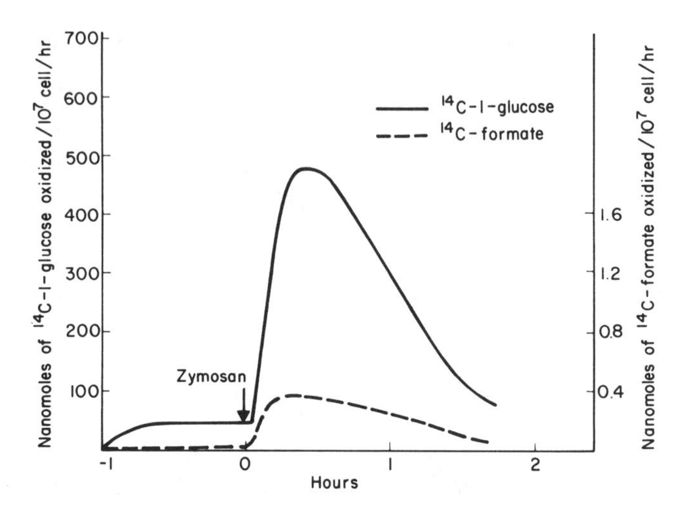

FIGURE 1. Oxidation of ^{14}C substrates by granulocytes. The curves represent a continuous measurement of $^{14}CO_2$ produced from the substrate by granulocyte suspensions using the ionization–electrometer system. After steady-state conditions were established, the addition of opsonized zymosan particles, as indicated by the arrow, results in a prompt increase in both [1-^{14}C]glucose and [^{14}C]formate oxidation. (From Sagone et al.[59])

simultaneously other functional and biochemical changes.[6,25] These include enzyme release,[10,11] chemotaxis,[80,81] the synthesis of prostaglandins,[34,82–84] leukotrienes, and other lipid peroxides.[81,85,86] These changes do not appear to require NADPH oxidase but potentially may be modified by the reactive oxygen species generated by this pathway.[34,83,87,88] The synthesis of prostaglandins can also be considered oxidative in that the reaction requires oxygen.[89] Therefore, part of the oxygen consumption occurring shortly after membrane stimulation may relate to the production of prostaglandins. However, the relative consumption of oxygen during prostaglandin biosynthesis compared with the amount consumed by NADPH oxidase pathway would appear small based on studies in chronic granulomatous cells. While these cells lack a metabolic burst,[36] they appear to have normal arachidonic acid metabolism.[83,85] It is theoretically possible that reactive oxygen species, particularly OH·, could be generated in PMN during prostaglandin biosynthesis.[41] Egan and Kuehl[89] postulated that a free radical. possibly OH· is produced from prostaglandin G_2 (PGG_2) during its conversion to PHG_2. This conclusion was derived following their studies of the metabolism of arachidonic acid by seminal vesicle microsomes.[89] Recent observations suggest that the arachidonic metabolism in PMNs is not associated with significant OH· release or the generation of other ROS.[34,83] However, the point remains controversial.[90]

All the ROS generated by granulocytes theoretically have the capacity to metabolize drugs. There is evidence that granulocytes metabolize drugs via the hydroxyl free radical (see Section 5), and the products generated from H_2O_2 by myeloperoxidase (see Section 6).

3. RELEVANT STUDIES CONCERNING THE METABOLISM OF DRUGS IN MODEL SYSTEMS

We shall focus primarily on the metabolism on xenobiotics by human phagocytic cells. Currently, a number of model systems have been used to study the metabolism of drugs, including pulse radiolysis. Studies in a number of these systems are relevant to the potential metabolism of drugs in vivo by human phagocytic cells. These enzyme systems include the horseradish peroxidase–H_2O_2 system, catalase, prostaglandin synthesis, the NADH–NADPH-dependent microsomal systems which have been isolated from a variety of tissues, and the xanthine oxidase enzyme system. For a comprehensive review of the studies in some of these enzyme systems, the reader is referred to several review articles.[91–101]

3.1. Horseradish Peroxidase–Hydrogen Peroxide Enzyme System

There have been relatively extensive studies using horseradish peroxidase (HRP)–hydrogen–peroxide enzyme system. This enzyme system appears to be analogous to the myeloperoxidase–H_2O_2 enzyme system known to be present in human phagocytic cells, particularly human granulocytes. These enzyme systems appear to have the capacity to metabolize a wide variety of drugs.[91–93,102] These include *p*-aminobenzoic acid (PABA), aniline, *p*-phenylenediamine, aminopyrine, phenothiazines such as chlorpromazine, aminophenol, phenols such as hydroquinones and morphine, and aromatic hydrocarbons. In addition, the HRP–H_2O_2 system oxidizes 4-hydroxylamino-quinoline-*N*-oxide, a known carcinogen to a nitroxide metabolite. Thiols may also be metabolized to thiol free radicals. In this regard, a number of thiol compounds are known to be metabolized by the myeloperoxidase–H_2O_2 enzyme system. These include oxidation of methionine to a sulfoxide,[64] oxidation of methional, and KTMB to ethylene[32]; the formation of the stable oxidant taurine chloramine from taurine[65,66] and oxidation of GSH.[68,103] Several hormones have also been demonstrated to be metabolized by the myeloperoxidase–H_2O_2 enzyme system. These include thyroid hormones,[104–106] estrogen,[107] *n*-formylpeptide (*N*-formyl methionyl-leucyl-phenylalanine),[87,88] and met^5-enkephalin (tyrosine, glycine, glycine, phenylalanine, methionine).[108] There is good evidence that a number of the reactions described in the isolation peroxidase–H_2O_2 enzyme system also occur in the intact cells.[104–109]

These background studies provide relatively solid evidence that the activation of the meyloperoxidase–H_2O_2 system in granulocytes may result in the metabolism of a variety of substrates in vivo, including drugs, carcinogens, and hormones.

3.2. Catalase–H_2O_2 Enzyme System

Almost all tissues contain substantial concentrations of catalase. While the primary role of this enzyme system in the mammalian cells appears to provide protection against oxidant injury by hydrogen peroxide,[101] the possibility that this enzyme system may serve a role in the metabolism of a variety of compounds in the phagocytic cell in vivo has not been explored in depth. The catalytic function of catalase is normally associated with the simultaneous oxidation and reduction of hydrogen peroxide to water and oxygen.[99,110] However, the formation of compound I (a complex of H_2O_2 and catalase) is also known to have peroxidative activity.[110] Compound I is able to oxidize a number of other substrates, including alcohols, aldehyde, and formate.[101,110] The peroxidase activity is favored by high concentrations of the substrate and low concentration levels of compound I.[101] In this regard, the oxidation of formate or methanol by the H_2O_2–catalase system represents a method for the detection of hydrogen peroxide formation in tissues.[49,99,111,112] Furthermore, it has been suggested that the H_2O_2–catalase system in liver may represent one mechanism for the degradation of ethanol and methanol in vivo.[101] In some cells, such as the red blood cells (RBC), hydrogen peroxide is degraded primarily by the glutathione peroxidase system, even under conditions in which the production of cellular H_2O_2 is increased substantially.[113] In spite of this physiology, the metabolism of alcohol by the H_2O_2–catalase system has been demonstrated in RBC. For example, stimulation of H_2O_2 production in RBC by the oxidant drug, methylene blue, results in metabolism of ethyl alcohol and methanol by the H_2O_2–catalase system.[114] The reaction is analogous to their capacity to oxidize formate.[113] In all cases, the reaction involves the oxidation of these substrates by catalase in the presence of hydrogen peroxide (compound I) and is enhanced by conditions in the RBC, which impair the glutathione activity peroxidase pathway.[113] In this regard, numerous studies have demonstrated that human phagocytic cells have the capacity to oxidize formate to CO_2.[1–6,49,74] This reaction also occurs in these cell tissues as a consequence of the oxidation of formate by the catalase–H_2O_2 system,[1–6,49,74] and the reaction is known to be markedly enhanced in the stimulated cell. Furthermore, the reaction may be facilitated in the stimulated granulocytes by a fall in the cellular reduced glutathione, limiting to some degree the activity of the glutathione peroxidase enzyme system.[76,78] These observations suggest that stimulated phagocytic cells acquire the capacity to oxidize a number of organic compounds such as alcohols and aldehydes during degradation of H_2O_2 by the catalase system.

3.3. Arachidonic Acid Metabolism

There has been an extensive study of the metabolism of arachidonic acid to prostaglandins in isolated enzyme systems.[115–118] Enzymes isolated from several tissues are able to oxidize arachidonic acid to prostaglandins. The initial reaction is the oxidation of arachidonic acid to a hydroperoxide (PGG_2) by the enzyme prostaglandin endoperoxide synthase, also known as cyclo-oxygenase. The PGG_2 is then reduced to its corresponding alcohol PGH_2. Both the oxygenase and peroxidase activity of the system appear to reside on the same enzyme.[115] The cooxidation of a variety of xenobiotics is known to occur during these reactions. Of interest, Egan et al.[89] suggested that a free radical might be generated during the degradation of PGG_2 to PGH_2. They observed that the addition of phenol to the incubation markedly enhanced the rate of the reaction. They postulated that a free radical was released during the degradation of PGG_2 to PGH_2 that had the capacity to inactivate cyclo-oxgenase and that this free radical might be hydroxyl radical. Egan and coworkers studied a number of compounds with the capacity to react with hydroxyl radical and demonstrated augmentation of the reaction similar to that observed with phenol.[115] These workers suggest that in addition to the production of prostaglandins that the free radical generated by this enzyme system in vivo could have inflammatory properties. A number of anti-inflammatory drugs have been studied in this system including indomethacin and aspirin.[41,89,115–118] These latter drugs block the cyclo-oxygenase activity, thereby impairing the production of prostaglandins and potentially the release of free radical by this enzyme pathway. A similar production of hydroxyl free radical has also been postulated during the metabolism of arachidonic acid by the lipoxgenase pathway.[41,114–119] These model systems have important implications. If the results from these systems can be extrapolated to the intact cell, almost every tissue would appear to have the capacity to generate hydroxyl radical. The production of hydroxyl free radical by this mechanism would imply that a number of compounds could be oxidized or hydroxylated by this pathway in vivo.

In addition, there may be other mechanisms. One may be the capacity of organic peroxides to react directly with drugs. Egan et al.[120] demonstrated that the oxidation of salindac sulfide to its sulfoxide (Sulindac) by 15-hydroperoxyprostaglandin E_2 (15-HPE_2) and endoperoxide synthase obtained from seminal vesicle microsomes. A similar reaction occurs if PGG_2 and 15-HPETE are used as substrates. In this reaction, there appears to be a direct transfer of an oxygen from the substrate to the drug.[115] Therefore, the biotransformation of drugs could occur in granulocytes as a consequence of the interaction of a drug with either the OH· or the organic peroxides produced by the arachidonic acid cascade.

3.4. The Glutathione Peroxidase Pathway

Glutathione peroxidase, similar to catalase, appears to have a primary role in the protection of the mammilian cell against oxidant injury. This enzyme which requires reduced glutathione as a co-factor is known to degrade the hydrogen peroxide.[20,75] This enzyme pathway also has the ability to metabolize organic peroxides.[99] The presence of the phagocytic cells at an inflammatory site could potentially reduce organic peroxides. However, the activity of glutathione peroxidase may be suboptimal in activated granulocytes due to a decrement in cellular GSH.[76–78]

3.5. Microsomal Enzyme System

Liver endoplasmic reticulum contains an important group of oxidative enzymes called mixed function oxidases or monoxygenases that require both a reducing agent (NADPH) and oxygen for the activity. These enzyme systems are involved in the metabolism of many drugs in vivo. This enzyme system is reported to consist primarily of cytochrome P-450 reductase, cytochrome P-450 and phospholipid.[48] These microsomal enzyme systems have been isolated from a variety of tissues including the nuclei of cells.[121,122] The microsomal mixed-function oxidase appears to catalyze a number of reactions.[48] These include *N*- and *O*-dealkylation, aromatic ring and side chain hydroxylation, sulfoxide formation, *N*-hydroxylation, and deamination of primary and secondary amines, and desulfuration, (replacement of a sulfur by an oxygen).[48] This enzyme system has been used to study the metabolism of a number of drugs in vitro, and there is extensive literature in this area.[48,91–101]

The consumption of oxygen by this enzyme system is known to be associated with the production of the reactive oxygen species O_2^-, H_2O_2, and OH·, as well as lipid peroxides.[99–101,121–126] This capacity of microsomes to produce ROS appears to be one mechanism for the metabolism of drugs. For example, hydroxyl free radical may relate to the metabolism of several compounds, particularly alcohols.[126–131] Cederbaum and co-workers demonstrated that several hydroxyl scavengers are oxidized by the liver microsomes presumably by OH·.[126,129–131] Methional and KTBA are oxidized to ethylene, DMSO to methane or formaldehyde, isopropyl alcohol to acetone, and tertiary butyl alcohol to formaldehyde and acetone.[130] The microsomal system has also been shown to oxidize the specific hydroxyl scavenger benzoate.[131,132] This compound is known to be oxidized by the OH· generated during the radiolysis of water[133,134] and by xanthine oxidase enzyme system.[36] Therefore, the oxidation of benzoate is evidence for the generation of OH· in microsomes. Oxidation of these substrates occurs by a different mechanism than the metabolism of aminopyrine to formaldehyde. The latter reaction can be impaired by two known inhibitors of mixed function oxidase–metyrapone and SKF-525A. By contrast, the oxidation of DMSO, KTBA, methanol, *t*-butyl alcohol, and benzoic acid are not impaired by these compounds, indicating that their metabolism by the microsomes system is not dependent on mixed-function oxidase activity.[130,131] The generation of OH· in microsomes appears dependent on the production of hydrogen peroxide, and not superoxide.[131,132] Therefore, a Haber–Weiss mechanism does not appear to be involved. Also, the addition of azide to the incubation enhances the rate of reaction, suggesting that iron is not required.[131,132] Therefore the mechanism for the generation of OH· in microsomes may differ from that described in other biological systems.[36]

One observation reported by Winston and Cederbaum further demonstrates the possible complex nature of the reactions occurring in microsomes.[131] This system is known to generate lipid peroxides as well as O_2^- and H_2O_2. This raises the possibility that alkoxy (RO·) radicals as well as OH· might explain the oxidation of compounds, such as benzoate, KTBA, and alcohols. This was evaluated further using a chemical system to generate OH· and RO·. Cumene or *t*-butyl hydroperoxide support the oxidation of KTBA and ethanol in the present Fe^{2+} chelate, but not the oxidation of benzoic acid. Rather, H_2O_2 is required as a substrate for the oxidation of benzoic acid to occur in this system. These data suggest that both alkoxy radicals (RO·) and OH· may oxidize compounds, such as methional, KTMB, and alcohols. By contrast, the oxidation of benzoic acid in biological systems may be mediated primarily by OH·. Activated human granulocytes acquire the capacity to metabolize benzoic acid in a manner similar to that observed in the liver microsomes system.[36] In both cases, the ROS mediating the reaction appears to be OH·. However, the biochemical mechanisms for the generation of OH· in these two biology systems appear to differ.

Another aspect of the microsomal metabolism of interest could relate to the metabolism of drugs by phagocytic cells. The metabolism of quinone drugs by cytochrome P-450 reductase is associated with the enhanced generation of reactive oxygen species.[96,121–124,132] The following mechanism has been suggested to explain the enhanced generation of ROS by the quinone drugs.[121,123] The drug is first reduced to a drug free radical (semiquinone) by a single electron transfer from NADPH, and cytochrome P-450 reductase. The drug free radical then transfers the electron to O_2 with the production O_2^-. During the reaction with O_2, the drug is reoxidized. The O_2^- produced becomes the source for the other ROS generated in this system—H_2O_2, OH·, and possibly singlet oxygen. Therefore, in the scheme, the quinone drug functions as an electron carrier between NADPH and O_2. The apparent similarities between the capacity of the NADPH–oxidase system of phagocytic cells to generate ROS and the microsomal system suggest that quinone drugs might similarly serve as electron carriers from NADPH to O_2 in phagocytic cells. This would include a large group of drugs such as the quinone type of anticancer drugs[121,122] and antibiotics.[96] In this regard, it has been established for some time that quinones markedly increase the rates of oxygen consumption and hexose monophosphate shunt activity of resting granulocytes but not stimulated ones.[49] Some of the drugs include naphthoquinone, menadione, phenazine methosulfate, methylene blue, and α-tocopherol quinone. However, this stimulation is not necessarily dependent on NADPH oxidase, since CGD cells that lack this enzyme are known to be stimulated with methylene blue.[1] While the metabolism of the quinone drugs has been evaluated in a variety of tissues, such as tumor cells, micro-organisms, and heart, their effects on the metabolism of phagocytic cells appear to have been less extensively studied.

3.6. Xanthine–Xanthine Oxidase Enzyme System

Currently, there is no evidence to indicate that human phagocytic cells have an enzyme system similar to the aldehyde–xanthine–xanthine oxidase found in a variety of tissues, particularly liver, kidney, and small bowel.[91–101] However, this enzyme system is known to generate three of the reactive oxygen species produced by granulocytes during a metabolic burst, including superoxide, hydrogen peroxide, and hydroxyl radical. This enzyme system may also generate singlet oxygen as well.[135–138]

Therefore, the metabolism of xenobiotics by this enzyme system is of interest because it provides a reasonable model system for the study of the potential metabolism of drugs by activated phagocytic cells. Of importance, however, in evaluating the studies done with the xanthine–xanthine oxidase enzyme system, is the fact that phagocytic cells also generate additional ROS. There are extensive studies concerning the metabolism of compounds by this enzyme system. These cannot all be discussed. However, it seems worthwhile to mention a few of these studies because they establish a few points. This enzyme system is known to oxidize methional to ethylene,[139] ethanol to acetaldehyde,[140] dimethyl sulfoxide to formaldehyde,[131] and *t*-butyl alcohol to both acetone and formaldehyde.[130] Presumably all these reactions are mediated by OH·. Recent studies of the metabolism of benzoic acid and salicylic acid by the X–XO enzyme are also of interest. We recently demonstrated that both benzoic acid and salicylic acid are oxidized by the xanthine–xanthine oxidase enzyme system.[36,141] Winterbourn[47] recently demonstrated that benzoic acid is hydroxylated by the xanthine-xanthine oxidase system. Floyd et al.[142] reported a similar hydroxylation of phenol and salicylate acid by this enzyme system. These reactions were predicted by the known metabolism of these compounds with OH· during pulse radiolysis.[133]

It is now known that a number of these compounds are also oxidized by stimulated granulocytes. However, the mechanism for these reactions may differ considerably. The metabolism of methional and KTMB by phagocytic cells was originally studied as a possible method for the detection of hydroxyl free radical production by the cells.[38,39] However, subsequent studies have suggested that the reactions are more complex[143] and that methional and KTMB are oxidized primarily as a consequence of the reactive oxygen species generated by the myeloperoxidase pathway (HOCl) rather than by OH·.[32] Ethanol may be oxidized by the H_2O_2–catalase system, but this point has not been studied.[49] Currently, the oxidation of benzoate, salicylate, and DMSO appears to be mediated primarily by the OH· produced by granulocytes.[16,36,102,141] Therefore, the oxidation of these substrates may represent methods for the measurement of the production of OH· by these cells.

The apparent marked differences in the mechanism for the oxidation of the compounds in the microsomal system (discussed in Section 3.5), in the xanthine–xanthine oxidase enzyme system, and in granulocytes raises an important consideration; i.e., in extrapolating the potential metabolism of a drug or compound from a model enzyme system to the intact cell, all the ROS generated by the cell system must be considered, as well as other factors such as the uptake and distribution of the drug in the cell and its rate constants for the variety of ROS produced.

3.7. Drug Studies in the Model System—Pulse Radiolysis

Recent studies concerning the effects of nonsteroidal anti-inflammatory drugs on the radiation damage induced in enzymes and bacteriophage have important implications concerning the possible metabolism of drugs in humans. I would, therefore, like to discuss some of these studies.

Studies by Samuni et al.[144] have provided important background information concerning the mechanism of radiation damage to bacteriophage T_4. These workers studied the radiation damage induced in the bacteriophage under different gas mixtures and scavenger conditions in order to evaluate the role of oxygen and its metabolites superoxide and H_2O_2 in the oxygen-enhancing ratio (OER). Viruses do not appear to have the repair mechanisms present in bacteria or more complex cells. Therefore, this system appears to provide a simple model for the study of the mechanism of radiation damage. Their results suggest that the indirect radiation effect mediated by exogenous water radical accounts for 90% of the radiolytic damage to the virus, while direct injury due to a direct excitation of biomolecules appears to account for only 10%. Furthermore, the hydroxyl free radicals produced during radiolysis appear to be responsible for the major portion of the indirect damage, while the role of the hydrated electrons, superoxide (formed by the reaction of hydrated electrons with oxygen), and H_2O_2 appears negligible. The role of oxygen in this system appears to be related to its capacity to react directly with target molecules causing irreversible peroxidation. The results are consistent with the previously proposed Howard-Flander's theory of radiation damage.[145] Of interest is the effect of formate in the system. Formate provided good protection of the virus against radiodamage even when the irradiation was done under an atmosphere which was oxygen saturated. Under these conditions, formate theoretically reacts with OH· forming an intermediate radical that then reacts with O_2 to produce CO_2 and O_2^-. This experiment is of importance for two reasons. First, it appears to exclude O_2^- and H_2O_2 as major mediators of radiolytic damage in this system. Second, it indicates that the capacity of the hydroxyl scavenger to provide protection against damage by OH· requires the formation of a product which cannot transmit the energy. In this case, the ROS was O_2^-.

Recently, Hiller and Wilson[146] demonstrated that a number of nonsteroidal anti-inflammatory (NSAIDS) drugs have relatively high rate constants for reaction with OH·. The drugs studied included aspirin, indomethacin, flurbiprofen, metiazinic acid, and D-penicillanine and were found to protect bacteriophage T_2 inactivation by the hydroxyl free radical produced during irradiation.

Studies that have involved the use of uric acid as a radioprotective agent are also of interest, since uric acid has been reported to have antioxidant properties in other biological systems.[147] Kittridge and Willson[148] studied the protective effect of uric acid and ascorbic acid against the damage occurring to lactic acid dehydrogenase (LDH) and yeast alcohol dehydrogenase (YADH) by the hydroxyl free radical generated during radiolysis.[147] Paradoxically, while uric acid protected LDH against inactivation, uric acid actually increased the degradation of yeast alcoholic dehydrogenase. This increased inactivation did not occur under anaerobic conditions indicating the importance of oxygen in the reaction. Subsequently, Gee et al.[149] studied further the capacity of several amino acids and a number of drugs and agents known to be hydroxyl scavengers in the inactivation of yeast alcoholic dehydrogenase by the hydroxyl radical produced by radiation (Table I). Compounds that were found to be protected were believed to scavenge hydroxyl radical with the formation of a new compound that was unable to transfer its energy to the target

TABLE I. Percent Remaining Activity of Alcohol Dehydrogenase after Irradiation with 20 Gy in Air-Saturated Solutions (0.02 mg/ml, pH = 7) Containing Other Solutes (10mM). The Data Are Presented as the Mean, Standard Error and the Number of Samples (*N*)[a]

	Mean	(SEM)	*N*		Mean	(SEM)	*N*
Control	22	(0.8)	16	Isoleucine	58	(2.9)	3
Alanine	31	(1.6)	5	Leucine	65	(0.9)	3
Arginine	43	(1.5)	5	Lysine	11	(1.5)	8
Asparagine	23	(1.2)	7	Methionine	92	(1.4)	2
Aspartate	28	(1.9)	3	Phenylalanine	27	(1.9)	4
Cysteine	100	(0.0)	4	Proline	34	(0.8)	4
Cystine	—			Serine	74	(5.6)	3
Glutamate	31	(0.0)	6	Threonine	58	(3.2)	3
Glutamine	1	(0.4)	11	Tryptophan	82	(0.9)	3
Glycine	71	(0.3)	4	Tyrosine	93	(0.3)	4
Histidine	93	(2.4)	4	Valine	46	(1.8)	3
Hydroxyproline	57	(2.1)	4	Salicylate	98	(2.5)	2
Thymine	5	(1.4)	3	Acetyl salicylate	17	(1.7)	5
ADP	39	(2.0)	2	Ethanol	87	—	1
Urate (2mM)	2	—	1	Mannitol	87	—	1
Formate	97	(1.0)	2	DMSO	86	—	1
Benzene	32	—	1	Glucose	68	—	1
Phenol	100	—	1	Benzoate	34	—	1

[a]From Gee et al.[149]

enzyme. Again, formate proved to be highly protective in this system. Formate presumably reacted with the hydroxyl radical to produce an intermediate CO_2 radical that was then reacted with molecular oxygen to produce CO_2 and superoxide. Superoxide appears unable to inactivate the enzyme system studied. In contrast to formate, thymine proved capable of accelerating the reaction in the presence of oxygen. The following hypothesis was raised to explain the results. The hydroxyl free radical reacted with the thymine to produce an intermediate radical. In the presence of oxygen, an additional compound was formed, a peroxy radical, which could then interact with sensitive sites in the enzyme causing inactivation. Gee et al. postulated that the specific sites were most likely the amino acids, cysteine and methionine. Their results also suggested that compounds with free hydroxyl groups appeared to afford much better protection in this model system than those without a free hydroxyl group. For example, the capacity of phenol and salicylic acid to protect the enzyme against radiation induced injury by hydroxyl radical proved to be much greater than benzene and acetylsalicylic acid (aspirin). In fact, aspirin appeared to actually enhance radiation-induced inactivation of the enzyme.

These data have significant implications in regard to the potential metabolism of drugs by granulocytes in vivo. First, the results indicate that these drugs have high rate constant for interaction with hydroxyl radical. Second, the compounds produced after reaction with OH· may be capable of further reactivity with molecular oxygen or another ROS. This new metabolite may be able to transfer its energy in some cases to a target system. Under other circumstances, however, the compound formed may be unable to react with a target system and thereby afford protection. As discussed above, uric acid has the capacity to protect LDH against inactivation by hydroxyl radical while actually enhancing radiation injury to yeast alcoholic dehydrogenase.

There is considerable evidence to suggest that activated granulocytes have the capacity to generate hydroxyl radicals (see Section 5). In view of what appears to be a relatively sustained period of production of this free radical by granulocytes, there would appear to be ample opportunity for a drug to interact with the hydroxyl radical produced by the granulocytes at a site of inflammation. While a major activity of the drugs might be to scavenge OH· and thereby deplete the production of hydroxyl radical by the granulocytes, the overall biological effect cannot be predicted with any certainty. If the compound formed following the interaction of the hydroxyl radical with the drug is capable of interacting with oxygen, and producing the peroxy radical, the metabolite might be capable of mediating enhanced injury if produced in the vicinity of a sensitive target. It is clear at this point that the potential interaction of the granulocyte with a variety of drugs needs considerable study. Furthermore, while the data discussed by the radiation model system are of interest, there are significant differences in the production of ROS by granulocyte, particularly the capacity to generate additional reactive oxygen species to those generated during pulse radiolysis. These additional ROS have the potential to react with the drugs as well.

4. EVIDENCE THAT THE ROS GENERATED BY THE MYELOPEROXIDASE SYSTEM IN PMN HAVE THE CAPACITY TO METABOLIZE DRUGS

The myeloperoxidase system present in the granulocytes results in the production of the unstable oxidant hypochlorous acid as well as stable oxidants produced by the interaction of HOCL with endogenous substrates such as taurine (see Section 2). In addition, monochloramine (NH_2Cl) may be produced from HOCl.[66] This compound is unstable. There is evidence that a large number of drugs, compounds, and hormones can be metabolized by the isolated enzyme system (see Section 3.1). Recently, Kalyanaraman and Sohnle[150] studied the potential interaction of a few of these drugs with the products of the MPO pathway. They evaluated the capacity of two products of the myeloperoxidase pathway, hypochlorite and taurine chloramine, to produce drug radicals. Both sodium hypochlorite solutions and taurine chloramine solutions were able to produce the drug free radical intermediates from chlorpromazine, aminopyrine, and phenylhydrazine. The drug radical was measured using electron spin resonance (ESR) or visible light spectroscopy. Similarly, they were able to demonstrate the production of drug radicals using the supernatants derived from zymosan-stimulated granulocytes suspensions. As expected, the formation of the drug radical was enhanced by supernatant derived from zymosan stimulated suspensions incubated with taurine. Under these conditions, the hypochlorous acid produced by the PMN is trapped in the supernatant as a stable oxidant–taurine chloramine. On the basis of their studies, Kalyanaraman and Sohnle[149] concluded that the oxidation of certain chemical compounds by neutrophils in vivo could produce potentially damaging electrophilic free radical forms. They suggested that the MPO/H_2O_2 enzyme system could represent a new metabolic pathway for the biotransformation of these substances in vivo and could be important in drug toxicity as well as chemical carcinogenesis. Therefore, these observations appear to have important implications. First, they provide evidence that the generation of reactive oxygen species by the myeloperoxidase pathway represents a potential biochemical mechanism in vivo for the metabolism of these drugs. Second, the implication of these studies is that the toxic compound produced by the granulocytes could result in tissue damage. One important area that they suggested for investigation is the potential interaction of a variety of compounds found in our diet with the reactive oxygen species produced by activated granulocytes in vivo. A number of these compounds could be toxic or carcinogenic, if biotransformed.[151]

However. while the results of the experiments discussed above are exciting, the situation in vivo could be more complex. Kalyanaraman and Sohnle did not study the metabolism of the drugs during the direct incubation with stimulated granulocytes but only the effects of the stable oxidants generated in the supernatant. The metabolism of the drugs under the former conditions might differ considerably. Several compounds appear to be metabolized similarly whether incubated directly with stimulated granulocytes or in the supernatants containing stable oxidants. A few include methionine, FMLP, and α-1-proteinase inhibitor, all of which are inactivated.[64–68]

By contrast, while thyroid hormones appear to be degraded by the ROS species produced by activated granulocyte suspensions,[104,105] triiodotyronine (T_3) is metabolized to T_4 by supernatants containing stable oxidants.[106] Also, I^- is oxidized to I_2 by the stable oxidant found in supernatants of activated granulocytes suspensions and is in fact a basis for the quantitation of the amount of chloramines produced.[65–68] However, the oxidation of I^-, when incubated with granulocytes, may be more complex and may involve reactions with OH· as well as HOCL,[79] since under these conditions, the substrate will be exposed to both ROS simultaneously.

5. EVIDENCE THAT ACTIVATED GRANULOCYTES CAN OXIDIZE AND HYDROXYLATE DRUGS

Several years ago, we became interested in developing a specific method for detection of hydroxyl radical in biological systems. After some consideration, we chose to evaluate an assay based on the oxidation of benzoic acid,[36] as there was already background information in the radiation literature supporting the specificity of the reaction of benzoic acid with OH·, compared with the other ROS known to be produced by phagocytic cells. Matthew and Sangster[133] had demonstrated that benzoic acid was oxidized and hydroxylated during the radiolysis of water. Furthermore, they demonstrated that the reaction was not mediated by either O_2^- or H_2O_2. Other investigators have reported similar results.[152,153] We found that while unstimulated granulocyte suspensions did not oxidize [^{14}C]carboxyl benzoic acid, those stimulated by opsonized zymosan particles rapidly oxidized this substrate (Fig. 2). Comparison of the pattern for benzoic oxidation with [1-^{14}C]glucose, and [^{14}C]formate shown in Fig. 1 suggests similar kinetics for the oxidation of all three substrates. The reaction could be inhibited by superoxide dismutase, somewhat by catalase, azide, and the hydroxyl scavengers, mannitol and DMSO. These results suggest that the decarboxylation of benzoic acid is related to the production of OH· and the OH· may be produced by the interaction of O_2^- and H_2O_2 in a Haber–Weiss-type reaction. The capacity of azide to inhibit the reaction suggests that heme is required in some way. The observation that the stimulated granulocytes of patients with CGD fail to oxidize benzoic acid indicates that NADPH oxidase is required for the reaction. In addition, the normal capacity of GCD granulocytes to metabolize arachidonic acid by the cyclo-oxygenase and lipo-oxygenase pathway appears to exclude a major production of OH· by these pathways.[83,85]

In order to provide firmer evidence that the oxidation of benzoic acid is mediated by hydroxyl radical, we recently demonstrated that benzoic is hydroxylated during incubation with zymosan activated granulocytes.[37] We were able to

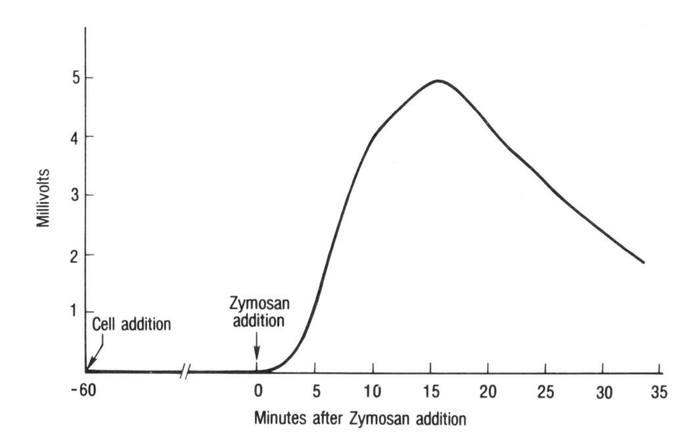

FIGURE 2. The oxidation of benzoic acid by granulocyte suspensions (2 × 10^7), both resting and during the phagocytosis of zymosan particles. The y axis represents the electric signal (in mV) generated from $^{14}CO_2$ passing through the ionization chamber and the x axis represents the time of incubation in minutes. As seen in the first part of the curve, unstimulated cells did not oxidize benzoic acid. However, as indicated by the arrow, the addition of opsonized zymosan particles is associated with a prompt oxidation of this substrate. (From Sagone et al.[42])

demonstrate the production of ortho, para, and meta benzoic acid. These products of benzoic were all predicted by the reaction known to occur between the benzoic acid and OH· during the radiolysis of water. However, we were unable to demonstrate the ratio of the products that were predicted from the radiation studies. We were also unable to detect the presence of phenol or resorcinol, even though these could be potential products.

One reason for the apparent differences in the ratio of the hydroxylated products produced from benzoic acid by OH· following the radiolysis of water compared to activated granulocytes may be the kinetics of the reactions. Following pulse radiolysis, there is a rapid and short-term production of OH·. This would leave relatively little opportunity for the secondary reaction of OH· with the products formed by OH· from benzoic acid (primary reaction). Matthews and Sangster[133] estimated that less than 2% of the G value, which they determined for benzoic acid was related to such secondary reactions. In contrast to pulse radiolysis, the production of reactive oxygen species by stimulated granulocytes appears to be sustained. Therefore, there would appear to be ample time for compounds produced from benzoic acid by OH· to react further with OH· or another ROS produced simultaneously as a consequence of the metabolic burst. This possibility has obvious implications concerning the potential capacity of granulocytes to metabolize drugs at the site of an inflammatory response.

The experiment with benzoic demonstrates several points:

1. Our results provide relatively firm evidence that OH· is produced by activated granulocytes as a consequence of the metabolic burst. It therefore seems likely that OH· free radical is produced at a site of inflammation by granulocytes in vivo.
2. We demonstrated that salicylic acid (*o*-hydroxybenzoic acid) is produced from benzoic acid, indicating that activated granulocytes acquire the capacity to hxdroxylate drugs.
3. If the ROS produced by activated granulocytes are related to the inflammatory response, the capacity of a drug to react with ROS could explain their anti-inflammatory properties. By contrast, the metabolite formed might be associated with significant toxic properties. This would be analogous to the formation of a toxic compound from carbon tetrachloride as a consequence of activation of this drug by liver microsomes.[95]

Recent observations by Floyd et al.[154] provide additional evidence indicating that granulocytes have the capacity to generate hydroxyl radicals. These investigators have been able to detect subpicomole levels of hydroxyl radicals in chemical systems with phenol and salicylic acid as radical traps using high-pressure liquid chromatography (HPLC) with electrochemical detection.[142] Using this approach, they were recently able to show that stimulation of human granulocytes with the tumor promoter, TPA, results in accumulation of 8-hydroxydeoxyguanosine. This observation appears to correlate with those of Birnhoim, which indicate that TPA activated granulocytes have DNA strand breakage.[155] The reaction could be significantly diminished by the addition of superoxide dismutase or catalase to the incubation, suggesting that the primary mechanism of injury was due to OH· produced by a Haber–Weiss-type reaction. These results are important for two reasons. They provide additional evidence for the generation of hydroxyl free radical by activated granulocytes. Also, the data provide evidence that hydroxyl free radical may be able to be generated or to permeate by some mechanism the nucleus of the cell, inducing DNA damage. The consequences of such injury include cellular death or transformation into malignant cells.[155]

6. EVIDENCE FOR THE METABOLISM OF NONSTEROIDAL ANTI-INFLAMMATORY DRUGS BY THE HYDROXYL RADICAL PRODUCED BY GRANULOCYTES

The myeloperoxidase system may represent one pathway that is virtually unexplored for the metabolism for a variety of xenobiotics by activated granulocytes. We recently extended our studies of the metabolism of the salicylate group of drugs by granulocytes for a number of reasons.[141] First, our prior experiments with the parent compound benzoic acid suggested that both aspirin (acetylsalicylic acid) and its major metabolite in vivo, salicylic acid, might be similarly oxidized by activated granulocytes. This raised the possibility that a major anti-inflammatory effect of this family of nonsteroidal anti-inflammatory drugs may relate to their capacity to scavenge hydroxyl free radical. If this proved to be the case, our ability to demonstrate that these drugs could be metabolized by the granulocytes would provide evidence that they could act as free radical scavengers in vivo as well as additional evidence that activated granulocytes have the capacity to metabolize drugs.

We compared the capacity of unstimulated and zymosan stimulated granulocytes suspensions with oxidize benzoic acid, acetylsalicylic acid and salicylic acid. The results of our experiments are given in Fig. 3. None of these substrates was oxidized by unstimulated granulocyte suspensions. However, the addition of oxidized zymosan particles resulted in a prompt oxidization of all three substrates. Similar results were found using PMA as a stimulus. The experiments were done using a similar final concentration of each substrate, so the relative rate of oxidation of the substrates could be compared. The mean peak rate of oxidation of salicylic acid was 3.65 $\pm$ 0.92 nmoles per 10^7 cells/hr, 16-fold higher than the rate of benzoate oxidation of 0.22 $\pm$ 0.03 nmoles per 10^7 cells/hr. The average rate of oxidation of aspirin was fourfold higher of benozate and approximately 25% of that rate, salicylic acid. The differences in the rates of benzoic and salicylic acid could be demonstrated over a range of concentration (Fig. 4). The oxidation of salicylic acid could be inhibited by superoxide dismutase, but not clearly by catalase (Table II). Further, the reaction could be impaired by several compounds known to be specific scavengers of hydroxyl radical. These included phenol, dimethylthiourea (DMTU), and benzoate. The reaction was oxygen dependent and was inhibited by the addition of azide, suggesting the possible importance of heme or a iron enzyme

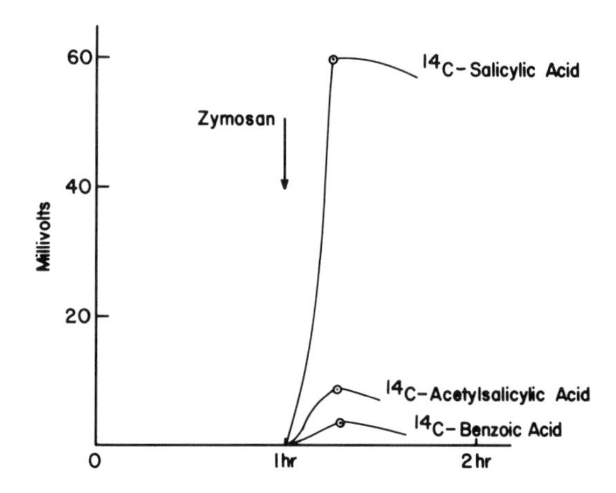

FIGURE 3. Decarboxylation of benzoic acid compounds during incubation with granulocytes. The curves represent the production of $^{14}CO_2$ from 5 μCi (carboxyl-^{14}C) salicylic acid, 2.5 μCi acetyl (carboxyl-^{14}C) salicylic acid, and 5 μCi (carboxyl-^{14}C) benzoic acid incubated with 2×10^7 granulocyte in 4 ml Dulbecco's phosphate-buffered saline with 2.5 mM glucose and 0.5 ml normal saline. The final concentrations of the substrates were similar and ranged from 0.019 to 0.022 mM. The electric signal in millivolts generated by the $^{14}CO_2$ in the ionization chamber is given on the *y* axis and time in hours on the *x* axis. After 1 hr, opsonized zymosan particles were added to the incubation as indicated by the arrow. The curves indicate representative experiments of 29 done with salicylic acid, 3 with benzoic acid, and 2 with ASA. The peak rate of CO_2 oxidation indicated by the dot in each curve was calculated from the observed millivolt signal and chamber constant and are given in the text. (From Sagone et al.[141])

in the reaction (Table III). The possible requirement of iron raised the question as to whether or not the reaction might require myeloperoxidase. Under these circumstances, the metabolism of salicylic acid by granulocytes would be analogous to the biotransformation of other xenobiotics as a consequence of the production of hypochlorous acid or stable oxidants such as taurine chloramine by the granulocytes. For this reason, we evaluated the effect of three compounds known to be potent scavengers of hypochlorous acid end chloramines (Table III). These include GSH, methionine, and taurine. None of these compounds significantly impaired the rate of the reaction. This observation appears to exclude the myeloperoxidase pathway in the metabolism of the salicylic acid. Furthermore, the reaction did not appear to represent a direct reaction of salicylate with superoxide or hydrogen peroxide, since the addition of high concentrations of salicylic acid to granulocyte suspensions did not impair the production of superoxide as determined by cytochrome *c* reduction or hydrogen peroxide production as measured by the oxidation of formate. Recently, dimethylthiourea has been proposed to be a potent and specific scavenger of hydroxyl radical and has protected lung injury in a model of adult respiratory distress syndrome (ARDS).[9] For this reason, we explored further the effects of this compound on the production of reactive oxygen species by zymosan stimulated granulocytes. DMTU did not impair the production of hypochlorous acid or stable

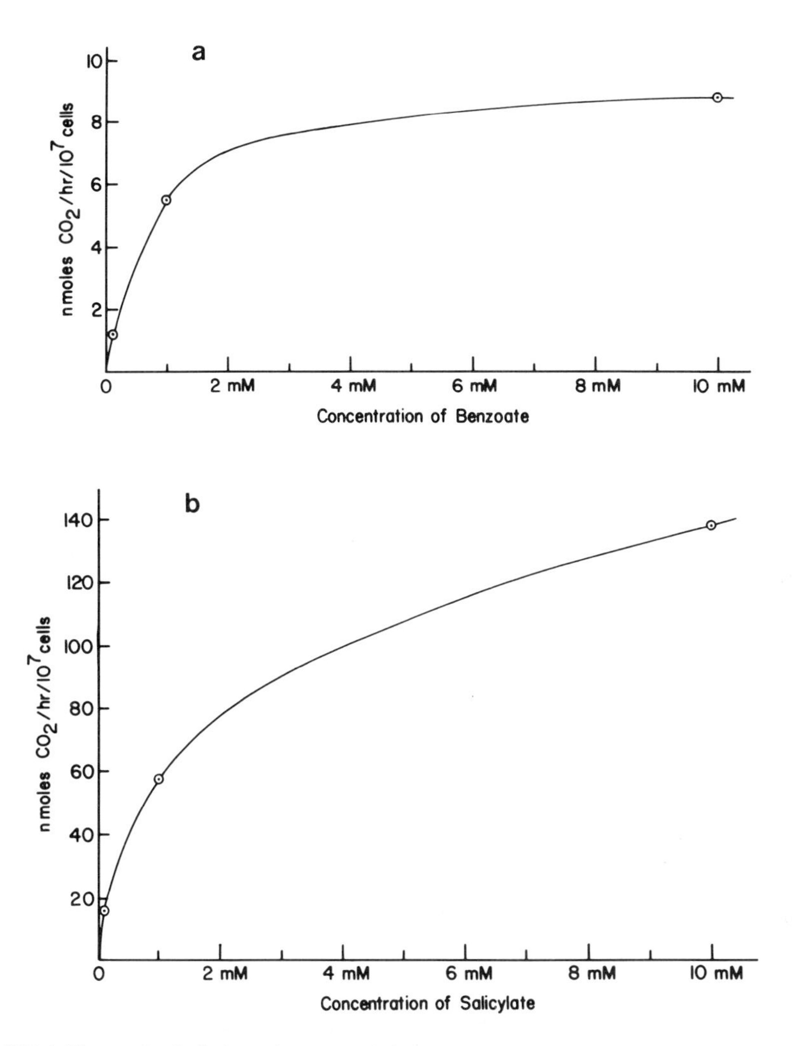

FIGURE 4. Figure 2a indicates the rate of the production of CO_2 from benzoic acid (in nmoles)/10^7 cells/hr over a range of substrate concentrations. Figure 2b indicated similar studies done with salicylic acid. The values indicate the peak values achieved after the addition of opsonized zymosan particles to the granulocyte incubation (see Fig. 3). Note the differences in the scales on the *y* axis. (From Sagone et al.[141])

oxidants by activated granulocytes as determined by the GSH method.[141] In addition, it did not impair the stimulation of the HMPS pathway or the production of hydrogen peroxide, but did impair the OH· production of granulocyte suspensions (Table IV). Our observations appear to support the results previously reported by Fox[9] and provide additional evidence that the metabolism of salicylic acid by

TABLE II. Effect of Scavengers of Reactive Oxygen Species on Salicylate (0.02 mM) Oxidation by Zymosan Stimulated Granulocytes[a,b]

Agent	Control	Experimental	Percent control
SOD 10 μg/ml (3)	3.63 ± .40	1.08 ± .15	30
Heat denatured SOD (3)		2.85 ± .16	78
Catalase 100 μg/ml (3)	3.45 ± 1.13	2.77 ± .15	80
Heat denatured catalase (3)		2.67 ± .11	77
Phenol 10^{-4} M (3)	4.15 ± .76	0.27 ± .06	6
DMTU 10 mM (2)[c]	1.26	0.09	7
Benzoate 1 mM (4)	2.85 ± .52	2.71 ± .30	95
Benzoate 20 mM (3)	3.70 ± .19	0.98 ± .08	27
DMSO 100 mM (6)[c]	1.19 ± .42	0.98 ± .35	83
Mannitol 40 mM (4)	4.24 ± .172	3.55 ± .18	84

[a]From Sagone et al.[141]
[b]Values are given in nanomoles/10^7 cells/hr and represent paired experiments done with each agent. The number of experiments is indicated in the parentheses. The mean ± SD for the controls and the experimental values are compared.
[c]The experiments were done with zymosan lot 71-F8086. The rest of the experiments were done with zymosan lot 24-09419.

activated granulocytes occurs by a mechanism that does not require myeloperoxidase. However, this point requires further study.

Our studies with salicylic acid support the following conclusions. First, salicylic acid is oxidized by granulocytes as a consequence of the production of hydroxyl free radical or a very potent ROS by these cells and the reaction is not mediated

TABLE III. Effect of Compounds which Impair Myeloperoxidase Activity on the Oxidation of Salicylate[a,b]

Agent	Percent of control value
Azide 10^{-4} M (3)	32 ± 5
GSH 1 mM (4)	134 ± 38
Methionine 1 mM (3)	123 ± 52
Taurine 5 mM (4)	116 ± 40

[a]From Sagone et al.[141]
[b]The results give the percentage of oxidation of ^{14}C-salicylic acid (0.02 mM) in the presence of the agent, compared to the control. The results were derived from paired experiments. The number of experiments done is given in the parentheses. The mean value for the controls ranged from 1.16 ± 0.23 (SD) to 1.99 ± 0.32 (SD) nanomoles/10^7 cells/hr. The experiments were done with zymosan particles lot 71F8086.

TABLE IV. Effect of 10 mM DMTU on the Oxidative Metabolism of Zymosan Stimulated Granulocytes[a,b]

	Control	DMTU treated
^{14}C-1-glucose oxidation (3)	543 ± 218	748 ± 128
^{14}C-formate oxidation (3)	186 ± 14	260 ± 21
^{14}C-benzoate oxidation (3)	0.104 ± 0.021	0.025 ± 0.005

[a] From Sagone et al.[141]
[b] Values are given in nanomoles/10^7 cells/hr. The experimental conditions used were identical to those described for the salicylate experiments. The 5 μCi of ^{14}C-substrates was added to the incubation. The final concentration of glucose in all experiments was 50 mg/dl. The final concentrations of formate and benzoate used were 5 mM and 0.02 mM, respectively. Catalase (25 μg/ml) was added to the experiments in which formate oxidation was studied, in order to enhance the efficiency of the reaction.

directly by superoxide, hydrogen peroxide, HOCl, or chloramines. The mechanism appears to involve the NADPH oxidase system and an additional as yet unidentified enzyme that may require iron. One possibility is lactoferrin.[28] Second, unstimulated granulocytes do not appear to have a major capacity to metabolize drugs. Therefore, the granulocytes would develop this capacity primarily following their activation in vivo by a number of stimuli. Third, the capacity of salicylic acid to interact with hydroxyl radical in the granulocyte suggests that oxidation of this drug provides an additional method for the quantitation of hydroxyl free radical by these cells. Finally, the capacity of this nonsteroidal anti-inflammatory agent to interact with hydroxyl radical may explain its anti-inflammatorv properties. This latter possibility seems attractive. It has been difficult to explain why salicylic acid would be anti-inflammatory if the major mechanism of action of salicylates in vivo is primarily by the inhibition of prostaglandin biosynthesis.

Previously, we reported that a number of other NSAID selectively impair the release of hydroxyl radical by zymosan-stimulated granulocytes.[42] The drugs studied included aspirin, indomethacin, and ibuprofen (Table V). We originally speculated that the capacity of these anti-inflammatory drugs to impair benzoate oxidation might relate to their ability to impair granulocyte cyclo-oxygenase activity.[41] However, the alternate explanation for these results, which now seems most plausible, is that the drugs react directly with hydroxyl radical. The reported high rate constants of these NSAID for OH· support this possibility.[146] Therefore, the capacity of other NSAID beside salicylates to scavenge OH· may relate to their anti-inflammatory properties and require further study.

We have demonstrated that the parent compound of the salicylate group of drugs, benzoate is both hydroxylated and oxidized by the OH· produced by stimulated granulocytes. Therefore, the oxidation of salicylate by stimulated granulocytes is evidence for the simultaneous hydroxylation of the compound as well. This consideration suggests that stimulated PMN may be able to hydroxylate numerous

TABLE V. Effect of Anti-inflammatory Agents on Hydroxyl Radical Generation by Granulocytes Stimulated by Opsonized Zymosan Particles[a,b]

	Control	Experimental	% Control
Indomethacin[c]			
(3) 10^{-5} M	19.1 ± 3.6	13.0 ± 3.5	68 ± 5.0
(3) 10^{-4} M	16.1 ± 3.9	4.6 ± 0.8	29 ± 2.6
Aspirin[c]			
(3) 10^{-3} M	16.0 ± 4.0	9.0 ± 2.0	57 ± 10.0
Phenol[c]			
(3) 10^{-5} M	18.9 ± 1.8	7.8 ± 1.5	42 ± 4.6
(3) 10^{-4} M	20.4 ± 1.1	2.0 ± 0.2	10 ± 1.8
Ibuprofen[c]			
(3) 10^{-4} M	29.6 ± 8.3	25.0 ± 9.6	82 ± 14.0
(3) 10^{-3} M	38.4 ± 4.9	11.7 ± 3.8	29 ± 5.9
DMSO[c]			
(3) 10^{-3} M	24.0 ± 1.9	18.2 ± 2.7	75 ± 11.8
(1) 10^{-2} M	20.0	7.9	40

[a]From Sagone[42]
[b]Values given indicate the peak rate of $^{14}CO_2$ production from (^{14}C) Carboxyl-benzoic acid in nanomoles per 10^9 PMN per hour from PMN suspensions during the phagocytosis of zymosan particles. Three paired experiments were done with most concentrations of the drug. The experimental rate (suspension with drugs) compared to the corresponding controls in each experiment is given. For convenience the percentage in the presence of the drug compared to the control is also given in the third column.

other ring compounds, such as tyrosine and dopamine. In this regard, ring hydroxylation has been demonstrated as a consequence of the incubation of dopamine with a hydroxyl radical-generating system.[156] Other studies that have used salicylic acid as an hydroxyl trap during the radiolysis of water,[133,152,153] in chemical systems,[142,157] in enzyme systems,[142] and in vivo[158] have demonstrated the production of catechol and dihydroxybenzoic acids. These include 2,5-dihydroxybenzoic acid (gentisic acid) and 2,3-dihydroxybenzoic acid. Both metabolites are both known to display biological activity.[159,160] It has been known for some time that 2,5-dihydroxybenzoic is a metabolite of salicylate in vivo.[48,141] In addition, Grootveld and Halliwell[158] were able to identify 2,3-dihydroxybenzoate in the serum of patients with arthritis and normal volunteers after administration of aspirin. Presumably, the hydroxylation of salicylate in vivo is evidence for the generation of OH· in vivo. While hydroxylation of salicylate may occur in the liver, or other tissues with microsomes or xanthine–xanthine oxidase, it seems likely that stimulated granulocytes also have the capacity to metabolize these compounds in vivo as well. Therefore, the relatively solid evidence that granulocytes produce hydroxyl radical and the demonstrated metabolism of the salicylic acid compounds by activated granulocytes suggests a general mechanism for the oxidation and/or hydroxylation of numerous drugs in vivo. This system, which would be independent of that known to occur in a variety of tissues as a consequence of microsomal and xanthine–xanthine oxidase enzyme systems, would be predicted to occur primarily in patients with stimulated/activated granulocytes.

7. METABOLISM OF DRUGS BY MONONUCLEAR CELLS

Currently, there are few data to permit a comparison of the capacities of monocytes, macrophages, and granulocytes to metabolize xenobiotics. Evidence has already been discussed indicating that the myeloperoxidase system may be able to activate certain drugs to free radicals due to its capacity to produce hypochlorous acid and stable oxidants. Similarly, studies with benzoic acid and salicylic acid indicate that PMNs may have the ability to hydroxylate a number of compounds when activated. Monocytes appear to have lower concentrations of myeloperoxidase than do granulocytes, and the macrophages derived from a variety of tissues appear to lack this enzyme entirely.[161,162] In fact, maturation of the monocytes to the macrophages appears to be associated with the deletion of this enzymatic activity.[161] These characteristics of mononuclear phagocytes suggest that their potential capacity to activate drugs to drug radical as a consequence of the production of hypochlorous acid or chloramines will be limited compared with granulocytes. We have studied the oxidation of benzoic acid as a measurement of the OH· produced by activated monocytes and macrophages under similar conditions to those used in our studies involving granulocytes.[162] Our experiments indicate that the rate of benozate oxidation by stimulated monocytes is significantly lower than that of granulocytes. Furthermore, in experiments involving human alveolar macrophages, we have been unable to detect the significant oxidation of benzoic acid. Our results parallel those of Repine et al.,[16] which quantitated the capacity of these cells to generate OH· using the oxidation of DMSO. These studies do not exclude the possibility that these phagocytic mononuclear phagocytes generate hydroxyl radical, since it is possible that there is a perferential reaction of any OH· generated by these cells with cellular substrates. However, our observations concerning the oxidation of benzoic acid by mononuclear phagocytes, suggest that little free hydroxyl radical is available for the metabolism of xenobiotics.

We have also studied the capacity of monocytes and alveolar macrophages to oxidize salicylic acid (unpublished data). The results parallel those with benzoic acid. The capacity of the monocytes to oxidize salicylic acid is lower than that of the granulocytes. Thus far, we have been unable to detect significant oxidation of salicylic acid by zymosan or PMA activated macrophages. The capacity of activated granulocytes to oxidize salicylic acid compared to benzoic acid appears to be 16-fold higher (see Fig. 3). When this fact is considered, the capacity of the macrophage to generate hydroxyl free radical must be significantly lower than that of the granulocytes.

Overall, these considerations predict that the capacity of the mononuclear phagocytes to oxidize or hydroxylate xenobiotics will be significantly lower than granulocytes. However, it has been reported that the animal macrophage particularly those which are activated may produce OH·.[30] Therefore, the possibility that activated macrophages may also have an enhanced capacity to oxidize drugs cannot be excluded. Furthermore, the metabolic characteristics of the reticuloendothelial cells from different tissues in humans has not been extensively evaluated. Therefore, the metabolism of RE cells from the liver, spleen or bone marrow may differ significantly from those of alveolar macrophages.

Presumably, the capacity of the eosinophil to metabolize xenobiotics will be similar to or superior to that of the granulocytes. The eosinophil may be the predominating cell type infiltrating the tissue of patients with a variety of infectious disorders, allergies, autoimmune disorders, and tumors.[163–168] Therefore, the capacity of the eosinophil to metabolize drugs would seem to be a fruitful area of investigation.

8. SUMMARY

This chapter briefly discusses evidence for the concept that stimulated phagocytic cells acquire the capacity to metabolize xenobiotics. Potentially, these reactions could occur at inflammatory sites or in circulating phagocytic cells and would represent a unique pathway for drug metabolism in a diverse group of patients compared to healthy individuals. There is now evidence that the biotransformation of xenobiotics can occur in granulocytes as a consequence of the myeloperoxidase–H_2O_2 system and their capacity to generate OH· free radicals. These two mechanisms appear to be relatively independent of each other.

Studies already done with the isolated horseradish peroxidase–H_2O_2 and myeloperoxidase–H_2O_2 enzyme predict that the myeloperoxidase system of granulocyte may have the capacity to metabolize a variety of drugs. These include hydroquinones, hydroaminoquinone, aromatic amines, phenothiazines, hydroxyaromatics, hydroxylamine, sulfhydryls, and hydrazines. Only a few of these reactions have been shown to occur in the intact granulocyte, and additional studies are clearly needed.

The capacity of granulocytes to oxidize and hydroxylate benzoate, ASA. and salicylic acid raise the possibility for a similar metabolism of numerous compounds. A few include barbiturates, tyrosine, phenol, dopamine, phenytoin (Dilantin), and phenothiazine.

The possibility that a xenobiotic could be metabolized by both pathways simultaneously, or sequentially, should also be considered. In addition, the biotransformation of the drug could permit further reaction with cellular substrates, other ROS known to be generated by the phagocytic cells, or molecular O_2. The biotransformation of the drug may be to an inactive metabolite or conversely to more active ones. Therefore, the overall biological effects of the transformation cannot be predicted with certainty.

Finally, it has been known for some time that quinone drugs stimulate the oxidative metabolism of phagocytic cells.[49] Presumably, this stimulation is due to the capacity of these drugs to serve as electron carriers and is associated with an enhanced cellular production of ROS.[49,169] However, few recent studies have characterized further the mechanism involved or the nature of the ROS produced particularly stable oxidants and OH·. It is interesting to speculate that under some circumstances the drug may have the capacity to react with ROS, thereby inducing its own biotransformation by PMNs. In addition, several compounds are known to stimulate the oxidative metabolism of granulocytes by other mechanisms. Examples include phorbol esters and hexachlorocyclohexanes,[170] a family of chlorinated hydrocarbons which have recently been shown to be potent stimulators of the oxidative metabolism.[170]

It seems reasonable to conclude that the capacity of activated granulocytes to biotransform xenobiotics represents as yet a relatively unexplored and important area for research.

REFERENCES

1. Baehner RL: The growth and development of our understanding of chronic granulomatous disease, in Bellanti JA, Dayton DH (ed): *The Phagocytic cell in Host Resistance*. New York, Raven, 1975, pp 173–200.
2. Sbarra AJ, Paul BB, Jacobs AA, et al: Biochemical aspects of phagocytic cells as related to bacterial function. *J Reticuloendothel Soc* **11:**492–502, 1972.
3. Klebanoff SJ: Antimicrobial mechanism in neutrophilic polymorphonuclear leukocytes. *Semin Hematol* **12:**117–142, 1975.
4. Badwey JA, Karnovsky ML: Active oxygen species and the functions of phagocytic leukocytes. *Annu Rev Biochem* **49:**695–726, 1980.
5. Babior BM: Oxygen-dependent microbial killing by phagocytes. *N Engl J Med* **298:**659–668, 1978.
6. Babior GM: The respiratory burst of phagocytes. *J Clin Invest* **73:**599–601, 1984.
7. Carp H, Janoff A: Potential mediator of inflammation phagocytic-derived oxidants suppress the elastase-inhibitory capacity of alpha$_1$-proteinase inhibitor in vitro. *J Clin Invest* **66:**987–995, 1980.
8. Fridovich I: The biology of oxygen radicals. *Science* **201:**875–880, 1978.
9. Fox RB: Prevention of granulocyte-mediated oxidant lung injury in rats by a hydroxyl radical scavenger, dimethylthiourea. *J Clin Invest* **74:**1456–1464, 1984.
10. Goldstein IM, Roos D, Kaplan HB, Weissman G: Complement and immunoglobulin stimulate superoxide production by human leukocytes independently of phagocytosis. *J Clin Invest* **56:**1155–1163, 1975.
11. Henson PM, Oades ZG: Stimulation of human neutrophils by soluble and immunoglobin aggregates—Secretion of granule constituents and increased oxidation of glucose. *J Clin Invest* **56:**1053–1061, 1975.
12. Johnston RB, Lehmeyer JE: Elaboration of toxic oxygen by-products by neutrophils in a model of immune complex disease. *J Clin Invest* **51:**836–841, 1976.
13. Klebanoff SJ: Oxygen metabolism and the toxic properties of phagocytes. *Ann Intern Med* **93:**480–489, 1980.
14. Krieger BP, Loomis WH, Czer GT, et al: Mechanism of interaction between oxygen and granulocytes in hyperoxic lung injury. *J Appl Physiol* **58:**1326–1330, 1985.

15. McCord J: Free radicals and inflammation: Protection of synovial fluid by superoxide dismutase. *Science* **185:**529–531, 1974.
16. Repine JE, Eaton JW, Anders MW, et al: Generation of hydroxyl radicals by enzymes, chemicals, and human phagocytes in vitro. Detection with the anti-inflammatory agent, dimethyl sulfoxide. *J Clin Invest* **64:**1642–1651, 1979.
17. Oxygen free radicals and tissue damage, in Fitzsimons DW (ed): *Ciba Foundation Symposium 65*. Amsterdam, Excerpta Medica, 1979, pp 1–381.
18. Sacks T, Moldow CF, Craddock PR, et al: Oxygen radicals mediate endothelial cell damage by complement-stimulated granulocytes an in vitro model of immune vascular damage. *J Clin Invest* **61:**1161–1167, 1978.
19. Sagone AL, Husney R, Guter H, Clark L: The effect of catalase on the proliferation of human lymphocytes to phorbol myristate acetate. *J Immunol* **133:**1488–1494, 1984.
20. Spielberg SP, Boxer LA, Oliver JM, et al: Oxidative damage to neutrophils in glutathione synthetase deficiency. *Br J Haematol* **42:**215–223, 1979.
21. Till GO, Johnson KJ, Kunkel R, Ward PA: Intravascular activation of complement and acute lung injury. *J Clin Invest* **69:**1126–1135, 1982.
22. Ward PA, Till GO, Kunkel R, et al: Evidence for role of hydroxyl radical in complement and neutrophil-dependent tissue injury. *J Clin Invest* **72:**789–801, 1983.
23. Weiss SJ: The role of superoxide in the destruction of erythrocyte targets by human neutrophils. *J Biol Chem* **255:**9912–9917, 1980.
24. Weiss SJ: Oxidative autoactivation of latent collangenase by human neutrophils. *Science* **227:**747–749, 1985.
25. Weissman G, Smolen JE, Korchak HM: Release of inflammatory mediators from stimulated neutrophils. *N Engl J Med* **303:**27–34, 1980.
26. Babior BM, Kipnes R, Curnette JT: Biological defense mechanisms. The production of leukocytes of superoxide, a potential bactericidal agent. *J Clin Invest* **52:**741–744, 1973.
27. Davidson WD, Tanaka KR: Instantaneous and continuous measurement of phagocytosis-stimulated glucose oxidation in human granulocytes by an ionization chamber method. *Br J Haematol* **25:**783–792, 1973.
28. Ambruso DR, Johnston RB, Jr: Lactoferrin enhances hydroxyl radical production by human neutrophils, neutrophil particulate fractions, and an enzyme generating system. *J Clin Invest* **67:**352–360, 1981.
29. Cheung K, Lark J, Robinson MF, Pomery PJ, et al: The production of hydroxyl radical by human neutrophils stimulated by arachidonic acid-measurements by ESR spectroscopy. *Aust J Exp Biol Med Sci* **64**(pt 2)**:**157–164, 1986.
30. Drath DB, Karnovsky ML, Huber GL: Hydroxyl radical formation in phagocytic cells of the rat. *J Appl Physiol* **46:**136–140, 1979.
31. Johnston RB Jr, Keele BB, Misra HP, et al: The role of superoxide anion generation in phagocytic bactericidal activity—Studies with normal and chronic granulomatous disease leukocytes. *J Clin Invest* **55:**1357–1372, 1975.
32. Klebanoff SJ, Rosen H: Ethylene formation by polymorphonuclear leukocytes, role of myeloperoxidase. *J Exp Med* **148:**490–506, 1978.
33. Kubo A, Sasada M, Yamamota K, et al: The role of phagosome formation in hydroxyl radical generation by human polymorphonuclear leukocytes: studies with normal and cytochalasin B-treated cell. *Acta Haematol Jpn* **49:**34–42, 1986.
34. Mallery SR, Zeligs BJ, Ramwell PW, Bellanti JA: Gender-related variations of human neutrophil cyclo-oxygenase and oxidative burst metabolites. *J Leukocyte Biol* **40:**133–148, 1986.
35. Rosen H, Klebanoff SJ: Hydroxyl radical generation by polymorphonuclear leukocytes measured by electron spin resonance spectroscopy. *J Clin Invest* **64:**1725–1729, 1979.
36. Sagone AL Jr, Decker MA, Wells RM, Democo C: A new method for the detection of hydroxyl radical production. *Biochim Biophys Acta* **628:**90–97, 1980.
37. Alexander MS, Husney RM, Sagone AL Jr: Metabolism of benzoic acid by stimulated polymorphonuclear cells. *Biochem Pharmacol* **35:**3649–3651, 1986.

38. Tauber AI, Babior B: Evidence for hydroxyl radical production by human neutrophils. *J Clin Invest* **60:**374–379, 1977.
39. Weiss SJ, Rustagi PK, LoBuglio AF: Human granulocyte generation of hydroxyl radical. *J Exp Med* **147:**316–323, 1977.
40. Repine JE, Fox RB, Berger EM: Hydrogen peroxide kills staphylococcus by reacting with staphylococcal iron to form hydroxyl radical. *J Biol Chem* **256:**7094–7096, 1981.
41. Sagone AL Jr, Wells RM, Democko C: Evidence that OH· production by human PMNs is related to prostaglandin metabolism. *Inflammation* **4:**65–71, 1980.
42. Sagone AL Jr: Effect of anti-inflammatory agents on the hydroxyl radical (OH·) production of zymosan stimulated human granulocytes, in Rodgers MA, Power EL (eds): *Oxygen and Oxyradicals in Chemistry and Biology*. New York, Academic, 1981, p 719.
43. Kaplan HB, Edelson HS, Korchak HM, et al: Effect of non-steroidal anti-inflammatory agents on human neutrophil functions in vitro and in vivo. *Biochem Pharmacol* **33:**371–378, 1984.
44. Lehmeyer JE, Johnston RB Jr: Effect of anti-inflammatory drugs and agents that elevate intracellular cyclic AMP on the release of toxic oxygen metabolites by phagocytes: Studies in a model of tissue-bound IgG. *Clin Immunol Immunopathol* **9:**482–490, 1978.
45. Duniec Z, Robak J, Gryglewski R: Antioxidant properties of some chemicals vs their influence on cyclooxygenase and lipoxidase activities. *Biochem Pharmacol* **32:**2283–2286, 1983.
46. Britigan BE, Rosen GM, Chai Y, et al: Do human neutrophils make hydroxyl radical? Determination of free radicals generated by human neutrophils activated with soluble or particulate stimulus using electron paramagnetic resonance. *J Biol Chem* **261:**4426–4431, 1986.
47. Winterbourn CC: Myeloperoxidase as an effect inhibitor of hydroxyl radical production. Implication for the oxidative reactions of neutrophils. J Clin Invest **78:**545–550, 1986.
48. Goodman GA, Goodman LS, Rall TW, Murad F (eds): *Goodman and Gilman's The Pharmacological Basis of Therapeutics,* ed 7. New York, Macmillan, 1985.
49. Iyer GY, Islam MF, Quastel JH: Biochemical aspect of phagocytosis. *Nature (Lond)* **192:**535–541, 1961.
50. Sbarra AJ, Karnovsky ML: The biochemical basis of phagocytosis I, Metabolic changes during the ingestion of particles by polymorphonuclear leukocytes. *J Biol Chem* **234:**1355–1362, 1959.
51. Karnovsky ML: Metabolic basis of phagocytic activity. *Physiol Rev* **42:**143–168, 1962.
52. Holmes B, Page AR, Good RA: Studies of the metabolic activity of leukocytes from patients with a genetic abnormality of phagocytic function. *J Clin Invest* **46:**1422–1443, 1967.
53. Evans W, Karnovsky ML: The biochemical basis of phagocytosis. IV. Some aspects of carbohydrate metabolism during phagocytosis. *Biochemistry* **1:**159–165, 1962.
54. Curnette JT, Babior B: Biological defense mechanisms, the effect of bacteria and serum on superoxide production by granulocyte. *J Clin Invest* **53:**1662–1672, 1974.
55. Homan-Muller J, Weening RS, Roos D: Production of hydrogen peroxide by phagocytosing human granulocytes. *J Lab Clin Med* **85:**198–207, 1975.
56. Allen RC, Stjernholm, Steele RH: Evidence for the generation of an electronic excitation state(s) in human polymorphonuclear leukocytes and its participation in bactericidal activity. *Biochem Biophys Res Commun* **47:**679–684, 1972.
57. Webb LS, Keele BB Jr, Johnston RB Jr: Inhibition of phagocytosis-associated chemiluminines-cence by superoxide dismutase. *Infect Immun* **9:**1051–1056, 1974.
58. Baehner RL, Murrman SK, Davis J, et al: The role of superoxide anion and hydrogen peroxide in phagocytosis-associated oxidative metabolic reactions. *J Clin Invest* **56:**571–576, 1975.
59. Sagone AL Jr, King GW, Metz EN: A comparison of the metabolic responses to phagocytosis in human granulocytes and monocytes. *J Clin Invest* **57:**1352–1358, 1976.
60. Klebanoff SJ, Pincus SH: Hydrogen peroxide utilization in myeloperoxidase-deficient leukocytes: A possible microbiocidal control mechanism. *J Clin Invest* **50:**2226–2229, 1971.
61. Weening RS, Roos P, Loos JA: Oxygen consumption of phagocytizing cells in human leukocyte and granulocyte preparations: A comparative study. *J Lab Clin Med* **83:**570–576, 1974.
62. Root RK, Metcalf J, Oshino N: H_2O_2 release from human granulocytes during phagocytosis. I. Documentation, quantitation, and some regulating factors. *J Clin Invest* **55:**945–955, 1975.

63. Ruch W, Cooper PH, Baggiolini M: Assay of H_2O_2 production by macrophages and neutrophils with homovanillic acid and horseradish peroxidase. *J Immunol Methods* **63:**347–357, 1983.
64. Tsan Min Fu, Chen JW: Oxidation of methionine by human polymorphonuclear leukocyte. *J Clin Invest* **65:**1041–1050, 1980.
65. Weiss SJ, Klein R, Slivka A, et al: Chlorination of taurine by human neutrophils. Evidence for hypochlorous acid generation. *J Clin Invest* **70:**598–607, 1982.
66. Thomas EL, Grisham MB, Jefferson MM: Myeloperoxidase-dependent effect of amines on function of isolated neutrophils. *J Clin Invest* **72:**441–454, 1983.
67. Weiss SJ, Lambert MD, Test ST: Long-lived oxidants generated by human neutrophils: Characteristics and bioactivity. *Science* **222:**625–628, 1983.
68. Sagone AL Jr, Husney RM, O'Dorisio MS, et al: Mechanisms for the oxidation of reduced glutathione by zymosan stimulated granulocytes. *Blood* **63:**96–104, 1984.
69. Pincus S, Klebanoff SJ: Quantitative leukocyte iodination. *N Engl J Med* **284:**744–750, 1984.
70. Rosen H, Klebanoff SJ: Formation of singlet oxygen by the myeloperoxidase-mediated antimicrobial system. *J Biol Chem* **252:**4803–4810, 1977.
71. Rosen H, Klebanoff S: Chemiluminescence and superoxide production by myeloperoxidase-deficient leukocytes. *J Clin Invest* **58:**50–60, 1976.
72. Segal AW, Cross A, Garcia R, et al: Absence of cytochrome b_{245} in chronic granulomatous disease. *N Engl J Med* **308:**245–251, 1983.
73. Gabig TG, Lefker B: Deficient flavoprotein component of the NADPH-dependent O_2^--generating oxidase in the neutrophils from three male patients with chronic granulomatous disease. *J Clin Invest* **73:**701–705, 1984.
74. Sagone AL Jr, Mendelson DS, Metz EN: Effect of sodium azide on the chemiluminescence of granulocytes. Evidence for the generation of multiple oxygen radicals. *J Lab Clin Med* **89:**1333–1340, 1977.
75. Roos D, Weening R, Voetman A, et al: Protection of phagocytic cells by endogenous glutathione: Studies in a family with glutathione reductase deficiency. *Blood* **53:**851–866, 1979.
76. Voetman AA, Loos JA: Roos D: Changes in the levels of glutathione in phagocytosing human neutrophils. *Blood* **55:**741–747, 1980.
77. Burchill BR, Oliver JM, Pearson JM, et al: Microtubular dynamics and glutathione metabolism in phagocytosing human polymorphonuclear leukocytes. *J Cell Biol* **76:**439–447, 1978.
78. Mendelson DS, Metz EN, Sagone AL Jr: Effect of phagocytosis on the reduced soluble sulfhydryl content of human granulocytes. *Blood* **50:**1023–1030, 1977.
79. Klebanoff SJ: Iodination catalyzed by the xanthine oxidase system: Role of hydroxyl radical. *Biochemistry* **21:**4110–4116, 1982.
80. Boxer LA, et al: Effect of a chemotactic factor, *N*-formylmethionyl peptide on adherence, superoxide anion generation, phagocytosis and microtubule assembly of human polymorphonuclear leukocytes. *J Lab Clin Med* **93:**506–514, 1979.
81. Weissman G, Serhan C, Korchak HM, et al: Neutrophils generate phosphatidic acid, an "endogenous calcium ionophore" before releasing mediators of inflammation. *Trans Assoc Am Phys* **94:**357–365, 1981.
82. Weksler BB, Goldstein IM: Prostaglandin: Interactions with platelets and polymorphonuclear leukocytes in Hemostasis and inflammation. *Am J Med* **68:**419–428, 1980.
83. Segal ML, Fertel RH, Kraut EH, Sagone AL: The role of reactive oxygen species in thromboxane B_2 generation by polymorphonculear leukocytes. *J Lab Clin Med* **102:**788–794, 1983.
84. Goldstein IM, Malmsten CL, Kindahl H, et al: Thromboxane generation by human neutrophils polymorphonuclear leukocytes. *J Exp Med* **148:**787–792, 1978.
85. Ingraham LM, Boxer LA, Baehner RL, et al: Chlorotetracycline fluorescence and arachidonic acid metabolism of polymorphonculear leukocytes of chronic granulomatous disease patients. *Clin Res* **29:**519A, 1981.
86. Smolen JE, Shohet SB: Remodeling of granulocyte membrane fatty acids during phagocytosis. *J Clin Invest* **53:**726–734, 1974.

87. Tsan M, Denison R: Oxidation of *n*-formyl methionyl chemotactic peptide by human neutrophils. *J Immunol* **126:**1387–1389, 1981.
88. Clark RA, Klebanoff: Chemotactic factor inactivation by the myeloperoxidase–hydrogen peroxide–halide system. *J Clin Invest* **64:**913–920, 1979.
89. Egan RW, Paxton J, Kuehl FA Jr: Mechanism of reversible self-deactivation of prostaglandin synthetase. *J Biol Chem* **251:**7329–7335, 1976.
90. Goldstein IM, Malmsten CL, Kaplan HB: Thromboxane generation by stimulated human granulocytes: Inhibition by corticosteroid and superoxide dismutase. *Clin Res* **25:**518, 1977.
91. Mason RP: Free-radical intermediates in chemical metabolism of toxic chemicals, in Pryor WA (ed): *Free Radicals in Biology*. New York, Academic, 1982, vol V, pp 161–221.
92. Mason RP, Chignell CF: Free radicals in pharmacology and toxicology selected topics. *Pharmacol Rev* **33:**189–211, 1982.
93. Mason RP, Harrelson B, Kalyanaraman B, et al: Free radical metabolites of chemical carcinogens, in McBrien D, Slater T (eds): *Free Radicals, Lipid Peroxidation and Cancer*. New York, Academic, 1982, pp 377–400.
94. Kappus H, Sies H: Toxic drug effects associated with oxygen metabolism: Redox cycling and lipid peroxidation. *Experientia* **37:**1233–1241, 1981.
95. Remmer H: The role of the liver in drug metabolism. *Am J Med* **49:**617–628, 1970.
96. Docampo R, Moreno S: Free-radical intermediates in the antiparasitic acid of drugs and phagocytic cells, in Pryor WA (ed): *Free Radicals in Biology*. Orlando, Florida, Academic, 1984, vol VI, pp 243–288.
97. O'Brien PJ: Multiple mechanisms of metabolic activation of aromatic amine carcinogens, in Pryor WA (ed): *Free Radicals in Biology*. Orlando, Florida, Academic, 1984, vol VI, pp 289–322.
98. Cavalieri EL, Rogan E: One electron and two-electron oxidation in aromatic hydrocarbon carcinogenesis, in Pryor WA (ed): *Free Radicals in Biology*. Orlando, Florida, Academic, 1984, vol VI, pp 323–369.
99. Freeman BA, Crapo JD: Biology of disease-free radicals and tissue injury. *Lab Invest* **47:**412–426, 1982.
100. Kappus H: Overview of enzyme systems involved in bioreduction of drugs and redox cycling. *Biochem Pharmacol* **35:**1–6, 1986.
101. Chance B, Sies H, Boveris A: Hydroperoxide metabolism in mammalian organs. *Physiol Rev* **59:**527–605, 1979.
102. Green T, Fellman J, Eicher A: Myeloperoxidase oxidation of sulfur-centered and benzoic acid hydroxyl scavengers. *FEBS Lett* **192:**33–36, 1985.
103. Turkall RM, Tsan MF: Oxidation of glutathione by the myeloperoxidase system. *J Reticuloendothel Soc* **31:**353–360, 1982.
104. Woeber KA, Ingbar S: Metabolism of L thyroxine by phagocytosing human leukocytes. *J Clin Invest* **52:**1796–1803, 1973.
105. Klebanoff SJ, Green WJ: Degradation of thyroid hormones by phagocytosing human leukocytes. *J Clin Invest* **52:**66–72, 1973.
106. Rao MK, Sagone AL Jr: Evidence for the extracellular conversion of triodothyroxine (T_3) to thyroxine (T_4) by phagocytosing granulocytes. *Infect Immun* **43:**846–849, 1984.
107. Klebanoff SJ: Effect of estrogens on the myeloperoxidase-mediated anti-microbial system. *Infect Immun* **25:**153–156, 1979.
108. Turkall RM, Denison R, Tsan MF: Degradation and oxidation of methionine enkaphalin by human neutrophils. *J Lab Clin Med* **99:**418–427, 1982.
109. Klebanoff SJ: Estrogen binding by leukocytes during phagocytosis. *J Exp Med* **145:**983–998, 1977.
110. Chance B: The primary and secondary compounds of catalase and methyl or ethyl hydrogen peroxide. II. Kinetics and activity. *J Biol Chem* **179:**1341–1369, 1949.
111. Mannering GJ, Van Harken DR, Makar TR: Role of intracellular distribution of hepatic catalase in peroxidative oxidation of methanol. *Ann NY Acad Sci* **168:**265–280, 1969.

112. Tephy TR, Parks RE, Mannering GJ: Methanol metabolism in the rat. *J Pharmacol Exp Ther* **143:**292–300, 1964.
113. Metz EN, Balcerzak SP, Sagone AL Jr: Mechanism of methylene blue stimulation of the hexose monophosphate shunt in erythrocytes. *J Clin Invest* **58:**797–802, 1976.
114. Tephy TR, Atkins M, Mannering GJ, et al: Activation of catalase peroxidative pathway for the oxidation of alcohols in mammalian erythrocytes. *Biochem Pharmacol* **14:**435–444, 1965.
115. Gale PH, Egan RW: Prostaglandin endoperoxide synthase-catalyzed oxidation reactions, in Pryor WA (ed): *Free Radicals in Biology.* Orlando, Florida, Academic, 1984, vol VI, pp 1–38.
116. Lands WEM, Kulmacz RJ, Marshall PJ: Lipid peroxide actions in the regulation of prostaglandin biosynthesis, in Pryor WA (ed): *Free Radicals in Biology.* Orlando, Florida, Academic, 1984, vol VI, pp 39–61.
117. Kalyanaraman B, Sivarajah K: The electron spin resonance study of free radicals formed during the arachidonic acid cascade and oxidation of xenobiotics by prostaglandin synthase, in Pryor WA (ed): *Free Radicals in Biology.* Orlando, Florida, Academic, 1984, vol VI, pp 149–198.
118. Marnett L: Hydroperoxide-dependent oxidations during prostaglandin biosynthesis, in Pryor WA (ed): *Free Radicals in Biology.* Orlando, Florida, Academic, 1984, vol VI, pp 63–94.
119. Singh DJ, Grenwald JE, Bianchine J, et al: Evidence for the generation of hydroxyl radical during arachidonic acid metabolism by human platelets. *Am J Hematol* **11:**233–240, 1981.
120. Egan RW, Gale PH, Vanden Heuvel JA, et al: Mechanism of oxygen transfer by prostaglandin hydroperoxidase. *J Biol Chem* **255:**323–326, 1980.
121. Bachur NR, Gordon SL, Gee MV: A general mechanism for microsomal activation of quinone anticancer agents to free radical. *Cancer Res* **38:**1745–1750, 1978.
122. Kennedy KA, Sligar SG, Polomski L, et al: Metabolic activation of mitomycin C by liver microsomes and nuclei. *Biochem Pharmacol* **29:**1–8, 1982.
123. Mimnaugh EG, Trush MA, Gram TE: Stimulation by adriamycin of rat, heart and liver microsomal NADPH-dependent lipid peroxidation. *Biochem Pharmacol* **30:**2797–2804, 1981.
124. Berlin V, Haseltine WA: Reduction of adriamycin to a semiquinone-free radical by NADPH cytochrome P-450 reductase produces DNA damage in a reaction mediated by molecular oxygen. *J Biol Chem* **256:**4747–4756, 1981.
125. Fong KL, McCay PB, Poyer JL: Evidence that peroxidation of lysosomal membrane is initiated by hydroxyl free radicals produced during flavin enzyme activity. *J Biol Chem* **248:**7792–7797, 1973.
126. Winston GW, Cederbaum AI: NADPH-dependent production of oxy radicals by purified components of the rat liver mixed function oxidase system. I. Oxidation of hydroxyl radical scavenging agents. *J Biol Chem* **258:**1508–1513, 1983.
127. Ohniskhi K, Lieber CS: Respective role of superoxide and hydroxyl radical in the activity of the reconstituted microsomal ethanol-oxidizing system. *Arch Biochem Biophys* **191:**798–803, 1978.
128. Ekström G, Chronholm T, Ingelman-Sunoberg: Hydroxyl-radical production and ethanol oxidation by liver microsomes isolated from ethanol-treated rats. *Biochem J* **233:**755–761, 1986.
129. Winston GW, Feierman DE, Cederbaum AL: The role of iron chelates in hydroxyl radical production in rat liver microsomes: NADPH–cytochrome P-450 reductase and xanthine oxidase. *Arch Biochem Biophys* **232:**378–390, 1984.
130. Cederbaum AI, Qureshi A, Cohen G: Production of formaldehyde and acetone by hydroxyl-radical generating systems during the metabolism of tertiary butyl alcohol. *Biochem Pharmacol* **32:**3517–3524, 1983.
131. Winston GW, Cederbaum AI: Oxidative decarboxylation of benzoate by microsomes: A probe for oxygen radical production during microsomal electron transfer. *Biochemistry* **21:**4265–4270, 1982.
132. Tobia AJ, Couri D, Sagone AL Jr: The effect of the quinone type chemotherapeutic drugs on the hydroxyl production of microsomes. *J Toxicol Environ Health* **15:**265–277, 1985.
133. Matthews RW, Sangster DF: Measurement of benzoate radiolytic decarboxylation of relative rate constants for hydroxyl radical reactions. *J Phys Chem* **69:**1938–1946, 1965.
134. Sagone AL Jr, Democko C, Basinger L, et al: Determination of hydroxyl radical production in aqueous solutions irradiated to clinical significant doses. *J Lab Clin Med* **101:**196–204, 1983.

135. Kellogg EW, Fridovich I: Superoxide, hydrogen peroxide, and singlet oxygen in the lipid peroxidation by a xnathine oxidase system. *J Biol Chem* **150:**8812–8817, 1975.
136. McCord JM, Fridovich I: The reduction of cytochrome C by milk xanthine oxidase. *J Biol Chem* **243:**5753–5760, 1968.
137. Porras A, Olson JS, Palmer G: The reaction of reduced xanthine oxidase with oxygen kinetics of peroxide and superoxide formation. *J Biol Chem* **256:**9096–9103, 1981.
138. Arneson RM: Substrate-induced chemiluminescence of xanthine oxidase and aldehyde oxidase. *Arch Biochem Biophys* **136:**352–360, 1970.
139. Beauchamp C, Fridovich I: A mechanism for the production of ethylene from methional. *J Biol Chem* **245:**4641–4646, 1970.
140. Cohen G, Thurman RG, Williamson JR et al, (eds): *Alcohol and Aldehyde Metabolizing Systems.* New York, Academic, 1977, pp 403–412.
141. Sagone AL, Husney PM: Oxidation of salicylates by stimulated granulocytes. Evidence that these drugs act as free radical scavengers in biological systems. *J Immunol* **138:**1–7, 1987.
142. Floyd RA, Watson JJ, Wong PT: Sensitive assay of hydroxyl free radical formation utilizing high pressure liquid chromatography with electrochemical detection of phenol and salicylate hydroxylates products. *J Biochem Biophys Methods* **10:**221–235, 1984.
143. Bors W, Saran N, Michel C: Assays of oxygen radicals: Methods and mechanisms, in Oberly LW (ed): *Superoxide Dismutase.* CRC Press, Boca Raton, Florida, 1982, vol II, pp 31–62.
144. Samuni A, Chevion M. Halpern YS, et al: Radiation-induced damage in T_4 bacteriophage: The effect of superoxide radicals and molecular oxygen. *Radiat Res* **75:**489–496, 1978.
145. Howard-Flanders P, Levin J, Theriot L: Reactions of DNA radicals with sulfhydryl in x-irradiated bacteriophage systems. *Radiat Res* **18:**593–606, 1963.
146. Hiller KO, Wilson RL: Hydroxyl-free radicals and anti-inflammatory drugs: Biological inactivation studies and reaction rate constants. *Biochem Pharmacol* **32:**2109–2111, 1983.
147. Ames BN, Cathcart R, Scheirs E, et al: Uric acid provides an antioxidant defense in humans against oxidant- and radical-caused aging and cancer. A hypothesis. *Proc Natl Acad Sci USA* **78:**6858–6862, 1981.
148. Kittridge KJ, Willson RJ: Uric acid substantially enhances the free radical-induced inactivation of alcohol dehydrogenase. *FEBS Lett* **170:**162–164, 1984.
149. Gee CA, Kittridge KJ, Wilson RL: Peroxy free radicals, enzymes, and radiation damage: Sensitisation by oxygen and protection by superoxide dismutase and antioxidants. *Br J Radiol* **58:**251–256, 1985.
150. Kalyanaraman B, Sohnle P: Generation of free radical intermediates from foreign compounds by neutrophil-derived oxidants. *J Clin Invest* **75:**1618–1622, 1985.
151. Ames B: Dietary carcinogens and anticarcinogens. Oxygen radicals and degenerative diseases. *Science* **221:**1256–1264, 1983.
152. Loeff I, Swallow AJ: On the radiation chemistry of concentrated aqueous solutions of sodium benzoate. *J Phys Chem* **68:**2470–2475, 1964.
153. Klein GW, Bhatia K, Schuler R: Reaction of OH· with benzoic acid isomer distribution in the radical intermediates. *J Phys Chem* **79:**1767–1774, 1975.
154. Floyd RA, Watson J, Harris J, et al: Formation of 8-hydroxydeoxyquanosine, hydroxyl free adduct of DNA in granulocytes exposed to the tumor promoter, tetradeconylphorbol acetate. *Biochem Biophys Res Commun* **137:**841–846, 1986.
155. Birnboim H: DNA strand breakage in human leukocytes exposed to a tumor promoter, phorbol myristate acetate. *Science* **215:**1247–1249, 1982.
156. Slivka A, Cohen G: Hydroxyl radical attack on dopamine. *J Biol Chem* **260:**15466–15472, 1980.
157. Grootveld MA, Halliwell B: An aromatic hydroxylation assay for hydroxyl radicals utilizing high performance liquid chromatography (HPLC). Use to investigate the effects of EDTA on the Fenton reaction. *Free Radiat Res Commun* **1:**243–250, 1986.
158. Grootveld M, Halliwell B: Aromatic hydroxylation as a potential measure of hydroxyl-radical formation in vivo. *Biochem J* **237:**499–504, 1986.

159. Shahidi N, Westring D: Acetylsalicylic acid induced hemolysis and its mechanism. *J Clin Invest* **49:**1334–1340, 1970.
160. Boxer L, Allen J, Baehner R: Potentiation of polymorphonuclear leukocyte motile functions by 2,3-dihydroxybenzoic acid. *J Lab Clin Med* **92:**730–736, 1978.
161. Cohn Z: Properties of macrophages, in Williams R, Fundenberg H, (eds): *Phagocytic Mechanisms in Health and Disease*. New York, International Medical Book Corporation, 1972, pp 39–49.
162. Papermaster-Bender G, Whitcomb M, Sagone AL Jr: Characterization of the metabolic responses of human pulmonary alveolar macrophage. *J Reticulendothel Soc* **28:**129–139, 1980.
163. Bass DA, Szejda P: Mechanisms of killing of newborn larvae of *Trichinella spiralis* by neutrophils and eosinophils. *J Clin Invest* **64:**1558–1564, 1979.
164. Bass DA, Grover W, Lewis J, et al: Comparison of human eosinophils from normals and patients with eosinophilia. *J Clin Invest* **66:**1265–1273, 1980.
165. DeChatelet L, Shirley P, McPahil L, et al: Oxidative metabolism of the human eosinophil. *Blood* **50:**525–535, 1977.
166. Klebanoff S, Dorack D, Rosen H, Clark R: Functional studies of human peritoneal eosophils. *Infect Immun* **17:**167–173, 1977.
167. Pincus S, Schooley W, DiNapoli A, et al: Metabolic heterogeneity of eosinophils from normal and hypereosinophilic patients. *Blood* **58:**1175–1181, 1981.
168. Prin L, Charon J, Capron M, et al: Heterogenity of human eosinophils. II. Variability of respiratory burst activity related to cell density. *Clin Exp Immunol* **57:**735–742, 1984.
169. Humbert JA, Gross G, Vatter A, et al: Nitroblue tetrazolium reduction by neutrophils: Biochemical and ultrastructrual effects of methylene blue. *J Lab Clin Med* **82:**26–30, 1973.
170. Kuhns P, Kaplan S, Basford R: Hexachlorocyclohexanes, potent stimuli of O_2^- production and calcium release in human polymorphonuclear leukocytes. *Blood* **68:**535–540, 1986.

13

The Respiratory Burst and Carcinogenesis

LEO I. GORDON and SIGMUND A. WEITZMAN

1. INTRODUCTION

Recent experimental evidence implicating oxygen-derived free radicals* in carcinogenesis has supported prior observations that link certain chronic inflammatory conditions to cancer. In this chapter we review data that support the notion that these radicals, many of which are derived from the respiratory burst of phagocytes, interact with target cells and thereby result in cellular changes that permit expression of the cancer phenotype. We first review the respiratory burst in phagocytes and then address the association between inflammation and cancer. The multistep model of carcinogenesis is discussed, followed by a presentation of the evidence linking oxygen-derived free radicals with the following events: (1) DNA and chromosomal alterations, (2) nucleoside modification, (3) activation of xenobiotics to genotoxic intermediates, (4) malignant transformation, and (5) tumor promotion (Fig. 1).

2. THE RESPIRATORY BURST IN PHAGOCYTES

The respiratory burst of leukocytes is reviewed in detail in Chapter 2. We describe briefly the products of the respiratory burst in phagocytic cells as they may relate to events associated with inflammation and carcinogenesis.

Oxygen may be metabolized by triggered inflammatory phagocytes (polymorphonuclear leukocytes, monocytes and eosinophils) to reactive oxidants.[1] These

*Free radicals are defined as compounds with one unpaired electron. However, for the purposes of this review, the term ''oxygen derived free radicals'' is used interchangably with ''reactive oxygen species'' and encompasses compounds which are not technically free radicals, such as H_2O_2 and HOCl.

LEO I. GORDON and SIGMUND A. WEITZMAN • Department of Medicine, Section of Hematology/Oncology and the Cancer Center, Northwestern University Medical School, Chicago, Illinois 60611.

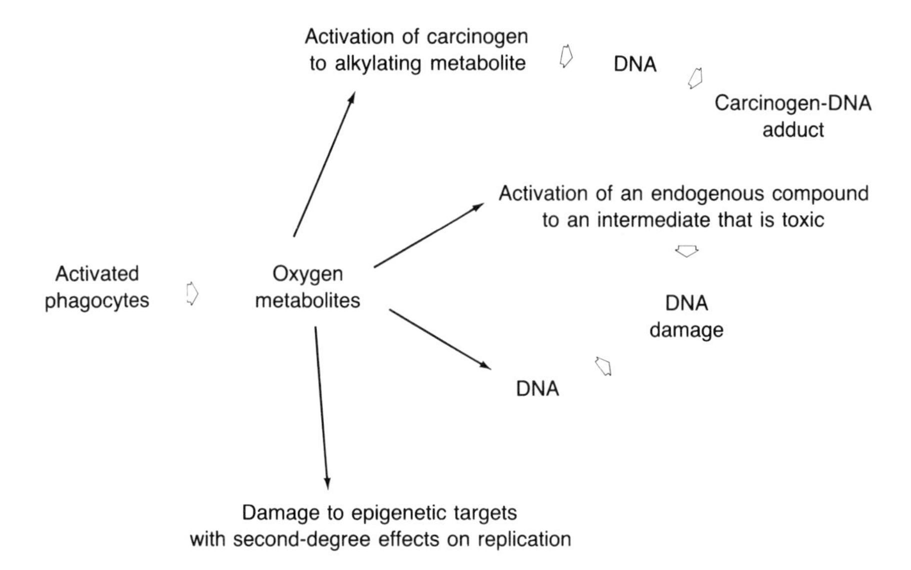

FIGURE 1. Possible pathways whereby products of the respiratory burst may influence targets and lead to subsequent alteration of DNA. (Partially adapted from Trush and Kensler.[121])

cells can take up molecular oxygen at a rate more than 50-fold their basal rate and in so doing produce superoxide anion radical (O_2^-) and hydrogen peroxide (H_2O_2). Initiation of this process depends on stimulation of the cellular membrane by any one of a number of stimuli that may be found at sites of inflammation, including microorganisms, activated complement, leukotriene B_4, and *N*-formyl-methylated oligopeptides[2,3] recently shown to be derived from bacteria.[4] Concomitantly, there is oxidation of glucose via the hexose monophosphate shunt, a process that provides NADPH as a source of electrons for the subsequent production of O_2^-. This process has been called the respiratory burst. In this review, we investigate its role in carcinogenesis. Critical to this is the concept that the immediate products of the respiratory burst in phagocytes, O_2^- and H_2O_2, may produce changes and influence events in target cells that result in expression of the cancer phenotype at sites of inflammation (Fig. 1).

The membrane of phagocytic cells, as the initial target of stimuli that can be found at inflammatory sites, has been the focus of the search for the enzyme(s) responsible for the respiratory burst. Indeed, a number of different oxidases have been identified,[5] but data now point convincingly to a membrane-associated pyridine nucleotide oxidase (first described in 1964[6]) as the most likely candidate.

List of Abbreviations: CGD, chronic granulomatous disease; BP-7,8 diol, 7,8-dihydroxy-7,8 dihydrobenzo[α]pyrene; CuDIPS, copper (II) (3,5-diisopropylsalicylate)$_2$; DMBA, demethylbenz-[α]anthracene; H_2O_2 hydrogen peroxide; HMdU, 5-hydroxymethyl-2-deoxyuridine; O_2^-, superoxide anion; ODC, ornithine decarboxylase; 8-OHdG, 8-hydroxyguanosine; SCE, sister chromatid exchange; SOD, superoxide dismutase; TPA (or PMA), tetradecanoylphorbol acetate; BHA, butylated hydroxyanisole; BHQ, butylated hydroxyquinolone; HMP shunt, hexose monophosphate shunt.

The enzyme is FAD requiring and catalyzes production of O_2^- using NADPH as an electron donor. The enzyme, which may or may not require a cytochrome *b* as an electron carrier,[7,8] has now been isolated in a dormant state from human polymorphonuclear leukocytes (PMN) and can be activated by detergent, sodium dodecyl sulfate (SDS), into its active form with a V_{max} of 2.1 μmoles of O_2^-/min/mg of membrane protein.[9]

The O_2^- and H_2O_2 produced in the initial steps of the respiratory burst are not the most reactive species formed. They are converted to oxidizing radicals and oxidized halides. Oxidizing radicals, e.g., hydroxyl radical (OH·),* are produced in part by the iron or copper catalyzed (Fenton reaction) Haber–Weiss reaction, such that

$$O_2^- + Fe^{3+} \rightarrow O_2 + Fe^{2+}$$
$$H_2O_2 + Fe^{2+} \rightarrow OH\cdot + OH^- + Fe^{3+}$$
$$\text{Net} \quad H_2O_2 + O_2^- \rightarrow OH\cdot + OH^- + O_2$$

Oxidized halides are produced by oxidation of chloride by H_2O_2, a reaction catalyzed by myeloperoxidase, an enzyme unique to certain phagocytic cells and recently cloned[10]:

$$CL^- + H_2O_2 \rightarrow HOCl + H_2O$$

The hypochlorous acid (HOCl) may react with other substances to form longer-lived oxidants, such as chloramines.[11,12]

When phagocytes are stimulated, a second phenomenon has been observed. This involves the release of unsaturated fatty acids from phosphatides. There are two phospholipases, A_2 (PLA_2) and C (PLC), which may be involved in this process, the end result of which is the liberation of arachidonic acid with subsequent oxidation by the lipoxygenase or cyclo-oxygenase pathways. The resulting hydroperoxides have the capability of initiating or sustaining free radical chains. When peroxides are reduced, a reaction catalyzed by prostaglandin hydroperoxidase, release of an oxidizing species[13] (perhaps singlet $^1O_2$[14]) is observed. Whether PLA_2 or PLC is primarily responsible for arachidonic acid release is uncertain. The former can cleave fatty acids at the C-2 position, to release arachidonic acid directly, while the latter hydrolyzes inositol phosphates to form diacylglycerol, and arachidonic acid is released by the subsequent action of a diacylglycerol lipase.[15] Both PLA_2 and PLC are Ca^{2+} dependent and have been described in inflammatory phagocytes.[15]

The processes by which phagocytic cells produce reactive oxidants are complex, and their products have been implicated in a variety of diseases.[16] We shall focus on the role of products of the respiratory burst, produced from stimulated phagocytes at sites of inflammation, in carcinogenesis.

*Some authors have recently questioned the production of hydroxyl radicals in vivo (Britigan et al.[9a]).

3. INFLAMMATION AND CANCER

The role of inflammation in carcinogenesis has been the subject of numerous investigations over the years. Boerhoave[17] and then Virchow[18] described the association of chronic inflammation with the subsequent development of cancer, and indeed attributed tumor formation to chronic irritation. Precancerous lesions are similar in many ways to chronic inflammatory conditions and have been described in burns,[19] prior trauma,[20] or chronic sinuses.[21,22] Malignant tumors have since been reported to occur at a variety of inflammatory sites. For example, in a review of 122 cases of Barrett's esophagus, Cameron et al.[23] found that adenocarcinoma of the esophagus and Barrett's esophagus were diagnosed simultaneously in 18, while adenocarcinoma developed in two additional patients over a mean interval of 8.5 years. Barrett's esophagus is associated with reflux esophagitis.[24] Furthermore, the association of lye stricture with subsequent neoplasia reinforces the role of chronic inflammation and the development of malignancy.

Chronic inflammatory conditions of the colon and small bowel have also been associated with the development of malignancy. The incidence of colon cancer is significantly increased in patients with ulcerative colitis,[25] so that prophylactic colectomy is sometimes recommended. Recently, the increased incidence of colonic carcinoma in patients with granulomatous colitis (Crohn disease) has been documented.[26] It is estimated that the incidence of colon cancer in patients with Crohn disease is approximately six times that of the general population, but approximately one half that seen in ulcerative colitis. Several features of these cases are of interest, including the multicentricity, the occurrence at sites of fistula formation, and the occurrence at sites of a preexisting benign stricture. This observation is especially interesting when considered in light of recent observations by Dolberg et al.,[27] who found that tumors in chickens induced by the Rous sarcoma virus appear to be localized to sites of wounding or inflammation despite the development of viremia in the host. It appears that inflammation, as well as mechanical puncture or laceration, must act as a cocarcinogen in Rous sarcoma virus-mediated carcinogenesis. Similarly, Konstantinides et al.[28] showed that tumors induced by systemically administered chemical carcinogens appear at sites of irritation or inflammation. Another type of viral oncogenesis may also depend on local irritation and inflammation, as papilloma formation following exposure to the Shope papillomavirus depends on concomitant skin abrasion.[29] These observations support the concept that the inflammatory process plays a role in carcinogenesis.

Indeed, in early models of experimentally induced cancer, wounding was found to enhance the development of tumors induced by chemicals.[30] In one experimental model, wounding could substitute for the diterpene tetradecanoylphorbol acetate (TPA) as a promoter.[31,32,33] Marks and Fürstenberger[34] found that a single wound was sufficient for tumor promotion.

The metaplastic and proliferative processes associated with wounding are also critical to an understanding of the wound response. Thus, regrowth of interfollicular epidermis from hair follicle cells and regeneration of lens from iris epithelium, occurring as a consequence of wounding, may play a role in expression of events associated with tumor promotion.

In 1928, Jean Nicholas Marjolin published the description of warty ulcers or carcinomatous changes arising in chronic ulcerative skin lesions and thus provided the first example of secondary squamous cell carcinomas.[35] Hawkins[36] described seven cases of warty tumors that arose in burns, flogging and asteomyelitis. These observations were extended by Da Costa,[37] who reported carcinomas developing in leg ulcers. More recently, similar observations have been made in other chronic inflammatory sites including chronic osteomyelitis,[38] perifolliculitis of the scalp,[39] lipoid pneumonia,[40] in the bladder in patients with shistosomiasis,[41] chronic indwelling Foley catheters,[42] chronic calculi,[43] and in the esophagus following lye stricture.[44] Adenocarcinoma of the gall bladder is more frequent in patients with chronic cholecystitis.[45] Thus, the role of inflammation in the development of cancer is established.[46] That asbestos can cause malignant change following an initial inflammatory reaction has been documented[47] and oxygen free radicals have now been implicated in the pathogenesis of these disorders.[48,49] Argyris and Slaga[50] showed that after initiation with dimethyl-benzanthracene, repeated abrasion without addition of other chemical compounds is sufficient stimulus for formation of both benign and malignant tumors in a mouse skin model.

Central to the concept of inflammation in cancer is the finding that chronic irritation of squamous or glandular epithelium will result in migration of inflammatory cells to the injured site.[51,52] These cells, stimulated to produce free radicals via the respiratory burst and NADPH oxidase activation or by activation of the arachidonic acid cascade with subsequent production of lipid hydroperoxides, then function as facilitators in the process of carcinogenesis. Lipid hydroperoxides also arise from nonenzymatic oxidation of membrane unsaturated fatty acids.[51] We shall review data that implicate products of the respiratory burst as well as products produced via the arachidonic acid cascade and derived predominantly in inflammatory phagocytes, as mediators of carcinogenesis as defined in the classic multistep model.

4. MULTISTEP CARCINOGENESIS

A number of recent epidemiological and experimental animal studies have supported the hypothesis of two-step carcinogenesis first advanced by Berenblum.[53] His mouse skin model has been further investigated and extended, and there are data that support the concept that chemical carcinogenesis can be divided into at least three steps or stages: (1) initiation, (2) promotion, and (3) progression (Table I). Although much of the experimental evidence derives from studies of early events in laboratory animals, the validity of these concepts has been established in certain instances in humans.[54] Recently the model was found to be applicable to a number of other systems including liver,[55,56] bladder,[57] colon,[58–60] mammary gland,[61] as well as to cells in culture (reviewed in Ref. 62–64). In order to discuss the role of

TABLE I. Multistep Carcinogenesis[a]

Initiation	Cellular phenomenon characterized by genetic changes that are irreversible
Promotion	Process of gene activation most often the result of the action of xenobiotics or endogenous substances—act on the entire tissue (initiated and noninitiated cells) by both epigenetic and genetic mechanisms to produce a benign tumor
Progression	Conversion of benign into malignant tumors, usually accompanied by more rapid growth, invasiveness, metastasis, and increased genetic instability

[a]From Cerutti.[168]

products of the respiratory burst on carcinogenesis, we shall review the model of multistage carcinogenesis (Table I) but focus on initiation and promotion.

4.1. Initiation

The initiation step requires application of an agent that is either a direct or indirect carcinogen with the capacity to alter the genome irreversibly. That initiators are both mutagenic and carcinogenic has been shown by correlation between the two upon exposure to a variety of chemical agents.[65,66] Most of the initiators described are electrophilic reactants[67] or may be converted into an electrophilic (activated) form that may bind covalently to DNA or other molecules.[68,69] An electrophile is a positively charged molecule that is able to react at certain electron-dense sites with a variety of cellular components, including DNA, RNA, proteins, and sugar groups, and it appears that only a minority of chemical carcinogens are able to function as initiators without undergoing metabolic conversion.[68] We shall review the process of metabolic conversion of initiators, as it appears that products of the respiratory burst are involved in this process.

4.2. Promotion

This second stage of carcinogenesis involves agents, termed promoters, that are not mutagenic and that for the most part do not bind DNA. However, recent data suggest that certain tumor promoters may not function totally by epigenetic mechanisms,[70] espeically when one considers that promotion may, in some cases, be irreversible and may alter the genome.[71] Tumor promotion usually involves a process of gene activation rather than alteration, in a process by which the altered (by initiation) phenotype becomes expressed over a defined latent period, a process that appears to correlate well with the clinical observations of a latent period during which progressive tissue and cellular changes take place.[72,73]

Many of the tumor promotors, and certainly the most studied, are diterpenes derived from croton oil. The most active promoter is phorbol myristate acetate (PMA), or TPA. This compound and others, such as mezerein and teleocidin, all are membrane-active agents that bind to cells with high affinity in a manner that is both reversible and saturable.[74,75] These compounds are able to send signals across cell membranes by a process mediated through a phospholipid dependent, calcium-dependent protein kinase (protein kinase C) by binding directly to it in place of the second messenger diacylglycerol.[15,76] This kinase may then catalyze phosphorylation of proteins that may alter a variety of cellular functions, including gene activation.[76] The phorbol esters can also stimulate respiratory burst activity in human phagocytes[2,3] and cause a general activation of the phagocytic cell membrane as measured by lysosomal enzyme release and aggregation. These compounds also can result in changes in cyclic nucleotide metabolism, calcium ion flux, and activation of the arachidonic acid cascade.[2,3] The resulting elaboration of free radicals from inflammatory phagocytes may result in a number of alterations of neighboring cells and biomolecules that have an impact on various steps in the multistep model of carcinogenesis.

The end result of the combination of initiation and promotion is most often the development of benign tumors. In the skin system, the interval between the applications of the promotor must be less than 1–2 weeks, and the tumors usually appear after 8 weeks, while a maximal response (5–15 tumors per mouse) occurs after 15 weeks. Recently, the process of promotion was further subdivided into two additional steps: (1) conversion, and (2) propagation. Conversion, or stage I promotion can occur after short-term treatment with a complete promoter (TPA), especially if this is followed by application of an incomplete promoter, such as mezerein. This model may seem relatively simple in the mouse skin system (it is not) but becomes even more complex in nonepidermal systems.[77] Tumor promoters have been found to cause apparently paradoxical effects in nonepidermal tissues, including either inhibition or induction of differentiation in fibroblast, neuroblastoma, and murine friend erythroleukemia cells and stimulation in some leukemia cell lines.[78] Differentiation and DNA synthesis of epidermal cells have been reported to exhibit divergent responses to phorbol esters.[79,80]

In nonepidermal in vitro systems, using phorbol ester promotion, a variety of criteria have been employed to investigate promoting activity, including formation of transformed cell foci, anchorage-independent growth in culture, or tumorigenicity as assessed by animal inoculation. Following promotion, a third stage of carcinogenesis, progression, is associated with a separate event involving alteration of DNA that will lead to the eventual production of a rapidly growing malignant tumor. Recent experiments have focused on this third stage of carcinogenesis.[81] For the purposes of this review, we confine our description of the role of products of the respiratory burst to the two well-described stages of carcinogenesis, initiation and promotion.

5. PHAGOCYTE-GENERATED OXIDANTS IN CARCINOGENESIS

5.1. Oxygen-Derived Free Radicals Cause DNA Strand Breaks

Critical to the initiation process as the first step in the multistep model of carcinogenesis is damage to DNA, induced by a variety of mechanisms. Birnboim et al.[82] suggested that O_2^-, in the absence of the more reactive H_2O_2, can induce DNA strand breaks in human inflammatory phagocytes. Catalase, an H_2O_2 inhibitor, prevented about 50% of the DNA damage measured by fluorimetric analysis of DNA unwinding (an extremely sensitive assay for DNA strand breakage), while SOD, an O_2^- scavenger, was more completely protective.[83]

There is additional evidence that phorbol ester-induced DNA damage in phagocytes is related to the respiratory burst. The HMP shunt inhibitor, 2-deoxyglucose, prevents DNA strand breaks, while *N*-ethylmaleimide, iodoacetamide, and iodoacetate (sulfhydryl inhibitors), as well as cysteamine and cysteine, block PMA-induced DNA strand breaks.[84] Furthermore, phagocytes from patients with CGD, which lack an activatable NADPH oxidase, are unable to induce DNA strand breaks after exposure to PMA.[85]

Birnboim also showed that DNA strand breaks could be induced by PMA stimulated phagocytes in murine erythroleukemia cells, providing evidence that the products of the phagocyte are diffusable into an adjacent target cell. The damage to the target cells could be blocked by catalase but not by SOD, while in phagocytes, both catalase and SOD could block the formation of strand breaks.[84,86] Experiments to explain this phenomenon have been performed, and suggest to some that O_2^- produced by stimulated phagocytes spontaneously dismuted to H_2O_2 as it diffused toward the target cell, so that the active substance in phagocyte–target cell experiments is H_2O_2. This oxidant, liberated from inflammatory phagocytes during the respiratory burst,[87] can also induce lysis of target cells.[88–93] Using murine macrophages (P388Dl cells) as targets, Schrauffstatter et al.[94] found that upon exposure to oxidant stress, there was a rapid fall in total intracellular NAD^+ levels at H_2O_2 concentrations of ≥ 40 μM, while at the same time there was stimulation of the repair enzyme poly-ADP–ribose polymerase, which resulted in NAD degradation and the subsequent depletion of ATP. The ADP–ribose polymerase enzyme inhibitors, 3-aminobenzamide or theophylline, prevented the H_2O_2-induced fall in NAD. Higher concentrations of H_2O_2 caused alteration of DNA (single-strand breaks), caused sustained depletion of NAD and ATP, and inhibited DNA synthesis, all within less than 30 sec of exposure. The observations that reactive oxygen species (produced via the respiratory burst by activated inflammatory phagocytes) can cause rapid-onset DNA strand breaks help interrelate the processes of inflammation, mutagenesis, and malignant transformation.

5.2. Oxygen-Derived Free Radicals Are Mutagenic

A number of groups have shown that human phagocytes can produce mutations in bacteria.[95–97] Macrophages isolated from established tumors have mutagenic activity in the Ames assay[98] and have the ability to induce the development of drug-resistant mutants in a mammary tumor line by a process that could be inhibited by the free radical scavengers SOD and catalase.[99] That phagocyte-generated oxidants were responsible for mutagenic events in bacteria was shown by the observation that while normal phagocytes produced mutations in bacteria, stimulated phagocytes from patients with CGD (which lack the NADPH oxidase) did not.[100]

Mutational events can also be induced by triggered phagocytes in cultured Chinese hamster ovary (CHO) cells.[101] It became apparent that in CHO cells and in a radiation-hypersensitive transformant A552, the radiation and chemical induced mutations are mediated by oxygen radicals, since streptomycin and bleomycin, which act via reactive oxygen species, and the oxidizing agents potassium superoxide and hydrogen peroxide, could all induce similar mutagenic responses in both cell types. Southern blot analysis showed that mutations caused by O_2^- and X-rays were of the deletion type, while other chemical mutagens cause primarily frame shift and nonsense mutations.[102]

5.3. Oxygen-Derived Free Radicals Cause Malignant Transformation

Further studies have demonstrated that cultured cells (C3H mouse fibroblasts of the 10T1/2 clone 8-line, C3H 10T1/2 cells), when exposed to human neutrophils stimulated to produce reactive oxygen species or to a cell free oxygen generating system[103,104] could form both benign and malignant tumors when injected into athymic mice.[105] Furthermore, these cells could be transformed in vitro, as measured by a standard transformation assay.[103] These data demonstrated that inflammatory phagocytes can produce malignant transformation by a process dependent on free radical generation by the respiratory burst. Although transformation may occur in a single step as a single event, the finding that in one animal a benign lesion eventually expressed the malignant phenotype is more consistent with a multistage process. Although both inflammatory phagocytes and a cell-free oxidant-generating system can cause these changes in target cells, the precise identity of the offending agent is unknown, since a variety of potential intermediates may be involved.

Borek[106] used superoxide dismutase to inhibit the oncogenic effects of X-rays and bleomycin to prevent enhancement of transformation by the tumor promoter TPA. Catalase, an enzyme that catalyzes the degradation of H_2O_2, did not prevent enhancement in this system. The action of SOD was probably at the cell surface, since SOD is known to penetrate cells poorly. Friedman et al. found similar results using reduction of ODC as an end point.[107] Later, Borek[108] extended her initial observations and, using the free radical scavenger vitamin E, further implicated free radicals in transformation.

5.4. Oxygen-Derived Free Radicals Cause Nucleoside Modification

That oxygen free radicals bind directly to DNA bases was first shown by Kasai and Nishimura,[109] who found that hydroxyl radical can react with deoxyguanosine to produce 8-hydroxyguanosine (8-OHdG). Using high-pressure liquid chromatography (HPLC) with electrochemical detection, Floyd[110] found that PMA-treated human polymorphonuclear leukocytes (PMN) accumulated 8-OHdG in DNA. Indeed, the amount of 8-OHdG in the DNA of activated cells could be decreased when SOD or SOD plus catalase were included in the incubation media. It is interesting that the 63% decrease in 8-OHdG in SOD-treated PMA stimulated phagocytes was similar to the observed decrease in DNA strand breaks reported by Birnboim when SOD was added to his system, supporting the notion that the strand breaks and base modification were caused by similar mechanisms. Likewise, Lewis[111] found that TPA-stimulated macrophages caused production of 5,6-ring thymine in 3T3 target cells co-incubated with phagocytes. Kasai and Nishimura found that 8-OHdG could also be produced in DNA by irradiation,[112] by co-incubation with H_2O_2 and asbestos,[113] and with a variety of auto-oxidizing agents.[114] The observations suggesting that asbestos and H_2O_2 produce DNA base changes are of importance in light of recent experimental evidence[49] linking the toxic effects of asbestos to the production of reactive oxygen metabolites in mouse macrophages and the association of asbestos, inflammation, and cancer.[48]

More recently, Frenkel[115] showed that the DNA of PMA-activated phagocytes has 5-hydroxymethyl-2-deoxyuridine (HMdU), a modified base also found in γ-irradiated cells, suggesting that the effects of phagocyte-generated oxidants on DNA mimic those of ionizing radiation. Similar effects were seen when exogenous DNA was co-incubated with various tumor promoters and phagocytes, and the amounts of modified DNA bases correlated with tumor-promoting activity and the amount of H_2O_2 generated.[116]

5.5. Oxygen-Derived Free Radicals Activate Xenobiotic Procarcinogens to Genotoxic Intermediates

The multistep model of carcinogenesis allows for two protocols commonly used in mouse skin: (1) the initiation–promotion model under discussion, and (2) a model of complete carcinogenesis wherein the offending agent is a complete carcinogen. This model calls for the repeated exposure to an agent that permits the development of autonomous malignant cells. In complete carcinogenesis, tumors may be induced by agents such as 7,12-dimethylbenz[α]anthracene (DMBA). There is evidence that phagocyte-generated free radicals may be involved in the metabolic activation of complete carcinogens.

We have reviewed the data describing the reactions of reactive oxidants with nucleic acids. The functional properties of the DNA molecule can be modified by covalent binding or by other mechanisms, such as strand scission and inaccurate repair. Products of arachidonic acid metabolism, which have been shown to influence tumor promotion with mediation of respiratory burst-generated free radicals,[117] have also been shown to lead to DNA damage by two mechanisms. In one, lipid peroxidation products, such as those generated by the respiratory burst, bind directly to DNA and induce mutation or cell transformation, while in the other, oxidizing agents, which are generated during hydroperoxide metabolism, can activate xenobiotic compounds so that they react with DNA. Examples of the first include peroxides, malondialdehyde, epoxides, and other aldehydes.[118] That epoxidation of BP-7,8 diol, the terminal activation step in benzpyrene carcinogenesis, occurs during prostaglandin synthesis, has been shown in the studies of Dix and Marnett.[119] Aromatic amines or radical cations[118] have also been shown to be activated during events associated with lipid peroxidation.

That inflammatory phagocytes are involved in the co-oxidation of xenobiotics to potentially genotoxic intermediates has also been shown by Trush et al.[120,121] Phorbol stimulation of polymorphonuclear leukocytes (PMN) results in both bleomycin A_2 and BP-7,9 diol oxidation, a process that leads to chemiluminescence, covalent binding to DNA, mutagenesis, and sister chromatid exchange. Indeed, data[122,123] suggest that the bay region diol epoxides are the ultimate carcinogens of a group of polycyclic aromatic hydrocarbons (PAH) that react with DNA and that these metabolites can be produced from the parent compounds in the presence of the products of the phagocytic respiratory burst. Thus tumor promoters, interacting with inflammatory phagocytes at sites of inflammation, may lead to genotoxic derivatives by a process independent of mixed-function oxidase systems, but is dependent on oxidation by cellular sources from the respiratory burst. In separate experiments, Epe[124] showed that under conditions of peroxidative metabolism, diethylstilbesterol (DES) can give rise to DNA strand breaks mediated by superoxide generation and that hydroxyl radical, produced in a Fenton-like reaction requiring iron, was the mediator of this process.

Recently, Eling et al.[125] showed that 7,8-dihydroxy-7,8-dihydrobenzo[α] pyrene (BP-7,8-diol) is oxidized to diolepoxides when incubated with arachidonic acid and prostaglandin synthetase, a process that appears to be mediated by two separate pathways. One is dependent on peroxyl radicals (apparently derived from lipid peroxidation) and is independent of mixed-function oxidases (cytochrome P-450), while the other is dependent on oxidation in a mixed-function oxidase system.

The contributions of the various endogenous oxidizing systems to cause oxidation of xenobiotics to substances with tumor-initiating or -promoting abilities in vivo has not been elucidated. However, evidence linking endogenous free radicals to liver cell death, proliferation, and carcinogenesis in a rat model has recently been presented.[126] Rats fed a choline-deficient and low-methionine diet develop a 100% incidence of preneoplastic hepatic nodules and a 50% incidence of hepatocellular carcinoma within 2 years. There is evidence of early triglyceride accumulation, nuclear lipid peroxidation, DNA alteration, mitochondrial lipid peroxidation, liver cell death and proliferation, and initiation of carcinogenesis. That these events are due to free radical generation is suggested by the presence of lipid peroxidation breakdown products (4-hydroxyalkenals).

Egner and Kensler[127,128] showed that a biomimetic superoxide dismutase, copper (II) $(3,4\text{-diisopropylsalicylate})_2$ (CuDIPS) that can permeate cells can inhibit tumor formation after weekly application of DMBA in a mouse skin model but cannot inhibit tumor formation when applied prior to a single initiating dose followed by addition of TPA in a typical two-stage carcinogenesis model. Thus, the data support the hypothesis that free radicals may mediate production of an ultimate carcinogen. Data supporting the role of oxygen free radicals and activation of xenobiotic compounds come from other sources as well. Certain antioxidants, such as butylated hydroxytoluene (BHT), butylated hydroxyanisole (BHA), flavones, and vitamins C and E, all function to decrease the level of polycyclic aromatic hydrocarbons (PAH) bound to epidermal DNA. All these compounds have been shown to inhibit PAH-induced tumor initiation.[77]

5.6. Oxygen-Derived Free Radicals Cause Chromosomal Abnormalities

Cerutti[129] advanced the concept that in situations where the intracellular concentration of activated oxygen species is high, initiated cells can be promoted to tumor growth. He described clastogenic factor(s) produced in human leukocytes that can be measured in the ultrafiltrate of supernatant from activated cells.[130,131] Clastogenic factors are substances that can cause breakage in chromosomes.

Antioxidants can reduce both the production and the activity of clastogenic factors. It appears that the tumor promoter PMA can induce chromosomal aberration in target cells by causing production of a low-molecular-weight (<10,000) clastogenic factor from human leukocytes.[129,131,132] This process is analogous to that seen in certain hereditary diseases, such as ataxia telangectasia, Fanconi anemia, and Bloom syndrome, all diseases characterized by an increased incidence of cancer and spontaneous chromosome breakage.[129] Skin fibroblasts from patients with these disorders are extremely sensitive to agents that also induce pro-oxidant states. Cerutti[129] stresses the role of superoxide anion in this process by citing the observation that SOD can inhibit the clastogenic factor in patients with Bloom syndrome. Thus, it appears that chromosomal changes can be induced indirectly from oxygenated products rather than O_2^- itself, but the process involves membrane-active agents that invoke inflammatory cells in the promotion process. However, Kinsella[133] found that although O_2^- production and reduction in SOD levels do occur in PMA-exposed cells under optimal media conditions, no chromosomal aberrations could be found—data that contradict those presented by Emerit and Cerutti.[131] An alternate mechanism for the observed effects on tumor promotion is proposed, one that implicates proliferation and differentiation by the effects of free radicals on the nucleotide pool, rather than an effect on chromosomes themselves.

Data implicating a cellular pro-oxidant state in cancer was recently provided by Reddy,[134] who found that certain nonmitogenic compounds that cause proliferation of peroxisomes and production of H_2O_2 lead to hepatocarcinogenesis. These peroxisome proliferators, which are lipid-lowering agents such as clofibrate, are complete carcinogens that function by increasing oxidative stress resulting from increased generation of intracellular hydrogen peroxide and other reactive oxygen species.

It has also been shown that stimulated human phagocytes can cause sister chromatid exchanges (SCEs) in cultured mammalian cells.[135,136] The production of these chromosomal alterations depends on oxygen metabolites. Hydroxyl radicals have been implicated as important, but they are quite unstable in aqueous media, so that they are unlikely to be the sole factor responsible for the changes. Hydroxyl radicals may, however, react with membrane lipids to produce secondary radicals, a notion supported by the finding that the hydrophobic antioxidant, vitamin E, when cultured with target cells, protected them from phagocyte-induced SCE.[135] Furthermore, it was observed that depletion of cellular glutathione levels makes the cell more susceptible to phagocyte-induced SCEs, an observation demonstrating the importance of cellular antioxidant defenses in prevention of genetic injury.[136]

5.7. Oxygen-Derived Free Radicals Act as Tumor Promoters

The evidence that free radicals derived from the respiratory burst in inflammatory phagocytes are important in tumor promotion comes from several lines of experimental evidence.[137] The most striking finding of a tumor promoter in a classic two stage skin model is its ability to induce a hyperplastic skin response. Since any irritant has the ability to induce this type of response without necessarily qualifying as a promoter, hyperplasia may likely be considered a necessary but insufficient condition for tumor promotion.[138]

In the mouse skin model, early events, such as development of dark basal keratinocytes, can be separated from late events, such as epidermal hyperplasia, or induction of ornithine decarboxylase (ODC) activity. Benzoyl peroxide may be involved at multiple sites in the promotion process and may, in fact, be important in mediating later events.[67,139]

Anthralin, an agent used commonly in the treatment of psoriasis, is capable of free radical production and is a tumor promoter in a mouse skin model.[140] Superoxide anion is formed upon oxidation of anthralin by air, and the tumor-promoting activity of this compound is inhibitable by the biomimetic superoxide dismutase, CuDIPS.[139,140]

Other free radical-generating systems are derived from cellular sources themselves and have been implicated in the tumor-promotion process. Important among these is the initiation of the arachidonic acid cascade by the calcium-dependent hydrolysis of membrane fatty acids by phospholipase A_2, a process that appears to occur concomitantly with phospholipid methylation.[141] Arachidonic acid is, in turn, oxidized by means of one of two pathways, using either cycloxygenase or lipoxygenase to form thromboxane A_2 or leukotrienes, respectively. The prostaglandin endoperoxides that occur as intermediates in this cascade can be compounds that serve as cellular sources of free radicals, and they function in the tumor-promotion process.

The interaction of inflammatory cells with tumor promoters, such as phorbol esters, has been well studied.[142,143] There is an almost immediate increase in oxygen consumption and production of O_2^- and H_2O_2. There is evidence that the phorbols bind to protein kinase C,[144] a phospholipid calcium-dependent kinase. These observations have led to the conclusion that the free radicals so generated are important contributors to the process of tumor promotion. These conclusions are supported by data that show that certain agents, which are able to block tumor promotion in a two-stage mouse skin model, are also potent inhibitors of the respiratory burst in human phagocytes. These include inhibitors of arachidonic acid metabolism, retinoids, protease inhibitors,[145] antioxidants, and dihydroxyepiandrosterone.[146]

Although it is postulated that most of the generated free radicals arise from inflammatory cells triggered to activate the respiratory burst,[147] Fisher showed that the target cells in the carcinogenesis model most often employed, basal keratinocytes, can also generate oxygen radicals in response to tumor promoters[148,149] and that inhibitors of oxygen radical production in the polymorphonuclear leukocytes also inhibited mouse epidermal oxygen radical production. The response, as measured by chemiluminescence, appears to be mediated by protein kinase C[150] and is strain dependent, with strains that are most sensitive to diterpene tumor promotion (Sencar mice) producing the maximal chemiluminescence response.[147] No measurements of O_2^- generation other than chemiluminescence were used. Similarly, ODC induction in mouse epidermal cells can be stimulated by phospholipase C incubation, presumably as a consequence of diacylglycerol activation of protein kinase C.[151]

Witz[152] has found that retinol can inhibit production of O_2^- in PMN induced by the promoters mezerein and teleocidin B. The mechanism of inhibition was thought secondary to alteration in membrane fluidity. Protease inhibitors have also been found to inhibit O_2 consumption and O_2-generation in human PMN, rat alveolar macrophages and rat PMN[153,154]

Promotion by TPA can also be inhibited by lipophilic analogues of BHA,[155] a free radical scavenger, as well as by 3-tert-butyl-4-hydroxyanisole and Cu DIPS.[156,157] Indeed, the BHA analogues were found to be capable of scavenging O_2^-, inhibiting lipid peroxidation and inhibiting chemiluminescence. Using a cell-free system, none of the analogues interacted with O_2^-, but the ± BHQ underwent auto-oxidation, resulting in increased O_2^- levels and causing reduction of molecular O_2. These data suggest that lipophilic BHA analogues inhibit ODC induction by scavenging free radicals other than O_2^- and that O_2^- may be converted to more reactive substances more directly involved in tumor promotion. The antioxidant was required at the time of exposure to the promoter, so that BHA was likely not to be exerting secondary effects. Agents which cause elevation of glutathione levels can protect against tumor promotion and ODC induction.[158,159]

Very few studies have been completed to further localize the influence of free radicals on one or another aspect of the tumor promotion process, but preliminary data suggest that antioxidants are most active in inhibiting stage 2 promotion[67,117,160,161] supporting the concept that respiratory burst derived free radicals may be involved in events associated with mitogenesis and terminal differentiation.

Evidence linking free radicals generated from the respiratory burst and tumor promotion also comes from the observations that host defense systems against oxidant stress are altered during the process of tumor promotion. The enzymes involved in reactions by which O_2^- and H_2O_2 are removed include superoxide dismutase, catalase, and glutathione peroxidases. Other natural antioxidants, including vitamins A, C, and E, are also protective both in vitro and in vivo.

Kensler and Taffe[162] showed that glutathione peroxidase activity is increased after multiple exposures to tumor promoters, while glutathione reductase levels are depressed. Oxidized glutathione levels are found to be increased following TPA exposure[157] providing evidence for a pro-oxidant state.

Krinsky and Deneke[163] found that carotenoids can scavenge singlet oxygen and are protective against other radical species that oxidize unsaturated fatty acids in membranes. Retinoic acid and RPA, structurally related to the carotenoids, did not inhibit stage I promotion,[164] suggesting that lipid peroxidation may not be involved in stage I promotion, although carotenoids can block SCE or bacterial mutations in cultured mammalian cells exposed to activated phagocytes.[136] Furthermore, carotenoids can inhibit O_2^- generation by neutrophils stimulated by phorbol esters.[165] The mechanism of carotenoid inhibition of lipid peroxidation is unclear, but may involve the inhibition of a peroxyl species by direct interaction with the carotenoid.[166] The role of peroxyl free radicals in carcinogenesis has recently been reviewed.[167]

6. SUMMARY

We have reviewed experimental data that link oxygen-derived free radicals produced by the respiratory burst in phagocytes to carcinogenesis. The historical association between inflammation and cancer was reviewed, and the classic multistep model of carcinogenesis was discussed. The evidence that oxygen-derived free radicals cause DNA-strand breaks, mutations, nucleoside alterations, and chromosomal abnormalities was reviewed, and data supporting the role of oxygen radicals derived from the respiratory burst and other sources in malignant transformation and activation of xenobiotics to genotoxic intermediates was provided. Finally, recent evidence that suggests that oxygen-derived free radicals play a role in tumor promotion was discussed. It seems clear that phagocyte-generated oxidants can play an important role in epithelial carcinogenesis, but further investigation is necessary in order to develop an understanding of the precise mechanisms by which this occurs in vivo.

REFERENCES

1. Weiss SJ: Oxygen, ischemia and inflammation. Acta Physiol Scand **126:**9–37, 1986.
2. Babior B: The respiratory burst of phagocytes. *J Clin Invest* **73:**599–607, 1984.
3. Babior B: Oxidants from phagocytes: Agents of defense and destruction. *Blood* **64:**959–966, 1984.
4. Marasco WA, Phan SH, Krutzsch H, et al: Purification and identification of formyl-methionyl-leucyl-phenylalanine as the major peptide neutrophil chemotactic factor produced by *E. coli. J Biol Chem* **259:**5430–5439, 1984.
5. Cheson BD, Curnutte JT, Babior BM: The oxidative killing mechanism of the neutrophil. *Prog Clin Immunol* **3:**1–65, 1977.
6. Rossi F and Zatti M: Biochemical aspects of phagocytosis in polymorphonuclear leukocytes. NADH and NADPH oxidation by the granules of resting and phagocytising cells. *Experientia (Basel)* **20:**21–23, 1964.
7. Gabig TG, Schervish EW, Santiaga JT: Functional relationship of the cytochrome b to the superoxide-generating oxidase of human neutrophils. *J Biol Chem* **257:**4114–4119, 1982.
8. Babior, BM: The nature of the NADPH oxidase. *Adv Host Defense Mech* **3:**91, 1983.
9. Curnutte ST, Kuver R, Babior BM: Activation of the respiratory burst oxidase in a fully soluble system from human neutrophils. *J Biol Chem* **262:**6450–6452, 1987.

9a. Britigan BE, et al: Do human neutrophils make hydroxyl radical? Determination of free radicals

generated by human neutrophils activated with a soluble or particulate stimulus using electron paramagnetic resonance spectroscopy. *J Biol Chem* **261:**4426–4431, 1986.
10. Weil S, Rosner C, Reid MS, et al: DNA cloning of human myeloperoxidase: Decrease in myeloperoxidase mRNA upon induction of HL-60 cells. *Proc Natl Acad Sci USA* **84:**2057–2061, 1987.
11. Test ST, Lampert MB, Ossana PJ, et al: Generation of nitrogen–chlorine oxidants by human phagocytes. *J Clin Invest* **74:**1341–1349, 1984.
12. Weiss SJ, Lampert MB, Test ST: Long-lived oxidants generated by human neutrophils: Characterization and bioactivity. *Science* **222:**625–628, 1983.
13. Kuehl FA, Humes JC, Hass EA, et al: Inflammation: The role of peroxidase-derived products. *Adv Prostaglandin Thromboxane Res* **6:**77–86, 1980.
14. Cadenas E, Sies H, Nastainczyk W, Ullrich V: Singlet oxygen formation detected by low-level chemiluminescence during enzymatic reduction of prostaglandin G_2 to H_2. *Hoppe Seylers Z Physiol Chem* **364:**519–528, 1983.
15. Majerus PW, Connolly TM, Deckmegn H, et al: The metabolism of phosphoinositide-derived messenger molecules. *Science* **234:**1519–1526, 1986.
16. Southorn PA, Powis G: Free radicals in medicine I. Chemical nature and biologic reactions II. Involvement in human disease. *Mayo Clin Proc* **63:**381–408, 1988.
17. Boerhave H, from Shimkin MB: Contrary to nature. DHEW publication No. (NIH) 76-720, Department of Health and Human Services, Washington, DC, 1977, pp 79–80.
18. Virchow R: *Cellular Pathology as Based upon Physiological and Pathological Histology*. Chance F (Trans). New York, Dover, 1971.
19. Lawrence EA: Carcinoma arising in the scars of thermal burns. *Surg Gynecol Obstet* **95:**579–588, 1952.
20. Gardner AW: Trauma and squamous skin cancer. *Lancet* **1:**760–761, 1959.
21. Menkin V: Role of inflammation in carcinogenesis. *Br Med J* **5186:**1585–1594, 1960.
22. Cruickshank AH, McConnell EM, Miller DG: Malignancy in scars, chronic ulcers and sinuses. *J Clin Pathol* **16:**573–580, 1963.
23. Cameron AJ, Oh BJ, Payne WS: The incidence of adenocarcinoma in columnar-lined (Barrett's) esophagus. *N Engl J Med* **313:**857–859, 1985.
24. Dahms BB, Rothstein FC: Barrett's esophagus in children: A consequence of chronic gastroesophageal reflux. *Gastroenterology* **86:**318–323, 1984.
25. Collins RH, Feldman M, Fordtran JS: Colon cancer, dysplasia and surveillance in patients with ulcerative colitis: A critical review. *N Engl J Med* **316:**1654–1658, 1987.
26. Korelitz BI: Carcinoma of the intestinal tract in Crohn's disease: Results of a survey conducted by the National Foundation for Ileitis and Colitis. *Am J Gastroenterol* **78:**44–46, 1983.
27. Dolberg DS, Hollingsworth R, Hertle M, Bissel MJ: Wounding and its role in RSV-mediated tumor formation. *Science* **230:**676–678, 1985.
28. Konstantinides A, Smulow JB, Sonnenschein C: Tumorigenesis at a predetermined oral site after one intraperitoneal injection of *N*-Nitroso-*N*-Methyurea. *Science* **216:**1235–1237, 1982.
29. Rogers S, Rous P: Joint action of a chemical carcinogen and a neoplastic virus to induce cancer in rabbits. *J Exp Med* **93:**459–488, 1951.
30. Friedwald UF, Rous P: Initiating and promoting elements in tumor production: Analysis of effects of tar, benzylpyrene, and methylcholanthrene on rabbit skin. *J Exp Med* **80:**101–126, 1944.
31. Hennings RH, Boutwell RK: Studies on the mechanism of skin tumor promotion. *Cancer Res* **30:**312–320, 1970.
32. Clark-Lewis I, Murray AW: Tumor promotion and the induction of epidermal ornithine decarboxylase activity in mechanically stimulated mouse skin. *Cancer Res* **38:**494–497, 1978.
33. Argyris TS: Tumor promotion by regenerative epidermal hyperplasia in mouse skin. *J Cutaneous Pathol* **9:**1–18, 1982.
34. Marks F, Fürstenberger G: Tumor promotion in skin: Are active oxygen species involved?, in Sies H (ed): *Oxidative Stress*. New York, Academic, 1985, pp 437–475.
35. Marjolin JN: *Dictionnaire de médicine*, ed 2. 1846.

36. Hawkins C: Cases of warty tumors in cicatrices. *Med Chir Trans Lond* **19:**19–34, 1835.
37. DaCosta JC: Carcinomatous changes in an area of chronic ulceration of Marjolin's ulcer. *Ann Surg* **37:**496–502, 1903.
38. Lovell WW, King RE, Alldredge R: Carcinoma of skin, sinuses and bone following chronic osteomyelitis. *South Med J* **50:**266–271, 1957.
39. Curry SS, Gaither DH, King LE: Squamous cell carcinoma arising in dissecting perifolliculitis of the scalp. A case report and review of secondary squamous cell carcinomas. *J Am Acad Dermatol* **37:**673–678, 1981.
40. Felson B, Ralaisomoy G: Carcinoma of the lung complicating lipoid pneumonia. *Am J Radiol* **141:**901–907, 1983.
41. Ferguson AR: Associated bilharziosis and primary malignant disease of the urinary bladder, with observations on a series of forty cases. *J Pathol Bacteriol* **16:**76–94, 1911.
42. Locke JR, Hill DE, Walzer Y: Incidence of squamous cell carcinoma in patients with long term catheter drainage. *J Urol* **133(6):**1034–1035, 1985.
43. Chapman WH, Kirchheim D, McRoberts JW: Effect of the urine and calculus formation on the incidence of bladder tumors in rats implanted with paraffin wax pellets. *Cancer Res* **33:**1225–1229, 1973.
44. Brandborg L: Neoplasms of the esophagus, in Beeson PB, McDermott W (eds): *Textbook of Medicine*. New York, WB Saunders, 1975, pp 1290–1293.
45. Way LW: Chronic cholecystitis, in Beeson PB, McDermott W (eds): *Textbook of Medicine*. New York, WB Saunders, 1975, pp 1313–1316.
46. Templeton AC: Acquired diseases, in Fraumeni JF Jr (ed): *Persons at High Risk of Cancer: An Approach to Cancer Etiology and Control*. New York, Academic, 1975, pp 175–183.
47. Craighead JE, Mossman BT: The pathogenesis of asbestos-associated diseases. *N Engl J Med* **306:**1446–1455, 1982.
48. Mossman BT, Marsh JP: Mechanisms of Toxic Inury by Asbestos Fibers: Role of Oxygen-free Radicals, in Beck EG, Bignon J (eds): *NATO ASI Series in Vitro Effects of Mineral Dusts*. Berlin, Springer-Verlag, 1985.
49. Goodglick LA, Kane AB: Role of reactive oxygen metabolites in crocidolite asbestos toxicity to mouse macrophages. *Cancer Res* **46:**5558–5566, 1986.
50. Argyris TS, Slaga TJ: Promotion of carcinomas by repeated abrasion in initiated skin of mice. *Cancer Res* **41:**5793–5795, 1981.
51. Diamond L, O'Brien TG, Baird WM: Tumor promoters and the mechanism of tumor promotion. *Adv Cancer Res* **32:**1–74, 1980.
52. Whittemore A, McMillan A: Lung cancer mortality among U.S. uranium miners: A reappraisal. *J Natl Cancer Inst* **71:**489–99, 1983.
53. Berenblum I: The carcinogenic action of croton resin. *Cancer Res* **11:**44–48, 1941.
54. Pitot HC: The natural history of neoplastic development: The relation of experimental models to human cancer. *Cancer* **49:**1206–1211, 1982.
55. Pitot HC: The stages of initiation and promotion in hepatocarcinogenesis. *Biochim Biophys Acta* **605(2):**191–215, 1980.
56. Pitot HC, et al: Properties of incomplete carcinogens and promoters in hepatocarcinogenesis, in Hecker E et al (eds): *Carcinogenesis*. New York, Raven, 1982, vol 7.
57. Chester JF, Gaissert HA, Ross JS, et al: *N*-[4-(5-nitro-2-furyl)-2-thiazolyl] formamide induced bladder cancer in mice: Augmentation by sutures through the bladder wall. *J Urol* **137:**769–771, 1987.
58. Lasser A, Acosta AE: Colonic neoplasms complicating uretero-sigmoidostomy. *Cancer* **35(4):**1218–1222, 1975.
59. Crissey MM, Steele GD, Gittes RF: Rat model for carcinogenesis in ureterosigmoidostomy. *Science* **207:**1079–1080, 1980.
60. Bristol JB, Williamson RCN: Ureterosigmoidostomy and colon carcinogenesis. *Science* **214:**351, 1981.
61. Pitot HC: Chemicals and cancer initiation and promotion. *Hosp Prac* **18(7):**101–113, 1983.

62. Slaga TJ, Sivak A, Boutwell RK (eds): *Carcinogenesis: A Comprehensive Survey*. New York, Raven, 1978, vol 2: *Mechanism of Tumor Promotion and Cocarcinogenesis*.
63. Slaga TJ (ed): *Mechanisms of Tumor Promotion*. Boca Raton, Florida, CRC Press, 1984, vol 1–4.
64. Hsia CC: Possible roles of fungal infection and mycotoxin in human esophageal carcinogenesis, in Harris CC, Autrup HN (eds): *Human Carcinogenesis*. New York, Academic, 1983, pp 883–911.
65. Ames BN, McCann J, Yamaski E: Methods for detecting carcinogens and mutagens with the *Salmonella*/mammalian–microsome mutagenicity test. *Mutat Res* **31**:347–364, 1975.
66. McCann J, Ames BN: Detection of carcinogens as mutagens in *Salmonella* microsome test: Assay of 300 chemicals. Discussion. *Proc Natl Acad Sci USA* **73**:950–954, 1976.
67. Slaga TJ: Overview of chemical carcinogenesis and anticarcinogenesis, in Nygaard OF, Simic MG (eds): *Radioprotectors and Anticarcinogens*. New York, Academic. 1983, pp 437–448.
68. Miller EC: Some current perspectives on chemical carcinogenesis in humans and experimental animals. *Cancer Res* **38**:1479–1496, 1978.
69. Miller EC, Miller JA: The metabolism of chemical carcinogens to reactive electrophiles and their possible mechanisms of action in carcinogenesis, in Searle CE (ed): *Chemical Carcinogens*. American Chemical Society, Washington, DC, 1976, pp 737–762.
70. Snyder RD: An examination of the DNA damaging and repair inhibitory capacity of phorbol myristate acetate in human diploid fibroblasts. *Carcinogenesis* **6**:1667–1670, 1985.
71. Kinzel V, Fürstenberger G, Loehoke H, Marks F: Three stage tumorigenesis in mouse skin. DNA synthesis as a prerequisite for the conversion stage induced by TPA prior to initiation. *Carcinogenesis* **7**:779–782, 1986.
72. Farber E, Cameron R: The sequential analysis of cancer development. *Adv Cancer Res* **31**:125–226, 1980.
73. Farber E: Chemical carcinogenesis. *N Engl J Med* **305**:1379, 1981.
74. Driedger PE, Blumberg PM: Specific binding of phorbol ester tumor promoters. *Proc Natl Acad Sci USA* **77**:567–71, 1980.
75. Shoyab M, Todaro GJ: Specific high affinity cell membrane receptors for biologically active phorbol and ingerol esters. *Nature (Lond)* **288**:451–455, 1980.
76. Nishizuka V: Studies and perspectives of protein kinase C. *Science* **233**:305–312, 1986.
77. Slaga TJ, Solanki V, Logani M: Studies on the Mechanism of Action of antitumor promoting agents: Suggestive evidence for the involvement of free radicals in promotion, in Nygaard OF, Simic MG (eds): *Radioprotectors and Anticarcinogens*. New York, Academic, 1983, p 471.
78. Mastro AM: Phorbol esters: Tumors promotion, cell regulation and the immune system. *Lymphokines* **6**:263–313, 1982.
79. Yuspa SH, Ben T, Patterson E, et al: Stimulated DNA synthesis in mouse epidermal cell cultures treated with 12-*O*-tetradecanoyl-phorbol-13-acetate. *Cancer Res* **36**:4062–4068, 1976.
80. Yuspa SH, Ben T, Hennings H, Lichti U: Divergent responses in epidermal basal cells exposed to the tumor promoter TPA. *Cancer Res* **42**:2344–2349, 1982.
81. Slaga TJ, Rotstein JB, O'Connell JF: The possible involvement of free radicals in skin tumor progression, in *Second International Conference on Anticarcinogenesis and Radiation Protection, Gaithersburg, Maryland, March 8–12, 1987*.
82. Birnboim HC: DNA strand breakage in human leukocytes exposed to a tumor promoter, PMA. *Science* **215**:1242–1249, 1982.
83. Birnboim HC, Kanabus-Kaminska M: The production of DNA strand breaks in human leukocytes by superoxide anion may involve a metabolic process. *Proc Natl Acad Sci USA* **84**:6820–6824, 1984.
84. Birnboim HC: Factors which affect DNA strand breakage in human leukocytes exposed to a tumor promotor, phorbol myristate acetate. *Can J Physiol Pharmacol* **60(11)**:1359–1366, 1982.
85. Birnboim HC, Biggar WD: Failure of phorbol myristate acetate to produce damage to DNA in leukocytes from patients with CGD. *Infection Immun* **38(3)**:1299–1300, 1982.
86. Birnboim HC: Importance of DNA strand break damage in tumor promotion, in Nygaard OF, Simic MG (eds): *Radioprotectors and Anticarcinogenesis*. New York, Academic, p 539.

87. Fridovich I: Superoxide radical: An endogenous toxicant. *Annu Rev Pharmacol Toxicol* **23:**239–257, 1983.
88. Spragg RF, Hinshaw DB, Hyslop PA, et al: Alteration in ATP and energy charge in cultured endothelial and P388D$_1$ cells following oxidant injury. *J Clin Invest* **76:**1471–1476, 1985.
89. Schraüfstatter IU, Hinshaw DB, Hyslop PA, et al: Glutathione cycle activity and pyridine nucleotide levels in oxidant induced injury of cells. *J Clin Invest* **76:**1131–1139, 1985.
90. Nathan CF, Silverstein SC, Brukner LH, Cohn ZA: Extracellular cytolysis by activated macrophages and granulocytes. II. Hydrogen peroxide as a mediator of cytotoxicity. *J Exp Med* **149:**100–113, 1979.
91. Nathan CF, Arrick BA, Murray HW, et al: Tumor cell antioxidant defenses: Inhibition of the glutathione redox cycle enhances macrophage mediated glycolysis. *J Exp Med* **158:**766–788, 1980.
92. Weiss SJ, Young J, LoBuglio AF, Slivka A: Role of hydrogen peroxide in neutrophil mediated destruction of cultured endothelial cells. *J Clin Invest* **68:**714–724, 1981.
93. Simon RH, Scroggin CH, Patterson D: Hydrogen peroxide causes fatal injury to human fibroblasts exposed to oxygen radicals. *J Biol Chem* **256:**7181–7186, 1981.
94. Schraüfstatter IU, Hinshaw DB, Hyslop PA, et al: Oxidant injury of cells. DNA strand breaks activate polyadenosine diphosphate–ribose polymerase and lead to depletion of NAD. *J Clin Invest* **77:**1312–1320, 1986.
95. Levin DE, Hollstein M, Christman MF, et al: A new *Salmonella* tester strain (CTA 102) with A.T. base pairs at the site of mutation detects oxidative mutagens. *Proc Natl Acad Sci USA* **79:**7445–7449, 1982.
96. Barak M, Ulitzur S, Merzbach D: Phagocytosis induced mutagenesis in bacteria. *Mutat Res* **121:**7–16, 1983.
97. Farr SP, D'Ari R, Touati D: Oxygen dependent mutagenesis in *E. coli* lacking SOD. *Proc Natl Acad Sci USA* **83:**8268–8272, 1986.
98. Fulton AM, Loveless SE, Heppner GH: Mutagenic activity of tumor associated macrophages in *Salmonella typhimeurium* strains TA98 and TA100. *Cancer Res* **44:**4308–4311, 1984.
99. Yamashima K, Miller B, Heppner GH: Macrophage-mediated induction of drug resistant variants in a mouse mammary tumor cell line. *Cancer Res* **46:**2396–2401, 1986.
100. Weitzman SA, Stossel TP: Mutation caused by human phagocytes. *Science* **212:**546–547, 1981.
101. Weitzman SA, Stossel TP: Phagocyte-induced mutation in Chinese Hamster ovary cells. *Cancer Lett* **22:**337–341, 1984.
102. Hsie AW, Resio L, Katz S, Lee CQ, et al: Evidence for reactive oxygen species inducing mutations in mammalian cells. *Proc Natl Acad Sci USA* **83:**9616–9620, 1986.
103. Zimmerman RA, Cerutti P: Active oxygen acts as a promoter of transformation in mouse embryo C3H/10T 1/2/C18 fibroblasts. *Proc Natl Acad Sci USA* **81:**2085–2087, 1984.
104. Lesko SA, Lorentzen RJ, Tsao PO: Role of superoxide in deoxyribonucleic acid strand scission. *Biochemistry* **19:**3023–3028, 1980.
105. Weitzman SA, Weitberg AR, Clark EP, Stossel TP: Phagocytes as carcinogens: Malignant transformation produced by human neutrophils. *Science* **227:**1231–1233, 1985.
106. Borek C, Troll W: Modifiers of free radicals inhibit in vitro the oncogenic actions of x-rays, bleomycin and the tumor promoter 12-*O*-tetradecanoyl phorbol 13-acetate. *Proc Natl Acad Sci USA* **80:**1304–1307, 1983.
107. Friedman J, Cerutti P: The induction of ornithine decarboxylase by phorbol 12-myristate 13-acetate or by serum is inhibited by antioxidants. *Carcinogenesis* **4:**1425–1427, 1983.
108. Borek C, Ong A, Mason H, et al: Selenium and vitamin E inhibit radiogenic and chemically induced transformation *in vitro* via different mechanisms. *Proc Natl Acad Sci USA* **83:**1490–1494, 1986.
109. Kasai H, Nishimura S: Hydroxylation of deoxyguanosine at the C-8 position by ascorbic acid and other reducing agents. *Nucleic Acid Res* **12:**2137–2145, 1984.
110. Floyd RA, Watson JJ, Horvis J, et al: Formation of 8-hydroxydeoxyguanosine, hydroxyl free

radical adduct of DNA in granulocytes exposed to the tumor promoter, tetradecanoylphorbolacetate. *Biochem Biophys Res Commun* **137**:841–846, 1986.

111. Lewis JG, Adams DO: Induction of 5,6-ring saturated thymine bases in NIH-3T3 cells by phorbol ester stimulated macrophoges, role of reactive oxygen intermediates. *Cancer Res* **45**:1270–1275, 1985.
112. Kasai H, Tanooka H, Nishimura S: Formation of 8-hydroxyguanosine by x-irradiation. *Gann* **75**:1037–1039, 1985.
113. Kasai H, Nishimura S: DNA damage induced by asbestos in the presence of hydrogen peroxide. *Gann* **75**:841–844, 1984.
114. Kasai H, Nishimura S: Hydroxylation of deoxyguanosine at the C-8 position by polysterols and aminophenols in the presence of hydrogen peroxide and ferric ion. *Gann* **75**:565–566, 1984.
115. Frenkel K, Chrzan K, Troll W, et al: Radiation-like modification of bases of DNA exposed to tumor promoter-activated polymorphonuclear luekocytes. *Cancer Res* **46**:5533–5540, 1986.
116. Frenkel K, Chrzan K: Hydrogen peroxide formation and DNA base modification by tumor promoter activated polymorphonuclear leukocytes. *Carcinogenesis* **8**:455–460, 1987.
117. Fischer SM: Arachidonate cascade and skin tumor promotion, in Thaler-Dao H, Crostes de Paulet GA, Paoletti R (eds): *Icosanoids and Cancer*. New York, Raven, 1984, p 79.
118. Marnett LJ: Arachidonic acid and tumor initiation, in Marnett LJ (ed): *Arachidonic Acid Metabolism and Tumor Initiation*. Boston, Martinus Niljhoff, 1985, pp 39–82.
119. Dix TA, Marnett LA: Metabolism of polycyclic aromatic hydrocarbon derivatives to ultimate carcinogens during lipid peroxidation. *Science* **221**:77–79, 1983.
120. Trush MA, Kensler TW, Seed JL: Activation of xenobiotics by human PMN via reactive oxygen dependent reactions, in Kocsis JJ, Jollow DJ, Witmer M, et al (eds): *Biological Reactive Intermediates*. New York, Plenum, 1986, vol III, p 311.
121. Trush MA, Seed JL, Kensler TA: Oxidant dependent metabolic activation of polycyclic aromatic hydrocarbons by phorbol ester-stimulated human polymorphonuclear leukocytes: Possible link between inflammation and cancer. *Proc Natl Acad Sci USA* **82**:5194–5198, 1985.
122. Jerina DM, Daly JW: Oxidation at carbon, in Parke DV, Smith RL (eds): *Drug Metabolism*. London, Taylor & Francis, 1977, pp 15–33.
123. Jerina DM, Lehr RE, Yogi H, et al: Mutagenicity of benzo[a]pyrene derivatives and the description of a quantum mechanical model which predicts the ease of carbonium ion formation from diol epoxides, in de Serres FJ, Fouts JR (eds): *In Vitro Metabolic Activation and Mutagenesis Testing*. Amsterdam, Elsevier/North-Holland Biomedical Press, 1976, pp 159–177.
124. Epe B, Schiffman D, Metzler M: Possible role of oxygen radicals in cell transformation by diethylstilbesterol and related compounds. *Carcinogenesis* **7**:1329–1334, 1986.
125. Eling T, Curtis J, Battista J, Marnett LJ: Oxidation of (+)-7,8-dihydroxy-7,8-dihydrobenzo[a]pyrene by mouse keratinocytes: Evidence for peroxyl radical—and monooxygenase dependent metabolism. *Carcinogenesis* **7**:1957–1963, 1986.
126. Ghoshol A, Rushmore IH, Ghazarian D, et al: Dietary choline deficiency as a new model to study the possible role of free radicals in acute cell injury and in carcinogenesis, in *Second International Conference on Anticarcinogenesis and Radiation Protection, Gaithersburg, Maryland, March 8–12, 1987*.
127. Egner PA, Kensler TW: Effects of a biomimetic superoxide dismutase on complete and multistage carcinogenesis in mouse skin. *Carcinogenesis* **6**:1167–1172, 1985.
128. Egner PA, Taffe BG, Kensler TW: Effects of copper complexes on multistage carcinogenesis, in Sorenson JRJ (ed): *Medicinal Uses of Copper*. Clifton, New Jersey, Human Press, 1986.
129. Cerutti P: Prooxidant states and tumor promotion. *Science* **227**:375–381, 1985.
130. Cerutti PA, Amstad P, Emerit I: Tumor promoter PMA induces membrane mediated chromosomal damage, in Nygaard OF, Simic MG (eds): *Radioprotectors and Anticarcinogenesis*. New York, Academic, 1983, pp 527–538.
131. Emerit I, Cerutti P: Icosanoids and chromosome damage, in Thaler-Dao H, Crostes dePaulet GA, Paoletti R (eds): *Icosanoids and Cancer*. New York, Raven, 1984, pp 127–138.

132. Kozumbo WJ, Muehlematter D, Jörg A, et al: Phorbol ester induced formation of a clastogenic factor from human monocytes. *Carcinogenesis* **8:**521–526, 1987.
133. Kinsella JR, St. C. Garner H, Butler J: Investigation of a possible role for superoxide anion production in tumor promotion. *Carcinogenesis* **4:**717–719, 1983.
134. Reddy JK, Reddy MK, Usman MI, et al: Comparison of hepatic peroxisome proliferative effect and its implications for hepatocarcinogenicity of pthalate esters di(2-ethylhexyl) pthalate and di(2-ethylhexyl) adipate with hypolipidemic drug. *Environ Health Persp* **65:**317–327, 1986.
135. Weitberg AB, Weitzman SA, Destremps M, et al: Stimulated human phagocytes produce cytogenetic changes in cultured mammalian cells. *N Engl J Med* **308:**26–29, 1983.
136. Weitberg AB, Weitzman SA, Clark EP, Stossel TP: Effects of antioxidants on oxidant induced sister chromatid exchange formation. *J Clin Invest* **75:**1835–1841, 1985.
137. Slaga TJ, Klein-Szanto AJP, Triplett LC, et al: Skin tumor promoting activity of benzoyl peroxide, a widely used free radical generating compound. *Science* **213:**1023–1025, 1981.
138. Marks F, Fürstenberger G, Kownatzki E: Prostaglandin E–mediated mitogenic stimulation of mouse epidermis in vivo by divalent cation ionophore A23187 and by tumor promoter 12-0-tetradecanoylphorbol-13-acetate. *Cancer Res* **41:**696–702, 1981.
139. O'Connell JF, Klein-Szanto AJP, DiGiovanni DM, et al: Enhanced malignant progression of mouse skin tumors by the free radical generator benzoyl peroxide. *Cancer Res* **46:**2863–2865, 1986.
140. Bock FG, Burns R: Tumor-promoting properties of anthralin (1,8,9-anthratriol). *J Natl Cancer Inst* **30:**393–397, 1963.
141. Hirata F, Axelrod J: Phospholipid methylation and biological signal transmission. *Science* **209:**1082–1090, 1980.
142. Troll W, Frenkel K, Teebor G: Free oxygen radicals: Necessary contributors to tumor promotion and cocarcinogenesis, in Fukiki H, et al (eds): *Cellular Interactions by Environmental Tumor Promoters.* Utrecht, VNU Science Press, 1984, pp 207–218.
143. Troll W, Witz G, Goldstein B, et al: The role of free oxygen radicals in tumor promotion and carcinogenesis, in Hecker F, Fusening NE, Kunz W, et al (eds): *Carcinogenesis—A Comprehensive Survey.* New York, Raven, 1982, vol 7, pp 593–597.
144. Tauber, AI, Brettler DB, Kennington EA, Blumberg PM: Relation of human neutrophil phorbol ester receptor occupancy and NADPH-oxidase activity. *Blood* **60:**333, 1982.
145. Troll W, Wiesner R: Protease inhibitors as anticarcinogens and radioprotectors, in Nygaard OF, Simic MG (eds): *Radioprotectors and Anticarcinogens.* New York, Academic, 1983, p 567.
146. Troll W, Wiesner R: The role of oxygen radicals as a possible mechanism of tumor promotion. *Annu Rev Pharmacol Toxicol* **25:**509–528, 1985.
147. Freeman BA, Crapo JD: Biology of disease: Free radical and tissue injury. *Lab Invest* **82:**412, 1982.
148. Fischer SM, Adams LM: Supression of tumor promoter-induced chemiluminescence in mouse epidermal cells by several inhibitors of arachidonic acid metabolism. *Cancer Res* **45:**3130–3136, 1985.
149. Fischer SM, Baldwin JK, Adams LM: Effects of anti-promoters and strain of mouse on tumor promoter-induced oxidants in murine epidermal cells. *Carcinogenesis* **7:**915–918, 1986.
150. Fisher SM, Baldwin JK, Adams LM: Phospholipase C mimics tumor promoter induced chemiluminescence in murine epidermal cells. *Biochem Biophys Res Commun* **131:**1103–1108, 1985.
151. Jeng AY, Lichti U, Strickland JE, Blumberg PM: Similar effects of phospholipase C and phorbol ester tumor promoters on primary mouse epidermal cells. *Cancer Res* **45:**5714–5721, 1985.
152. Witz G, Goldstein BD, Amoruso M, et al: Retinoid inhibition of superoxide anion radical production by human PMN stimulated with tumor promoters. *Biochem Biophys Res Commun* **97:**883–888, 1980.
153. Goldstein BD, Witz G, Amoruso M, Troll W: Protease inhibitors antagonize the activation of PMN oxygen consumption. *Biochem Biophys Res Commun* **88:**854–860, 1979.
154. Goldstein B, Witz G, Zimmerman J, Gee C: Free radicals and reactive oxygen species in tumor

promotion, in Greenwald PA, Cohen G (eds): *Oxyradicals and Their Scavenger Systems*. New York, Elsevier, 1983, vol II: *Cellular and Medical Aspects*, pp 321–325.
155. Kozumbo WJ, Trush MA, Kensler TW: Are free radicals involved in tumor promotion? *Chem Biol Interac* **54**:199–207, 1985.
156. Kozumbo WJ, Seed JL, Kensler TW: Inhibition by 3-tert-butyl-4-hydroxyanisole and other antioxidants of epidermal ODC activity induced by TPA. *Cancer Res* **43**:2555–2559, 1983.
157. Kensler TW, Bush DM, Kozumbo WJ: Inhibition of tumor promotion by a biomimetic superoxide dismutase. *Science* **221**:75–77, 1983.
158. Perchellet JP, Owen MD, Posey TD, et al: Inhibitory effects of glutathione level-raising agents and D-tocopherol on ornithine decarboxylase induction and mouse skin tumor promotion by 12-*O*-tetradecanoylphorbol-13-acetate. *Carcinogenesis* **6**:567–573, 1985.
159. Perchellet JP, Perchellet EM, Orten DK, Schneider BA: Decreased ratio of reduced/oxidized glutathione in mouse epidermal cells treated with tumor promoters. *Carcinogenesis* **7**:503–506, 1986.
160. Schwartz F, Peres G, Kunz W, et al: On the role of superoxide anion radicals in skin tumor promotion. *Carcinogenesis* **5**:1663–1670, 1984.
161. Czerniecki B, Gad SC, Reilly C, et al: Phorbol diacetate inhibits superoxide anion radical production and tumor promotion by mezerein. *Carcinogenesis* **7**:1637–1641, 1986.
162. Kensler TW, Taffe BG: Free radicals in tumor promotion. *Adv Free Radical Biol Med* **2**:347–387, 1986.
163. Krinsky NI, Deneke SM: Interaction of oxygen and oxyradicals with carotenoids. *J Natl Cancer Inst* **69**:205–209, 1982.
164. Gschwendt M, Horn F, Kittstein W, et al: Calcium and phospholipid dependent protein kinase activity in mouse epidermis cytosol. Stimulation by complete and incomplete tumor promoters and inhibition by various compounds. *Biochem Biophys Res Commun* **124**:63–68, 1984.
165. Krinksy NI: Mechanisms of inactivation of oxygen species by carotenoids, in *Second International Conference on Anticarcinogenesis and Radiation Protection, Gaithersburg, Maryland, March 8–12, 1987.*
166. Burton GW, Ingold KU: Beta-carotene an unusual type of lipid antioxidant. *Science* **224**:569–573, 1984.
167. Marnett LJ: Peroxyl free radicals: Potential mediators of tumor initiation and promotion. *Carcinogenesis* **8**:1365–1373, 1987.
168. Cerutti PA: Multistep carcinogenesis, in *Second International Conference on Anticarcinogenesis and Radiation Protection, Gaithersburg, Maryland, March 8–12, 1987.*

14

The Respiratory Burst and Mechanisms of Oxygen Radical-Mediated Tissue Injury

JEFFREY S. WARREN, PETER A. WARD, and KENT J. JOHNSON

1. INTRODUCTION

Although it has been recognized since the nineteenth century that the influx of phagocytic cells into sites of tissue injury is critical to the development of the inflammatory response, the recognition that neutrophils, monocytes, and macrophages are complex cells possessing diverse biological capabilities is relatively recent. Inflammatory cells are controlled by an extensive array of biochemical mediators. The concept that leukocytic proteases contribute to the tissue injury observed in acute inflammation has considerable supporting evidence but has been extensively broadened in recent years.[1,2] Although numerous studies have demonstrated that lysosomal proteases can cause or enhance tissue injury, other studies have shown that tissue injury cannot be completely prevented by antiproteases.[3,4] In addition, tissue injury still occurs in strains of mice whose leukocytes are deficient of proteases suggesting that other mechanisms are important in phagocyte-mediated tissue injury.[5] Recent studies have provided compelling evidence that leukocyte-derived oxygen radicals and their metabolites are important mediators of inflammation and tissue injury.

Phagocytic cells, when activated with a variety of soluble and particulate stimuli, exhibit a sharp increase in consumption of molecular oxygen and produce several biologically active oxygen-derived radicals and metabolites. The focus of this chapter will be how phagocytes, particularly neutrophils, generate these sub-

JEFFREY S. WARREN, PETER A. WARD, and KENT J. JOHNSON • Department of Pathology, University of Michigan Medical School, Ann Arbor, Michigan 48109-0650.

stances and an examination of in vitro and in vivo evidence that oxygen radicals and other oxygen-derived metabolies are important mediators of tissue damage.

2. OXYGEN-DERIVED METABOLITE PRODUCTION BY PHAGOCYTIC CELLS

Membrane perturbation by a variety of particulate and soluble stimuli (Table I) results in the release of lysosomal contents and an abrupt increase (2 to 20-fold) in oxygen consumption by phagocytic leukocytes. The increased oxygen consumption (respiratory burst) is associated with the production of several reactive oxygen-derived metabolites and an increase in glucose catabolism due to activation of the hexose monophosphate shunt.[6,7] The primary oxygen-derived species generated by phagocytes are superoxide anion ($O_2^- \cdot$) and hydrogen peroxide (H_2O_2). When molecular oxygen (O_2) accepts a single electron from a reducing agent, the principal species generated is $O_2^- \cdot$. The divalent reduction product of molecular oxygen is H_2O_2. $O_2^- \cdot$ and H_2O_2 participate in several chemical reactions resulting in formation of additional oxidant species (Table II). These oxygen-derived species are currently thought to have direct or indirect roles in tissue injury and modulation of the inflammatory response.

Superoxide anions can be generated from microsomal and mitochondrial electron transport systems and as by-products of enzymatic reactions such as the xanthine-xanthine oxidase reaction but these sources appear to play a minor role in phagocyte-mediated processes.[8] The NADPH oxidase system appears to be the most important source of phagocyte-derived superoxide anion. $O_2^- \cdot$ is generated following activation of a membrane-associated pyridine nucleotide oxidase which

TABLE I. Stimuli for Phagocyte-Derived Oxygen Metabolite Production

Soluble stimuli
C5a
Calcium ionophore A23187
Platelet activating factor (PAF)
Tumor necrosis factor (TNF)
N-formyl-methionyl-leucyl-phenylalanine (FMLP)
Soluble immune complexes
Phorbol myristate acetate (PMA)
Particulate stimuli
Bacteria (opsonized, heat-killed)
Opsonized zymosan particles
Insoluble immune complexes
Immunoglobulin G (heat-aggregated)
Bacterial endotoxin

TABLE II. Reactions Resulting in Formation of Oxygen-Derived Metabolites

1.	Reduction of molecular oxygen
	$O_2 + e^- \longrightarrow O_2^- \cdot$ (superoxide anion)
2.	Dismutation of superoxide anion
	$2O_2^- \cdot + 2H^+ \longrightarrow H_2O_2 + O_2$ (or $^1O_2$) (singlet oxygen)
	$O_2^- \cdot + HO_2\cdot$ (perhydroxyl radical)[a] $\longrightarrow H_2O_2 + O_2$
3.	Haber–Weiss reaction
	$O_2^- \cdot + H_2O_2 \longrightarrow O_2$ (or $^1O_2$) $+ HO^- + HO\cdot$ (hydroxyl radical)
4.	Fenton reaction
	$O_2^- \cdot + Fe^{3+} \longrightarrow Fe^{2+} + O_2$
	$Fe^{2+} + H_2O_2 \longrightarrow Fe^{3+} + HO^- + HO\cdot$
	$O_2^- \cdot + H_2O_2 \rightarrow O_2 + HO^- + HO\cdot$
5.	Myeloperoxidase (MPO)—hydrogen peroxide–halide system
	$H_2O_2 + Cl^- \xrightarrow{MPO} H_2O + HOCl$ (hydrochlorous acid)

[a]In aqueous media, the perhydroxyl radical exists in equilibrium with $O_2^- \cdot$. Formation of this species is favored by decrease in pH toward 4.8.

can utilize either NADH or NADPH. The components and assembly of this enzyme system are unclear, but several studies suggest that this enzyme system functions as an electron-transport chain containing a flavoprotein, ubiquinone, a *b*-cytochrome, and a hydrophobic oxidase.[9,10] Oxidation of pyridine nucleotides (NADH or NADPH) accompanies reduction of molecular oxygen to produce superoxide anion by the following reaction:

$$2O_2 + NAD(P)H \longrightarrow 2O_2^- \cdot + NADP^+ + H^+$$

By definition, superoxide anion ($O_2^- \cdot$) is a free radical, since it contains an extra unpaired electron. Because of the general reactivity of free radicals, O_2^-PD was considered a likely mediator of phagocyte-dependent cytotoxicity. In recent years it has become clear that although $O_2^- \cdot$ can mediate a variety of chemical reactions. it is not particularly reactive in aqueous media.[11] In the aqueous medium of biological systems, $O_2^- \cdot$ undergoes spontaneous dismutation as noted by the following equivalent reactions:

$$O_2^- \cdot + O_2^- \cdot + 2H^+ \longrightarrow H_2O_2 + O_2$$

or

$$O_2^- \cdot + HO_2\cdot \longrightarrow H_2O_2 + O_2$$

This reaction is accelerated by a decrease in pH to 4.8, where $O_2^- \cdot$ and $HO_2\cdot$ (perhydroxyl radical, the pronated form of $O_2^- \cdot$) concentrations are equal.[12] The dismutation of $O_2^- \cdot$ can also be accelerated by a group of enzymes called superoxide dismutases.[13] In activated neutrophils, approximately 90% of the consumed oxygen is reduced to form $O_2^- \cdot$.[14] In turn, nearly 80% of the H_2O_2 produced by phagocytic cells is derived from $O_2^- \cdot$ via dismutation reactions. $O_2^- \cdot$ and H_2O_2 production by neutrophils are increased when the cells have been pretreated (primed) with a different agonist, when the cells have been allowed to adhere to a surface prior to stimulation, or in some cases, when the cells have been treated with cytochalasin B.[15] Despite the limited reactivity of $O_2^- \cdot$ in biological conditions, its capacity to dismutate into H_2O_2, either spontaneously or aided by the catalytic influence of superoxide dismutase, makes it an important intermediate among the oxygen-derived metabolites of inflammation.

In contrast to $O_2^- \cdot$, hydrogen peroxide is a relatively potent oxidant. Approximately four fifths of the H_2O_2 generated by human neutrophils arises from the dismutation of $O_2^- \cdot$.[14] It should be noted that there are experimental systems in which this stoichiometric relationship between $O_2^- \cdot$ and H_2O_2 cannot be demonstrated. Glucose oxidase directly reduces O_2 to H_2O_2 without production of $O_2^- \cdot$.[12] In addition, some macrophage cell lines produce $O_2^- \cdot$ in the absence of detectable H_2O_2.[16] In addition to direct effects as an oxidant, hydrogen peroxide can participate in several chemical reactions resulting in the formation of additional reactive metabolites (Table III).

Singlet oxygen ($^1O_2$) is formed when one of the two unpaired orbital electrons in molecular oxygen absorbs enough energy to provoke spin inversion.[17] It is difficult to quantify production of $^1O_2$ in biological systems. The assays that have been used to detect $^1O_2$ are indirect and employ radical scavengers or electron-spin traps. Singlet oxygen reacts with molecules that are electron-rich or contain unsaturated double bonds. Singlet oxygen may be responsible for some chemiluminescent reactions that occur in biological systems, but the mechanisms are unclear. Hydrogen peroxide–peroxidase–halide–derived oxidants such as OCl^- can also react with H_2O_2 to produce singlet oxygen, but there is no convincing evidence that this occurs within phagocytes.[18]

TABLE III. Formation of Oxidants from Hydrogen Peroxide

1. Haber–Weiss reaction
2. Fenton reaction
3. Myeloperoxidase (MPO)–hydrogen peroxide–halide system

$$H_2O_2 + Cl^- \xrightarrow{MPO} H_2O + OCl^- \text{ (hydrochlorous acid)}$$
$$OCl^- + H_2O_2 \rightarrow H_2O + Cl^- + {}^1O_2 \text{ (singlet oxygen)}$$
$$OCl^- + RNH_2 \rightarrow RNHCl \text{ (monochloramine)}^a + HO^-$$
$$OCl^- + RNHCl \rightarrow RNCl_2 \text{ (dichloramine)}^a + HO^-$$

[a] Monochloramines and dichloramines are long-lived, hydrophilic nitrogen–chlorine derivatives thought to modulate inflammatory reactions by reacting with a variety of biological molecules.

The hydroxyl radical (OH·) is a potent oxidizing agent.[19] The bulk of evidence suggesting that phagocytes or phagocyte-derived oxygen metabolites can generate hydroxyl radicals (OH·) is indirect. Beauchamp and Fridovich described the production of ethylene gas from methional when the latter was incubated with xanthine and xanthine oxidase.[20] These investigators demonstrated that neither $O_2^-\cdot$ alone nor H_2O_2 alone (products of the xanthine–xanthine oxidase system) could oxidize methional. They suggested that these two oxygen-derived species reacted via the Haber–Weiss reaction (Table II) to form ·OH. This hypothesis was supported by the ability of OH· scavengers to retard ethylene formation. The Fenton reaction (see Table II) was invoked to explain the generation of OH· by $O_2^-\cdot$ and H_2O_2 when it became apparent that the Haber–Weiss reaction occurred too slowly to account for the OH· or OH·-like oxidant production observed in a variety of cell-free systems.[21] The Haber–Weiss and Fenton reactions are identical net chemical reactions, but the later requires the presence of a transition metal to occur. There is emerging evidence that iron is the most important transition metal involved in in vivo reactions. Iron, in conjunction with several biological compounds (e.g., ferritin, transferrin or lactoferrin), can catalyze the formation of hydroxyl radicals.[22,23] Methods employing quantification of oxidized exogenous compounds have been used to assay OH· generation by phagocytes. Neutrophils can oxidize thioethers, dimethylsulfoxide and benzoate, but the mechanisms are unclear.[24–26] In several systems, oxidation of these three substances requires myeloperoxidase and can be blocked by myeloperoxidase inhibitors.[24–27] A more specific myeloperoxidase-independent technique to detect OH· generation by phagocytes has been described. Interaction of OH· with 5,5′-dimethyl-1-pyrroline-*N*-oxide (DMPO) results in the formation of a semistable free radical that can be detected by a characteristic electron-spin resonance (ESR) spectrum.[28,29] Although the mechanisms of OH· generation by phagocytes are unclear, it appears that phagocytic cells can produce OH· or OH·-like compounds by a myeloperoxidase-independent mechanism.

A number of stimuli cause the release of myeloperoxidase into phagocytic vacuoles or the extracellular surroundings. Under appropriate conditions, myeloperoxidase can catalyze the oxidation of halides by H_2O_2, resulting in the generation of hypohalous acids (HOCl, HOI, HOBr), which can oxidize biological molecules. Target molecules can also be directly halogenated by phagocytes which release sufficient quantities of H_2O_2 and peroxidase. In vivo, Cl^- is found in substantially greater concentrations that I^- or Br^-, although phagocytes appear to have access to sufficient quantities I^- or Br^- for utilization by peroxidase.[30]

In addition to the intensively studied oxidative capabilities of hypochlorous acid, it can react with other phagocyte-derived oxygen metabolites to form addition reactive products[31] (Table III). The precise chemical mechanisms by which these oxygen-derived metabolites inflict cell and tissue injury are complex. There is experimental evidence that suggests that lipid peroxidation reactions involving cell membranes are important. Changes in membrane function have been described and a variety of peroxidation by-products have been detected and quantitated.[32] Although measurements of lipid peroxidation products such as conjugated dienes, malonyl dialdehyde, fluorescent lipid products, and volatile hydrocarbons have been employed to quantitate oxidant-mediated injury,[33–35] it should be emphasized that much of this evidence is indirect and controversy persists. Reactive oxygen-derived metabolites have also been shown to modify carbohydrates and proteins. Protein polymerization may occur secondary to incorporation of fragments of lipid oxidation products derived from the lipid peroxides or as the result of a direct reaction with free radicals.[36,37] In another system, hydrogen peroxide has been shown to oxidize amino acid residues in a target protein, hemoglobin, thereby rendering it more susceptible to subsequent proteolytic degradation.[38] Oxygen-derived free radicals have also been shown to damage carbohydrates by an iron-dependent mechanism.[39,40]

It is apparent from the foregoing discussion that a large armamentarium of oxygen-derived metabolites can be produced by phagocytic inflammatory cells. In turn, there are several well characterized antioxidant defense systems which protect the host from the untoward effects of oxidants.

3. HOST ANTIOXIDANT DEFENSE MECHANISMS

The toxic properties of oxygen-derived radicals and their metabolites are counterbalanced by an elaborate antioxidant defense system. A variety of nonbiological compounds such as dimethyl sulfoxide (DMSO), dimethyl-thiourea (DMTU) and desferoxamine, have proved useful probes in studies of oxidant-mediated tissue injury, because of their relatively specific antioxidant functions. Physiological antioxidant defense systems can be divided into two categories: intracellular enzymes and extracellular scavengers of oxygen radical electrons (Table IV). Superoxide dismutase accelerates the conversion of $O_2^-\cdot$ into H_2O_2, while catalase catalyzes the degradation of H_2O_2 into O_2 and water. Glutathione (GSH) peroxidase and reductase are complementary enzymes that function to maintain stable levels of GSH within the cytosol. If this enzyme system is impaired (as in the case of selenium deficiency or by inhibitors of other enzymes that influence GSH levels) the resulting decrease in GSH levels results in increased susceptibility of the cells to oxidant injury.[41] Since these antioxidant enzymes are cytosolic, they may have limited ability to protect cells against toxic oxygen-derived metabolites diffusing toward them from the outside. Extracellular substances that have been shown to have oxygen radical scavenging potential include ascorbic acid, vitamin E (α-tocopherol) and ceruloplasmin.[42,43] A variety of amino acids and carbohydrates can also function as oxygen radical scavengers, but only at high concentrations.

4. IN VITRO EVIDENCE FOR PHAGOCYTE-DERIVED OXYGEN METABOLITE-MEDIATED CELL AND TISSUE INJURY

Numerous in vitro studies have shown that phagocyte-derived oxygen metabolites can damage eukaryotic cells and various biologically important molecules.[44–53] Different oxygen-derived metabolites, including $O_2^-\cdot$, H_2O_2, $OH\cdot$, HOCl, and

TABLE IV. Physiological Antioxidant Defense Systems

Intracellular enzymes
Superoxide dismutase
Catalase
Glutathione peroxidase/reductase
Extracellular antioxidants
Ascorbic acid
Ceruloplasmin
α-Tocopherol (vitamin E)
Amino acids
Carbohydrates

$^1O_2$, have been implicated in these inflammatory processes. In vitro studies using specific oxidant-inactivating enzymes such as superoxide dismutase and catalase, scavengers such as DMSO or DMTU, or MPO inhibitors such as azide, suggest that products of the myeloperoxidase–H_2O_2–halide system or products of $O_2^-\cdot$–H_2O_2 reactions are the most powerful cytolytic species. Despite these observations, it is apparent that cells vary in their susceptibility to specific oxidant species. For example, endothelial cells and red blood cells (RBC) appear to be lysed by $O_2^-\cdot$ or H_2O_2 (or their reaction products) independently of myeloperoxidase. Despite relatively clear-cut data in some systems, quantitative inflammatory potencies cannot be assigned to each oxygen-derived metabolite. The ability of activated phagocytes to injure specific cells or damage biomolecules appears to be dependent on various factors, such as the species of target cell or molecule, effector cell, activating stimulus, physical relationship of mediator substance(s) and target, and modulating influences such as antioxidant defenses or other inflammatory mediators.

Keeping in mind that oxidant-mediated cell and tissue injury have been studied in diverse in vitro model systems, this chapter examines data from some of the in vitro studies of endothelial cell injury. Additional in vitro studies have provided insights into regulatory mechanisms that appear to play a role in oxygen metabolite mediated tissue injury.

More than 10 years ago, it was proposed that the acute lung injury which periodically occurs in hemodialysis, burn or trauma patients is the result of activation of plasma mediator systems (including complement) and subsequent sequestration of neutrophils within pulmonary capillaries.[54,55] There are considerable in vitro data suggesting that both oxygen-derived metabolites and nonoxidant mechanisms may be important in endothelial cell injury.

It must be emphasized that endothelial cell injury may be reversible or irreversible and is defined by the particular assay system. Sacks et al.[44] showed that human neutrophils, in the presence of serum fractions enriched for C5a anaphylatoxin cause endothelial cell lysis. Endothelial cell lysis induced by activated neutrophils can be blocked by the addition of catalase alone or together with SOD, but not consistently by SOD alone. When xanthine and xanthine oxidase (an enzymatic system which produces $O_2^- \cdot$ and H_2O_2) are added instead of neutrophils, similar results are observed. These data suggest that either H_2O_2 or metabolites derived from H_2O_2 mediate lysis of endothelial cells. The inconsistent protective effects of SOD may be explained by the fact that it catalyzes the conversion of $O_2^- \cdot$ to H_2O_2. Superoxide anion may participate not only as a precursor for H_2O_2 but also as an electron donor for the reduction of Fe^{3+} to Fe^{2+}, allowing a Fenton reaction to occur (see Table II). Endothelial cell lysis also occurred following the addition of opsonized zymosan particles to neutrophils. Supernatants from neutrophil–endothelial cell culture preparations in the presence of activated serum, purified C5a, and opsonized zymosan particles were assayed for myeloperoxidase and lysozyme activities. Significant enzyme release occurred only in neutrophil–endothelial cell culture preparations incubated with opsonized zymosan. These studies suggest that endothelial cell lysis does not require the release of measurable amounts of granule proteases.

Using a similar system, Weiss et al.[56] showed that neutrophils stimulated with phorbol myristate acetate (PMA) can lyse endothelial cells obtained from human umbilical veins. Neutrophils derived from patients with chronic granulomatous disease (CGD) neutrophils (unable to produce oxygen-derived metabolites) do not lyse endothelial cells, again suggesting an important role for H_2O_2 or an H_2O_2-derived metabolite. Weiss et al. also demonstrated that hypochlorous acid scavengers and myeloperoxidase inhibitors (azide and cyanide) do not prevent lysis of endothelial cells incubated with PMA-stimulated neutrophils. Other in vitro studies suggest that the H_2O_2 metabolite, OH·, has an important role in endothelial cell killing.

Varani et al.[57] showed that neutrophils incubated with PMA, immune complexes or opsonized zymosan mediate catalase-suppressible endothelial cell lysis. Incubation of neutrophil-endothelial cell preparations with *N*-formyl-methionyl-leucyl-phenylalanine (FMLP) or platelet activating factor does not result in lysis of endothelial cells even though these agonists are effective secretagogues for lysosomal enzymes. Endothelial cell injury produced by PMA-stimulated neutrophils can be prevented by addition of OH· scavengers (*N,N*-dimethylthiourea and D-mannitol) or deferoxamine mesylate (an iron chelator) but not by the addition of iron-saturated deferoxamine. These data suggest that neutrophil-derived OH· may be involved in endothelial cytolysis. The data also suggest that iron is important in neutrophil-mediated endothelial cell lysis, since chelation of iron by deferoxamine would prevent formation of OH· via the Fenton reaction. While these studies implicate neutrophil-derived H_2O_2 (and its metabolite, OH·) in oxygen radical mediated endothelial cell injury, there is also evidence that nonoxidant mechanisms may be important.

Harlan et al.[58] showed that human neutrophils incubated with opsonized zymosan results in detachment of endothelial cells from a culture monolayer without causing cell lysis. The addition of reagent H_2O_2 to endothelial cell cultures causes detachment and cell lysis, while addition of catalase or SOD prevents only lysis. CGD neutrophils also cause detachment of endothelial cells suggesting an oxygen-metabolite-independent mechanism. Because endothelial cell detachment can be mediated by cell-free postsecretory media from zymosan stimulated neutrophils and by partially purified lysosomal granule components, and because detachment can be blocked by neutral protease inhibitors, it has been proposed that the activated neutrophils secrete elastase-like enzymes.

In vitro studies using cultured human microvascular endothelial cells have shown that pretreatment of neutrophils with lipopolysaccharide, followed by stimulation with C5a or FMLP results in endothelial cell injury.[59] In this model, the time course of injury is similar to that produced by the addition of human neutrophil elastase and unlike that produced by the addition of H_2O_2. Neutrophil-mediated injury is not blocked by oxygen radical scavengers but can be largely prevented by methoxysuccinyl-alanyl-alanyl-prolyl-valyl-chloromethyl ketone, a specific elastase inhibitor. It should be emphasized that these neutrophil mediated mechanisms of disrupting endothelium are not mutually exclusive, since C5a and PMA are weak primary granule secretagogues in neutrophils. In addition, serum was present in the H_2O_2-dependent models described by Sacks et al.[44] and by Weiss et al.,[56] while protease-mediated endothelial cell detachment can be blocked with serum. These studies emphasize the diverse array of mediators and mechanisms which appear to be involved in neutrophil-mediated endothelial cells injury.

Oxygen metabolites and leukocytic proteases appear to participate in numerous chemical reactions involving cellular and noncellular targets. Oxygen radical mediated lysis of a variety of eukaryotic cells including endothelial cells has been well documented.[60] Reactions directly attributed to toxic oxygen metabolites include lipid peroxidation, covalent crosslinking of proteins via amino acid or sulfide moieties, and DNA damage. Synergism between leukocytic proteases and oxidants has been demonstrated in several in vitro models which have direct relevance to tissue injury. Neutrophil derived oxidants can activate latent neutrophil collagenase and gelatinase, which in turn can digest connective tissue proteins.[61,62] Exposure of fibronectin, a basement membrane protein, to H_2O_2 results in increased susceptibility to subsequent proteolytic degradation.[38] Schraufstätter et al.[63] showed breakage of DNA strands in rabbit pulmonary endothelial cells following exposure to millimolar concentrations of H_2O_2. These workers postulate that H_2O_2 may mediate endothelial cell injury (DNA damage and cytolysis) through activation of poly-ADP-ribose polymerase and subsequent intracellular depletion of NAD and ATP.[64] While many facets of this process are uncertain, this model allows detailed analysis of a possible lytic mechanism. It is also clear that this is only one potential mechanism of oxidant-induced cytolysis since red blood cells (which are subject to oxidant damage by neutrophils) do not contain poly ADP ribose polymerase.

Additional in vitro studies have shown that oxygen metabolites can mediate proinflammatory events unrelated to lysis of target cells or degradation of structural proteins. A heat-labile albumin-bound chemotactic lipid can be generated from arachidonic acid following incubation with an $O_2^{-}\cdot$ and H_2O_2-generating system.[65,66] This oxidant-derived lipid is distinct from arachidonic acid and is active at nanomolar concentrations.

Another mechanism through which oxygen metabolites appear to amplify inflammatory response is by inactivation of serum protease inhibitors. In vitro studies have shown that oxygen metabolites generated by immune complex-stimulated human neutrophils can inhibit α_1-antiprotease activity.[67,68] Phagocyte-generated hydroxyl radicals oxidize methionyl residues within the antiprotease molecules, rendering them inactive.

These studies illustrate the diversity of mechanisms by which phagocyte-derived oxyradicals may mediate tissue injury. There is substantial evidence that oxygen-derived metabolites, sometimes in conjunction with lysosomal enzymes, damage both cellular and stromal tissue components and that oxyradicals also modulate the inflammatory response by altering biochemical mediators such as chemotactic lipids and α_1-antiprotease.

5. IN VIVO MODELS OF OXYGEN METABOLITE-MEDIATED TISSUE INJURY

Several in vivo models of tissue injury have been developed to delineate the role of toxic oxygen-derived metabolites in the inflammatory response. Much of the evidence implicating oxyradicals in tissue injury has hinged on studies in which specific antioxidant enzymes and chemical have been used to diminish quantifiable tissue damage. Reduction of carrageenan-induced pleural inflammation and footpad edema by pretreatment of animals with SOD are early examples in which this type of specific pharmacologic intervention was used to implicate oxygen radicals (O_2^-RPD) in tissue injury.[69,70] Johnson et al.[71] characterized two models of acute lung injury in which inhibitor studies have provided evidence that $O_2^- \cdot$ contributes to the development of tissue damage. In the first model, Johnson et al.[71] demonstrated that lung injury resulting from intratracheal instillation of xanthine and xanthine oxidase could be blocked by the addition of SOD to the reaction mixture. Xanthine oxidase, in the presence of xanthine, generates $O_2^- \cdot$ and H_2O_2. Lung injury is quantitated by counting the accumulation of ^{125}I-labeled bovine serum albumin ([^{125}I]-BSA) in the lungs. The total lung content of [^{125}I]-BSA, after flushing blood from the vasculature, serves as an index of lung injury since it reflects increased alveolar–capillary permeability and leakage of the radioisotope-labeled protein from the vascular space into adjacent tissue. Acute lung injury can also be elicited by IgG immune complexes which form in situ at the alveolar–capillary interface. This model is complement and neutrophil dependent and resembles the reverse passive Arthus reaction in the skin[72]; 4–6 hr after intratracheal administration of antibody and intravenous injection of antigen, immune complexes and complement components can be demonstrated in alveolar septae by immunofluorescence. In both of these lung injury models, treatment with SOD decreases the permeability increases and appears to decrease the number of neutrophils that accumulate at the site of inflammation. In the IgG immune complex model, SOD-mediated inhibition of injury decreases with time.[3] These studies suggest that $O_2^- \cdot$

can contribute to development of tissue injury by increasing the influx of neutrophils into the inflammatory site. This conclusion is supported by the previously discussed in vitro studies that show that $O_2^- \cdot$ is capable of generating chemotactic factors.[65,66]

In contrast to $O_2^- \cdot$, which appears to contribute to proinflammatory reactions as a precursor to other oxidants and as participant in chemotactic lipid formation, in vivo studies suggest that H_2O_2 or its conversion products can directly mediate tissue damage. Analogous to the inactivation of $O_2^- \cdot$ by SOD, much of the in vivo data suggesting a role for H_2O_2 in tissue injury is derived from the specific ability of catalase to inactivate H_2O_2 by catalyzing its conversion into water and molecular oxygen. Among lung injury models that have been studied in our laboratory, the most direct example of H_2O_2-mediated tissue injury involves the use of glucose and glucose oxidase, an enzyme-substrate system that generates H_2O_2.[71] Glucose and glucose oxidase are instilled directly into the tracheas of rats. After 3–4 hr, lung injury is quantitated by indexing permeability changes as described previously. Similar to the lung injury induced by xanthine–xanthine oxidase and by IgG-immune complexes, there is alveolar edema and fibrin deposition as well as dense neutrophil infiltration. Lung injury can be markedly suppressed by catalase but not by SOD. The specificity of H_2O_2 generation by the glucose–glucose oxidase system and the ability of catalase to suppress this injury provide convincing evidence that H_2O_2 is integral to the development of tissue damage.

The presence of lactoperoxidase or myeloperoxidase with glucose and glucose oxidase (which would result in production of hypoiodous and hypochlorous acids, respectively) results in an accentuated form of lung damage whereas horseradish peroxidase (which does not have a halide requirement and does not produce hypohalous acids) does not enhance the injury.[71] These data are consistent with the conclusion that hypohalous acids from H_2O_2 are toxic to the lung parenchyma. In this case, extensive parenchymal damage and interstitial fibrosis are the outcomes.

Additional in vivo evidence implicating H_2O_2 in lung injury comes from the IgG-immune complex model described above. Administration of increasing quantities of catalase results in up to 80% suppression of lung injury. Morphologically, there is extensive neutrophil infiltration but little edema, hemorrhage, or cell death. While these studies provide convincing evidence that H_2O_2 is critical to the full development of oxidant-mediated IgG-immune complex lung injury, it is less clear whether the toxic metabolite is H_2O_2 or substances derived from H_2O_2, such as $OH\cdot$, HOCl, or other less well characterized metabolites.

In contrast to immune complex lung injury. in which the role of H_2O_2 metabolites is unclear, there is in vivo evidence that $OH\cdot$ may mediate acute microvascular injury of the lung. Bolus infusion of cobra venom factor (CVF), a potent complement activator, results in the rapid development of acute lung injury.[73] In this model, infusion of CVF results in complement activation, appearance of C5a, and the activation of neutrophils. Activated neutrophils become adherent to capillary endothelium of the lung, resulting in sequestration of neutrophils within the lung and neutropenia. The essential role of C5 has been demonstrated by the inability of

CVF to produce acute lung injury in mice that are genetically deficient of C5, in contrast to lung damage induced in congenic mice that are C5 sufficient.[74] The requirement for neutrophils in the CVF model of lung injury has been established by neutrophil depletion procedures. CVF-induced lung injury involves destruction of endothelial cells in the interstitial capillaries, and the appearance of gaps in the endothelial lining of interstitial capillaries. The result of endothelial injury and gap formation is interstitial edema and alveolar hemorrhage. Ward et al.[75] provided in vivo evidence that CVF-triggered lung injury may be mediated at least in part by OH· formation. Pretreatment of CVF-injected rats with iron chelators (apolactoferrin and deferoxamine) or dimethyl-sulfoxide (an OH· scavenger) results in reduction in lung injury. Pretreatment of CVF injected animals with iron-saturated lactoferrin does not prevent lung injury and addition of iron, especially Fe^{3+}, increases lung injury. These data corroborate the in vitro studies which suggest that OH· plays a role in neutrophil-mediated endothelial cell injury. Suppression of CVF-triggered lung injury with iron chelators or catalase in conjunction with the injury enhancing effect of supplemental Fe^{3+} again suggests an important role for OH· generated from H_2O_2 via the Fenton reaction (see Table II).

In the in vivo models described above, tissue injury appears to be mediated primarily by oxygen metabolites produced by neutrophils. Recently, we characterized an acute lung injury model in which tissue injury appears to be mediated by oxyradical derived from macrophages. When IgA-immune complexes are formed in rat lungs, injury occurs that morphologically is associated with increased numbers of alveolar macrophages.[76] When lung cells are retrieved by bronchoalveolar lavage, the alveolar macrophages show evidence of activation by their increased spontaneous generation of $O_2^-\cdot$ and their hyperresponsiveness following stimulation with PMA.[77] Full development of IgA-induced lung injury requires complement but still occurs after neutrophil depletion. Catalase, HO· scavengers such as dimethyl sulfoxide and dimethylthiourea, and deferroxamine are protective. Taken together, the data suggest that HO· may be a key toxic product involved in the tissue injury and that these oxygen derivatives are probably generated from lung macrophages.

6. SUMMARY AND CONCLUSIONS

These in vivo studies provide compelling evidence that phagocyte-derived oxygen metabolites contribute substantially to the development of tissue injury. It should be emphasized that while the data discussed in this section underscore the variety of mechanisms that appear to be involved in oxidant mediated tissue injury, considerable work has also been carried out in other organ systems.

The large body of in vitro and in vivo data has provided insight into the pathogenesis of tissue injury. Despite these advances, many questions remain to be answered before an integrated and detailed picture emerges.

REFERENCES

1. Senior RM, Tegner H, Kuhn C, et al: The induction of pulmonary emphysema with human leukocyte elastase. *Am Rev Respir Dis* **116:**469–475, 1977.
2. Janoff A, Sloan B, Weinbaum G, et al: Experimental emphysema produced by purified human neutrophil elastase: Tissue localization of the instilled protease. *Am Rev Respir Dis* **115:**461–478, 1977.
3. Johnson KJ, Ward PA: Role of oxygen metabolites in immune complex injury of lung. *J Immunol* **126:**1365–2369, 1981.
4. Cochrane CG, Janoff A: The Arthus reaction: A model of neutrophil, complement-mediated injury, in Zweifach B, Grant L, McClusky (eds): *The Inflammatory Process.* ed 2. New York, Academic, 1974, vol 3, p 85.
5. Johnson KJ, Varani J, Oliver J, et al: Immunologic vasculitis in beige mice with deficiency of leukocytic neutral protease. *J Immunol* **122:**1807–1812, 1979.
6. Babior BM: Oxygen-dependent microbial killing by phaogycytes. *N Engl J Med* **298:**659–668, 721–725, 1978.
7. Klebanoff SJ: Oxygen metabolism and the toxic properties of phagocytes. *Ann Intern Med* **93:**480–490, 1980.
8. Fantone JC, Ward PA: *Oxygen-Derived Radicals and Their Metabolites: Relationship to Tissue Injury.* Kalamazoo, Michigan, The UpJohn Company, 1981.
9. Gabig TG, Lefker BA: Catalytic properties of the resolved flavoprotein and cytochrome B components of the NADPH dependent $O_2^-\cdot$ generating oxidase from human neutrophils. *Biochem Biophys Res Commun* **118:**430–436, 1984.
10. Crawford DR, Schneider DL: Ubiquinone content and respiratory burst activity of latex-filled phagolysosomes isolated from human neutrophils and evidence for the probable involvement of a third granule. *J Biol Chem* **258:**5363–5367, 1983.
11. Valentine JS: The chemical reactivity of superoxide anion in aprotic versus protic media: A review, in Caughey WS (ed): *Biochemical and Clinical Aspects of oxygen.* New York, Academic, 1979, p 659.
12. Fridovich I: Oxygen radicals, hydrogen peroxide and oxygen toxicity, in Pryor WA (ed): *Free Radicals in Biology.* New York, Academic, 1976, vol I, p 239.
13. Fridovich I: The biology of oxygen radicals. *Science* **201:**875–880, 1978.
14. Root RK, Metcalf JA: H_2O_2 release from human granulocytes during phagocytosis: Relationship to superoxide anion formation and cellular catabolism of H_2O_2: Studies with normal and cytochalasin B-treated cells. *J Clin Invest* **60:**1266–1279, 1977.
15. Dahinden CA, Fehr J, Hugli T: Role of cell surface contact in the kinetics of superoxide production by granulocytes. *J Clin Invest* **72:**113–121, 1983.
16. Nathan CF, Brukner LH, Silverstein SC, et al: Extracellular cytolysis by activated macrophages and granulocytes. I. Pharmacologic triggering of effector cells and the release of hydrogen peroxide. *J Exp Med* **149:**84–99, 1979.
17. Koppenol WH: Reactions involving singlet oxygen and superoxide anion. *Nature (Lond)* **262:**420–421, 1976.
18. Foote CS: Detection of singlet oxygen in complex systems: A critique, in Caughey WS (ed): *Biochemical and Clinical Aspects of Oxygen.* New York, Academic, 1979, p 603.
19. Willson RL: Hydroxyl radicals and biological damage *in vitro:* What relevance *in vivo?,* in *Oxygen Free Radicals and Tissue Damage,* Ciba Foundation Symposium 65, Amsterdam, Excerpta Medica, 1979, p 19.
20. Beauchamp C, Fridovich I: A mechanism for the production of ethylene from methional. *J Biol Chem* **245:**4641–4646, 1970.
21. Halliwell B, Gutteridge JM: Formation of a thiobarbituric acid reactive substance from deoxyribose in the presence of iron salts. *FEBS Lett* **128:**347–352, 1981.

22. Ambruso DR, Johnston RB: Lactoferrin enhances hydroxyl radical production by human neutrophils, neutrophil particulate fractions, and an enzymatic generating system. *J Clin Invest* **67:**352–360, 1981.
23. Rosen H, Klebanoff SJ: Role of iron and ethylenediaminetetracetic acid in the bactericidal activity of a superoxide anion-generating system. *Arch Biochem Biophys* **208:**512–519, 1981.
24. Weiss SJ, Rustagi PK, LoBuglio AF: Human granulocyte generation of hydroxyl radical. *J Exp Med* **147:**316–323, 1978.
25. Repine JE, Eaton JW, Anders MW, et al: Generation of hydroxyl radical by enzymes, chemicals and human phagocytes *in vitro*. *J Clin Invest* **64:**1642–1651, 1979.
26. Sagone AL, Decker MA, Wells RM, et al: A new method for the detection of hydroxyl radical production by phagocytic cells. *Biochim Biophys Acta* **628:**90–97, 1980.
27. Klebanoff SJ, Rosen H: Ethylene formation by polymorphonuclear leukocytes: Role of myeloperoxidase. *J Exp Med* **148:**490–506, 1978.
28. Green MR, Hill AD, Okolow-Zubkowska MJ, et al: The production of hydroxyl and superoxide radicals by stimulated human neutrophils-measurements by EPR spectroscopy. *FEBS Lett* **100:**23–26, 1979.
29. Rosen H, Klebanoff SJ: Hydroxyl radical generation by polymorphonuclear leukocytes measured by electron spin resonance spectroscopy. *J Clin Invest* **64:**1725–1729, 1979.
30. Ramsey PG, Martin T, Chi E, et al: Arming of mononuclear phagoyctes by eosinophil peroxidase bound to *Staphylococcus aureus*. *J Immunol* **128:**415–420, 1982.
31. Stelmaszynska T, Zgliczynski JM: Myeloperoxidase of human neutrophilic granulocytes as colorinating enzyme. *Eur J Biochem* **45:**305–312, 1974.
32. Mead JF: Free radical mechanisms of lipid damage and consequences for cellular membranes, in Pryor WA (ed): *Free Radicals in Biology*. New York, Academic, 1976, vol 1, p 51.
33. Fletcher BL, Dillard CJ, Tappel AL: Measurement of fluorescent lipid peroxidation products in biological systems and tissues. *Anal Biochem* **52:**1–9, 1973.
34. Riley CA, Cohen G, Lieberman M: Ethane evolution: A new index of lipid peroxidation. *Science* **183:**208–210, 1974.
35. Tappel AL, Dillard CJ: *In vivo* lipid peroxidation: Measurement via exhaled pentane and protection by vitamin E. *Fed Proc* **40:**174–178, 1981.
36. Roubul WT, Tappel AL: Polymerization of protein induced by free radical lipid peroxidation. *Arch Biochem Biophys* **173:**150–155, 1966.
37. Roubul WT, Tappel AL: Damage to protein, enzymes and amino acids by peroxidizing lipids. *Arch Biochem Biophys* **113:**5–8, 1966.
38. Fligiel SEG, Lee EC, McCoy JP, et al: Protein degradation following treatment with hydrogen peroxide. *Am J Pathol* **115:**418–425, 1984.
39. Gutteridge JM: thiobarbituric acid reactivity following iron dependent free radical damage to amino acids and carbohydrates. *FEBS Lett* **128:**343–346, 1981.
40. Halliwell B: The biological effects of the superoxide radical and its products. *Bull Eur Physiopathol Respir* **17**(suppl):21–28, 1981.
41. Doroshow JH, Locker GY, Myers CE: Enzymatic defenses of the mouse heart against reactive oxygen metabolites: Alterations produced by doxorubicin. *J Clin Invest* **65:**128–137, 1980.
42. Lucy JA: Functional and structural aspects of biological membranes: A suggested structural role of vitamin E in the control of membrane permeability and stability. *Ann NY Acad Sci* **203:**4–21, 1972.
43. Denko CW: Protective role of ceruloplasmin in inflammation. *Agents Actions* **9:**333–341, 1979.
44. Sacks T, Moldow CF, Craddock PR, et al: Oxygen radical mediated endothelial cell damage by complement-stimulated granulocytes: An *in vitro* model of immune vascular damage. *J Clin Invest* **50:**327–335, 1978.
45. Slivka A, LoBuglio AF, Weiss S: A potential role for hypochlorous acid in granulocyte mediated tumor cell cytotoxicity. *Blood* **55:**347–350, 1980.
46. Simon RH, Scoggin CH, Patterson D: Hydrogen peroxide causes the fatal injury to human fibroblasts exposed to oxygen radicals. *J Biol Chem* **256:**7181–7186, 1981.

47. Nathan CF, Brukner L, Silverstein SC, et al: Extracellular cytolysis by activated macrophages and granulocytes. *J Exp Med* **149:**100–113, 1979.
48. Weiss SJ: Neutrophil generated hydroxyl radicals destroy RBC targets. *Clin Res* **27:**466A, 1979.
49. Klebanoff SJ, Clark RA: Hemolysis and iodination of erythrocyte components by a myeloperoxidase-mediated system. *Blood* **45:**699–707, 1975.
50. Hafeman DD, Lucas ZJ: Polymorphonuclear leukocyte mediated antibody dependent cellular cytotoxicity against tumor cells: Dependent on oxygen and the respiratory burst. *J Immunol* **123:**55–62, 1979.
51. Baehner RL, Boxer LA, Allen JM, et al: Autooxidation as a basis for altered function by polymorphonuclear leukocytes. *Blood* **50:**327–335, 1977.
52. Clark RA, Klebanoff SJ: Neutrophil-mediated tumor cell cytotoxicity: Role of the peroxidase system. *J Exp Med* **141:**1442–1447, 1975.
53. Clark RA, Klebanoff SJ, Einstein AB, et al: Peroxidase–H_2O_2–halide system: Cytotoxic effect on mammalian tumor cells. *Blood* **45:**161–170, 1975.
54. Craddock PR, Fehr J, Brigham K, et al: complement and leukocyte mediated pulmonary dysfunction in hemodialysis. *N Engl J Med* **196:**769–774, 1977.
55. Rinaldo JE, Rogers RM: Adult respiratory distress syndrome: Changing concepts of lung injury and repair. *N Engl J Med* **306:**900–909, 1982.
56. Weiss SJ, Young J, LoBuglio AF, et al: Role of hydrogen peroxide in neutrophil-mediated destruction of cultured endothelial cells. *J Clin Invest* **68:**714–721, 1981.
57. Varani J, Fligiel SEG, Till GO, et al: Pulmonary endothelial cell killing by human neutrophils. Possible involvement of hydroxyl radical. *Lab Invest* **53:**656–633, 1985.
58. Harlan JM, Killen PD, Harker LA, et al: Neutrophil-mediated endothelial injury *in vitro:* Mechanism of cell detachment. *J Clin Invest* **68:**1394–1403, 1981.
59. Smedly LA, Tonnesen MG, Sandhaus RA, et al: Neutrophil-mediated injury to endothelial cells. Enhancement by endotoxin and essential role of neutrophil elastase. *J Clin Invest* **77:**1233–1243, 1986.
60. Weiss SJ: Oxygen as a weapon in the phagocytic armamentarium, in Ward PA (ed): *Immunology of Inflammation.* New York, Elsevier, 1983, p 37.
61. Weiss SJ, Peppin G, Ortiz X, et al: Oxidative autoactivation of latent collagenase by human neutrophils. *Science* **227:**747–749, 1985.
62. Peppin GJ, Weiss SJ: Activation of the endogenous metalloproteinase, gelatinase, by triggered human neutrophils. *Proc Natl Acad Sci USA* **83:**4322–4326, 1986.
63. Schraufstätter IU, Hyslop PA, Jackson J, et al: Induction of DNA strand breaks by H_2O_2 and pMA. *Fed Proc* **45:**451, 1986.
64. Schraufstätter IU, Hinshaw DB, Hyslop PA, et al: Oxidant injury of cells. DNA strand-breaks activate polyadenosine diphosphate–ribose polymerase and lead to depletion of nicotinamide adenine dinucleotide. *J Clin Invest* **77:**1312–1320, 1986.
65. Perez HD, Weksler BB, Goldstein I: Generation of a chemotactic lipid from arachidonic acid by exposure to a superoxide generating system. *Inflammation* **4:**313–328, 1980.
66. Petrone WF, English DK, Wong K, et al: Free radicals and inflammation: the superoxide dependent activation of a neutrophil chemotactic factor in plasma. *Proc Natl Acad Sci USA* **77:**1159–1163, 1980.
67. Matheson NR, Wong DS, Travis J: Enzymatic inactivation of human alpha-1-proteinase inhibitor by neutrophil myeloperoxidase. *Biochem Biophys Res Commun* **88:**402–409, 1979.
68. Carp H, Janoff A: *In vitro* suppression of serum elastase inhibitory capacity by reactive oxygen species generated by phagocytosing polymorphonuclear leukocytes. *J Clin Invest* **63:**793–797, 1979.
69. Oyanagui Y: Participation of superoxide anions at the prostaglandins phase of carrageenin foot edema. *Biochem Pharmacol* **25:**1465–1471, 1976.
70. Huber W, Saifer MG: Orgotein. the drug version of bovine Cu–An superoxide dismutase. I. A summary account of safety and pharmacology in laboratory animals, in Michelson AM, McCord J, Fridovich I (eds): *Superoxide and Superoxide Dismutases.* New York, Academic, 1977, p 517.

71. Johnson KJ, Fantone JC, Kaplan J, et al: *In vivo* damage of rats lungs by oxygen metabolites. *J Clin Invest* **67**:983–993, 1981.
72. Johnson KJ, Ward PA: Acute immunologic pulmonary alveolitis. *J Clin Invest* **54**:349–357, 1974.
73. Till GO, Johnson KJ, Kunkel R, et al: Intravascular activation of complement and acute lung injury. Dependency on neutrophils and toxic oxygen metabolites. *J Clin Invest* **69**:1126–1135, 1982.
74. Tvedten HW, Till GO, Ward PA: Mediators of lung injury in mice following systemic activation of complement. *Am J Pathol* **119**:92–100, 1985.
75. Ward PA, Till GO, Kunkel R, et al: Evidence for the role of hydroxyl radical in complement and neutrophil-dependent tissue injury. *J Clin Invest* **72**:789–801, 1983.
76. Johnson KJ, Wilson BS, Till GO, et al: Acute lung injury in rat caused by immunoglobulin A immune complexes. *J Clin Invest* **74**:358–369, 1984.
77. Johnson KJ, Ward PA, Kunkel RG, et al: Mediation of IgA induced lung injury in the rat. Role of macrophages and reactive oxygen products. *Lab Invest* **54**:499–508, 1986.

15

The Respiratory Burst of Neutrophilic Granulocytes and Its Influence on Infected Tissues

Indirect Consequences

JAN MACIEJ ZGLICZYŃSKI and
TERESA STELMASZYŃSKA

1. REACTIVE SPECIES PRODUCED DUE TO THE RESPIRATORY BURST

1.1. Triggering the Respiratory Burst

1.1.1. Steps of Activation of Neutrophilic Granulocytes

The respiratory burst of neutrophilic granulocytes is a metabolic response of these cells to signals of infection. This metabolic pathway starts with markedly increased oxygen uptake,[1,2] resulting in production of primary and secondary products, harmful to invaders.[3,4]

Such a description of the respiratory burst raises some questions and requires more explanations especially concerning sources of the signals and a possible sequence of their action.

Signals of infection may originate from bacteria bodies. The bacterial metabolism's being distinct from the host processes opens a possibility that some of its products may be recognized by neutrophils as signals of invasion. An example of such substances is one of the chemotactic *N*-formylpeptides, *N*-formylmethionyl-leucyl-phenylalanine,[5] initiating the respiratory burst of neutrophils.[6]

The second step of the activation of neutrophils may result from the interaction of microbial substances with a ubiquitous complement system, present in the host

---

JAN MACIEJ ZGLICZYŃSKI and TERESA STELMASZYŃSKA • Institute of Medical Biochemistry, Nicolaus Copernicus Academy of Medicine, 31-034 Kraków, Poland.

body fluids. Activation of the complement cascade results in the subsequent generation of the chemotactic fragments C3a and C5a, also stimulating the respiratory burst of neutrophils[7] and of opsonizing protein, C3b.[8] The complement cascade can be activated even to a higher extent via the classical pathway by bacteria coated with specific antibodies, but it is noteworthy that specific antibodies are not necessary for the information "emergency-infection."

The next step of the activation of neutrophils occurs as a consequence of a close contact of bacteria with the neutrophil membrane, followed by ingestion of the particle.[2,9] Phagocytosis may be developed by the nonspecific contact of particles with neutrophils, as is the case for hydrophobic polystyrene beads or the microbes with hydrophobic surface. The optimal phagocytosis and maximal stimulation of the respiratory burst, however, require opsonization of particles by C3b and/or specific immunoglobulins involving expression of specific receptors for C3b and Fc fragments on neutrophil membrane.[8,10,11]

Finally, activated neutrophils produce powerful chemoattractant species that can further stimulate phagocytes and enhance inflammatory response. Examples of these agents are leukotriene B_4[12] and tuftsin,[13] both augmenting the respiratory burst of neutrophils in the presence of other soluble or particulate stimuli.[14,15]

It seems that all activating agents, chosen arbitrarily in this review, should act on neutrophils in vivo in a defined time sequence.

1.1.2. Extent of Activation of the Neutrophil Respiratory Burst by Different Stimuli

Neutrophils in vivo seem to be activated gradually not only with respect to the stimuli sequence, but also with respect to the temporal pattern and size of the evoked effect. It can be expected that the metabolic response of neutrophils is triggered in a different way by chemotactic peptides, chemotactic complement fragments, and intimate contact of the cells with particulate matter.

In other words, the respiratory burst of neutrophils stimulated by chemoattractant should be shorter and lower than that resulting from final excitation of the cells—phagocytosis. Otherwise, the respiratory burst, devoted to bacteria killing, would be developed long before neutrophils come into contact with bacteria.

It is tempting to consider from this point of view data presented in papers dealing with initiation of the respiratory burst in neutrophils by various stimuli. First, it is known that neutrophils develop their ability to respond to the stimuli by increasing the availability of receptors for chemoattractants,[16,17] and immunoglobulins.[18] This occurs when neutrophils adapt to foreign environment or when they have to migrate.[16–18] Second, it was found that the respiratory burst stimulated by chemoattractant peptide is short-lasting and lower, even at optimal concentration of the stimulus, than that initiated by particulate matter or nonspecific agent phorbol myristate acetate (PMA).[19,20] It means that the chemoattractant-mediated activity is rapidly suppressed, while continuous contact of particles with neutrophil or phorbol ester persistently embedded in the neutrophil membrane give longer-lasting stimulation.

The activity of the respiratory burst system of neutrophils is thus regulated not only by different stimulating mechanisms[19] but also by deactivating processes which seem to be unique for various stimuli.

1.1.3. The Metabolic States of Neutrophils

The above considerations and available data, suggesting that the blood of patients with active bacterial infection may contain normal, activated, and deactivated neutrophils,[21,22] prompted us to examine also the metabolic (functional) states of these cells. The essential difficulty involved is the choice of an appropriate experimental procedure. It is well known that the course of isolation of purified neutrophils influences their various functions. Thus, we have realized that the metabolic activity of neutrophils measured by chemiluminescence (CL) technique in whole blood[23,24] might reflect some nuances in physiological states of the cells.

We chose the luminol-enhanced CL method, worked out by Selvaraj et al.,[23] as indirect means to detect formation of reactive oxygen species by stimulated neutrophils.[25] According to this method,[23] CL was measured in whole (diluted 1:250) blood after stimulation with polystyrene latex particles. The latex particles appear to be an appropriate stimulant of phagocytic and respiratory burst activity of neutrophils independent of the opsonic influence of serum factors.[23] This method obviates the need of cell isolation and permits measurement of activity of unchanged neutrophils quickly after venipuncture.

Using this experimental approach, four metabolic (functional) states of neutrophils in blood can be distinguished:

1. There is no spontaneous CL and low and long-lasting response after challenge with the polystyrene particles. The particle-stimulated CL is inhibited by plasma. This type of behavior may reflect the resting state of neutrophils (Fig. 1a, lower curve).
2. There is still no spontaneous CL, but the time pattern and the extent of CL changes. There is a distinct peak of the CL after the stimulation with the particles, and the CL is not inhibited by plasma. The neutrophils seem to be preactivated (or devoid of an inhibitor?). Thus, this state may be considered as a standby state (Fig. 1a, upper curve; Fig. 1b, lower curve).
3. Some, very low and long-lasting spontaneous CL is observed. The particle-stimulated CL is higher than in corresponding state 2, with shortened time lag between phagocytic stimulus and the CL initiation. This state may be defined as an activation state (Fig. 1b, upper curve).
4. The high- and moderate-duration spontaneous CL was observed, but after the particle addition, CL decreases rapidly. The neutrophils are unable to develop high and efficient respiratory burst after phagocytic stimulus and may be considered exhausted.

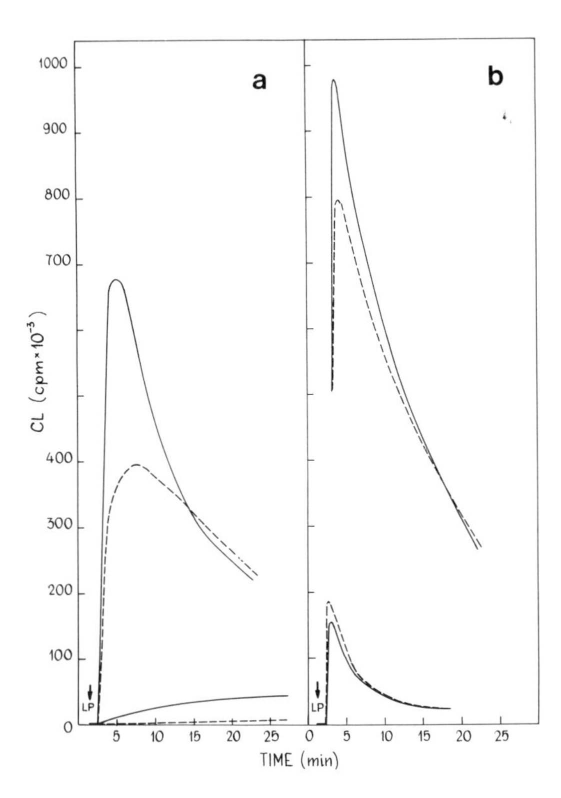

FIGURE 1. Examples of the chemiluminescence (CL) patterns of freshly drawn (a) and preincubated (b) blood for 1 hr at room temp. Lower curves from control healthy subject, upper curves from patient with purulent appendicitis. Sample composition: 2.5 ml luminol (1.4 mM), 0.8 ml 5× concentrated Krebs–Ringer solution, 1.5 ml glucose (18 mM) and 20 μl blood (———), + 20 μl autologous plasma (- - - -). To this 200 μl ($\sim 10^{10}$) latex particles (LP) was added (arrows). The light emission was measured immediately in a Packard Tri-Carb liquid scintillation spectrometer, in the IN-coincidence mode, at ambient temp. Unstimulated CL was observed only for preincubated blood from infected patient (2000 cpm). (From J. M. Zgliczyński, E. Kwasnowska, S. Olszowski, E. Olszowska, and T. Stelmaszyńska, unpublished data.)

It is noteworthy that some of the states might be observed, using the blood of one donor, depending on the blood's initial pre-treatment. The resting features of the CL response (state 1) could be observed only using freshly drawn blood from healthy subjects[26] (Fig. 1a, lower curve); 60–90 min after venipuncture, the CL pattern of blood reflected the standby behavior of neutrophils[26] (Fig. 1b, lower curve). The blood of some of the infected patients exhibited just after venipuncture the standby pattern of particle-stimulated CL that, after preincubation at room temperature, changed to the activation state (Figs. 1a,b, upper curves), with low and long-lasting spontaneous CL. The CL activity of these patients returns to normal (i.e., characteristic of the resting state) after therapy and recovery. In two cases of patients with bacteriemia who died a few days after the examination of blood, a CL pattern typical of the exhausted state was found.

It seems necessary to mention that changes in CL responsiveness of blood may be caused also by simple isolation procedures of the cells. Washing out of the blood cells with isotonic NaCl slightly enhances ability to the stimulation of CL (Fig. 2a). Hemolysis of erythrocytes by washing out of the blood with hypotonic NaCl solution (0.3%) produces the standby state (Fig. 2b) and, using isotonic NH_4Cl, can activate the cell to exhibit low spontaneous and the high, not inhibited by plasma particle-provoked CL (Fig. 2c). These observations indicate that purification of

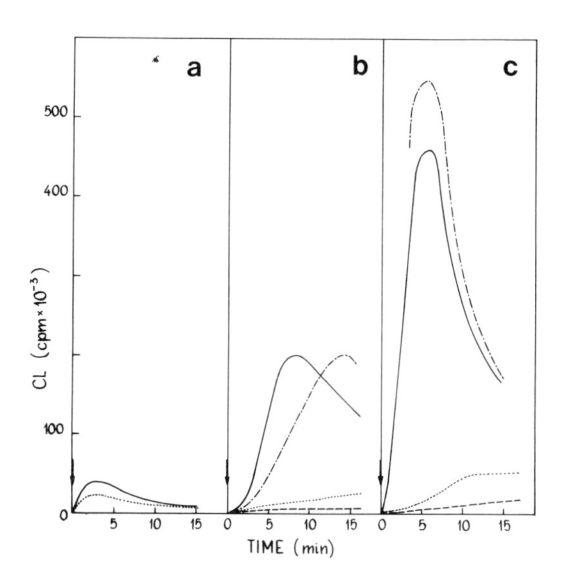

FIGURE 2. Effect of the pretreatment of blood on its chemiluminescence (CL) pattern. CL of whole blood without pretreatment (······) + 20 μl plasma (- - - - -); CL of blood pretreated with isotonic NaCl (a), hypotonic (0.3%) NaCl (b), and isotonic NH_4Cl (c) (———) + 20 μl plasma (—·—·—). Arrows indicate addition of Latex particles. Sample composition and CL measurement as described in Fig. 1. (From J. M. Zgliczyński, E. Kwasnowska, S. Olszowski, E. Olszowska, and T. Stelmaszyńska, unpublished data.)

neutrophils before evaluating their respiratory burst extent may cause their preactivation or even exhaustion.

There are at least two aspects of the effects discussed. The first is that neutrophils in circulating blood seem to be at the really dormant state, reluctant to express their function. It is beneficial for the host, since easily stimulated neutrophils would be the menace because of releasing toxic products of the respiratory burst. It is tempting to speculate that the neutrophils in the resting state are in fact inhibited by an unknown compound present in plasma of healthy subjects.

On the other hand, blood neutrophils may enhance responsiveness to stimuli by receiving earlier signal(s) of something happening. Then, initiation of the respiratory burst is easier, and its primary and secondary products may leak from the cells, acting on surrounding tissues instead of on bacteria. This may depend on the magnitude and duration of the signal(s). One may assume, not only that the neutrophils in circulating blood can be activated but also those marginated and infiltrating the tissues, even far away from infection focus. Next, to gain insight into the consequences of the respiratory burst initiation, we will survey its primary and secondary products.

1.2. Reaction Pathways Originating from the Increased Oxygen Uptake by Neutrophils

1.2.1. Superoxide Anion and Hydrogen Peroxide Formation

It is commonly accepted that the enhanced oxygen uptake triggered by the contact of the neutrophil surface with an appropriate stimulus results from the activation of the enzyme (system) catalyzing the reduction of oxygen with NADPH as an electron donor.[27–30] The oxidation of NADPH drives in turn the oxidation of glucose via the hexose monophosphate (HMP) shunt.[2,31] In this pathway, glucose is oxidized to CO_2 and a five-carbon sugar with $NADP^+$ serving as the electron acceptor. As the rate of HMP shunt is limited by the $NADP^+$/NADPH ratio, the oxidation of the reduced coenzyme by the oxidase system makes HMP go faster and produce in recycling mechanism NADPH as a substrate for the oxidase. The O_2 consumption increases to the point at which the entire process can be called the respiratory burst.

The plasma membrane-bound NADPH oxidase,[32,33] unique to phagocytes, seems to be a multicomponent system, but its structure and mechanism underlying the activation are far from being clear. It has been reported that this enzyme complex contains a *b*-type cytochrome[34,35] and flavoprotein,[36,37] but the soluble NADPH oxidase preparation obtained recently from bovine neutrophils does not contain spectroscopically detectable amounts of cytochrome or flavin.[30]

NADPH oxidase is dormant in resting cells and is stimulated by at least two transductional mechanisms[19,38] with the soluble stimuli of the neutrophils. The active enzyme catalyzes the formation of superoxide anion radical by the one electron reduction of oxygen[28]:

$$2O_2 + NADPH \rightarrow 2O_2^{-}\cdot + NADP^+ + H^+ \tag{1}$$

Superoxide radicals were detected in the medium of phagocytosing neutrophils for the first time in 1973 by Babior et al.[39] They used cytochrome *c* as an oxidant for $O_2^{-}\cdot$ and the specificity of this reaction was established by the fact that the reduction of ferricytochrome *c* could be inhibited by the addition of superoxide dismutase (SOD)[40]:

$$2O_2^{-}\cdot + 2H^+ \xrightarrow{\text{SOD}} H_2O_2 + O_2 \tag{2}$$

In neutrophils, SOD seems responsible for formation of H_2O_2 mainly in the cytosol,[41] but noncatalyzed dismutation accelerated by H^+ ions is thought to take place in the phagocytic vacuole.[42]

Some authorities believed that the spontaneous dismutation of superoxide gives rise to singlet oxygen,[43] but this point is controversial[44]:

$$HO_2 + O_2^{-}\cdot + H^+ \rightarrow {}^1O_2 + H_2O_2 \tag{3}$$

Thus, H_2O_2 is the next oxygen species produced as a result of the respiratory burst. In fact, this compound was detected in the medium of phagocytosing neutrophils[31] prior to the discovery of $O_2^{-}\cdot$ as its origin.[39] Several assays were performed to quantitate hydrogen peroxide formed during phagocytosis.[45–48] The quantities of $O_2^{-}\cdot$ and H_2O_2 estimated in surrounding medium of phagocytosing neutrophils reflect only this part of the compounds that escape from compartments of the cell in which they are formed (phagosomes).

Using cytochalasin B-treated neutrophils stimulated by bacteria, Root and Metcalf[48] demonstrated that almost all H_2O_2 produced during contact of bacteria with the cell has $O_2^{-}\cdot$ as a precursor and is formed by the spontaneous dismutation. Moreover, they found that all the oxygen taken up in the respiratory burst was reduced to $O_2^{-}\cdot$ and that about 80% of this $O_2^{-}\cdot$ was converted to H_2O_2. This evaluation was possible by employing cytochalasin B, which inhibits phagocytosis but permits the respiratory burst and almost quantitative delivery of the oxygen to extracellular medium.[48]

NADPH oxidase is most probably a component of the cell membrane.[32,33] When primary phagosome is formed from the cell membrane and the enzyme begins to operate, $O_2^{-}\cdot$ is discharged into phagocytic vacuole,[49] where it reacts directly or dismutates to H_2O_2. The simultaneous process of fusion between primary phagosome and the cell granules (degranulation) permits discharge of the granule enzymes into the phagosome. One of these enzymes, myeloperoxidase (MPO),[50] is especially abundant in the azurophil granules of neutrophils.[51] The interaction of the MPO with H_2O_2 in the presence of Cl^- ions, which are also abundant in these cells,[52,53] is responsible for the unusual features of the respiratory burst in neutrophils (and monocytes).

1.2.2. The Specialty of the Neutrophil Respiratory Burst—Hypochlorite

The neutrophils, the most potent host-defensive cells launch the respiratory burst to kill and destroy foreign agents. This process is in some way amplified and its products broadly diversified dut to the MPO, which uses moderately antibacterial products of the respiratory burst, hydrogen peroxide, to form more reactive and toxic compounds. The enzyme catalyzes oxidation of halides ($X^- = Cl^-$, Br^-, and I^-) to hypohalite ions [54]:

$$X^- + H_2O_2 \rightarrow XO^- + H_2O \qquad (4)$$

Hypohalites are oxidizing as well as halogenating species. The importance of MPO as halogenating antimicrobial agent was first reported by Klebanoff.[55] Early experiments with the MPO–H_2O_2–I^- system and with the neutrophils stimulated in the presence of I^- showed incorporation of iodine into a protein-bound form.[55]

Since chloride, not iodide, is present in the leukocytes in considerable amounts,[52,53] fulfilling the requirement of the MPO–H_2O_2–halide system to operate,[56] it is most likely that Cl^- is the physiological substrate and $HOCl/OCl^-$ species are formed in the neutrophil phagosome.

In fact, a direct proof of the hypochlorite formation in neutrophils has not been obtained. This is understandable, taking into account its high reactivity. However, as this compound has an ability to chlorinate, its transient existence in the cells has been traced in the more stable chlorinated products[57–60] (Fig. 3).

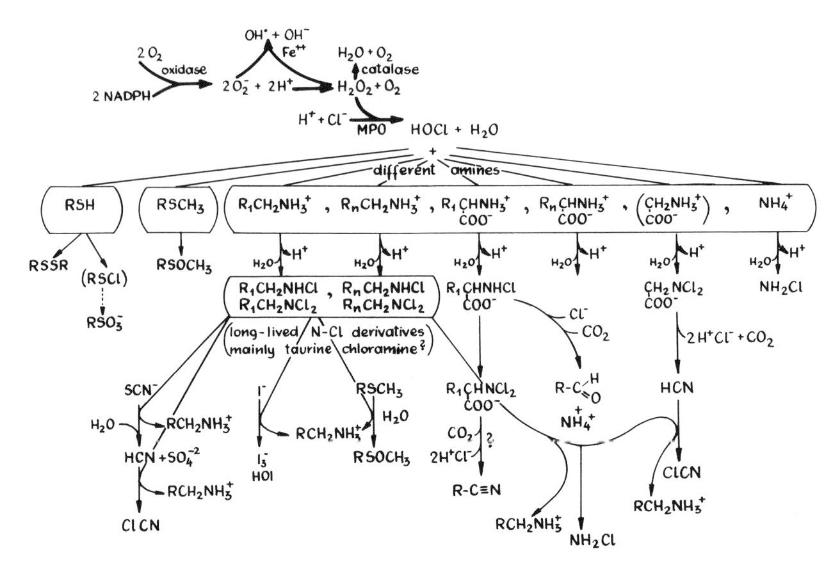

FIGURE 3. Some well-established and tentative primary and secondary products of neutrophil respiratory burst.

1.2.3. The Chloramines as Detectable Markers of Hypochlorite Formation in Neutrophils

$HOCl/OCl^-$ formed due to the action of the MPO system may react with many target compounds in the living cell.[56–59,61–71] Among them are primary and secondary amines that form relatively stable N–Cl compounds, retaining the Cl^+ moiety of hypochlorite:

$$RNH_2 + HOCl \rightarrow RNHCl + H_2O \tag{5}$$

The chlorinating MPO system may react with any amino groups available in the medium, i.e., in low-molecular-weight amines, amino acids, proteins, and macromolecular structures of bacteria.[56–59,61–68] The stability and chlorinating and oxidizing ability of the chloramines are determined by their overall structure rather than by the N–Cl moiety alone. For example, *N*-monochloro-α-amino acids undergo oxidative decarboxylation to yield aldehyde, ammonia, and CO_2[62,71]:

$$\begin{array}{c} R \\ | \\ HC\text{-}NHCl \\ | \\ COOH \end{array} \xrightarrow{H_2O} \begin{array}{c} R \\ | \\ HC{=}O \end{array} + NH_4^+ + CO_2 + Cl^- \tag{6}$$

whereas N–Cl derivatives of β-amino acids and of amines are relative stable.[56,62,66]

Leukocytes contain unusually high amount of amino acids and amines, especially taurine.[72] This latter compound deserves special attention as a target of chlorination. Although its intracellular distribution and site of chlorination are unknown, taurine chloramine is released from stimulated neutrophils.[73–75] Its accumulation in the environment proves that the rate of its formation exceeds consumption by components of the neutrophil.

Taurine seems to have a dual biological role in neutrophils. It may be regarded as a trap for toxic HOCl, thus protecting cells from its cytotoxic and cytolytic action.[75–78] On the other hand, taurine chloramine may exert prolonged oxidative and chlorinating effect long after the initiation of the respiratory burst and in some distance from a cell of origin.[73–75,77]

1.2.4. Tentative Secondary Reactions of Respiratory Burst Intermediates

The reactions starting at univalent reduction of oxygen (Fig. 3) may proceed in phagocytic vacuoles and/or in surrounding medium of stimulated neutrophils when products of the respiratory burst are released from these cells.

1.2.4a. Formation of HCN and Its Chlorination to ClCN. *N*-Monochloramines seem to be the main products of amine chlorination in the neutrophils. In some circumstances, however, dichloramines may be formed.[56,66] This occurs at moderately low pH due to dismutation of monochloramine:

$$RNHCl + RNHCl + H^+ \rightarrow RNH_3^+ + RNCl_2 \quad (7)$$

and by the extensive chlorination, when the ratio of HOCl to amine is much higher than 1:

$$RNH_2 + 2HOCl \rightarrow RNCl_2 + 2H_2O \quad (8)$$

N-Dichloroamines have higher chlorinating power, are less stable and display different reactivity as compared with that of monochloramines.[56,64,66,77] *N*-Dichloroamino acids decompose to form nitriles[79]:

$$Cl_2N-\underset{COOH}{\overset{R}{CH}} \longrightarrow R-C\equiv N + CO_2 + 2H^+ + 2Cl^- \quad (9)$$

N-Dichloroglycine and *N*-dichloroglycyl peptides and proteins with *N*-terminal dichloroglycyl residue decompose releasing HCN and CO_2[64,80]:

$$Cl_2N-CH_2-COOH \xrightarrow[2H^+ 2Cl^-]{} \left[\begin{array}{c} C\equiv N \\ | \\ COOH \end{array}\right] \rightarrow HCN + CO_2 \quad (10)$$

$$Cl_2N-CH_2-CO-NH-R \xrightarrow[2H^+ 2Cl^-]{} N\equiv C-\underset{\underset{O}{\|}}{C}-NH-R \xrightarrow{H_2O} HCN + CO_2 + RNH_2 \quad (11)$$

Reactions (10) and (11) have been shown to proceed in the cell-free MPO–H_2O_2–Cl^- system[64,80] as well as in the phagocytosing neutrophils.[81] *Staphylococcus epidermidis* bacteria were used as phagocytosed objects and the target for the enzymatic chlorination.[81] These bacteria were chosen because they contain polyglycylseryl crosslinkages in the peptidoglycan of their cell wall. Several percent of glycyl residues in the peptidoglycan are not crosslinked[82] and may serve as a substrate for chlorination.

The percentage of free amino groups of glycine is higher in mucopeptide of the bacteria grown in the presence of sublethal doses of penicillin.[83] As might be expected, penicillin pretreatment of the bacteria markedly increased efficiency of their chlorination (amount of the N–Cl groups formed) and subsequent HCN production by the MPO–H_2O_2–Cl^- system as well as in phagocytosing neutrophils.[81]

This is an interesting aspect of the interaction of neutrophils and cell-wall-active antibiotics in the bactericidal action. In fact, a cooperative effect between penicillin at low concentration and neutrophils was shown.[84] It may be assumed that *Staphylococcus* with the damaged peptidoglycan, having more free amino groups, undergoes more efficient chlorination. HCN is then formed as a flux in the close vicinity of the bacteria membrane and by itself exerts a potent bactericidal effect.

HCN originated from bacterial structures (as has been judged using [2-^{14}C]glycine-labeled bacteria) contributes only to 20% of the total HCN formation during phagocytosis. It means that there are other cyanogenic substrates in the neutrophils or in their environment.

Indeed, it was shown that the bulk of HCN released during phagocytosis originates from thiocyanate[85] (10^7 neutrophils/ml challenged by 2×10^9 bacteria in the presence of $S^{14}CN^-$ at a concentration of 50 μM produces 2–3 nmoles $H^{14}CN$).

Thiocyanate is ubiquitous in human body fluids. Its concentration ranges from 80–90 μM to 135–150 μM in plasma of nonsmokers and smokers, respectively.[86] As plasma leaks with neutrophils to inflammatory foci, the cells develop the respiratory burst usually in the presence of SCN^- in the surrounding medium.

SCN^- may be oxidized directly by the MPO–H_2O_2 system as pseudohalogen to hypothiocyanous acid[87,88] and other products.[88] The MPO–H_2O_2–Cl^- system and/or the products of its action, chloramines, oxidize SCN^- to HCN[85]:

$$SCN^- + 3RNHCl + 4H_2O \rightarrow HCN + H^+ + SO_4^{2-} + 3RNH_3^+ Cl^- \quad (12)$$

In physiological conditions, Cl^- ion concentration is about 1000 times higher than that of SCN^-. Thus, most likely SCN^- is a secondary substrate oxidized by the HOCl and/or long-lived chloramines formed by the MPO. This assumption is confirmed by the increase in HCN production by the activated neutrophils in the presence of SCN^- and taurine.[85]

It is not, however, the end of the HCN story. Although HCN is known as a potent inhibitor of hemoproteins, including MPO, chloride competes with HCN in binding to the MPO,[80,89–91] especially at low pH. Because of this competition, HCN is readily chlorinated by the MPO–H_2O_2–Cl^- system.[80,89] The extent of enzymatic chlorination of HCN depends on the relative concentrations of Cl^-, HCN, H_2O_2, and H^+.[89] In other words, HCN can be chlorinated by the MPO system even at a concentration of 2 mM, provided that Cl^-, H^+, and H_2O_2 concentrations are properly adjusted.[89] HCN is also readily chlorinated by chloramines.[92]

Formation of ClCN by the stimulated neutrophils from HCN as well as from SCN^- was shown only indirectly employing the reaction of ClCN with thiol compounds[93] (thiol groups of albumin[85]).

1.2.4b. Formation of Ammonia Chloramine (NH_2Cl). Stimulated neutrophils in the presence of NH_4^+ ions (at mM concentration) increase the rate of inactivation of their own function, and decrease O_2 uptake and glucose metabolism.[59] Thomas et al.[59] showed that NH_2Cl formed by the MPO–chlorinating system is responsible for this phenomenon.

NH_2Cl is one of the lipophilic chloramines that penetrate the cell membrane and oxidize intramembranous and intracellular components.[59,94,95] This compound is bactericidal[94] and has a multiple toxic effect on eukaryotic cells.[95]

Supply of $NH_4{}^+$ to the neutrophils is a rate-limiting step for the production of NH_2Cl.[75,95] All cells produce $NH_4{}^+$ as a result of catabolic reactions of amino compounds, but neutrophils have an additional source of $NH_4{}^+$ due to the MPO-mediated deamination and decarboxylation of *N*-monochloro-α-amino acids[71] (Fig. 3).

It is noteworthy that the frequently employed washing out of neutrophils from erythrocytes by isotonic NH_4Cl, in view of Thomas et al.[59,75,95] and our results (Fig. 2) seems to affect the neutrophils and change their function.

1.2.4c. Formation of Secondary Oxygen Species ($OH\cdot$ and $^1O_2$). Oxygen species ($O_2^-\cdot$, H_2O_2, and OCl^-) produced due to the respiratory burst may react one with another forming much more reactive oxygen products—hydroxyl radicals ($OH\cdot$) and singlet oxygen ($^1O_2$).[96,97]

Several such reactions has been observed in model experiments.[96] First, H_2O_2 may be converted by ferrous or ferric ions:

$$H_2O_2 + Fe^{2+} \rightarrow Fe^{3+} + OH\cdot + OH^- \tag{13}$$

$$H_2O_2 + Fe^{3+} \rightarrow Fe^{2+} + O_2^{-}\cdot + 2H^+ \tag{14}$$

$O_2^{-}\cdot$ and $OH\cdot$ react, yielding excited oxygen[98]:

$$O_2^{-}\cdot + OH\cdot \rightarrow {}^1O_2 + OH^- \tag{15}$$

H_2O_2 reacts with $O_2^{-}\cdot$ in the Haber–Weiss process catalyzed by transition metal ions[99]:

$$H_2O_2 + O_2^{-}\cdot \rightarrow OH\cdot + OH^- + O_2 \tag{16}$$

In reaction (16), the formation of $^1O_2$ was also proposed.[100] The main source of singlet oxygen in neutrophils, however, was supposed to be the reaction between HOCl and H_2O_2[96]:

$$HOCl + H_2O_2 \rightarrow {}^1O_2 + H_2O + Cl^- \tag{17}$$

There is also a possibility that $O_2^{-}\cdot$ reacts with HOCl[101]:

$$O_2^{-}\cdot + HOCl \rightarrow O_2 + Cl^- + OH\cdot \tag{18}$$

All these reactions may deliver oxygen species harmful to biological environment, $OH\cdot$ and $^1O_2$.

So far, reactions (13) and (16) are thought to be the most likely origins of $OH\cdot$ radicals in phagocytic cells.[97] The formation of hydroxyl radicals by neutrophils has been observed using more or less specific trapping reagents.[102–105]

There is, however, no clear evidence for the production of singlet oxygen by neutrophils.[106–108]. Moreover the reaction of HOCl with H_2O_2 (17) is rather unlikely as a source of $^1O_2$ in neutrophils because HOCl reacts preferably with amines (5), which are abundant in phagosomes or in extracellular space. It has been shown that HOCl added to an equimolar mixture of amine and H_2O_2 reacts quantitatively with amine.[109]

1.2.4d. Possible Involvement of Chloramines in Excited Oxygen Species Formation. Consumption of HOCl in the reaction with amines does not exclude, however, the formation of excited oxygen species, which are thought to be responsible for light emission (chemiluminescence, CL). Chloramines of taurine and of proteins or chlorinated whole bacteria when reacting with H_2O_2 exhibit prolonged CL[109] (Fig. 4). The N–Cl derivatives of proteins and chlorinated bacteria appeared to be especially efficient light-producing substrates, which may indicate that specific nature of local environment of excited moiety can influence its reactivity. The mechanism of CL in these reactions is unknown, but the involvement of excited oxygen ($^1O_2$) or chloramine radical(s) is not unlikely.

The CL of the chloramine–H_2O_2–system is markedly increased by luminol—

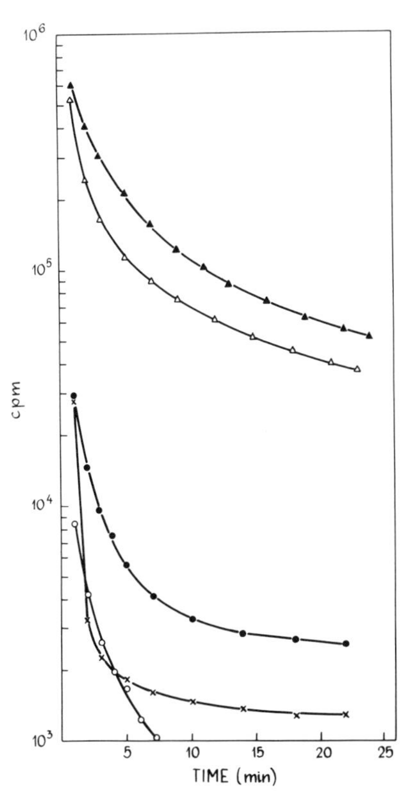

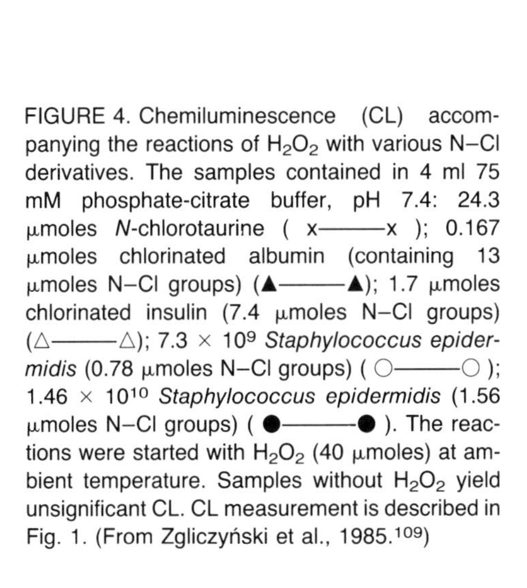
FIGURE 4. Chemiluminescence (CL) accompanying the reactions of H_2O_2 with various N–Cl derivatives. The samples contained in 4 ml 75 mM phosphate-citrate buffer, pH 7.4: 24.3 μmoles *N*-chlorotaurine (x——x); 0.167 μmoles chlorinated albumin (containing 13 μmoles N–Cl groups) (▲——▲); 1.7 μmoles chlorinated insulin (7.4 μmoles N–Cl groups) (△——△); 7.3 × 10^9 *Staphylococcus epidermidis* (0.78 μmoles N–Cl groups) (○——○); 1.46 × 10^{10} *Staphylococcus epidermidis* (1.56 μmoles N–Cl groups) (●——●). The reactions were started with H_2O_2 (40 μmoles) at ambient temperature. Samples without H_2O_2 yield unsignificant CL. CL measurement is described in Fig. 1. (From Zgliczyński et al., 1985.[109])

about three orders of magnitude—and in this feature resembles the MPO-dependent CL of stimulated neutrophils. It is known that the luminol-enhanced CL of these cells is totally dependent on MPO-mediated reactions.[110,111]

Chloramines, too, deliver light reacting with tryptophan.[112] Again, in this reaction, excited oxygen species are likely to be involved, since anaerobic conditions markedly decrease light emission. This reaction may be of biological relevance, too, as it is known that indole derivatives amplify the CL of phagocytosing neutrophils.[113]

1.3. Reactivity of the MPO-Mediated Products of the Respiratory Burst

1.3.1. General Remarks

There is no doubt that the major biological role of the respiratory burst products is to act on and kill invading bacteria, but at least in three circumstances they may act on host cells.

First, this may occur when neutrophils reach sites infiltrated by bacteria, engulf them, and discharge their granule content to the phagocytic vacuole; then, some part of the respiratory burst products and granule enzymes leaks to the extracellular space. This is true especially after the inevitable disruption of neutrophils, which occurs as a consequence of their bactericidal action.

Second, this may happen when neutrophils are excited with immune complexes, especially by large, inappropriate to phagocytose immune complex-opsonized surfaces; then the vacuoles fail to close and the primary respiratory burst products ($O_2^- \cdot$ and H_2O_2) and granule enzymes are secreted to the opsonized autologous cell or to the environment.

Third, this has been supposed to happen (see Section 1.1.3) when neutrophils due to bacterial infection or chronic inflammation are in excited state in blood or in tissues even far from the infection focus.

Reactivity of the respiratory burst products released allows them to interact with almost all vital cell components.

Many recent papers are dealing with the experimental model systems in which activated neutrophils exert cytotoxic and cytolytic effects on surrounding target cells. Some reports are emphasizing the contribution of hydroxyl radicals in the cell injury,[97,114–116] while other reports are proving action of the MPO-mediated products.[76–78,117,118] Most likely, in vivo all these compounds act simultaneously, although the MPO-mediated reactions and $OH\cdot$ generation compete for H_2O_2 [eqs. (4), (13), and (16)].

Since it is a broad field of investigations, this survey focuses on the reactivity of the MPO-mediated products.

1.3.2. Reactions with Amino Acids, Peptides, and Proteins

Deamination and decarboxylation of free amino acids through *N*-monochloramine intermediates [eq. (6)] is one of the most characteristic reactions mediated by the chlorinating MPO system.[71,119] *N*-terminal amino groups of peptides and proteins, when chlorinated, yield more stable *N*-monochloramines than those of the free amino acids but also decompose to ammonia and *N*-oxoacyl derivative, retaining carbonyl group in the peptide.[64]

N-Terminal dichloramines of peptides and proteins decompose to nitriles corresponding to *N*-terminal amino acids and to the shortened peptides.[64]

ϵ-Amino groups of lysine, when chlorinated, give relatively stable chloramine. Recently, however, it has been shown, that lysyl residues can be deaminated by the MPO system yielding aldehyde residues.[120] This renders possible a spontaneous formation of crosslinkages in proteins due to aldol condensation or due to the reaction of aldehyde with amino groups.[120] In fact, it was found that polymerization of protein occurs as a result of the MPO system action.[121]

Thiol[68] and thioether[122,123] side residues of amino acids appeared to be the most susceptible moieties to the attack of the MPO system. Cysteine and methionine are about 100 times more reactive with the MPO system than taurine.[69] Thiol groups may be oxidized to disulfide or (through chlorination to sulfenyl chloride[124]) even to sulfonic acid residues, and thioether groups to sulfoxides.[123] The oxidation of cysteinyl or methionyl residue functioning at the reactive site of protein is of special importance because it leads to the loss of biological function of the protein.[68,125,126]

Modification of indole rings by the MPO system has been followed by ultraviolet (UV) absorption changes. An increase in absorption at 250 nm and a decrease at 280 nm with isosbestic points, characteristic of oxidation to oxindole (2-indolone),[127] could be observed in the case of tryptophan, *N*-acetyltryptophan, indoleacetic acid, and protein.[112,121]

Interesting results have been obtained in the case of protein chlorination by chloramines. Namely, it was reported that *N*-chloro-D-leucine specifically chlorinated the apoenzyme of D-amino acid oxidase from pig kidney.[128] The chlorination was highly specific with respect to the structure of the chlorinating agent and of the chlorinated site in the enzyme. It appeared that the target residue for chlorination was tyrosine of the enzyme-active site and that this was converted to 3,5-dichlorotyrosine. This chlorination decreased the catalytic activity of D-amino acid oxidase.[128]

N-Dichloramines, in addition to their chlorinating and oxidizing properties, may be incorporated to proteins through histydyl, tyrosyl, or lysyl residues.[66]

The formation of ClCN in stimulated neutrophils permitted assumption of incorporation of the -CN moiety into proteins as a consequence of the action of the MPO–H_2O_2–Cl^- system in the presence of cyanide.[85] Since alkylation (with iodoacetamide) of albumin, used as a $Cl^{14}CN$ acceptor, reduced the amount of $Cl^{14}CN$ bound, most likely the -SH groups of the protein were involved in this reaction, yielding thiocyanoalanyl residues.[93]

In conclusion, modification of proteins by various chlorinating species may be broadly diversified, especially when the dose and time dependence of the process is envisaged. From the biological point of view, this leads to inactivation of proteins.

A characteristic example of this modification effect is autoinactivation of the MPO. In the presence of H_2O_2 and Cl^-, but in the absence of any acceptor of chlorine or reducing substrate, this enzyme loses its catalytic properties mainly due to destruction of the porphyrin ring.[71] MPO appeared to be a suicidal enzyme, which may be regarded as a special biological adaptation aimed at switching off chlorination irreversibly in the absence of target substrates.

It cannot be excluded, however, that some proteins after moderate oxidation (chlorination ?) become more active. This may be true of the proenzymes, which can be activated or rendered more susceptible for activation by oxidative modification.[129]

1.3.3. Reactions with Nucleotides and Nucleic Acids

Purine and pyrimidine nucleotides have been found to be good targets for the action of the respiratory burst products. Adenine and guanine are damaged due to chlorination by HOCl[68] or the MPO–H_2O_2–Cl^- system (Zgliczyński *et al*, unpublished data). The reactions were followed by UV absorption changes.[68] The initial reaction involved purine chlorination, with subsequent loss of UV chromophore, which indicated the destruction of the ring.[68] Both adenine and guanine protect MPO from autoinactivation, which indicates that they are better acceptors of chlorine than the porphyrin of the MPO itself.

Pyrimidine bases act as chlorine-accepting substrates of the MPO–H_2O_2–Cl^- system and undergo chlorination at position 5 (A. Lis and J. M. Zgliczyński, unpublished data).

As free bases and nucleotides are modified by the MPO system, it is likely that nucleic acids, when available, can be a target for MPO-mediated HOCl.

Whereas the MPO system seems to act preferably on bases of nucleic acids, hydroxyl radicals are thought also to cause fragmentation of DNA.[130]

1.3.4. Reactions with Glycosaminoglycans

Sugars and *N*-acetylated amino sugars are not reactive with HOCl.[68] Thus, it is conceivable that glycosaminoglycans are poor substrates for this system.

Nevertheless, neutrophils stimulated by PMA in the presence of FeEDTA chelate depolymerize hyaluronic acid.[131] The contribution of iron-EDTA chelate in this process indicates that OH· radicals are involved mainly[97] in damage to hyaluronic acid.

1.3.5. Reactions with Lipids

The studies of the reactions of the respiratory burst products with structural lipids are of a special importance, since they can give insight into the biochemical mechanism of cell injury and resistance to the neutrophil-derived oxidants. This was recently accomplished by developing a model system in which liposomes of known composition were employed as targets for the attack of activated neutrophils.[132,133]

Exposure of liposomes composed of egg phosphatidylcholine (PC) to the cell-free MPO–H_2O_2–Cl^- system results in the disruption of liposomes. The same was observed when the MPO–H_2O_2–$^{125}I^-$ system was used. In this case, a close correspondence between liposome lysis and incorporation of ^{125}I into liposomes was found.[132] Since liposomes composed of dipalmitoyl PC were resistant to attack by the MPO–H_2O_2–halide system, it has been suggested that this system oxidized or halogenated structural lipid at the C—C double bond.[132,133]

Disruption of PC liposomes by stimulated neutrophils required not only chloride but also iodide in the medium.[133] This phenomenon may be explained by the fact that stimulated neutrophils release mainly N–Cl compounds (*N*-chlorotaurine).[73–75] These compounds are incapable of affecting lipids but may oxidize iodide to active iodinating species (I_3^-, HOI), which would iodinate unsaturated fatty acids.

The liposomal membrane injury by the MPO-halogenating system required continued MPO activity, which argues against a free radical chain reaction within this model lipid bilayer.[132]

Free radicals generated by the activated neutrophils might be a potent source of polyunsaturated fatty acid peroxidation.[134] This process yields a variety of products, including peroxides, hydroperoxides, aldehydes, unsaturated aldehydes, and hydroxyalkenals.[135] Reviewing this broad and complicated subject is not a task that can be attempted here.

1.3.6. Reactions with Selected Biologically Active Compounds

The diversity of the target substrates affected by the MPO–H_2O_2–halide system is conspicuous. It results clearly from the generation of chemically reactive species by this system such as HOCl, Cl_2, chloramines, I_3^-, HOI, and $^1O_2$ (?). The occurrence of such a "predacious" enzyme in the defensive cells is amazing from the teleological point of view.

The MPO–H_2O_2–halide system reacts with many vital compounds of the cell. The vulnerability of adenine and guanine to the HOCl attack[68] may diminish the nucleotide coenzyme pool. It has been found that the MPO–H_2O_2–Cl^- system chlorinates and inactivates NADH and NADPH.[136] The components of the respiratory chain—cytochromes (cytochrome *c*)[137] and especially iron–sulfur centers[137,138]—are susceptible to the MPO system action. Ferredoxins are also oxidized by chloramines.[68] It is noteworthy that oxidation of iron–sulfur centers and heme of cytochrome *c* in the presence of chelating agent is accompanied by iron release,[137] which renders possible in turn the generation of OH· radicals [eqs. (13) and (16)].[97]

The MPO-chlorinating system oxidizes and chlorinates steroids (estrogens),[139] ascorbic acid,[68,69] and β-carotene.[68] The unspecific action of halogenating and oxidizing species produced by the MPO also inactivates the stimulants of inflammation—agents responsible directly or indirectly for initiation of the respiratory burst by neutrophils. These include prostaglandins,[140,141] leukotriens C_4, D_4, and E_4,[142,143] and chemoattractants C5a and *N*-formyl-methionyl-leucyl-phenyl-alanine.[144,145] Thus, in spite of its predominantly damaging activity, MPO may also be considered an anti-inflammatory agent.[146]

2. EFFECTS OF RESPIRATORY BURST PRODUCTS ON METABOLIC REACTIONS IN THE INFECTED TISSUES

2.1. Proteolysis

2.1.1. Introduction

Stimulation of neutrophils by particles or soluble stimuli is accompanied by partial release of both granule enzymes and oxidative species into the extracellular space.[147] The secreted proteases have an opportunity to act on extracellular proteins in the presence of MPO and the oxidative products of respiratory burst. A question arises as to whether there is synergy or antagonism between proteolytic and oxidating (chlorinating) systems.

It has been reported that the neutrophil respiratory burst products inactivate lysosomal enzymes in phagolysosomes, resulting in an inhibition of the lysosomal enzyme release.[148] On the other hand, it has been shown that intraphagosomal digestion is impaired in CGD neutrophils, which are unable to respond to stimulation with a respiratory burst.[42]

In the light of recent findings, there are four ways by which oxidative systems of neutrophil may enhance proteolytic processes.

2.1.2. Inactivation of α_1-Protease Inhibitor (α_1-PI) by the MPO–H_2O_2–Cl^- System of Neutrophils

It has been shown that α_1-PI, the major plasma inhibitor of elastolytic activity, is efficiently inactivated by the stimulated neutrophils[149,150] or the MPO–H_2O_2–Cl^- system.[125,126,149] The loss of activity results from the oxidation of the methionine residue occupying a part of the reactive site of α_1-PI.[125,126] The thioether moiety is one of the most sensitive groups to the MPO–H_2O_2–Cl^- system as well as to chloramines[73] and its oxidation to form sulfoxide impairs elastase-binding capacity of α_1-PI. The activity of elastase is thereby potentiated.

Under the influence of the MPO–H_2O_2–Cl^- system, the rate of α_1-PI activity loss is much higher than that of other proteinase inhibitors and proteases.[125]

The oxidative inactivation of α_1-PI resulting in enhancement of elastase activity is widely believed to be a pathophysiological event leading to the tissue damage[126,149] for example in lung emphysema.[151] In individuals exposed to the external pollutants, neutrophils become massively sequestered in the lung and release elastase, MPO, and oxygen species. Inactivation of α_1-PI that has penetrated the lung causes elastase to develop an uncontrolled hydrolysis of lung elastin. It may be suggested that in this disease, the enormous extent of the oxidative processes plays a role in decontrolling tissue degradation.

A question arises as to whether the influence of the oxidative species on α_1-PI has physiological significance. One may speculate that the process may regulate connective tissue turnover when the oxidation of α_1-PI can be controlled by reducing agents (enzyme?).

2.1.3. Direct Fragmentation of Protein by Halogenating or Oxygen Radical Attack on Proteins

The study by Selvaraj et al.[63] showed the direct cleavage of bacterial proteins by the cell-free MPO–H_2O_2–Cl^- system. In these experiments, [1,7-^{14}C]diaminopimelic acid (DAP) was incorporated into *Escherichia coli* proteins. It was observed that the MPO–H_2O_2–Cl^- system released as $^{14}CO_2$ about 94% of the radioactivity from heat-killed labeled bacteria. This would indicate that the decarboxylation had also involved [^{14}C]carboxyl groups engaged in the peptide bond due to the chlorination of both the free amino and amide group of DAP residue.[63]

Oxidative peptide cleavage of the bacterial proteins by the MPO chlorinating system was also suggested by Thomas[65] and was considered by Adeniyi-Jones and Karnovsky,[152] who have found decarboxylation of uniformly labeled ^{14}C-particulate protein after phagocytosis by guinea pig neutrophils. In the two latter cases, however, peptide scission might be a result of the action of proteases present in the systems studied.[65,152]

It has also been observed that tryptophanyl peptide bonds are oxidatively cleaved by a peroxidase–H_2O_2–I^- system, by chloramine alone, or by chloramine in the presence of iodide.[127] This reaction proceeds at pH 3–5 via oxidation of the tryptophan residues to oxindole derivatives, subsequent cyclization to iminolactones, and hydrolysis to lactones and new free *N*-terminal residues.[127]

The fragmentation of proteins may also proceed under the influence of the oxygen radicals, especially hydroxyl radicals in the presence of oxygen ($OH\cdot$–O_2).[153] Thus, several mechanisms of the protein degradation have been proposed, one leading to hydrolysis of the peptide bond, another to the deamination of the peptide bond via protein carbon peroxyradicals.[153]

It has also been found that scission of proteins occurs at prolyl residues; i.e., oxidation of the proline ring by $OH\cdot$–O_2 results in subsequent spontaneous hydrolysis of the 2-pyrrolidone formed and the production of the new *N*-terminal glutamate residue.[153]

2.1.4. Enhancement of Enzymatic Proteolysis as an Effect of Protein Modification

It is now claimed that even limited oxidative modification of proteins can render them more susceptible to enzymatic proteolysis.[153,154] This phenomenon is regarded as an important step in protein turnover in cells. These proteins, which are covalently marked, become more compliant substrates for hydrolytic enzymes.[154] Several marking reactions have now been identified: the oxidations of methionine residues,[155] -SH groups,[153] and nonheme iron–sulfur centers[156] and the formation of carbonyl groups[157] and of mixed disulfide derivatives with cysteine residues.[158] It is noteworthy that all these modifications of proteins are readily catalyzed by the MPO–H_2O_2–halogenating system (see Sections 1.3.2 and 1.3.6).

The latter process, i.e., formation of mixed disulfides, may be also expected to be a result of the MPO action. Namely, it has been shown[159] that GSH is readily oxidized to GSSG by the stimulated neutrophils due to the activity of the MPO system. GSSG, in turn, is thought to be responsible for the formation of mixed disulfides with protein thiol groups.[158]

Pretreatment of bovine albumin with the MPO–H_2O_2–Cl^- system or NaOCl was found to result in faster hydrolysis by trypsin as compared with that of native proteins (Olszowska et al.[159a]). Moreover, the modified (chlorinated) albumin competes more efficiently with trypsin artificial substrate (*N*-benzoyl-D,L-arginine-*p*-nitroanilide) than the native albumin.

The susceptibility of the modified albumin to trypsin attack depends on the extent of chlorination (oxidation). It is dose dependent, reaches maximum, and decreases as a result of the high concentration of chlorinating species even below proteolysis level of untreated albumin. This latter result is compatible with observations[153] that in some cases oxidatively modified proteins become more resistant to proteolysis and accumulate in tissues.

2.1.5. Activation of Proteolysis by Conversion of Latent to Active Proteolytic Enzyme(s)

Stimulated neutrophils were shown to release and simultaneously activate latent collagenase.[129] Activation of the proenzyme was almost completely suppressed by catalase, but HOCl added to the neutrophil supernatant containing inactive collagenase in the presence of catalase abolished its inhibitory effect. Thus, the production of HOCl by the MPO–H_2O_2–Cl^- system operating after triggering neutrophils has been considered critical in the formation of active collagenase.[129] The mechanism underlying this activation is unknown, but it may be supposed that it proceeds through oxidation and dissociation of the inhibitor–enzyme complex, changes in the conformation of the modified proenzyme, or splitting the proenzyme at a proper site.[129]

It is of a particular interest whether the case of activation of latent collagenase is unique or there are other enzymes whose activities are potentiated by the action of the respiratory burst products.

2.2. Changes in Nucleic Acid Metabolism

Purine and pyrimidine bases may be affected by the MPO–H_2O_2–Cl^- system,[67,68] and fragmentation of nucleic acid proceeds under the influence of oxygen radicals[130] (see Section 1.3.3). Modification of ribonucleic acids does not seem to have such pronounced biological effects as changes in DNA structure.

Under subtle modification, DNA may still be biologically active but may express changed properties that, when propagated to daughter cells, may be important in tumorigenesis. In fact, certain chronic inflammatory conditions associated with continuous leukocyte infiltration cause a markedly increased risk of cancer in the affected tissues.[160]

Mutagenic abilities of chlorinating species, i.e., hypochlorite[161] and ammonia chloramine[162,163] were established. Recently, it was found that stimulated neutrophils produced mutations in bacteria[164] and elicited cytogenetic changes in cultured mammalian cells.[165] It was also shown[166] that DNA of human neutrophils underwent extensive damage with maximum at 40–45 min after addition of respiratory burst stimulus. In all three cases,[164–166] the active oxygen products of the respiratory burst were thought to be responsible for the genetic disorders.

2.3. Degradation of Proteoglycans

It is believed that the degradation of proteoglycans is mediated by both oxygen radicals and proteases released from phagocytic cells, i.e., from neutrophils in acute and macrophages in chronic phase of inflammation. Oxygen radicals depolymerize hyaluronic acid[131] as well as affect protein components of proteoglycan.[167,168] This may be responsible for profound changes in the matrix of connective tissue, especially in chronic inflammatory conditions.

2.4. Tentative Mechanisms of Cytotoxic and Cytolytic Activities of Neutrophils toward Target Cells

In experimental model systems, activated neutrophils exert unspecific cytotoxic and cytolytic effects on bystander cells.[76–78,114–118,169] Alternatively, a specific mode of cytotoxicity involves attachment of neutrophil through its surface receptor for the Fc portion of IgG to the antibody-coated target cell.[170–174]

In the first case, cytotoxic properties of neutrophils result from a partial release or leak of the respiratory burst products and granule enzymes to the extracellular space.[76–78,114–118,169] Under other circumstances, effector–neutrophil cells discharge toxic respiratory burst products directly into the tightly attached plasma membrane of the target cells.[170–174]

It is widely believed that the respiratory burst is essential for the maximal cytotoxic activities of neutrophils in the both cases.[76–78,114–118,169,172–174] There is, however, a difference in the availability and involvement of toxic products of respiratory burst in these two types of neutrophil cytotoxicity.

The MPO chlorinating system and its long-lived products—chloramines were reported to play an important role in the cytotoxicity of neutrophils to bystander cells.[76–78,117,118,169] The first object of the target cell to be attacked, plasma membrane, may be injured by the MPO–H_2O_2–Cl^- system.[132,133] This may result from the reaction with lipids[132,133] and/or most probably with membrane proteins. At this point, it is noteworthy that lipid composition and participation of antioxidants in the membrane are believed to determine resistance or susceptibility of the membrane to injury.[132,133]

In light of the former consideration (see Section 1.3.2), almost all enzymes can be affected by the MPO–H_2O_2–Cl^- system. It is expected, however (taking example from α_1-PI) that there are substantial differences in the susceptibility of particular proteins to inactivation. For example, respiration of target cells seems to be easily impaired because of the vulnerability of nonheme iron–sulfur centers,[137,138] hemes,[137] and at least reduced pyridine coenzymes[136] to MPO–H_2O_2–Cl^- attack.

Lipophilic chloramines, such as those formed from ammonia or spermidine produced by the MPO system penetrate plasma membrane and exert intracellular toxic effects.[59,75,95] It has been reported that NH_2Cl inhibits HMP shunt in erythrocytes[175] and is mutagenic.[162,163]

Although generally the MPO system acts unspecifically, it is conceivable that its products, *N*-chloro-analogues of amino acids, behave as active site-directed inhibitors of enzymes critical in amino acid metabolism. There is a precedent for such assumption: *N*-chloro-D-leucine inhibits D-amino acid oxidase, specifically binding to and chlorinating critical tyrosine residue of the enzyme-active site.[128]

The antibody-dependent cellular cytotoxicity (ADCC) of neutrophils has been reported to be independent of the $MPO–H_2O_2–Cl^-$ system.[172,173] Instead, it has been proposed that oxygen radicals ($O_2^- \cdot$, $OH \cdot$) act directly on plasma membrane of target cell, tightly attached to neutrophil. This leads to the lipid peroxidation, denaturation, and destruction of the membrane proteins, which result in lysis of the target cell.

3. MODULATION OF RESPIRATORY BURST BY ITS PRODUCTS

3.1. The $MPO–H_2O_2–Cl^-$ System as a Modulator of the Respiratory Burst

The MPO system appears to be involved in the regulation of the respiratory burst. This statement is based on findings indicating that genetically MPO-deficient neutrophils display increased and longer-lasting $O_2^- \cdot$ production induced by phagocytosis as compared with normal cells.[176–178] This phenomenon can result from the MPO system-destructive action on the first steps of the neutrophil activation. The MPO system, in fact, has been shown to damage Fc and C3b receptors of neutrophils[179] as well as reduce the opsonizing activity of immunoglobulin G (IgG) and complement component C3b.[180] Since the MPO system also affects other neutrophil-stimulating agents as C5a,[144] formyl-methionyl-leucyl-phenyl-alanine[144,145] and leukotrienes,[1421],[143] it can modulate in several ways the metabolic response of neutrophils.

In addition, some products of the MPO system-lipophilic chloramines inhibit the respiratory burst, impairing glucose utilization and oxygen uptake enzyme systems of neutrophils.[59]

3.2. Influence of Cyanide on Respiratory Burst and Other Neutrophil Activities

It has been shown that the respiratory burst of neutrophils is not only insensitive to, but also stimulated by, cyanide.[176,181–184] Cyanide stimulates HMP shunt,[176,182,185] oxygen uptake,[181,184] and superoxide production[183] (NADPH oxidase acivity[182]) in triggered neutrophils.

An indirect mechanism of the stimulatory effect of cyanide on HMP shunt has been proposed.[185,186] Cyanide inhibiting catalase spares endogenous H_2O_2, which, in turn, via the glutathione redox system, regenerates the NADP-rate-limiting substrate of the HMP shunt in neutrophils (Fig. 5). At this point, it is noteworthy that

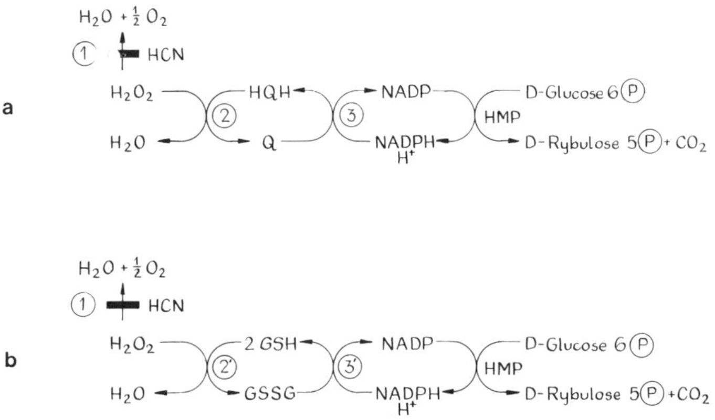

a

$H_2O + \frac{1}{2} O_2$
(1) HCN
H_2O_2 — (2) — HQH — (3) — NADP — HMP — D-Glucose 6(P)
H_2O — Q — NADPH H^+ — D-Rybulose 5(P) + CO_2

b

$H_2O + \frac{1}{2} O_2$
(1) HCN
H_2O_2 — (2′) — 2 GSH — (3′) — NADP — HMP — D-Glucose 6(P)
H_2O — GSSG — NADPH H^+ — D-Rybulose 5(P) + CO_2

FIGURE 5. Possible contribution of cyanide to the activation of HMP shunt accompanying (a) seed transformation from dormant to germinating state[187] and (b) neutrophil stimulation from resting to activated state.[185,186] In both cases, peroxidases (2) and (2′) (less sensitive to cyanide than catalase) limit the process. H_2O_2 as oxidizing substrate of peroxidase is spared by the inhibition of catalase (1) by cyanide. (a) Peroxidase (2) oxidizes hydroquinone substrate to quinone; (b) glutathione peroxidase (2′) oxidizes GSH to GSSG. In turn, oxidoreductases NADPH: quinone (3) and NADPH:GSSG (3′) regenerate NADP required as oxidant in HMP shunt.

there is a striking similarity between the proposed mechanism of cyanide influence on the neutrophil HMP shunt[185,186] and possible involvement of this compound in events accompanying the transformation of dormant seed to a germinating form[187] (Fig. 5).

The higher O_2 consumption and $O_2^{-}\cdot$ production in the presence of cyanide may result at least partly from an increase of HMP shunt, i.e., increase of NADPH concentration,[188] but other explanations have been also presented. First, it has been reported[189] that cyanide alters $O_2^{-}\cdot$ formation by increasing the amount of the activated NADPH oxidase. Second, it has been suggested that cyanide acts due to chelation of metal ions which could catalyze the dismutation of $O_2^{-}\cdot$ or could be involved in other way in the respiratory burst.[184]

The effect of cyanide on the respiratory burst seems to be of interest since this compound is generated by stimulated neutrophils.[81,85] In other words, a question arises as to whether endogenous cyanide is of a physiological significance in the neutrophil function.

It has been shown in several model experiments that addition of cyanide at 1 mM concentration inhibits bactericidal and cytotoxic activities of neutrophils.[76,117,118,190] But in some other experimental systems[114,173] cyanide potentiates lysis of target cells.

The former effect of cyanide, inhibition of neutrophil cytotoxicity, has been explained by its inhibitory action on myeloperoxidase.[76,117,118,190] Since it was found recently[85] that cyanide even at about 0.1 mM concentration was chlorinated to ClCN by PMA-stimulated neutrophils, it is conceivable that cyanide also acts as the scavenger of HOCl or chloramines formed due to MPO activity. It is noteworthy, that ClCN formed by neutrophils may exert also toxic effects on neutrophils themselves and/or target cells due to its reactivity toward thiol compounds.[85,93] It is not known, however, if the MPO–H_2O_2–Cl^- system in cells, in the presence of cyanide at commonly employed 1 mM concentrations still retains the ability to produce HOCl (see Section 1.2.4a).[89]

The second effect—potentiation of neutrophil cytotoxicity by cyanide may result from inhibiting catalase thus sparing H_2O_2 for the generation of oxygen radicals (OH·).

These two different effects of cyanide would indicate that efficacy of various secondary products of the respiratory burst in the cytotoxic action depends in a high degree on type of target cells—their resistance or susceptibility to versatile toxic products of neutrophils.

There may be another explanation of the stimulatory effect of cyanide on the cytotoxicity of neutrophils. Namely, it is known[191] that the development of respiratory burst depends on the availability of oxygen. Cyanide inhibiting the mitochondrial respiration of the target cells would save oxygen for the neutrophil respiratory burst. Such effect of cyanide was shown[192] in the case of the interaction of neutrophils and aerobic bacteria (*Neisseria gonorrhoeae* and *Staphylococcus aureus*). These bacteria inhibited the respiratory burst of neutrophils by reducing the local concentration of oxygen. Cyanide restored the respiratory burst of neutrophils by impairing the respiration of bacteria.[192] A possible contribution of cyanide and/or cyanogen chloride to the activity of stimulated neutrophils remains to be clarified.

4. CONCLUSIONS

The bird's-eye view on the neutrophil factory of oxidative and halogenating compounds shows more and more clearly that the defensive action of these cells against invaders is not neutral for the host. The inconvenient consequences of neutrophil respiratory burst can be seen. Modified junk proteins difficult to remove, decontrolled proteolysis of connective tissue, and mutations all lead to disease and aging.

In spite of this, the beneficial action of neutrophils predominates, they defend host against foreign agents and is not impossible that take part in tissue renovation and healing.

ACKNOWLEDGMENTS. The authors' research described in this chapter was supported by a grant (CPBP04.01.2.08) from the Polish Academy of Sciences.

REFERENCES

1. Baldridge CW, Gerard RW: The extrarespiration of phagocytosis. *Am J Physiol* **103:**235–236, 1933.

2. Sbarra AJ, Karnovsky ML: The biochemical basis of phagocytosis. I. Metabolic changes during the ingestion of particles by polymorphonuclear leukocytes. *J Biol Chem* **234:**1355–1362, 1959.
3. Babior BM: Oxygen-dependent microbial killing by phagocytes. *N Engl J Med* **298:**659–668, 1978.
4. Badwey JA, Karnovsky ML: Active oxygen species and the functions of phagocytic leukocytes. *Annu Rev Biochem* **49:**695–726, 1980.
5. Schiffman E, Corcoran BA, Wahl SM: *N*-formyl-methionyl peptides as chemoattractants for leukocytes. *Proc Natl Acad Sci USA* **72:**1059–1065, 1975.
6. English D, Rozoff JS, Lukens JN: Regulation of human polymorphonuclear leukocyte superoxide release by cellular responses to chemotactic peptides. *J Immunol* **126:**165–173, 1981.
7. Goldstein JM, Ross D, Kaplan HB, et al: Complement and immunoglobulins stimulate superoxide production by human leukocytes independently of phagocytosis. *J Clin Invest* **56:**1155–1163, 1975.
8. Klebanoff SJ, Clark RA: *The Neutrophil: Function and Clinical Disorders.* Amsterdam, Elsevier/North-Holland, 1978, pp 172–183.
9. Curnutte JT, Babior BM: Biological defense mechanisms. The effect of bacteria and serum on superoxide production by granulocytes. *J Clin Invest* **53:**1662–1672, 1974.
10. Messner RP, Jelinek J: Receptors for human γG globulin on human neutrophils. *J Clin Invest* **49:**2165–2171, 1970.
11. Sajnani AN, Ranadive NS, Movat HZ: The visualization of receptors for the Fc portion of the IgG molecule on human neutrophil leukocytes. *Life Sci* **14:**2427–2430, 1974.
12. Borgeat P. Samuelsson B: Transformation of arachidonic acid by rabbit polymorphonuclear leukocytes. Formation of a novel dihydroxyeicosatetraenoic acid. *J Biol Chem* **254:**2643–2646, 1979.
13. Najjar VA: Biochemistry and physiology of tuftsin Thr-Lys-Pro-Arg, in Sbarra AJ, Strauss RR (eds): *The Reticuloendothelial System, A Comprehensive Treatise.* New York, Plenum, 1980, vol 2, pp 45–71.
14. Gay JC, Beckman JK, Brash AR, et al: Enhancement of chemotactic factor-stimulated neutrophil oxidative metabolism by leukotriene B_4. *Blood* **64:**780–785, 1984.
15. Spirer Z, Zakuth V, Golander A, et al: The effect of tuftsin on the nitrous blue tetrazolium reduction of normal human polymorphonuclear leukocytes. *J Clin Invest* **55:**198–201, 1975.
16. Fletcher MP, Galin JI: Degranulating stimuli increase the availability of receptors on human neutrophils for the chemoattractant f Met-Leu-Phe. *J Immunol* **124:**1585–1588, 1980.
17. Dahlgren C, Magnusson KE, Stendahl O, et al: Modulation of polymorphonuclear leukocyte chemiluminescent response to the chemoattractant f-Met-Leu-Phe. *Int Arch Allergy Appl Immunol* **68:**79–83, 1982.
18. Targowski SP, Niemialtowski M: Appearance of Fc receptors on polymorphonuclear leucocytes after migration and their role in phagocytosis. *Infect Immun* **52:**798–802, 1986.
19. McPhail LC, Snyderman R: Activation of the respiratory burst enzyme in human polymorphonuclear leukocytes by chemoattractants and other soluble stimuli. Evidence that the same oxidase is activated by different transductional mechanisms. *J Clin Invest* **72:**192–200, 1983.
20. Harber MJ, Topley N: Factors affecting the measurement of chemiluminescence in stimulated human polymorphonuclear leukocytes. *J Bioluminescence Chemiluminescence* **1:**15–27, 1986.
21. Barbour AG, Allred CD, Solberg CO, et al: Chemiluminescence by polymorphonuclear leukocytes from patients with acute bacterial infections. *J Infect Dis* **141:**14–26, 1980.
22. Bass DA, Olbrantz P, Szejda P, et al: Subpopulations of neutrophils with increased oxidative product formation in blood of patients with infection. *J Immunol* **136:**860–866, 1986.
23. Selvaraj RJ, Sbarra AJ, Thomas GB, et al: A microtechnique for studying chemiluminescence response of phagocytes using whole blood and its application to the evaluation of phagocytes in pregnancy. *J Reticuloendothel Soc* **31:**3–16, 1982.
24. Tono-Oka T, Ueno N, Matsumoto I, et al: Chemiluminescence in whole blood. 1. A simple and rapid method for the estimation of phagocytic function of granulocytes and opsonic activity in whole blood. *Clin Immunol Immunopathol* **26:**66–75, 1983.
25. Allen RC, Stjernholm RL, Steele RH: Evidence for the generation of an electronic excitation

state(s) in the human polymorphonuclear leukocytes and its participation in bactericidal activity. *Biochem Biophys Res Commun* **47:**679–684, 1972.
26. Zgliczyński JM, Olszowski S, Olszowska E, et al: The influence of plasma factors on chemiluminescence of polymorphonuclear granulocytes. *Postępy Biologii Komórki* **11:**459–462, 1984.
27. Patriarca P, Cramer R, Moncalvo S, et al: Enzymatic basis of metabolic stimulation in leukocytes during phagocytosis: The role of activated NADPH oxidase. *Arch Biochem Biophys* **145:**255–262, 1971.
28. Babior BM, Curnutte JT, McMurrich BJ; The particulate superoxide-forming system from human neutrophils. Properties of the system and further evidence supporting its participation in the respiratory burst. *J Clin Invest* **56:**1035–1042, 1976.
29. Suzuki H, Kakinuma K: Evidence that NADPH is the actual substrate of the oxidase responsible for the "respiratory burst" of phagocytosing polymorphonuclear leukocytes. *J. Biochem* **93:**709–715, 1983.
30. Doussiere J. Vignais PV: Purification and properties of an $O_2^-\cdot$ generating oxidase from bovine polymorphonuclear neutrophils. *Biochemistry* **24:**7231–7239, 1985.
31. Iyer GYN, Islam MF, Quastel JG: Biochemical aspects of phagocytosis. *Nature (Lond)* **192:**535–541, 1961.
32. Roos D. Voetman AA and Meerhof LJ: Functional activity of enucleated human polymorphonuclear leucocytes. *J Cell Biol* **97:**368–377, 1983.
33. Korchak HM, Roos D, Giedd KN, et al: Granulocytes without degranulation: Neutrophil function in granule-depleted cytoplasts. *Proc Natl Acad Sci USA* **80:**4968–4972.
34. Segal AW, Jones OTG: Novel cytochrome b system in phagocytic vacuoles of human granulocytes. *Nature (Lond)* **276:**515–517, 1978.
35. Segal AW, Jones OTG: Absence of cytochrome b reduction in stimulated neutrophils from both female and male patients with chronic granulomatous disease. *FEBS Lett* **110:**111–114, 1980.
36. Babior BM, Kipness RS: Superoxide-forming enzyme from human neutrophils: Evidence for a flavin requirement. *Blood* **50:**517–524, 1977.
37. Gabig GT: The NADPH-dependent $O_2^-\cdot$ generating oxidase from human neutrophils: Identification of a flavoprotein component that is deficient in a patient with chronic granulomatous disease. *J Biol Chem* **258:**6352–6356, 1983.
38. Gerard C, McPhail LC, Marfat A, et al: Role of protein kinase in stimulation of human polymorphonuclear leukocyte oxidative metabolism by various agonists. *J Clin Invest* **77:**61–65, 1986.
39. Babior MB, Kipnes RS and Curnutte JT: Biological defense mechanisms. The production by leukocytes of superoxide, a potential bactericidal agent. *J Clin Invest* **52:**741–744, 1973.
40. McCord JM, Fridovich I: Superoxide dismutase: An enzymatic function for erythrocuprein (hemocuprein). *J Biol Chem* **244:**6049–6055, 1969.
41. DeChatelet LR, McCall CE, McPhail LC, et al: Superoxide dismutase activity in leukocytes. *J Clin Invest* **53:**1197–1201, 1973.
42. Segal AW, Geisgow M, Garcia R, et al: The respiratory burst of phagocytic cells is associated with a rise in vacuolar pH. *Nature (Lond)* **290:**406–409, 1981.
43. Khan AU: Singlet molecular oxygen from superoxide anion and sensitized fluorescence of organic molecules. *Science* **168:**476–477, 1970.
44. Arudi RL, Bielski BHJ, Allen AO: Search for singlet oxygen luminescence in the disproportionation of $HO_2/O_2^-\cdot$. *Photochem Photobiol* **39:**703–706, 1984.
45. Paul B, Sbarra AJ: The role of the phagocyte in host parasite interactions. XIII. The direct quantitative estimation of H_2O_2 in phagocytozing cells. *Biochim Biophys Acta* **156:**168–178, 1968.
46. Root RK, Metcalf J, Oshino N, et al: H_2O_2 release from human granulocytes during phagocytosis. *J Clin Invest* **55:**945–955, 1975.
47. Homan-Muller JWT, Weening RS, Roos D: Production of hydrogen peroxide by phagocytizing human granulocytes. *J Lab Clin Med* **85:**198–207, 1975.
48. Root RK, Metcalf JA: H_2O_2 release from human granulocytes during phagocytosis: Relationship to

superoxide anion formation and cellular catabolism of H_2O_2: Studies with normal and cytochalasin B-treated cells. *J Clin Invest* **60:**1266–1279, 1977.

49. Kakinuma K, Kaneda M: Apparent Km of leukocyte $O_2^- \cdot$ and H_2O_2 forming enzyme for oxygen, in Rossi F, Patriarca P (eds): *Biochemistry and Function of Phagocytes.* New York, Plenum, 1982, pp 351–359.
50. Agner K: Verdoperoxidase. A ferment isolated from leukocytes. *Acta Physiol Scand* **2:**1–62, 1941.
51. Bainton DF, Farquhar MG: Differences in enzyme content of azurophil and specific granules of polymorphonuclear leukocytes. II. Cytochemistry and electron microscopy of bone marrow cells. *J Cell Biol* **39:**299–317, 1968.
52. Wilson DL, Manery JF: The permeability of rabbit leukocytes to sodium, potassium and chloride. *J Cell Comp Physiol* **34:**493–519, 1949.
53. Baron DN, Ahmed SA: Intracellular concentration of water and of the principal electrolytes determined by the analysis of isolated human leukocytes. *Clin Sci* **37:**205–219, 1969.
54. Klebanoff SJ, Clark RA: *The Neutrophil: Function and Clinical Disorders.* New York, Elsevier/North-Holland Biomedical Press, 1978, pp 410–434.
55. Klebanoff SJ: Iodination of bacteria: A bactericidal mechanism. *J Exp Med* **126:**1063–1978, 1967.
56. Stelmaszyńska T, Zgliczyński JM: Myeloperoxidase of human neutrophilic granulocytes as chlorinating enzyme. *Eur J Biochem* **45:**305–312, 1974.
57. Zgliczyński JM, Stelmaszyńska T: Chlorinating ability of human phagocytosing leukocytes. *Eur J Biochem* **56:**157–162, 1975.
58. Weiss SJ, Klein R, Slivka A, et al: Chlorination of taurine by human neutrophils. Evidence for hypochlorous acid generation. *J Clin Invest* **70:**598–607, 1982.
59. Thomas EL, Grisham MB, Jefferson MM: Myeloperoxidase-dependent effect of amines on functions of isolated neutrophils. *J Clin Invest* **72:**441–454, 1983.
60. Foote CS, Goyne TE and Lehrer RI: Assessment of chlorination by human neutrophils. *Nature (Lond)* **301:**715–716, 1983.
61. Agner K: Biological effects of hypochlorous acid formed by "MPO" peroxidation in the presence of chloride ions, in Åkeson A, Ehrenberg A (eds): *Structure and Function of Oxidation-Reduction Enzymes,* Oxford, Pergamon, 1972, pp 329–335.
62. Zgliczyński JM, Stelmaszyńska T, Domański J, et al: Chloramines as intermediates of oxidation reaction of amino acids by myeloperoxidase. *Biochim Biophys Acta* **235:**419–424, 1971.
63. Selvaraj RJ, Paul BB, Strauss RR, et al: Oxidative peptide cleavage and decarboxylation by the MPO–H_2O_2–Cl^- anti-microbial system. *Infect Immun* **9:**255–260, 1974.
64. Stelmaszyńska T, Zgliczyński JM: N-(2-oxoacyl)amino acids and nitriles as final products of dipeptide chlorination mediated by the myeloperoxidase/H_2O_2/Cl^- system. *Eur J Biochem* **92:**301–308, 1978.
65. Thomas EL: Myeloperoxidase, hydrogen peroxide, chloride antimicrobial system: Nitrogen–chlorine derivatives of bacterial components in bactericidal action against *Escherichia coli. Infect Immun* **23:**522–531, 1979.
66. Thomas EL, Jefferson MM, Grisham MB: Myeloperoxidase-catalyzed incorporation of amines into proteins: Role of hypochlorous acid and dichloramines. *Biochemistry* **24:**6299–6308, 1982.
67. Zgliczyński JM: Characteristics of myeloperoxidase from neutrophils and other peroxidases from different cell types, in Sbarra AJ, Strauss RR (eds): *The Reticuloendothelial System. A Comprehensive Treatise.* New York, Plenum, 1980, pp 255–278.
68. Albrich JM, Mc Carthy CA, Hurst JK: Biological reactivity of hypochlorous acid: Implications for microbicidal mechanisms of leucocyte myeloperoxidase. *Proc Natl Acad Sci USA* **78:**210–214, 1981.
69. Winterbourn CC: Comparative reactivites of various biological compounds with myeloperoxidase–hydrogen peroxide–chloride, and similarity of the oxidant to hypochlorite. *Biochim Biophys Acta* **840:**204–210, 1985.
70. Onishi M, Odajima T: On the product of prostaglandin E_1 oxidized by the myeloperoxidase–H_2O_2–chloride system. *Jpn J Oral Biol* **27:**291–298, 1985.

71. Zgliczyński JM, Stelmaszyńska T, Ostrowski W, et al: Myeloperoxidase of human leukaemic leukocytes: Oxidation of amino acids in the presence of hydrogen peroxide. *Eur J Biochem* **56:**157–162, 1968.
72. McMenamy RW, Lund CC, Neville GJ, et al: Studies of unbound amino acid distributions in plasma, erythrocytes, leukocytes and urine of normal human subjects. *J Clin Invest* **39:**1675–1687, 1960.
73. Weiss SJ, Lampert MB, Test ST: Long-lived oxidants generated by human neutrophils: Characterization and bioactivity. *Science* **222:**625–628, 1983.
74. Test ST, Lampert MB, Ossanna PJ, et al: Generation of nitrogen–chlorine oxidants by human phagocytes. *J Clin Invest* **74:**1341–1349, 1984.
75. Grisham MB, Jefferson MM, Melton DF, et al: Chlorination of endogenous amines by isolated neutrophils. Ammonia-dependent bactericidal, cytotoxic, and cytolytic activities of the chloramines. *J Biol Chem* **259:**10404–10413, 1984.
76. Slivka A, LoBuglio AF, Weiss SJ: A potential role for hypochlorous acid in granulocyte-mediated tumor cell cytotoxicity. *Blood* **55:**347–350, 1980.
77. Thomas EL, Grisham MB, Melton DF, et al: Evidence for a role of taurine in the in vitro oxidative toxicity of neutrophils towards erythrocytes. *J Biol Chem* **260:**3321–3329, 1985.
78. Dallegri F, Patrone F, Ballestrero A, et al: Inhibition of neutrophil cytolysin production by target cells. *Blood* **67:**1266–1272, 1986.
79. Pereira WE, Hoyano Y, Summons RE, et al: Chlorination studies. II. The reaction of aqueous hypochlorous acid with α-amino acids and peptides. *Biochim Biophys Acta* **313:**170–180, 1973.
80. Zgliczyński JM, Stelmaszyńska T: Hydrogen cyanide and cyanogen chloride formation by the myeloperoxidase–H_2O_2–Cl^- system. *Biochim Biophys Acta* **567:**309–314, 1979.
81. Stelmaszyńska T: Formation of HCN by human neutrophils. 1. Chlorination of *Staphylococcus epidermidis* as a source of HCN. *Int J Biochem* **17:**373–379, 1985.
82. Tipper DJ, Berman MF: Structures of the cell wall peptidoglycans of *Staphylococcus epidermidis* Texas 26 and *Staphylococcus aureus* Copenhagen I. Chain length and average sequence of cross-bridge peptides. *Biochemistry* **8:**2183–2191, 1969.
83. Tipper DJ, Strominger JL: Mechanism of action of penicillins: A proposal based on their structural similarity to acyl-D-alanyl-D-alanine. *Proc Natl Acad Sci USA* **54:**1133–1141, 1965.
84. Root RK, Isturiz R, Molavi A, et al: Interactions between antibiotics and human neutrophils in the killing of *staphylococci:* Studies with normal and cytochalasin B-treated cells. *J Clin Invest* **67:**247–259, 1981.
85. Stelmaszyńska T: Formation of HCN and its chlorination to ClCN by stimulated human neutrophils. 2. Oxidation of thiocyanate as a source of HCN. *Int J. Biochem* **18:**1107–1114, 1986.
86. Weuffen W, Kramer A, Jülich WD: Vorkommen bei Mensch und Tier, in Weuffen W (ed): *Medizinische und biologische Bedeutung der Thiocyanate (Rhodanide).* Berlin, VEB Verlag Volk und Gesundheit, 1982, p 123.
87. Wever R, Kast WM, Kasinoedin JH, et al: The peroxidation of thiocyanate catalysed by myeloperoxidase and lactoperoxidase. *Biochim Biophys Acta* **709:**212–219, 1982.
88. Thomas EL: Peroxidase-catalysed oxidation of thiocyanate, in Weuffen W (ed): *Medizinische und biologische Bedeutung der Thiocyanate (Rhodanide).* Berlin, VEB Verlag Volk und Gesundheit, 1982, pp 89–102.
89. Stelmaszyńska T, Zgliczyński JM: The role of myeloperoxidase in phagocytosis, with special regard to HCN formation, in Vennesland B, Conn E, Knowles CJ, et al (eds): *Cyanide in Biology.* London, Academic, 1981, pp 371–383.
90. Zgliczyński JM, Stelmaszyńska T, Olszowska E, et al: Peroxidative oxidation of halides catalysed by myeloperoxidase. Effect of fluoride on halide oxidation. *Acta Biochim Biophys* **30:**213–222, 1983.
91. Bolscher BG, Wever R: A kinetic study of the reaction between human myeloperoxidase, hydroperoxides and cyanide. Inhibition by chloride and thiocyanate. *Biochim Biophys Acta* **788:**1–10, 1984.
92. Epstein J: Estimation of microquantities of cyanide. *Anal Chem* **19:**272–274, 1947.

93. Aldridge WN: The conversion of cyanogen chloride to cyanide in the presence of blood proteins and sulphhydryl compounds. *Biochem J* **48:**271–276, 1951.
94. Thomas EL: Myeloperoxidase–hydrogen peroxide–chloride antimicrobial system: Effect of exogenous amines on antibacterial action against *Escherichia coli. Infect Immun* **25:**110–116, 1979.
95. Grisham MB, Jefferson MM, Thomas EL: Role of monochloramine in the oxidation of erythrocyte hemoglobin by stimulated neutrophils. *J Biol Chem* **259:**6757–6765, 1984.
96. Allen RC: Free-radical production by reticuloendothelial cells, in Sbarra AJ, Strauss RR (eds): *The Reticuloendothelial System. A Comprehensive Treatise.* New York, Plenum, 1980, vol 2, pp 309–338.
97. Halliwell B: Metal ions, free radical reactions and inflammation; an assessment of current status, in Venge P, Lindbom A (eds): *Inflammation.* Stockholm, Almquist & Wiksell, 1985, pp 271–287.
98. Arneson RM: Substrate-induced chemiluminescence of xanthine oxidase and aldehyde oxidase. *Arch Biochem Biophys* **136:**352–360, 1970.
99. Haber F, Weiss J: The catalytic decomposition of hydrogen peroxide by iron salts. *Proc R Soc Lond A* **147:**332–351, 1934.
100. Kellog EW, Fridovich I: Superoxide, hydrogen peroxide and singlet oxygen in lipid peroxidation by a xanthine oxidase system. *J Biol Chem* **250:**8812–8817, 1975.
101. Long CA, Bielski BHJ: Rate of reaction of superoxide radical with chloride-containing species. *J Phys Chem* **84:**554–557, 1980.
102. Tauber AI, Babior BM: Evidence for hydroxyl radical production by human neutrophils. *J Clin Invest* 60:374–379, 1977.
103. Weiss SJ, Rustagi PK, LoBuglio AF: Human granulocyte generation of hydroxyl radical. *J Exp Med* **147:**316–323, 1978.
104. Green MR, Hill HAO, Okolow-Zubkowska MJ, et al: The production of hydroxyl and superoxide radicals by stimulated human neutrophils—Measurements by EPR spectroscopy. *FEBS Lett* **100:**23–27, 1979.
105. Sagone AL, Decker MA, Wells RM, et al: A new method for the detection of hydroxyl radical production by phagocytic cells. *Biochim Biophys Acta* **628:**90–97, 1980.
106. Cheson BD, Christensen RL, Sperling R, et al: The origin of the chemiluminescence of phagocytosing granulocytes. *J Clin Invest* **58:**789–796, 1976.
107. Held AM, Hurst JK: Ambiguity associated with use of singlet oxygen trapping agents in myeloperoxidase-catalyzed oxidations. *Biochem Biophys Res Commun* **81:**878–885, 1978.
108. Klebanoff SJ, Rosen H: The role of myeloperoxidase in the microbicidal activity of polymorphonuclear leukocytes, in *Oxygen Free Radicals and Tissue Damage.* Ciba Foundation Symposium 65. Amsterdam, Excerpta Medica, 1979, pp 263–282.
109. Zgliczyński JM, Olszowska E, Olszowski S, et al: A possible origin of chemiluminescence in phagocytosing neutrophils. Reaction between chloramines and H_2O_2. *Int J Biochem* **17:**515–519, 1985.
110. DeChatelet WR, Long GD, Shirley PS, et al: Mechanism of the luminol-dependent chemiluminescence of human neutrophils. *J. Immunol* **129:**1589–1593, 1982.
111. Dahlgren C, Stendahl O: Role of myeloperoxidase in luminol-dependent chemiluminescence of polymorphonuclear leukocytes. *Infect Immun* **39:**736–741, 1983.
112. Zgliczyński JM, Olszowska E, Olszowski S, et al: A possible origin of chemiluminescence in phagocytosing neutrophils. Myeloperoxidase-mediated chlorination of proteins and tryptophan. *Int J Biochem* **17:**393–397, 1985.
113. Ushijima Y, Nakano M: Excitation of indole analogs by phagocytosing leukocytes. *Biochem Biophys Res Commun* **82:**835–858, 1978.
114. Weiss SJ, Young J, LoBuglio, et al: Role of hydrogen peroxide in neutrophil-mediated destruction of cultured endothelial cells. *J Clin Invest* **68:**714–719, 1981.
115. Ward PA, Till GO, Kunkel R, et al: Evidence for role of hydroxyl radical in complement and neutrophil-dependent tissue injury. *J Clin Invest* **72:**789–795, 1983.
116. Varani J, Fligiel SEG, Till GO, et al: Pulmonary endothelial cell killing by human neutrophils. Possible involvement of hydroxyl radical. *Lab Invest* **53:**656–663, 1985.

117. Clark RA, Klebanoff SJ: Neutrophil-mediated tumor cell cytotoxicity. Role of the peroxidase system. *J Exp Med* **141:**1442–1450, 1975.
118. Weiss SJ, Slivka A: Monocyte and granulocyte-mediated tumor cell destruction. A role for the hydrogen peroxide–myeloperoxidase–chloride system. *J Clin Invest* **69:**255–262, 1982.
119. Jacobs AA, Paul BB, Strauss RR, et al: The role of the phagocyte in host–parasite interactions. XXIII. Relation of bactericidal activity to peroxidase-associated decarboxylation and deamination. *Biochem Biophys Res Commun* **39:**284–289, 1970.
120. Clark RA, Szot S, Williams MA, et al: Oxidation of lysine side chains of elastin by the myeloperoxidase system and by stimulated human neutrophils. *Biochem Biophys Res Commun* **135:**451–457, 1986.
121. Drożdż R: Modification of proteins by the MPO–H_2O_2–Cl^- system, doctoral thesis, 1985.
122. Tsan MF, Chen JW: Oxidation of methionine by human polymorphonuclear leukocytes. *J Clin Invest* **65:**1041–1050, 1980.
123. Tsan MF: Myeloperoxidase-mediated oxidation of methionine and amino acid decarboxylation. *Infect Immun* **36:**136–141, 1982.
124. Silverstein RM, Hager LP: The chloroperoxidase-catalyzed oxidation of thiols and disulphides to sulfenyl chloride. *Biochemistry* **13:**5069–5073, 1974.
125. Matheson NR, Wong PS, Travis J: Enzymatic inactivation of human alpha-1-proteinase inhibitor by neutrophil myeloperoxidase. *Biochem Biophys Res Commun* **88:**402–409, 1979.
126. Matheson NR, Wong PS, Schuyler M, et al: Interaction of human α-1-proteinase inhibitor with neutrophil myeloperoxidase. *Biochemistry* **20:**331–336, 1981.
127. Alexander NM: Oxidative cleavage of tryptophanyl peptide bonds during chemical and peroxidase catalyzed iodinations. *J Biol Chem* **249:**1946–1952, 1974.
128. Rudie NG, Porter DJT, Bright HJ: Chlorination of an active site tyrosyl residue in D-amino acid oxidase by *N*-chloro-D-leucine. *J Biol Chem* **255:**498–506, 1980.
129. Weiss SJ, Peppin G, Ortiz X et al: Oxidative autoactivation of latent collagenase by human neutrophils. *Science* **227:**747–749, 1985.
130. Mello Filho AC, Meneghini R: In vivo formation of single strand breaks in DNA by hydrogen peroxide is mediated by the Haber–Weiss reaction. *Biochim Biophys Acta* **781:**56–61, 1984.
131. Greenwald RA, Moy WW: Effect of oxygen-derived free radicals on hyaluronic acid. *Arthritis Rheum* **23:**455–458, 1980.
132. Sepe SM, Clark RA: Oxidant membrane injury by the neutrophil myeloperoxidase system. I. Characterization of a liposome model and injury by myeloperoxidase, hydrogen peroxide, and halides. *J Immunol* **134:**1888–1895, 1985.
133. Sepe SM, Clark RA: Oxidant membrane injury by the neutrophil myeloperoxidase system. II. Injury by stimulated neutrophils and protection by lipid-soluble antioxidants. *J Immunol* **134:**1896–1901, 1985.
134. Slater FF, Benedetto C: Free radical reactions in relation to lipid peroxidation, inflammation and prostaglandin metabolism, in Berti F, Velo GP (eds): *The Prostaglandin System.* New York, Plenum, 1981, pp 109–126.
135. Schauenstein E, Esterbauer H, Zollner R: *Aldehydes in Biological Systems.* London, Pion, 1977.
136. Selvaraj RJ, Zgliczyński JM, Paul BB et al: Chlorination of reduced nicotinamide adenine dinucleotides by myeloperoxidase: A novel bactericidal mechanism. *J Reticuloendothel Soc* **27:**31–38, 1980.
137. Rosen H, Klebanoff SJ: Oxidation of *Escherichia coli* iron centers by the myeloperoxidase-mediated microbicidal system. *J Biol Chem* **257:**13731–13735, 1982.
138. Rosen H, Klebanoff SJ: Oxidation of microbial iron sulfur centers by the myeloperoxidase–H_2O_2–halide antimicrobial system. *Infect Immun* **47:**613–618, 1985.
139. Odajima T, Onishi M, Sato N: Metabolism of steroids by myeloperoxidase. *Jpn J Oral Biol* **23:**489–497, 1981.
140. Paredes JM, Weiss S: Human neutrophils transform prostaglandins by a myeloperoxidase-dependent mechanism. *J Biol Chem* **257:**2738–2740, 1982.

141. Onishi M, Odajima T: On the product of prostaglandin E_1 oxidized by the myeloperoxidase–H_2O_2–chloride system. *Jpn J Oral Biol* **27:**291–298, 1985.
142. Henderson WR, Klebanoff SJ: Leukotriene production and inactivation by normal, chronic granulomatous disease and myeloperoxidase deficient neutrophils. *J Biol Chem* **258:**13522–13527, 1983.
143. Lee CW, Lewis RA, Tauber AI, et al: The myeloperoxidase-dependent metabolism of leukotrienes C_4, D_4 and E_4 to 6-trans leukotriene B_4 diastereoisomers and the subclass-specific S-diastereoisomeric sulfoxides. *J Biol Chem* **258:**15004–15009, 1983.
144. Clark RA, Klebanoff SJ: Chemotactic factor inactivation by the myeloperoxidase–H_2O_2–halide system. *J Clin Invest* **64:**913–919, 1979.
145. Tsan MF, Denison RC: Oxidation of *n*-formyl methionine chemotactic peptide by human neutrophils. *J Immunol* **126:**1387–1389, 1981.
146. Håkanson L: Antiinflammatory effects of myeloperoxidase and eosinophil peroxidase, in Venge P, Lindbom A (eds): *Inflammation.* Stockholm, Almquist & Wiksell, 1985, pp 319–323.
147. Klebanoff SJ, Clark RA: *The Neutrophil: Function and Clinical Disorders.* Amsterdam, Elsevier-North Holland, 1978, pp 217–488.
148. Kobayashi M, Tanaka T, Usui T: Inactivation of lysosomal enzymes by the respiratory burst of PMN leukocytes. Possible involvement of myeloperoxidase–H_2O_2–halide system. *J Lab Clin Med* **100:**896–907, 1982.
149. Clark RA, Stone PJ, ElHag A, et al: Myeloperoxidase-catalyzed inactivation of α_1-protease inhibitor by human neutrophils. *J Biol Chem* **256:**3348–3353, 1981.
150. Carp H, Janoff A: In vitro suppression of serum elastase-inhibitory capacity by reactive oxygen species generated by phagocytosing polymorphonuclear leukocytes. *J Clin Invest* **63:**793–797, 1979.
151. Lieberman J: Elastase, collagenase, emphysema and α_1-antitrypsin deficiency. *Chest* **70:**62–67, 1976.
152. Adeniyi-Jones SK, Karnovsky ML: Oxidative decarboxylation of free and peptide-linked amino acids in phagocytizing guinea pig granulocytes. *J Clin Invest* **68:**365–373, 1981.
153. Wolff SP, Garner A, Dean RT: Free radicals, lipids and protein degradation. *Trends Biochem Sci* **11:**27–31, 1986.
154. Stadtman ER: Oxidation of proteins by mixed-function oxidation systems: implication in protein turnover, ageing and neutrophil function. *Trends Biochem Sci* **11:**11–12, 1986.
155. Payne JW, Tuffnell JM: Cleavage of peptide-bound methionine sulphoxide and methionine sulphone: Possible role in enhanced proteolysis in *Escherichia coli. FEMS Microbiol Lett* **12:**279–282, 1981.
156. Bernlohr DA, Switzer RL: Reaction of *Bacillus subtilis* glutamine phosphoribosylpyrophosphate amidotransferase with oxygen. Chemistry and regulation by ligands. *Biochemistry* **20:**5675–5681, 1981.
157. Rivett AJ: Preferential degradation of the oxidatively modified form of glutamine synthetase by intracellular mammalian proteases. *J Biol Chem* **260:**300–305, 1985.
158. Offerman MK, McKay MJ, Marsh MW, et al: Glutathione disulphide inactivates, destabilizes and enhances proteolytic susceptibility of fructose 1,6-bis-phosphate aldolase. *J Biol Chem* **259:**8886–8891, 1984.
159. Sagone AL, Husney RM, O'Dorisio MS, et al: Mechanisms for the oxidation of reduced glutathione by stimulated granulocytes. *Blood* **63:**96–104, 1984.
159a. Olszowska E, Olszowski S, Zgliczyński JM: Enhanced Susceptibility of chlorinated proteins to proteolysis, unpublished.
160. Devraede GJ, Taylor WF, Sauer WG, et al: Cancer risk and life expectancy of children with ulcerative colitis. *N Engl J Med* **285:**17–21, 1971.
161. Wlodkowski TJ, Rosenkranz HS: Mutagenicity of sodium hypochlorite for salmonella typhimurium. *Mutat Res* **31:**39–43, 1975.
162. Shih KL, Lederberg J: Chloramine mutagenesis in *Bacillus subtilis. Science* **192:**1141–1143, 1976.

163. Cheh AM, Skochdopole J, Koski P, et al: Nonvolatile mutagens in drinking water: Production by chlorination and destruction by sulfite. *Science* **207**:90–92, 1980.
164. Weitzman SA and Stossel TP: Mutation caused by human phagocytes. *Science* **212**:546–547, 1981.
165. Weitberg AB, Weitzman SA, Destrempes M, et al: Stimulated human phagocytes produce cytogenetic changes in cultured mammalian cells. *N Engl J Med* **308**:26–30, 1983.
166. Birnboim HC: DNA strand breakage in human leukocytes exposed to a tumor promoter, phorbol myristate acetate. *Science* **215**:1247–1249, 1982.
167. Dean RT, Roberts CR, Forni LG: Oxygen-centered free radicals can efficiently degrade the polypeptide of proteoglycans in whole cartilage. *Biosci Rep* **4**:1017–1021, 1984.
168. Dean RT, Wolff SP: Free radical damage to proteoglycans and proteins, in Venge P, Lindbom A (eds): *Inflammation*. Stockholm, Almquist & Wiskell, 1985, pp 289–297.
169. Clark RA, Szot S: The myeloperoxidase–hydrogen peroxidase–halide system as effector of neutrophil-mediated tumor cell cytotoxicity. *J Immunol 126:*1295–1301, 1982.
170. Gale RP, Zighelboim J: Polymorphonuclear leukocytes in antibody-dependent cellular cytotoxicity. *J. Immunol* **114**:1047–1052, 1975.
171. Zighelboim J, Gale RP, Kedar E: Polymorphonuclear leukocyte Fc receptors in antibody-dependent cellular cytotoxicity (ADCC). *Transplantation* **21**:524–530, 1976.
172. Clark RA, Klebanoff SJ: Studies on the mechanism of antibody-dependent polymorphonuclear leukocyte-mediated cytotoxicity. *J Immunol* **119**:1413–1418, 1977.
173. Hafeman DG, Lucas ZJ: Polymorphonuclear leukocyte-mediated, antibody-dependent, cellular cytotoxicity against tumor cells: Dependence on oxygen and the respiratory burst. *J Immunol* **123**:55–62, 1979.
174. Dallegri F, Frumento G, Minervini F: Role of the oxidative metabolic burst in the antibody-dependent cellular cytotoxicity mediated by neutrophil polymorphonuclears. *Exp Hematol* **10**:859–866, 1982.
175. Eaton JW, Kolpin CF, Swofford HS, et al: Chlorinated urban water: A cause of dialysis-induced hemolytic anemia. *Science* **181**:463–464, 1973.
176. Klebanoff SJ and Pincus SH: Hydrogen peroxide utilization in myeloperoxidase-deficient leucocytes: A possible microbicidal control mechanism. *J Clin Invest* **50**:2226–2229, 1971.
177. Rosen H, Klebanoff SJ: Chemiluminescence and superoxide production by myeloperoxidase-deficient leukocytes. *J Clin Invest* **58**:50–60, 1976.
178. Dri P, Soranzo MR, Cramer R, et al: Role of myeloperoxidase in respiratory burst of human polymorphonuclear leukocytes. Studies with myeloperoxidase-deficient subjects. *Inflammation* **9**:21–31, 1985.
179. Stendahl O, Coble BI, Dahlgren C, et al: Myeloperoxidase modulates the phagocytic activity of polymorphonuclear neutrophil leukocytes. Studies with cells from a myeloperoxidase-deficient patient. *J Clin Invest* **73**:366–373, 1984.
180. Coble BI, Dahlgren C, Hed J, et al: Myeloperoxidase reduces the opsonizing activity of immunoglobulin G and complement component C3b. *Biochim Biophys Acta* **802**:501–505, 1984.
181. Klebanoff SJ and Hamon CB: Role of myeloperoxidase-mediated antimicrobial systems in intact leucocytes. *J Reticuloendothel Soc* **12**:170–196, 1972.
182. DeChatelet LR, McPhail LC, Shirley PS: Effect of cyanide on NADPH oxidation by granules from human polymorphonuclear leucocytes. *Blood* **49**:445–454, 1977.
183. Cohen HJ, Chowaniec ME: Superoxide production by digitonin-stimulated guinea pig granulocytes. The effects of *N*-ethylmaleimide, divalent cations and glycolytic and mitochondrial inhibitors on the activation of the superoxide generating system. *J Clin Invest* **61**:1088–1096, 1978.
184. Beswick PH, Slater TF: Modification by metals, sulphydryl reagents and cyanide of the particle stimulated enhancement of oxygen consumption in bovine granulocytes. *Chem Biol Interact* **20**:373–382, 1978.
185. Reed PW: Glutathione and the hexose monophosphate shunt in phagocytizing and H_2O_2-treated rat leucocytes. *J Biol Chem* **244**:2459–2464, 1969.

186. Strauss RR, Paul BB, Jacobs AA, et al: The role of the phagocyte in host–parasite interactions. XIX. Leucocytic glutathione reductase and its involvement in phagocytosis. *Arch Biochem Biophys* **135:**265–271, 1969.
187. Hendricks SB, Taylorson RB: Breaking of seed dormancy by catalase inhibition. *Proc Natl Acad Sci USA* **72:**306–309, 1975.
188. Suzuki H, Kakinuma K: Evidence that NADPH is the actual substrate of the oxidase responsible for the "respiratory burst" of phagocytosing polymorphonuclear leukocytes. *J Biochem* **93:**709–715, 1983.
189. Cohen HJ, Chowaniec ME, Davies WE: Activation of the guinea pig granulocyte NAD(P)H-dependent superoxide generating enzyme: Localization in a plasma membrane enriched particle and kinetics of activation. *Blood* **55:**355–363, 1980.
190. Klebanoff SJ: Myeloperoxidase: Contribution to the microbicidal activity of intact leukocytes. *Science* **169:**1095–1097, 1970.
191. Edwards SW, Hallett MB, Campbell AK: Oxygen-radical production during inflammation may be limited by oxygen concentration. *Biochem J* **217:**851–854, 1984.
192. Britigan BE, Cohen MS: Effects of human serum on bacterial competition with neutrophils for molecular oxygen. *Infect Immun* **52:**657–663, 1986.

16

The Respiratory Burst and Psoriasis

RUDOLF E. SCHOPF

1. INTRODUCTION

Psoriasis is a disease affecting about 1–2% of the population worldwide, similar to diabetes mellitus.[1] It occurs somewhat more frequently in the white population than among Asians and Eskimos.[1] The disease has two peaks of onset, an early one occurring around the age of 20 and a later one around 60.[2] In early-onset psoriasis, 85% of patients exhibit the HLA Cw6 antigen; in the late-onset group, no such pattern of inheritance can be found.[2]

Clinically, the disease can be divided into two groups: psoriasis vulgaris and psoriasis nonvulgaris. The former is characterized by red plaques with scales occurring with a predilection on the elbows and knees but also the trunk and scalp with a striking symmetry of lesions similar to the appearance of a Rorschach test[3] (Figs. 1 and 2). Psoriasis nonvulgaris comprises lesions occurring only in the genitocrural region, palms, and soles, but the term also includes psoriatic erythroderma and pustular psoriasis (Fig. 3). The pustules may occur generalized or may be confined to the palms, soles and tips of the toes and fingers. Psoriatic arthritis is not considered here.

Histologically, psoriasis is characterized by regular acanthosis and papillomatosis, i.e., thickening of the upper parts of the skin, in particular the dermoepidermal region (Fig. 4). Papillary vessels are surrounded by edema and a mononuclear cell infiltrate. Mononuclear cells move into the epidermis, where slight spongiosis develops. Apparently after the formation of parakeratosis, polymorphonuclear leukocytes (PMN) are discharged from the papillary capillaries, a process termed the squirting papilla.[4] PMN then move into the interstitial spaces in the epidermis to form small unique spongiform pustules of Kogoj[5] that later enlarge to become Munro microabscesses in the upper epidermis.[6] The epidermis contains,

RUDOLF E. SCHOPF • Department of Dermatology, Johannes Gutenberg University, D-6500 Mainz, West Germany.

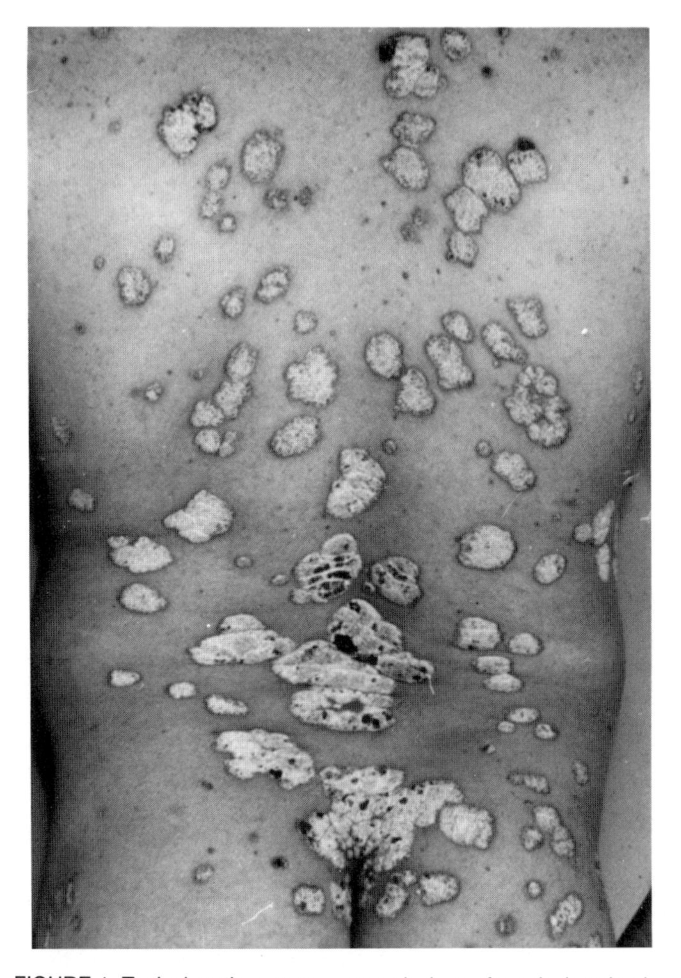

FIGURE 1. Typical erythematosquamous lesions of psoriasis vulgaris.

at its outer layer, cells that still have nuclei as a sign of disturbed cell maturation, termed parakeratosis. In pustular psoriasis, the histological features are similar, but the microabscesses are grossly enlarged to become macropustules visible by the naked eye.[6]

Pathophysiologically, psoriasis exhibits four characteristic features[7]: (1) an inherited predisposition, (2) increased proliferation of keratinocytes, (3) a correspondingly decreased keratinocyte differentiation, and (4) a marked inflammatory

List of abbreviations: ADCC, antibody-dependent cellular cytotoxicity; aggIg, aggregated immunoglobulin; Con A, concanavalin A; ETAF, epidermal cell-derived thymocyte-activated factor; G6PD, glucose 6-phosphate dehydrogenase; IL-1, interleukin-1; LPS, lipopolysaccharide; Mϕ, monocytes/Macrophages; PMA, phorbol myristate acetate; PMN, polymorphonuclear leukocytes.

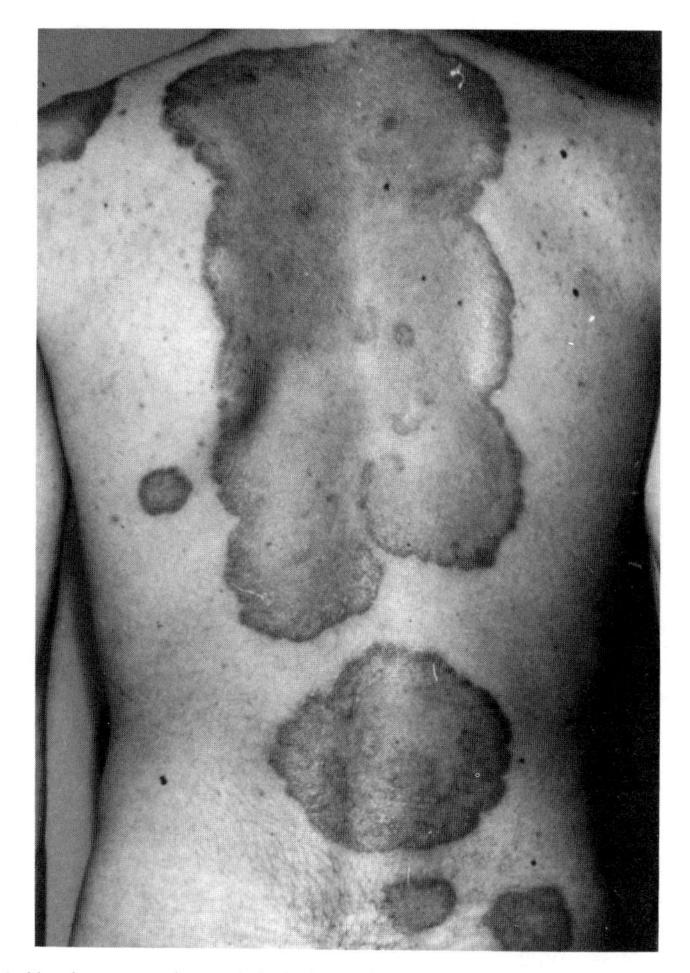

FIGURE 2. Nearly symmetric psoriatic lesions after removal of scales. Note demarcated and prominent borders of lesions indicating disease activity.

reaction, clinically evident as erythema. This chapter deals with the inflammatory reaction in psoriasis, in particular phagocytes with their respiratory burst.

2. THE IMPORTANCE OF PHAGOCYTES IN PSORIASIS

2.1. Vulgar Psoriasis

The cell infiltrate around the papillary vessels in psoriasis contains monocytes/macrophages (Mϕ) as well as PMN. Furthermore, the spongiform pustules and the Munro microabscesses are truly diagnostic of psoriasis; in their absence, the

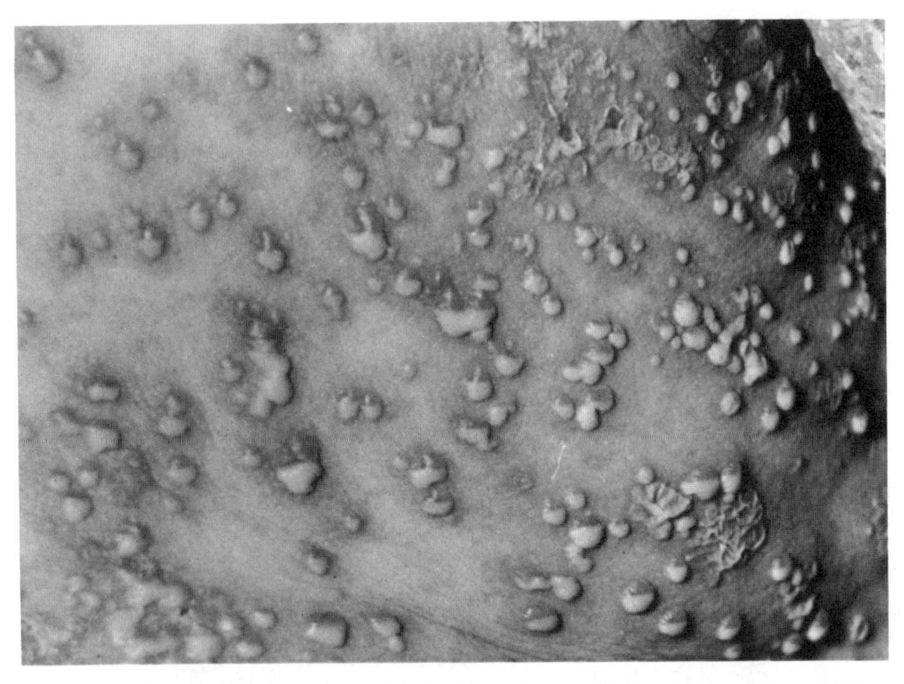

FIGURE 3. Generalized pustular psoriasis with visible pus-filled vesicles on truncal skin.

diagnosis can rarely be made with certainty on a histological basis.[6] Neutrophils migrating through the epidermis could in some way damage it and lead to reparative proliferation of epidermal cells as suggested.[8] This is supported by the Koebner phenomenon (isomorphic response), i.e., the occurrence of new psoriatic lesions at sites of skin injury. Wound healing that cannot shut itself off has thus been likened to psoriasis in individuals genetically programmed to develop psoriasis.[8] PMN and Mϕ are involved in wound repair, as is epidermal cell division, providing a working hypothesis for psoriasis. Although the Koebner penomenon is not entirely specific for psoriasis and occurs in only up to 50% of patients,[9] it provides an important clue for pathogenesis in predisposed individuals. Moreover a streptococcal infection of the tonsils may also elicit a Koebner phenomenon resulting in guttate or droplike psoriatic skin lesions, as well known clinically. This concept can be expanded in that all types of mechanical or inflammatory injuries may elicit a Koebner phenomenon. As the early cellular infiltrate in skin trauma and psoriasis consists of PMN and Mϕ, it is only natural to study phagocytes in psoriasis.

Indeed, phagocyte abnormalities have been found in psoriasis both in the reticuloendothelial system and in the peripheral blood. Enlarged bone marrow space and enhanced fixed phagocyte function of liver and spleen have been reported.[10,11] Peripheral blood PMN exhibit a more ruffled cell membrane.[12] Phagocytes have

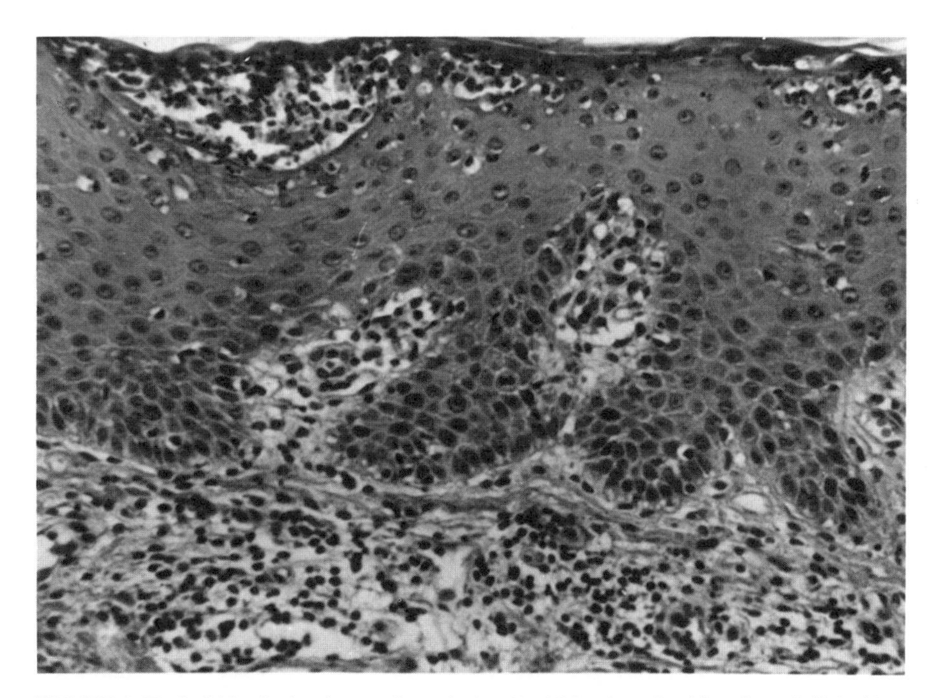

FIGURE 4. Typical histologic picture of psoriasis with thickening of epidermis and dilated tortuous papillar vessels embedded in edema, surrounded by an inflammatory infiltrate. The PMN interspersed in the epidermis form spongiform pustules that enlarge and become Munro microabscesses (top). The stratum corneum has been removed by topical therapy. (Hematoxylin & eosin, ×280.)

been demonstrated by several investigators to exhibit a generally enhanced function as summarized.[13,14] Owing to phagocytic proinflammatory properties, it was of considerable interest to study the respiratory burst and closely related events in psoriasis.

2.2. Pustular Psoriasis

What has been said for vulgar psoriasis is even more important for pustular psoriasis. A mononuclear infiltrate precedes the appearance of PMN-forming spongiform pustules, which enlarge up to several millimeters in diameter.[6] As the size of the pustule increases, degenerated and thinned epidermal cells undergo complete cytolysis, forming a large cavity.[6,17] Also, a chemotactic factor from scales in pustular psoriasis indentical to that in psoriasis vulgarus has been found.[18] Moreover a functional Fc-receptor defect was recently described in a patient with pustular psoriasis of the palms and soles.[19] It follows that phagocytes, in particular the respiratory burst with its rapid response, may be significant in psoriasis.

3. RESPIRATORY BURST-RELATED PERIPHERAL BLOOD PHAGOCYTE FUNCTION IN PSORIASIS

The interaction of phagocytes with appropriate stimuli results in markedly increased uptake of oxygen, an activation of the hexose monophosphate shunt, and the generation of highly reactive oxygen intermediates, including superoxide anion, hydrogen peroxide, hydroxyl radical, and singlet oxygen, as summarized by Babior.[20] These events are collectively termed the respiratory burst. Stimuli may be particulate, such as bacteria, zymosan particles, insoluble immune complexes, or soluble like complement components, including C5a, C3b, and notably factor H (β1H),[21] lectins, phorbol myristate acetate (PMA), or soluble immune complexes. The energy released in the resulting oxygenation of substrates can be detected by sensitive photomultipliers in the form of light, i.e., chemiluminescence.[22] A large increase of the signal can be obtained by the introduction of a chemiluminigenic probe such a luminol.[23,24] In the experiments described below, the chemiluminescenct responses of PMN and Mϕ of patients with active psoriasis and normal control individuals were compared. Chemiluminescence was chosen, as previous experiments had disclosed it to be more rapid and more sensitive than quantitative nitroblue tetrazolium reduction, another measure of the respiratory burst.[25] Furthermore, glucose 6-phosphate dehydrogenase (G6PD) activity was determined in

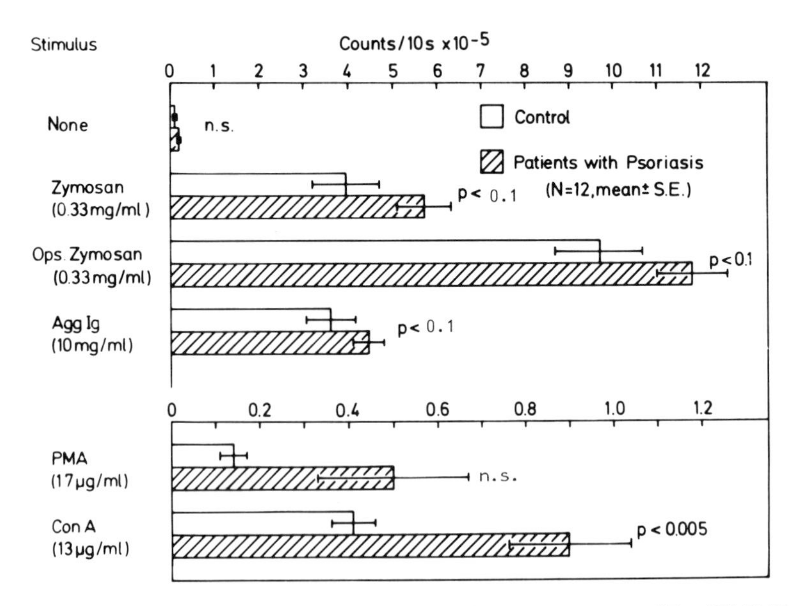

FIGURE 5. Luminol (2×10^{-4} M)-enhanced chemiluminescent response of 5×10^5 PMN upon stimulation with various stimuli. Peak counts were recorded from bottom to top after 4, 30, 30, 25, 30, and 30 min, respectively. Statistical comparisons were made by the two-tailed Student's *t*-test. (For details of method, see Schopf et al.[25])

mononuclear leukocytes in psoriasis. G6PD is the key enzyme regulating glucose oxidation by the hexose monophosphate shunt.

3.1. Chemiluminescent Response of PMN in Psoriasis

These cells were isolated from the peripheral blood of 12 patients with psoriasis and 12 age- and sex-matched healthy controls. After centrifugation over a density gradient, the cells were allowed to sediment in a dextran solution.[26] The cell suspensions obtained contained >98% PMN. In order to elicit a respiratory burst, the following stimuli were used: (1) zymosan particles, (2) zymosan particles opsonized with autologous serum (as C3b stimulant), (3) aggregated immunoglobulin (aggIg) (interacting with Fc receptors), (4) PMA for which specific receptors have been identified on phagocytes (activating protein kinase C),[27] and (5) concanavalin A (Con A) binding to mannose moieties on the cell surface. In addition, the generation of oxygen intermediates of quiescent unstimulated PMN was measured by chemiluminescence. Figure 5 displays the results.

Although the stimulation of PMN with zymosan, opsonized zymosan, aggIg, and PMA in all cases brought about a higher stimulatory response in the patient group the values failed to reach statistical significance. By contrast, incubation with ConA effected a markedly higher respiratory burst activity in PMN from patients with psoriasis. Unstimulated PMN exhibited no difference in their generation of oxygen intermediates between patients and controls.

3.2. Chemiluminescent Response of Mononuclear Leukocytes in Psoriasis

After isolation of mononuclear leukocytes,[26] Mϕ were identified by esterase staining.[28] For respiratory burst measurements, Mϕ did not have to be separated from lymphocytes, as the chemiluminescent activity had been shown to result from Mϕ and not lymphocytes.[21] Figure 6 shows that zymosan and notably zymosan opsonized with autologous serum induced a significantly higher generation of chemiluminescence in psoriasis patients compared with controls. By contrast, aggIg, PMA, and Con A did not elicit a higher respiratory burst in the Mϕ of patients. Quiescent, unstimulated Mϕ also behaved similar to the controls.

From these findings, it was concluded that different stimuli may regulate the extent of the respiratory burst in PMN and Mϕ in psoriasis. Furthermore, the data indicate that particulate stimuli causing phagocytosis elicit an enhanced respiratory burst in Mϕ. In addition, as the chemiluminescent activity of resting phagocytes was not enhanced, in vivo prestimulation of phagocytes is unlikely in psoriasis.

3.3. Effects of Psoriatic Serum on the Respiratory Burst

In order to assess further the effects of psoriatic serum on the respiratory burst of phagocytes, the sera of patients with psoriasis and healthy controls were incu-

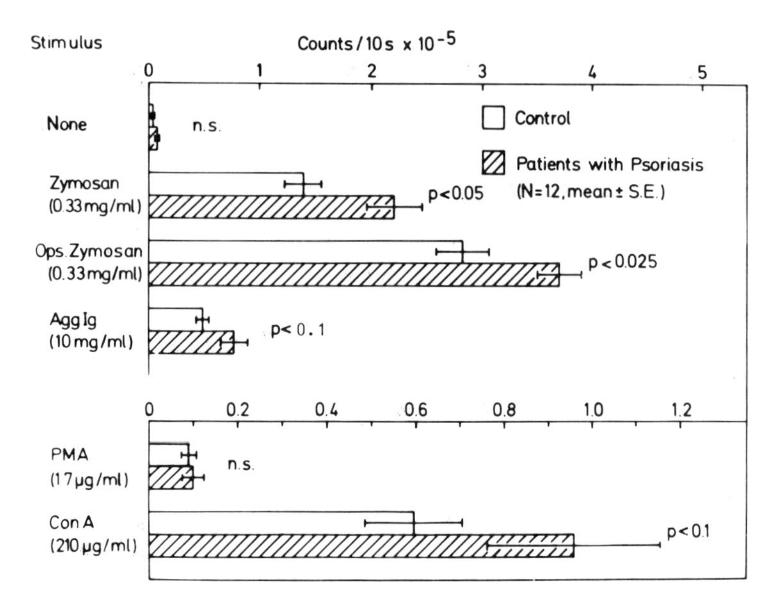

FIGURE 6. Effect of different stimuli on the generation of luminol (2×10^{-4} M)-amplified chemiluminescence of 2×10^5 Mϕ. Counts were recorded from bottom to top after 2, 30, 20, 25, 30, and 30 min, respectively. The two-tailed Student's *t*-test served for statistical comparisons.

bated with normal PMN in order to determine the chemiluminescent response. 50 μl of serum, stored at −20°C, thawed once only was mixed with 5×10^5 PMN and 2×10^{-4} M luminol in a final volume of 300 μl. Every 40 sec up to 6 min, chemiluminescent activity was determined. Maximum activities were measured between 800 and 1600 sec. Psoriatic sera, however, reached peak activity earlier than controls, the difference being small but statistically significant (Fig. 7). No such difference could be found comparing sera from patients with pyodermic infecfions, leg ulcers, or atopic dermatitis. Furthermore, the sums of the counts measured, i.e., the integrated counts, were higher in the psoriasis group (220 ± 46 versus 160 ± 33 ($\times 10^3$) in the control; mean ±SEM). These data suggested that psoriatic serum contains more proinflammatory activity for the respiratory burst than does normal serum or serum of other skin diseases.

3.4. Activity of G6PD Activity in Mononuclear Leukocytes in Psoriasis

Although different stimuli may regulate the extent of augmented leukocytic activity in psoriasis, the underlying biochemical mechanisms have not been determined. We therefore compared psoriatics and controls regarding concentrations of G6PD activity in mononuclear leukocytes. After isolation, the cells were lysed and

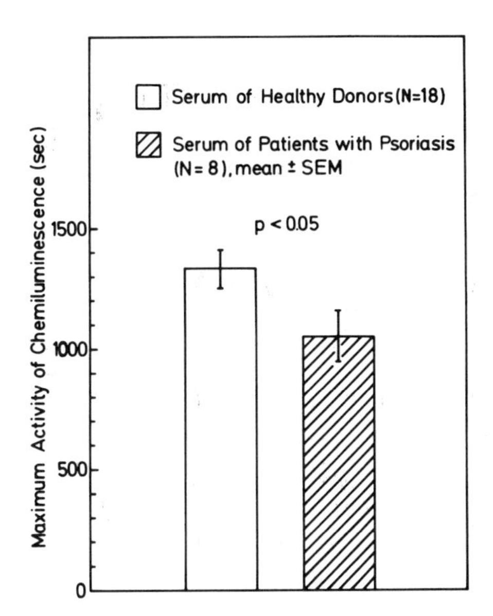

FIGURE 7. Influence of serum on the peak luminol-enhanced chemiluminescent response of 5×10^5 normal PMN (statistical comparison by the Wilcoxon test).

subjected to enzyme determinations by a photometric test. G6PD activity was expressed in terms of both enzyme per cell number and per total protein. Table I summarizes the results. In both cases, we found a statistically significant augmented activity of G6PD activity in the mononuclear leukocytes of 16 patients with active psoriasis compared with 16 controls. Psoriatics exhibited 52% more G6PD activity based on cell number and 34% more based on total soluble protein. In common with previous findings,[29] the number of Mϕ among the mononuclear leukocytes was in the range of 9–21% (as judged by esterase staining) and did not differ between patients and controls, as reported elsewhere.[30] Our finding of 4.16 ± 1.44 (mean ±

TABLE I. Comparison of G6PDH Activity in Mononuclear Leukocytes in 16 Patients with Active Psoriasis Vulgaris Compared with 16 Healthy Control Individuals[a,b]

Group	MNL ($mU/10^6$)	Protein ($mU/10^{-6}$ g)
Controls	4.16 ± 1.44	45.7 ± 11.6
Psoriasis	6.32 ± 3.23^c	61.1 ± 22.4^c

[a] From Schopf et al.[63]
[b] Mean ±SD.
[c] $p < 0.05$, two-tailed Wilcoxon test.

SD) mU G6PD/10^6 mononuclear cells in the control individuals compares well with the 3.37 ± 0.33 mU/10^6 cells reported in the literature.[31]

The activity of the hexose monophosphate shunt follows that of G6PD. Other studies have indicated enhanced hexose monophosphate shunt activity and thus of G6PD in mononuclear leukocytes based on rather indirect methods, such as increased nitroblue tetrazolium reduction[32] or antibody-dependent cellular cytotoxicity (ADCC).[33] The present findings, however, demonstrate the enzyme activity directly under well-defined conditions, comprising excess of substrate (glucose 6-phosphate, G6P) and coenzyme (NADP).

4. INFLUENCE OF EPIDERMAL CELL-CULTURE SUPERNATANTS OF PSORIATIC SKIN ON THE RESPIRATORY BURST OF PHAGOCYTES

In order to test whether products of epidermal cells exert an influence on the respiratory burst, we isolated epidermal cells from lesional and uninvolved psoriatic skin using a suction blister technique.[34] Thus we obtained five 7-mm-diameter blisters that were removed with scissors and foreceps, incubated on trypsin solution, and washed several times as described.[35] 1×10^6 viable epidermal cells were incubated for 48 hours in the presence of culture medium alone, with the addition of silica or lipopolysaccharide (LPS). After the culture period, cell culture plates were frozen and thawed to release total cytokines, Next, the supernatants were passed through a 0.22-μm sterile filter; 50 μl of the resulting solution was added to 5×10^5 normal PMN and zymosan (0.33 mg/ml) and 2×10^{-4} M luminol solution to a final volume of 0.3 ml. Every 10–90 min, the chemiluminescent response was recorded and plotted on graph paper. Computer-assisted planimetry determined the integrals of the curves. In order to compare results obtained with different experiments, the positive zymosan control was arbitrarily set at 100%. Figure 8 exhibits the results. Supernatants from unstimulated epidermal cells were all inhibitory, regardless of lesional or uninvolved epidermis. Similarly, supernatants of silica-stimulated psoriatic epidermal cells also inhibited chemiluminescence; notably, no difference emerged between uninvolved and involved skin. LPS-stimulated epidermal cells, however, produced supernatants that were able, at least in part, to overcome the inhibitory effect on the respiratory burst or even to induce a stimulation in four of nine cases studied. Moreover, compared with uninvolved skin, epidermal cells from lesional sites effected a higher chemiluminescent response, the difference reaching statistical significance.

When the above-described epidermal cell supernatants were tested on the chemiluminescent response of resting PMN, no stimulatory effect was observed and the values did not exceed the negative buffer control (data not shown). Furthermore, no difference could be ascertained between uninvolved and involved epidermal cells from the patients.

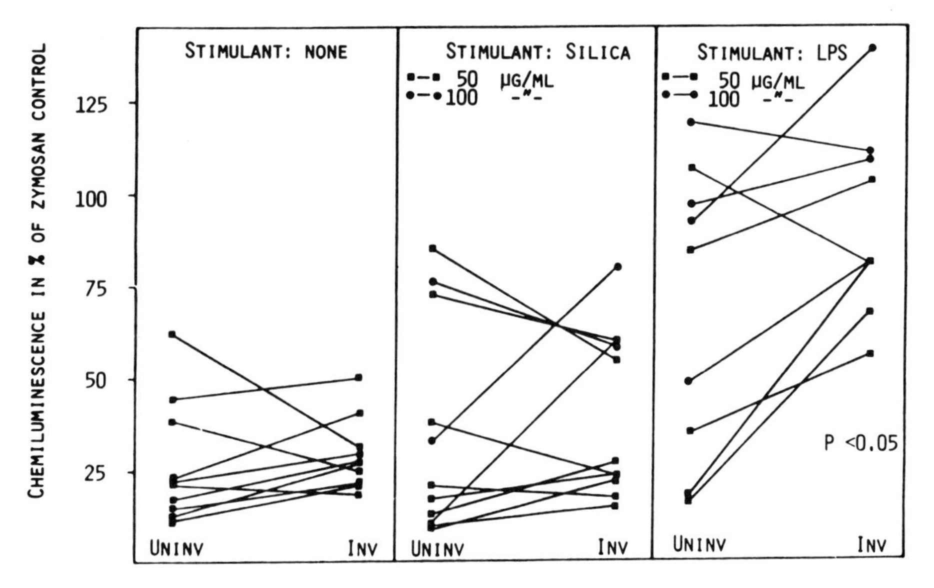

FIGURE 8. Effect of epidermal cell (1×10^6) culture supernatants (after 48 hr) of uninvolved (Uninv) and involved (Inv) psoriatic skin on zymosan (0.33 mg/ml)-induced chemiluminescence of 5×10^5 normal PMN (statistical comparison by the Wilcoxon test for matched pairs).

The supernatants tested in the zymosan-induced chemiluminescence of PMN were also examined for their activity of interleukin-1 (IL-1) of epidermal cell-derived thymocyte-activating factor (ETAF) as described.[36] We found that epidermal cells from lesional sites in psoriasis produced more IL-1-like activity than from uninvolved skin, the statistical significance reaching levels of $p < 0.01$ without further stimulus, $p < 0.05$ with silica as stimulus, and $p < 0.02$ under stimulation with LPS. There was no correlation between IL-1-like activity of epidermal cell supernatants on one hand and effect the zymosan-induced chemiluminescence on the other. This is consonant with the fact that purified IL-1 has only marginal influence on the respiratory burst of both resting and zymosan-stimulated PMN (Schopf, unpublished data) as measured by chemiluminescence.

These findings indicate that supernatants of epidermal cell cultures modify the zymosan-induced respiratory burst of PMN; i.e., tend to suppress the respiratory burst. LPS-stimulated epidermal cell supernatants of lesional psoriatic skin are able to cause a higher response on the respiratory burst than from uninvolved psoriatic skin. This finding suggests that lesional psoriatic epidermal cells possess a higher stimulatory potential for the respiratory burst than do cells from uninvolved skin. Others[37] characterized a factor derived from LPS-stimulated epidermal cells. This factor causes stimulation of PMN chemiluminescence. It may well be that our results with LPS-stimulated psoriatic epidermal cells are attributable to this factor.

5. EFFECTS OF PSORIATIC SCALES ON THE RESPIRATORY BURST

Psoriatic scales have been known for about 20 years to be chemotactic.[38] More recent papers have confirmed these findings.[39] In addition, complement split products were identified in psoriatic scales[40,41] as well as other chemotactic peptides or leukotriene B_4.[40,42–44] Superoxide generation of PMN has been achieved with aqueous extracts of psoriatic scales pooled from nine patients.[40] As pooling of scales does not permit insight into pathogenesis in individual patients, we studied extracts that were not pooled. Moreover, we compared both water-soluble and ether-soluble extracts of psoriatic scales and as a control, normal callus in order to study the effects on the respiratory burst of normal phagocytes.

5.1. Aqueous Scale Extracts

Psoriatic scales or callus from healthy individuals were lyophilized, weighed, and brought in 1 : 10 (w/v) solution with sterile pyrogen-free water to which 0.1 mg/ml penicillin–streptomycin was added. After centrifugation, the supernatants were aspirated and passed through a 0.22-μm sterile filter. Protein was adjusted to 1 mg/ml, using a commercial kit (BIO-RAD). Aliquots were added to PMN and Mϕ suspensions, and the respiratory burst activity was determined by luminol-enhanced chemiluminescence. Figure 9 displays the results. The respiratory burst in PMN and notably Mϕ was markedly activated by aqueous scale extracts. It should be noted that the extracts alone were able to cause the activation of the respiratory burst; zymosan was not added in these experiments.

In order to test the opsonizing properties, the extracts were co-incubated with zymosan, PMN, or Mϕ. Figure 10 shows that psoriatic scales effected an opsonization of the respiratory burst of PMN. No such effect was found, however, with Mϕ. From these data, it was concluded that aqueous scale extracts of psoriasis markedly stimulate the respiratory burst of phagocytes.

5.2. Organic Extracts

Ether extracts (pH 3.0) were prepared as described elsewhere.[44] In brief, freeze-dried scales or callus suspended as mentioned above were acidified to pH 3.0 with 0.5 N HCl, 3 ml (i.e., 300 mg) was extracted three times with 2 vol diethylether. After evaporation of ether, samples were suspended in physiological saline and adjusted to pH 7.4 with 0.05 N NaOH; 50 μl was used for respiratory burst measurements and added to PMN and Mϕ. Figure 11 shows that the ether extracts of psoriatic scales elicit a marked respiratory burst in both PMN and Mϕ. When the organic extracts were incubated together with zymosan-stimulated phagocytes, no potentiation of the chemiluminescent response ensued. Likewise. the extracts were not inhibitory either. We concluded that the organic extracts contained no opsonins, in contrast the findings with aqueous extracts described above.

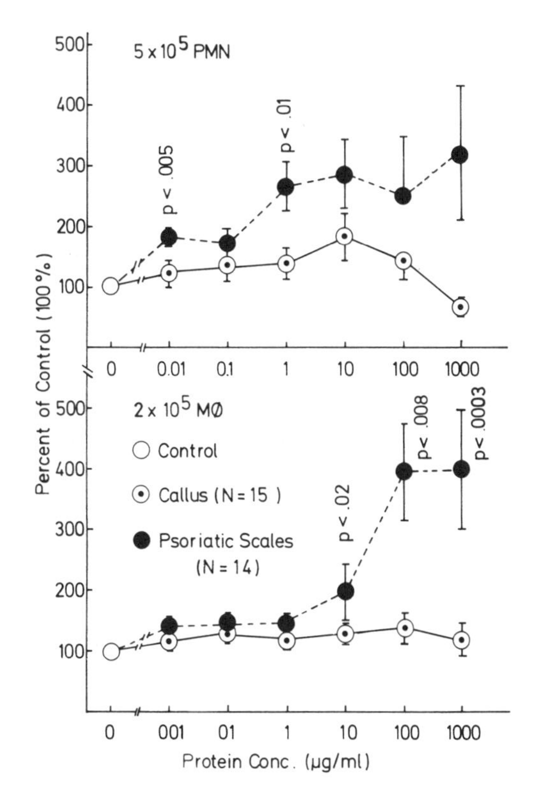

FIGURE 9. Comparison of aqueous extracts of psoriatic scales and callus on phagocyte chemiluminescence. The values are derived from the sums of the means of duplicate samples of counts every 10–60 min. The control consisted of incubation in buffer (Dulbecco's MEM for chemiluminescence, Boehringer Mannheim).

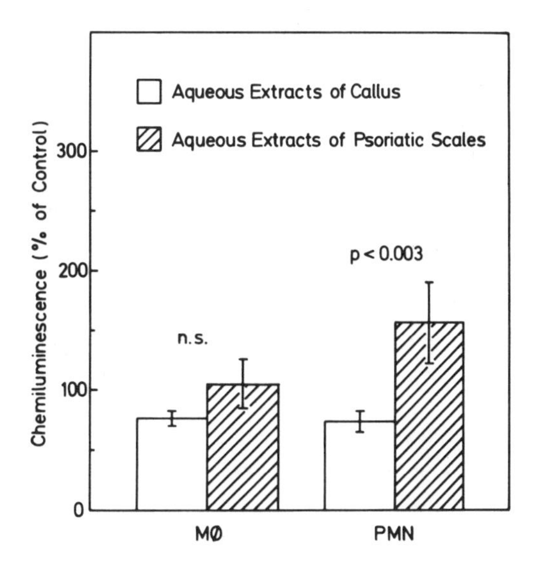

FIGURE 10. Effects of aqueous extracts (1 mg/ml) of callus and psoriatic scales on zymosan-induced (0.33 mg/ml) chemiluminescence. The 100% control consisted of incubation in zymosan (cf. Fig. 9).

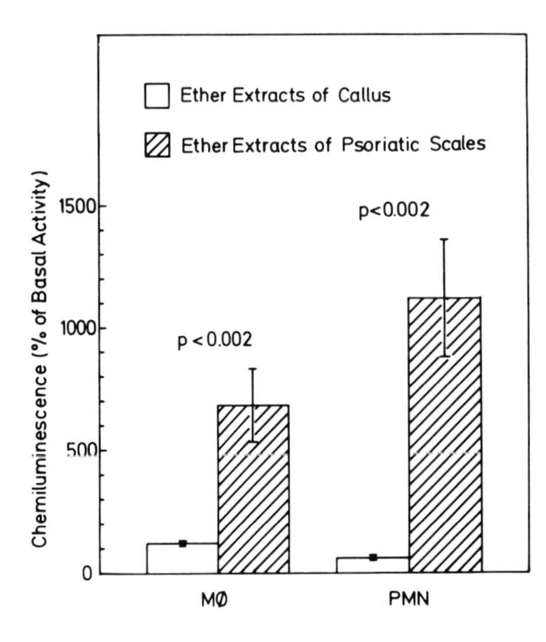

FIGURE 11. Action of organic extracts (pH 3.0) of callus ($N = 15$) and psoriatic scales ($N = 12$) (1 mg protein/ml) on the chemiluminescence of 2×10^5 Mϕ and 5×10^5 PMN. Measurements of duplicate samples were made every 5–30 min. The values represent the means ±SEM; 100% corresponds to buffer.

6. EFFECTS OF PMN ON KERATINOCYTES

Lysis of keratinocytes by stimulated phagocytes was recently demonstrated.[45] Isolated normal or neoplastic keratinocytes were labeled with ^{51}Cr and incubated in a 4-hr cytotoxicity assay with normal PMN stimulated with zymosan or PMA. After centrifugation, released ^{51}Cr was determined in the supernatants as a measure of target cell lysis. Figure 12 summarizes some of the results. When catalase (4000 U/ml) was added to the samples, cytotoxicity could be abrogated, suggesting a major involvement of H_2O_2 in cell lysis. Furthermore, 0.1 M H_2O_2 caused complete lysis of target keratinocytes, indicating that it was both essential and sufficient for cytotoxicity. In common with previous findings on the cytotoxicity of various target cell types, including lymphoma cells, fibroblasts, spleen cells, or red blood cells (RBC),[46] this suggests that consequences of or events closely associated with the respiratory burst of PMN include the killing of target cells, in particular keratinocytes.

6.1. Effect of Extracts of Scales on PMN-Mediated Cytotoxicity of Keratinocytes

Next, we addressed the question of whether psoriatic scale extracts would also exert an influence on the lysis of ^{51}Cr-labeled normal keratinocytes. We incubated ^{51}Cr-labeled keratinocytes with ether extracts (pH 3.0) of callus and psoriatic scales (0.25 mg/ml) with normal PMN with and without the addition of zymosan. Table II

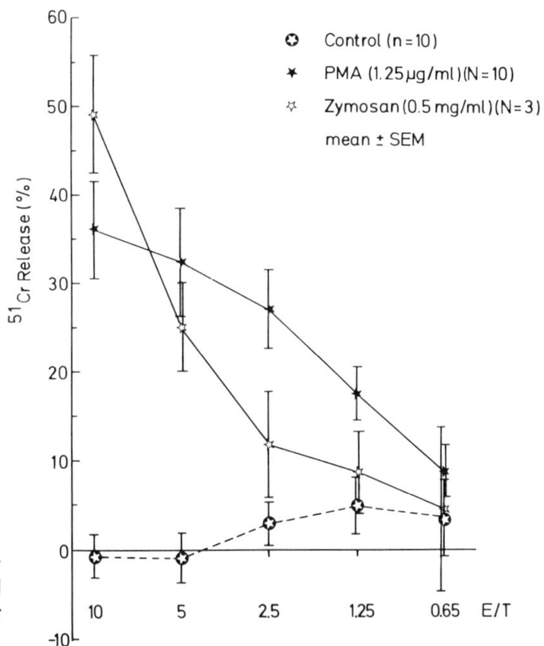

FIGURE 12. Killing of PMN effectors (E) of 8 × 10^4 normal human keratinocyte targets (T) after 4 hr. (From Schopf et al.[45])

shows that the callus extracts led to inhibition of lysis of keratinocytes. By contrast, psoriatic scale extracts did not cause suppression of lysis. Although this difference was small, it was statistically significant. Furthermore, there was no opsonizing effect of organic scale extracts when incubated together with zymosan, as expected from chemiluminescent measurements. From these data, it was concluded that organic extracts of normal callus exert a protective influence on the PMN-mediated lysis of keratinocytes. This effect is lost with psoriatic scale extracts.

TABLE II. Effect of Organic Extracts of Psoriatic Scales and Callus (0.25 mg/ml) on PMN (40×10^4)-Mediated ^{51}Cr Release of Keratinocytes (8×10^4) after 4 hr[a,b]

		Percentage of control (100%)	
Group	*N*	Without zymosan	With zymosan (0.5 mg/ml)
Callus	8	86.8±4.5	95.4±4.1
Psoriatic scales	13	101.5±3.1[c]	101.2±2.7

[a]From Schopf et al.[45]
[b]Mean ±SEM.
[c]$p < 0.01$, Wilcoxon test.

7. ROLE OF THE RESPIRATORY BURST IN PSORIASIS

The characteristic histological picture of psoriasis with the spongiform pustules and Munro microabscesses present in the epidermis provide a challenge to define the role of the respiratory burst in psoriasis. This is reinforced by the generally augmented phagocytic functions in psoriasis, including enlarged bone marrow space,[10] increased function of sessile phagocytes in liver and spleen,[11] increased locomotion of peripheral blood phagocytes,[47-54] adherence,[55,56] enzyme content,[57–63] chemiluminescence,[29,64] phagocytosis,[65,66] and killing activity,[49,67] although normal or subnormal phagocyte function has also been reported.[15,16,30,65,68–71] There is, however, wide agreement that increased phagocyte function correlates with disease activity; i.e., augmented function reverts to normal in remission.[30,51,54,62,67]

During active psoriasis, there is increased monocytopoiesis, which is indicative of consumption by pathological processes.[72] Furthermore, mononuclear leukocytes in psoriasis exhitit increased G6PD activity[63] important for the respiratory burst as G6PD is the key enzyme regulating the hexose monophosphate shunt. Permanent or at least long-lasting activation of G6PD has been repeatedly observed under strongly oxidizing conditions consuming NADPH.[73] For instance, it has been found that the administration of paraquat, a redox cycler using up NADPH, leads to enhanced G6PD activity in the lungs of rats and dogs, as summarized.[73] Similar processes could be relevant in psoriasis, since the enhanced turnover of arachidonic acid with intermediate formation of hydroperoxides has been described in the skin of psoriatics.[74] Such processes use NADPH as a coenzyme such as for glutathione transferase in the formation of leukotriene C.[75] Thus, the enhanced G6PD and respiratory burst activities in leukocytes in patients with psoriasis may reduce oxidant stress by virtue of their enhanced reductive power as also suggested by the increased reduction of nitroblue tetrazolium in PMN[76,77] and Mϕ.[32]

Serum factors in psoriasis have also generally been reported with increased proinflammatory potential. The formation of superoxide, hydrogen peroxide and the rate of opsonization have all been found enhanced in psoriasis.[78-80] In addition, our studies confirm that the maximum chemiluminescent activity of PMN under the influence of psoriatic serum occurs earlier than with normal serum. Specifically, complement split products such as C3a, C3c, C4, C4a, and C5 have been reported in increased concentration in psoriatic serum.[80–83] Immunoglobulins G, A, and E have also been measured in increased quantities in psoriatic serum.[81,84–86] Furthermore immune complexes have been detected in psoriatic serum,[87–89] in particular those containing IgA.[90] In addition to normal activation of the classic pathway of complement activation, the properdin activity has been reported to be reduced,[91] suggesting in vivo activation. The alternative pathway of complement can be activated by IgA, immune complexes, products of gram-negative bacteria (endotoxin), and streptococcal products. The latter finding is of considerable importance as streptococcal infections are well known to provoke psoriasis in susceptible individuals. Thus, these findings provide an ample explanation for the proinflammatory properties of psoriatic serum. In addition, it deserves mention that increased concentrations of leukotriene B_4-like activity have been measured in psoriatic serum.[52]

Scales in psoriasis contain a Pandora's box of pharmacologically and immunologically active components. Identified products include C3a, C4a, C5a,[41] chemotactic peptides,[42] leukotriene B_4[42–44], IgG, IgM, C3c,[92] and immune complexes.[93,94] Remarkably, attempts to find IL-1 in psoriatic scales have not proved fruitful.[95] Other components include plasminogen activator[96,97] and a catheptic endopeptidase.[98] The abnormalities involving proteolytic processes in psoriatic scales are believed to affect inflammation by providing more proinflammatory power on one hand and less inhibition of inflammation on the other. It follows that psoriatic scales offer ideal sources of potent mediators of inflammation. The data presented in this chapter stress the biological effect of these on the respiratory burst.

The multiple possibilities to stimulate the phagocytes in psoriasis may provide an explanation for the well-known clinical experience that patients with psoriasis hardly ever harbor pyodermic infections, despite the undoubltedly altered barrier function of the epidermis.

It should further be mentioned that nonsteroidal anti-inflammatory drugs affecting arachidonate metabolism may profoundly affect the course of psoriasis. This appears important, since leukocytes, in particular PMN, are a rich source of arachidonate metabolites. Exacerbation of psoriasis has been reported under treatment with indomethacin[99] and phenylbutazone.[100] By contrast, clearing of psoriasis has been achieved under treatment with benoxaprofen,[101] a drug inhibiting presumably the 5-lipoxygenase pathway of arachidonic acid oxidation, important in PMN. It follows that there is at least a relative contraindication for indomethacin and similar drugs in psoriasis, as supported by clinical experience.

Other drugs known to exacerbate psoriasis include lithium salts[102,103] and β-adrenergic blockers.[104,105] This side effect is thought to be mediated by the action of these drugs on the adenylate cyclase system, as both are able to block the enzyme system, although at different sites.[106] A decreased concentration of cyclic adenosine monophosphate (cAMP) is thought to promote psoriatic lesions, whereas an increase should lead to clearing of lesions.[107] The same mechanisms are apparently effective in phagocytes; i.e., stimulation of the adenylate cyclase system decreases phagocyte activity.[108] Furthermore, lithium increases the mass of PMN in the peripheral blood.[109] In addition lithium salts have been reported to increase the release of β-glucuronidase of PMN in psoriasis.[57] It follows that agents that potentially enhance phagocyte activity, including the respiratory burst, enhance the disease activity in psoriasis as well.

An important question to be considered is whether phagocytes are merely innocent bystanders. Thus far, only indirect evidence can answer this question. Peritoneal dialysis removing PMN and leukopheresis have been reported to clear psoriasis.[110] In addition, methotrexate, an effective medication used in psoriasis, also depresses PMN production. Together with the increased numbers of PMN in the peripheral blood in psoriasis,[96] this would suggest that phagocytes are important in pathogenesis.

Arguments against a role of phagocytes in psoriasis include the inability by some investigators to confirm the generally enhanced phagocyte function in psoriasis.[16,68–70] This may be due to the patient's state of disease activity, aside from technical problems. Appropriate patient selection will probably indicate differences in phagocyte function. Another, morphological, aspect is that the spongiform pustule does not occur in all cases of psoriasis. A further argument against a specific role of phagocytes in psoriasis may be that inflammatory diseases may generally exhibit increased phagocytic activity. This does not seem to be the case, as we have evidence that in rheumatic arthritis, a disease with considerable inflammatory activity, the respiratory burst of PMN is decreased (manuscript in preparation). Although a causal role for phagocytes is speculative, the importance for PMN and Mϕ in psoriasis is well established as also suggested elsewhere.[8]

8. SUMMARY

Psoriasis is a disease characterized by salmon pink squamous plaques of the skin. Pathogenetically, there is a hereditary determination, an increased epidermal proliferation coupled with decreased keratinocyte differentiation, and a pronounced inflammatory reaction visible as erythema and histologically characterized by dermal and epidermal cellular infiltration, capillary alteration, and exudation.[7] The association of hyperproliferative epidermis with inflammatory cell infiltration, including phagocytes forming the characteristic spongiform pustule and Munro microabscesses, has prompted investigations of phagocyte function, in particular the respiratory burst.

In active psoriasis, generally enhanced phagocytic functions have been found. The data presented indicate that augmented respiratory burst activities of peripheral blood phagocytes are regulated by different stimuli and that psoriatic serum is able to induce a more rapid peak of the chemiluminescent response of normal PMN than can normal serum or the serum of other inflammatory dermatoses. Furthermore, compared with normal controls, mononuclear leukocytes in psoriasis contain higher concentrations of G6PD, the key enzyme for hexose monophosphate shunt. LPS-stimulated epidermal cell-culture supernatants of lesional psoriatic skin possess a higher stimulatory potential for the respiratory burst of PMN than that from uninvolved skin. Furthermore, in contrast to callus, both aqueous and organic extracts of psoriatic scales exert strong stimulating activity for the respiratory burst of both PMN and Mϕ. When normal PMN stimulated with either PMA or Zymosan were incubated together with ^{51}Cr-labeled normal keratinocytes, cytotoxicity of keratinocytes as indicated by ^{51}Cr release resulted. The incubation of organic callus extracts with labeled keratinocytes led to a diminished spontaneous ^{51}Cr release. By contrast, psoriatic scale extracts failed to suppress lysis, indicating that the protective effect of callus extracts is lost with psoriatic scale extracts. This may facilitate cell necrosis in the neighborhood of PMN in lesional psoriatic epidermis.

9. CONCLUSION

Multiple stimuli in psoriasis are able to elicit an enhanced respiratory burst of phagocytes. As inflammatory stimuli are well known to exacerbate psoriasis, the respiratory burst may provide a trigger factor and may thus determine disease activity.

ACKNOWLEDGMENTS. This work was supported by grant Scho 273/2 from Deutsche Forschungsgemeinschaft and in part by Deutscher Psoriasis-Bund. The contribution of the following individuals in carrying out the experiments is gratefully acknowledged: Elke Straussfeld, Franz-Josef Müller, Doris Hoffmann, Anja Bruchhausen, Thomas Klaas, Claudia Ziegler, Marianne Rehder, and Peter Benes.

REFERENCES

1. Christophers E, Krueger GG: Psoriasis, in Fitzpatrick TB, Eisen AZ, Wolff K, Freedberg IM, Austen KF (eds): *Dermatology in General Medicine*, ed 3. New York, McGraw-Hill, 1987, p 461.
2. Henseler T, Christophers E: Psoriasis of early and late onset. Characterization of two types of psoriasis vulgaris. *J Am Acad Dermatol* **13:**450–456, 1985.
3. Farber EM, Nickoloff BJ, Recht B, Fräki JE: Stress, symmetry, and psoriasis: Possible role of neuropeptides. *J Am Acad Dermatol* **14:**305–311, 1986.
4. Pinkus H, Mehregan AH: The primary histologic lesion of seborrheic dermatitis and psoriasis. *J Invest Dermatol* **46:**109–116, 1966.
5. Kogoj F: Un cas de maladie de Hallopeau. *Acta Derm Venereol (Stockh)* **8:**1–12, 1927–1928.
6. Lever WF, Schaumburg-Lever G: *Histopathology of the Skin.* ed 6. Philadelphia, Lippincott, 1983.
7. Braun-Falco O, Scherer R: Immunoinflammatory phenomena in psoriasis, in Beutner EH (ed): *Autoimmunity in Psoriasis.* Boca Raton, Florida, CRC Press, 1982, p 165.
8. Dahl MV: Immunology and papulosquamous diseases, in Stone J (ed): *Dermatologic Immunology.* St. Louis, CV Mosby, 1985, p 619.
9. Krueger GG, Eyre RW: Trigger factors in psoriasis, in Weinstein GD, Voorhees JJ (eds): *Dermatologic Clinics.* Philadelphia, WB Saunders, 1984, vol 2, p 373.
10. Altmeyer P, Munz D, Holzmann H, Hör G: Functional studies of sessile macrophages in liver and spleen of psoriatics. *Dermatologica* **166:**15–22, 1983.
11. Altmeyer P, Munz DL, Chilf G, et al: Morphological and functional findings of fixed phagocytes in psoriatics. *Arch Dermatol Res* **275:**95–99, 1983.
12. Sedgwick JB, Hurd ER, Bergstresser PR: Abnormal granulocyte morphology in patients with psoriasis. *Br J Dermatol* **107:**165–172, 1982.
13. Greaves MW: Neutrophil polymorphonuclears, mediators and the pathogenesis of psoriasis. *Br J Dermatol* **109:**115–118, 1983.
14. Glinski W, Mansbridge J: Polymorphonuclear leukoctes in psoriasis. *Cutis* **30:**441–442, 1985.
15. Fräki JE, Jakoi L, Davies AO, et al: Polymorphonuclear leukocyte function in psoriasis. *J Invest Dermatol* **81:**254–257, 1985.
16. Bork K, Holzmann H: Phagozytoseleistung der neutrophilen Leukozyten bei der psoriatischen Entzündung. *Arch Dermatol Forsch* **251:**11–17, 1974.
17. Rupec M: Zur Ultrastruktur der spongiformen Pustel. *Arch Klin Exp Dermatol* **239:**30–49, 1970.
18. Tagami H, Ofuji S: A leukotactic factor in the stratum corneum of pustulosis palmaris et plantaris. *Acta Derm Venereol (Stockh)* **58:**401–405, 1978.

19. Schopf RE, Rehder M, Benes P, et al: Impaired function of numerically augmented Fc-receptors on granulocytes in a HLA B8$^+$ patient with palmoplantar pustulosis. *Arch Dermatol Res* **279:** 444–448, 1987.
20. Babior BM: Oxygen-dependent microbial killing by phagocytes. *N Engl J Med* **298:**659–668, 721–725, 1978.
21. Schopf RE, Hammann KP, Scheiner O, et al: Activation of human monocytes by both human β1H and C3b. *Immunology* **46:**307–312, 1982.
22. Allen RC, Stjernholm RL, Steele RH: Evidence for the generation of an electronic excitation state(s) in human polymorphonuclear leukocytes and its participation in bactericidal activity. *Biochem Biophys Res Commun* **47:**679–684, 1972.
23. Allen RC, Loose LD: Phagocytic activation of a luminol-dependent chemiluminescence in rabbit alveolar and peritoneal machrophages. *Biochem Biophys Res Commun* **69:**245–252, 1976.
24. Allen RC: Biochemiexcitation, in Adam W, Cilento G (eds): *Chemical and Biological Generation of Excited States.* New York, Academic, 1982, p 309.
25. Schopf RE, Mattar J, Meyenburg W, et al: Measurement of respiratory burst in human monocytes and polymorphonuclear leukocytes by nitro blue tetrazolium reduction and chemiluminescence. *J Immunol Method* **67:**109–117, 1984.
26. Böyum A: Isolation of mononuclear cells and granulocytes from human blood. *Scand J Clin Lab Invest* **97 (suppl):**77–89, 1968.
27. Lehrer RI, Cohen L: Receptor-mediated regulation of superoxide production in human neutrophils stimulated by phorbol myristate acetate. *J Clin Invest* **68:**1314–1320, 1981.
28. Tucker SB, Pierre RV, Jordon RE: Rapid identification of monocytes in a mixed mononuclear cell preparation. *J Immunol Method* **14:**267–269, 1977.
29. Schopf RE, Straussfeld E: Stimulus-dependent increased generation of oxygen intermediates in monocytes and polymorphonuclear leukocytes in psoriasis. *J Invest Dermatol* **84:**73–76, 1985.
30. Geerdink JPM, de Muldner PHM, Franzen MR, et al: Monocyte function is normal in quiescent psoriasis. *J Invest Dermatol* **82:**122–125, 1984.
31. Bergmeyer H-U: *Methoden der enzymatischen Analyse.* Weinheim, Verlag Chemie, 1970, p 599.
32. Krueger GG, Jederberg WW, Ogden BE, Reese DL: Inflammatory and immune cell function in psoriasis. *J Invest Dermatol* **71:**195–201, 1978.
33. Herlin T, Borregaard N, Kragballe K: On the mechanisms of enhanced monocyte and neutrophil cytotoxicity in severe psoriasis. *J Clin Invest* **79:**104–108, 1982.
34. Kiistala U, Mustakallio KK: Dermo-epidermal separation with suction. *J Invest Dermatol* **48:**466–477, 1967.
35. Schopf RE, Hoffman A, Jung M, et al: Stimulation of T cells by autologous mononuclear leukocytes and epidermal cells in psoriasis. *Arch Dermatol Res* **279:**89–94, 1986.
36. Luger TA, Stadler BM, Katz SI, Oppenheim JJ: Epidermal cell (keratinocyte) derived thymocyte activating factor (ETAF). *J Immunol* **127:**1493–1498, 1981.
37. Kapp A, Luger TA, Schöpf E: Modulation of granulocyte oxidative response by mediators derived from human monocytes and epidermal cells. Abstracted *J Invest Dermatol* **89:**446, 1987.
38. Langhof H, Müller H: Leukotaktische Eigenschaften von Psoriasis-Schuppen. *Hautarzt* **17:**101–104, 1966.
39. Schwartze G, Wozniak K-D, Wohlrab W, Höfer H-W: Migration and chemotaxis of donor leukocytes under the influence of psoriatic serum and psoriatic scale extracts. *Dermatol Monatsschr* **170:**637–641, 1984.
40. Schröder J-M, Christophers E: Identification of C5a desarg and an anionic neutrophil-activating peptide (ANAP) in psoriatic scales. *J Invest Dermatol* **87:**53–58, 1986.
41. Takematsu H, Ohkohchi K, Tagami H: Demonstration of anaphylatoxins C3a, C4a and C5a in the scales of psoriasis and inflammatory pustular dermatoses. *Br J Dermatol* **114:**1–6, 1986.
42. Takematsu H, Terni T, Ohkohchi K, et al: Presence of chemotactic peptides other than C5a anaphylatoxin in scales of psoriasis and sterile pustular dermatoses. *Acta Derm Venereol (Stockh)* **66:**93–97, 1986.

43. Brain SD, Camp RDR, Cunningham FM, et al: Leukotriene B4-like material in scale of psoriatic skin lesions. *Br J Pharmacol* **83**:313–317, 1984.
44. Grabbe J, Czarnetzki BM, Rosenbach T, Mardin M: Identification of chemotactic lipoxygenase products of arachidonate metabolism in psoriatic skin. *J Invest Dermatol* **82**:477–479, 1984.
45. Schopf RE, Rehder M, Bork K, Morsches B: Killing of normal and neoplastic epidermal cells by stimulated polymorphonuclear leukocytes in man. *J Invest Dermatol* **87**:165, 1986.
46. Nathan CF, Brukner LH, Silverstein SC, Cohn ZA: Extracellular cytolysis by activated macrophages and granulocytes. *J Exp Med* **149**:84–99, 1979.
47. Silny W, Pehamberger H, Zielinsky C, Gschnait F: Effect of PUVA treatment on the locomotion of polymorphonuclear leukocytes and mononuclear cells in psoriasis. *J Invest Dermatol* **75**:187–188, 1980.
48. Preissner WC, Schröder J-M, Christophers E: Altered PMN responses in psoriasis. *Br J Dermatol* **109**:1–8, 1983.
49. Csató M, Dobozy A, Hunyadi J, et al: PMN chemotaxis and chemotactic factor generation by Con A-stimulated peripheral blood mononuclear cells in patients with psoriasis. *Arch Dermatol Res* **271**:259–264, 1981.
50. Wahba A, Cohen HA, Bar-Eli M, Gallily R: Enhanced chemotactic and phagocytic activities of leukocytes in psoriasis vulgaris. *J Invest Dermatol* **71**:186–188, 1978.
51. Langner A, Chorzelski TP, Fraczykowska S, et al: Is chemotactic activity of polymorphonuclear leukocytes increased in psoriasis? *Arch Dermatol Res* **275**:226–228, 1983.
52. Fogh K, Ternowitz T, Kragballe K, Herlin T: Chemotactic lipoxygenase products in sera from patients with psoriasis. *Arch Dermatol Res* **278**:173–176, 1986.
53. Ternowitz T, Therstrup-Pedersen K: Neutrophil and monocyte chemotaxis in pustulosis palmoplantaris and pustular psoriasis. *Br J Dermatol* **113**:507–513, 1985.
54. Czarnetzki B: Increase monocyte chemotaxis towards leukotriene B4 and platelet activating factor in patients with inflammatory dermatoses. *Clin Exp Immunol* **54**:486–492, 1983.
55. Sedgwick JB, Bergstresser PR, Hurd ER: Increased granulocyte adherence in psoriasis and psoriatic arthritis. *J Invest Dermatol* **74**:81–84, 1980.
56. Majewski S, Jablonska S, Langner A, Pawinska M, Szmurto A: The adherence to human endothelium and plastic of neutrophils from psoriatic patients, and the effects of psoriatic patients' sera on normal neutrophils. *Br J Dermatol* **112**:655–662, 1985.
57 Bloomfield FJ, Young MM: Enhanced release of inflammatory mediators from lithium-stimulated neutrophils in psoriasis. *Br J Dermatol* **109**:9–13, 1983.
58. Glinski W, Barszcs D, Janczura E, et al: Neutral proteinases and other neutrophil enzymes in psoriasis, and their relation to disease activity. *Br J Dermatol* **111**:147–154, 1984.
59. Moszczynski P, Legiec C, Lisiewicz J: Acid phosphatase of neutrophils in peripheral blood of patients with psoriasis. *Dermatol Monatsschr* **170**:371–375, 1984.
60. Jain VK, Gupta V, Verma KC, Saini AS: NADPH oxidase and myeloperoxidase activity in psoriasis leukocytes. *J Dermatol* **12**:425–428, 1985.
61. Koh MS, Majewski BJ, Rhodes EL: Increased macrophage activity in psoriasis. *Acta Derm Venereol (Stockh)* **65**:194–198, 1985.
62. Lundin A, Hakansson L., Hällgren R, et al: Increased in vivo secretory activity of neutrophil granulocytes in patients with psoriasis and palmoplantar pustulosis. *Arch Dermatol Res* **277**:179–184, 1985.
63. Schopf RE, Müller F-J, Benes P, Morsches B: Augmented G-6-PDH activity and normal penetration of metabolism of dehydroepiandrosterone in mononuclear leukocytes in psoriasis. *Arch Dermatol Res* **278**:393–397, 1986.
64. Krueger GG, Jederberg WW: Monocyte function in patients with psoriasis as assessed by chemiluminescence and monokine production. *Clin Res* **27**:137, 1979.
65. Bar-Eli M, Gallily R, Cohen HA, Wahba A: Monocyte function in psoriasis. *J Invest Dermatol* **73**:147–149, 1979.
66. Lundin A, Hakansson L, Hällgren R, et al: Studies on the phagocytic activity of the granulocytes in psoriasis and palmoplantar pustulosis. *Br J Dermatol* **109**:539–547, 1983.

67. Herlin T, Kragballe K; Enhanced monocyte and neutrophil cytotoxicity and normal cyclic nucleotide levels in severe psoriasis. *Br J Dermatol* **105:**405–410, 1981.
68. Breathnach SM, Carrington P, Black MM: Neutrophil leukocyte migration in psoriasis vulgaris. *J Invest Dermatol* **76:**271–274, 1981.
69. Guillot B, Guilhou J, Vendrel JP, Meynadier J: Neutrophil chemotaxis in psoriasis before and after PUVA therapy. *Arch Dermatol Res* **275:**19–22, 1983.
70. Cunningham FM, Wong E, Woollard PM, Greaves MW: The chemokinetic response of psoriatic and normal polymorphonuclear leukocytes to arachidonic acid lipoxygenase products. *Arch Dermatol Res* **278:**270–273, 1986.
71. Geerdink JPM, Troost PW, Schalwijk J, et al: The metabolic burst in polymorphonuclear leukocytes from patients with quiescent psoriasis. *Br J Dermatol* **112:**387–392, 1985.
72. Meuret G, Schmitt E, Hagedorn M: Monocytopoiesis in chronic exzematous diseases, psoriasis vulgaris and mycosis fungoides. *J Invest Dermatol* **66:**22–28, 1976.
73. Brigelius R: Mixed disulfides: Biological functions and increase in oxidative stress, in Sies H (ed): *Oxidative Stress*. London, Academic, 1985, p 243.
74. Voorhees JJ: Leukotrienes and other lipoxygenase products in the pathogenesis and therapy of psoriasis and other dermatoses. *Arch Dermatol* **119:**541–547, 1983.
75. Higgs GA: Arachidonic acid metabolism in leukocytes, in Karnovsky ML, Bolis L (eds): *Phagocytosis—Past and Future*. New York, Academic, 1982, p 105.
76. Csató M, Dobozy A, Simon N: Nitroblue tetrazolium reductive activity of polymorphonuclear leukocytes in the blood and in the skin in psoriasis. *Z Hautkr* **58:**1731–1739, 1983.
77. Cotterill JA, Roberts MM, Freeman R, et al: Aspects of polymorph function in psoriasis. *Proc R Soc Med* **67:**874–875, 1974.
78. Miyachi Y, Niwa Y: Effects of psoriatic sera on the generation of oxygen intermediates bz normal polymorphonuclear leukocytes. *Arch Dermatol Res* **275:**23–26, 1983.
79. Herlin T, Borregaard N: Increased rate of opsonization of zymosan by serum from patients with psoriasis. *Arch Dermatol Res* **273:**343–345, 1982.
80. Sedgwick JB, Bergstresser PR, Hurd ER: Increased superoxide generation by normal granulocytes incubated in sera from patients with psoriasis. *J Invest Dermatol* **76:**158–163, 1981.
81. Marghescu S, Braun-Falco O: Quantitatives Verhalten von Immunoglobulinen and Komplement bei Psoriasis. *Arch Klin Exp Dermatol* **238:**417–427, 1970.
82. Kapp A, Wokalek H, Schöpf E: Involvement of complement in psoriasis and atopic dermatitis. *Arch Dermatol Res* **277:**359–361, 1985.
83. Ottkohchi K, Takematsu H, Tagami H: Increased anaphylatoxins (C3a and C4a) in psoriatic sera. *Br J Dermatol* **113:**189–196, 1985.
84. Guilhou J-J, Clot J, Meynadier J, Lapinski H: Immunological aspects of psoriasis. *Br J Dermatol* **94:**501–507, 1976.
85. Negrosanti M, Fanti PA, Gasponi A, et al: IgE serum levels in psoriasis. *Dermatologica* **163:**474–455, 1981.
86. Chen Z-Y, Aninsworth K, Khan T, et al: Immunoglobulin E in psoriasis is elevated by paper radioimmunosorbent and paper enzyme-immunosorbent tests. *Arch Derm Venereol (Stockh)* **65:**14–18, 1985.
87. Kapp A, Kemper A, Schöpf E, Deicher H: Detection of circulation immune complexes in patients with atopic dermatitis in psoriasis. *Acta Derm Venereol (Stockh)* **66:**121–126, 1986.
88. Westphal H-J, Schütt C, Hartwich M, et al: Evidence of circulating immune complexes in the sera of patients with psoriasis. *Dermatol Monatsschr* **168:**624–630, 1982.
89. Laurent MR, Panayi GS, Shepherd P: Circulating immune complexes, serum immunoglobulins, and acute phase proteins in psoriasis and psoriatic arthritis. *Ann Rheum Dis* **40:**66–69, 1981.
90. Hall RP, Peck GL, Lawley TJ: Circulating IgA immune complexes in patients with psoriasis. *J Invest Dermatol* **80:**465–468, 1983.
91. Marley WM, Belew PW, Rosenberg EW, et al: Abnormalities in the alternative pathway of complement in psoriasis. *J Invest Dermatol* **76:**303, 1981.

92. Krogh HK, Tønder O: Antibodies in psoriatic scales. *Scand J Immunol* **2:**45–51, 1973.
93. Kaneko R, Muramatsu R, Takahashi Y, Miura Y: Extractable immune complex in soluble substances from psoriatic scales. *Arch Dermatol Res* **276:**45–51, 1984.
94. Jones P, Kumar V: A solid-phase chemiluminescence immunoassay for determination of IgG in eluates of psoriatic scales. *J Immunol Method* **70:**185–192, 1984.
95. Takematsu H, Suzuki R, Tagami H, Kumagai K: Interleukin-1-like activity in horny layer extracts. *Dermatologica* **172:**236–240, 1986.
96. Fräki JE, Lazarus GS, Gilgor RS, et al: Correlation of epidermal plasminogen activator activity with disease activity in psoriasis. *Br J Dermatol* **108:**39–44, 1983.
97. Grøndahl-Hansen J, Nielsen LS, Kristensen P, et al: Plasminogen activator in psoriatic scales is of the tissue-type PA as identified by monoclonal antibodies. *Br J Dermatol* **113:**257–263, 1985.
98. Stüttgen G, Würdemann I: Zur Darstellung der katheptischen Endopeptidase in der normalen menschlichen Haut und bei Dermatosen. *Arch Klin Exp Dermatol* **208:**192–198, 1957.
99. Katayama H, Kawada A: Exacerbation of psoriasis induced by indomethacin. *J Dermatol (Tokyo)* **8:**323–337, 1981.
100. Reshad H, GK Hargreaves GK, Vickers CFH: Generalized pustular psoriasis precipitated by phenylbutazone and oxyphenbutazone. *Br J Dermatol* **108:**111–113, 1983.
101. Kragballe K, Herlin T: Benoxyprofen improves psoriasis. *Arch Dermatol* **119:**548–552, 1983.
102. Skott A, Mobacken H, Starmark JE: Exacerbation of psoriasis during lithium treatment. *Br J Dermatol* **96:**445–448, 1977.
103. Skoven I, Thormann I: Lithium compound treatment and psoriasis. *Arch Dermatol* **115:**1185–1187, 1979.
104. Søndergaard J, Wadskov S, Aerenlund Jensen H, Mikkelsen HI: Aggravation of psoriasis and occurrence of psoriasiform cutaneous eruptions induced by practolol. *Acta Derm Venereol (Stockh)* **56:**239–243, 1976.
105. Halevy S, Reuerman EJ: Psoriasiform eruption induced by propranolol. *Cutis* **24:**95–99, 1979.
106. Forn J: Lithium and cyclic AMP, in Johnson FN (ed): *Lithium Research and Therapy*. London, Academic, 1975, p 485.
107. Voorhees JJ, Duell EA: Psoriasis as a possible defect of the adenyl cyclase-cyclic AMP cascade. *Arch Dermatol* **104:**352–358, 1971.
108. Bourne HR, Lichtenstein LM, Melmon KL, et al: Modulation of inflammation and immunity by cyclic AMP. *Science* **184:**19–28, 1974.
109. Rothstein G, Clarkson DR, Larsen W, et al: Effect of lithium on neutrophil mass and production. *N Engl J Med* **298:**178–180, 1978.
110. Glinski W, Barszcz D, Jablonska S, et al: Leukophoresis for treatment of psoriasis. *Arch Dermatol Res* **278:**6–12, 1985.

17

The Respiratory Burst and Diabetes Mellitus

JÓZSEF T. NAGY, TAMÁS FÜLÖP, Jr., GEORGE PARAGH, and GABRIELLA FÓRIS

1. INTRODUCTION

Diabetes mellitus is known as a general disturbance of glucose utilization at the cellular level. However, the precise pathomechanism is unclear, according to the 1985 classification of the WHO study Group[1]: Two (clinical) types of the disease—insulin-dependent diabetes mellitus (IDDM) and noninsulin-dependent diabetes mellitus (NIDDM)—have been used. The discovery and accessibility of the hormone insulin have fundamentally changed the outcome of the patients, but complications of the disease have become conspicuous; such as increased sensitivity of the organism against different infectious illnesses.

Insulin depletion at the intracellular level leads to serious metabolic abnormalities and a consequent disturbance of many cellular functions, including immunological activity.[2,3] This is one possible explanation why the relative immunodeficiency, which is considered to be of metabolic origin, was not widely studied by immunologists. Recently, with the advance in the therapy of diabetes mellitus (e.g., continuous insulin pumping), an adequate control of the carbohydrate metabolism is possible. Episodes of excessive hyperglycemia or ketosis are not frequent in diabetics. Therefore, the direct effect of metabolic abnormalities on the immunological status is negligible.[4] Moreover, during the past two to three decades, IDDM has come to seem to be a more complicated disease than was thought with a possibility of autoimmune origin suggested by identification in the sera of patients of different autoantibodies against pancreatic islet cells,[5,6] as well as against the hormone insulin,[7,8] and in special cases against the insulin receptor. The genetics of diabetes was also studied, and a higher incidence of HLA-DR3 and -DR4 haplotypes was

JÓZSEF T. NAGY, TAMÁS FÜLÖP, Jr., GEORGE PARAGH, and GABRIELLA FÓRIS • First Department of Medicine, University Medical School, 4012 Debrecen, Hungary.

observed.[10,11] By contrast, the phenomenon of decreased resistance against infectious diseases also exists in NIDDM patients in whom evidence of an autoimmune origin or any connection with genetic factors are undemonstrable.[12,13]

According to our present knowledge, diabetes mellitus appears to be a disorder of carbohydrate metabolism, with different abnormalities of the specific and nonspecific immune response present to some degree independent of the metabolic disturbances.

The role of the reactive oxygen species (ROS)-generating system in the host defense has also been intensively studied during the past 20 years,[14,15] but it seems that the advances in the study of the diabetes, and of the respiratory burst, were not synthesized by the different teams. However, it is evident that superoxide, hydrogen peroxide, and other short-lived oxygen derivatives have significant importance in the bactericidal killing of phagocytic cells,[16] and disturbed resistance of the diabetics against infectious disease is also well known[17,18]; the number of reports in which the ROS of diabetics was studied is relatively small. A possible explanation is that the interest of immunologists turned first to the immunological processes involved in the pathomechanism of the disease and the biochemists studied especially the so-called pure forms of ROS disturbances (e.g., chronic granulomatous disease).

We believe that a special connection between the so-called immunological recognition and the effector function of phagocytic cells exists in diabetes mellitus. The aim of this short review is to give a possible view of this connection, including the special role of the ROS in it.

2. ALTERATION OF THE IMMUNE SYSTEM IN DIABETES MELLITUS

The confusion in the classification of diabetes makes evaluation of the different immunological findings difficult. Inaccuracies of correct and routine methods appear to have a common origin; i.e., the recent classification based on the insulin requirement of the patients has nothing to do with the pathomechanism of the disease. Although is abundant enough in everyday clinical practice, it is insufficient for immunological and biochemical measurements, in general.

A review of theories on the pathophysiology of diabetes mellitus is far beyond the dimensions of this chapter, but several facts must be taken into account:

1. There is an existing form of diabetes mellitus of juvenile onset, characterized by absolute insulin need, but the so-called immunological findings are very different. Autoantibodies against pancreatic iselet cells are present only in about 50% of the sera.[19,20] Another small proportion of patients has antibodies against insulin[21] and, very rarely, antibodies against insulin receptors[22,23] are detectable as well. It is significant that a great number of cases show no laboratory signs of autoimmune diseases. All these patients appear in the literature as IDDM or as type 1 DM.

2. The other primary type of the disease is presented in reports as the non-insulin-dependent form (NIDDM, or type 2 DM) and the term is associated with the elderly.[24,25] Doubtless no antibodies against islet cells, insulin, or insulin receptors are present in the sera of these patients, but other immunological disturbances, alteration of the cytotoxic activity of killer[26] or natural killer (NK)[27] cells, and pathological interleukin -2 (IL-2) synthesis[28] have recently been reported. On the other hand, a part of so-called NIDDM patients becomes temporarily or definitively insulin dependent over the course of the disease. Some of the above-mentioned immunological findings are characteristic as well for the aging, which this makes the question more complicated.

After these critical remarks, let us examine some immunological properties of the IDDM the former type 1 DM.

Characteristic change in the lympocyte T subsets (OKT4–OKT8) was found[29] and, in connection with the relatively great incidence of HLA-DR3 and -DR4 surface antigen determinants,[30] the importance of genetic factors in the pathomechanism is widely discussed.[31,32] Most investigators point out that abnormal distribution of the T-cell subsets appear only during the early stage of the disease,[29,33-35] and in remission or after accurate insulin treatment, the OKT4/OKT8 ratio is normalizing. In other laboratories, it was found that the proportion of helper and suppressor T cells is independent of the presence of pancreatic islet cell autoantibodies.[35] The positivity of the islet cell autoantibody test is also an object of discussion (recent efforts toward standardization). The increased number of activated (4F2 and Tac positive) cells seems to be a standard marker of autoimmune, DM but only if the patients are HLA-DR3 positive.[31] By contrast, disturbance of the complement-activating system with a measurable increase in the amounts of C3a and C5a fragments exists only in the early stage of IDDM.[36] The decrease in suppressor T-cell number as well as the reduction of circulating (Leu 7+) NK cell number and diminished cytotoxicity against K565 tumor cells was observed, while increased cytotoxicity against pancreatic iselet cells was described.[27]

Another, controversial, observation was reported by Fövényi et al.,[26] i.e., ADCC activity of killer cells increases in type 1 DM. These investigators found this phenomenon characteristic of type 2 DM as well, but only in the case of a transitional episode of insulin requirement.

Investigation of the lymphoid system in IDDM was focused first of all on its role in the pathomechanism of the disease.[33,35] Surprisingly, alteration of the lymphocyte function was found in NIDDM as well. Increased ADCC activity and disturbed synthesis of IL-2[26,28] were described independently, later relating to the deep injury of the immune system in the nonautoimmune form of diabetes mellitus.

Whereas comprehensive studies on lymphocytes focused on the role of autoimmunity or immunity in the pathomechanism of IDDM, investigation of the granulocyte–monocyte system was carried out chiefly to explain the decreased host defense.

Diminished glycolysis in the granulocytes of patients suffering from diabetes mellitus was among the first elucidations.[2] Evidence of decreased synthase–phosphatase activity and measurable loss of intracellular ATP level was also demonstrated. Energy depletion of phagocytic cells is a possible cause of reduced adherence and chemotactic activity.[13]

The role of the extracellular milieu is questionable: serious disturbances of granulocyte functions were found in vitro in the presence of extremely high concentrations of glucose. Hyperosmolarity was found responsible for decreased synthesis of a protein that has an essential role in phagocytosis.[3] Some reports pointed out that the granulocytic disturbances could be transmitted by autologous serum,[3,4,12] but a series of abnormalities were observed in patients with normoglycemia; increased chemiluminescence, which seems to be characteristic as well of the resting phagocytic cells of patients with diabetes, was independent of the presence or absence of autologous serum.[37]

Decreased intracellular killing was also demonstrated in some laboratories,[3,4,12] and this abnormality was quite pronounced in the presence of autologous serum or glucose. The early phase of killing seems to be damaged, an NADPH oxidase-dependent process.[38,39] Diminished conversion of glucose to sorbitol is considered responsible for this alteration.[3]

There is a relatively integrated view of ROS production of resting phagocytic cells; i.e., increased chemiluminescence, superoxide, or H_2O_2 production is detectable in general. Despite production of these cells among high reactive oxygen species, no sufficient respiratory burst could be measured after different stimuli (phagocytosis, PMA stimulation). Cells obtained from type 1 DM patients as well as those of experimental rat diabetes has reduced SOD activity,[40,41] which is a possible explanation for the increased superoxide production of the resting cells.[42,43] Elevated activity of glutathion peroxidase enzyme was demonstrated in diabetes, which is probably responsible for the reduced GSH level of thrombocytes obtained from diabetic patients.[38]

The more data on the disturbed immunological processes in diabetes we have, the less integrated our view. What are the disturbances of insulin depleted hyperglycemia? Is IDDM really an autoimmune disease? What are the main pathogenetic and immunological differences in IDDM and NIDDM? Our attention focuses only on the last question. The following discussion demonstrates a series of experiments we performed in which we studied the reactive oxygen species-generating system of two patient groups with diabetes and an age-matched control group.

3. CASE STUDIES

IDDM and NIDDM were diagnosed according to the criteria of the WHO Study Group.[1] In order to form relatively homogeneous groups in both diseases, additional criteria were used as follows: In IDDM, all the patients initiated in the study have near-normal or even higher serum insulin levels, indicating that their own hormonal production is not totally disturbed. In addition, this observation made it probable that in these cases a receptor- or postreceptor-coupling disturbance exists. As a further restriction, only patients with decreased receptor activity were chosen. A proportion of these cases had a serum factor that was able to inhibit the [^{125}I]-insulin-binding capacity of PMNL obtained from healthy subjects.

Similarly in NIDDM, patients with near-normal serum insulin levels were chosen, indicating that not (or not only) the hormonal production but its effect is impaired. In this group, neither a decrease in the receptor activity nor blocking serum factors against insulin binding were detected.

IDDM patients received monocomponent insulin therapy; subjects with NIDDM were only on diet. In the first group, insulin administration was omitted for at least 24 hours before the investigations were performed.

All other parameters (e.g., fasting glucose level, HbAlc, relative weight, lack of chronic serious hepatic and renal failure, lack of episode of acute infection, standard diet) in the two patient groups were comparable, apart from the average age, which was 22.6 ± 1.8 years in IDDM and 60.3 ± 4.8 year in NIDDM.

The results of the diabetics were also compared with those of two age-matched control groups selected on the basis of the following criteria: good physical and mental health confirmed by clinical, radiological, and biochemical examinations.

4. REACTIVE OXYGEN SPECIES-GENERATING SYSTEM AND THE EFFECTOR FUNCTIONS OF GRANULOCYTES IN DIABETES MELLITUS

As the first step in our study, we demonstrated the decreased intracellular killing activity of granulocytes in both groups of patients suffering from diabetes mellitus (Table I). According to our results and in agreement with the clinical data, a significant impairment of this effector function was found; however, slight decrement of intracellular killing could be measured in the healthy aged group as well. To clarify the mechanism of this decreased intracellular killing activity, the ROS of granulocytes was studied in the subsequent experiments.

TABLE I. Intracellular Killing Activity of PMNL Obtained from Patients with IDDM and NIDDM as Well as from Age-Matched Controls[a]

Groups	No. of patients	^{51}Cr release % of control
IDDM	5	32.8± 7.6[b]
Controls for IDDM	10	61.7±12.3
NIDDM	8	27.2± 6.3[b]
Controls for NIDDM	10	50.1± 8.7

[a]Intracellular killing activity was measured by the use of ^{51}Cr-*Candida albicans* in the presence of 5% homologous sera, according to the method of Yamamura et al.[44]

[b]Values differ significantly from control values ($p < 0.01$).

A clear-cut distinction must be drawn between the ROS of resting cells and the respiratory burst.

Table II demonstrates alterations in the resting granulocytes in both types of diabetes mellitus measuring some parameters of the ROS as well as superoxide and H_2O_2 production, oxygen consumption, glutathione peroxidase, (GPx) and reductase (GR) enzymatic activities, levels of reduced (GSH), and oxidized (GS–SG) forms of glutathione. The following results were obtained:

1. Significant activation of the ROS of resting granulocytes was measured, which could be characterized by increased oxygen consumption, superoxide and H_2O_2 production. Similar, but not significant, elevation of these parameters was measured with age.
2. Similarities between changes among the aged and the NIDDM groups are pronounced.
3. Despite similar changes in both diabetes mellitus groups, a marked difference was measured, i.e., GPx enzymatic activity, and the total glutathione content of cells obtained from NIDDM patients was decreased in comparison with those of IDDM granulocytes.

Our results suggest several conclusions: Resting phagocytic cells of patients suffering from diabetes mellitus have activated ROS generation, independent of the type of the disease. It seems that the process that resulted in elevation of the level of toxic short-lived oxygen derivatives differs in IDDM and NIDDM. This difference is most striking at the level of the glutathione system. The behavior of resting granulocytes of NIDDM patients and aged healthy subjects is almost comparable, but there are quantitative differences as well. The increased ROS activity displayed by phagocytic cells may have pathophysiological consequences if the possible damaging effect of toxic oxygen derivatives to endothelial cells is taken into account.[49,50]

These results are in good agreement with those published by Shah et al,[37] who demonstrated the increased chemiluminescence of resting granulocytes obtained from patients with diabetes mellitus. Reasonable changes in the oxidative products of healthy aged subjects were recently reported from our laboratory.[51] Similar alterations were found in the granulocytes of patients suffering from chronic renal failure by Rhee et al.[52] These observations apparently contradict the clinical experience, i.e., with the tendency to infectious diseases in these conditions. Still, it must not be forgotten that respiratory burst, and not oxidative products of resting granulocytes, is responsible for intracellular killing and extracellular cytotoxicity.

Next, we studied the respiratory burst of granulocytes obtained from IDDM patients, NIDDM patients, and age-matched healthy subjects. IgG-coated zymosan particles and carbachol were used as stimuli acting through receptors. The effect of $PGF_{2\alpha}$ was studied, since post-receptor coupling of this agent is not comparable to any other receptor-stimulating drugs known so far.[3] Calcium ionophore A23187 (which elevates the intracellular free calcium,) and PMA (which stimulates PKC) were used as direct stimuli. Superoxide production was measured as the product of the respiratory burst.

TABLE II. Parameters of Oxidative Processes in Resting PMNL Obtained from Patients with IDDM and NIDDM as Well as from Age-Matched Control Subjects[a]

Parameters	Control for IDDM 10 persons	IDDM 5 persons	Control for NIDDM 10 persons	NIDDM 8 persons
O_2 consumption nmoles/5 min per 10^7 cells	15.1±1.1	31.4±2.9[b]	21.8±1.42	30.4±2.85[b]
O_2^- generation nmoles/5 min per 10^7 cells	6.1±0.3	15.6±1.2[b]	5.2±0.36	13.6±1.5[b]
H_2O_2 generation nmoles/5 min per 10^7 cells	1.08±0.08	4.48±0.3[b]	1.08±0.1	1.71±0.1[b]
GPxmU/10^7 cells	586.2±41	524±42	648±42	404±48[b]
GR U/10^7 cells	10.8±0.41	3.6±1.1[b]	10.7±0.95	11.8±1.3
GSH μg/10^7 cells	3.86±0.5	2.94±0.3	3.71±0.21	2.27±0.28[b]
GS-SG μg/10^7 cells	0.052±0.004	0.182±0.02[b]	0.078±0.02	0.105±0.02[b]

[a] O_2 consumption was determined by Tanabe et al.,[45] O_2^- generation according to the method of Cohen and Chovaniek,[46] H_2O_2 production by Pick and Keisari,[47] and GPx activity by Paglia and Valentine.[48]

[b] Values differ significantly from age-matched control values ($p < 0.01$).

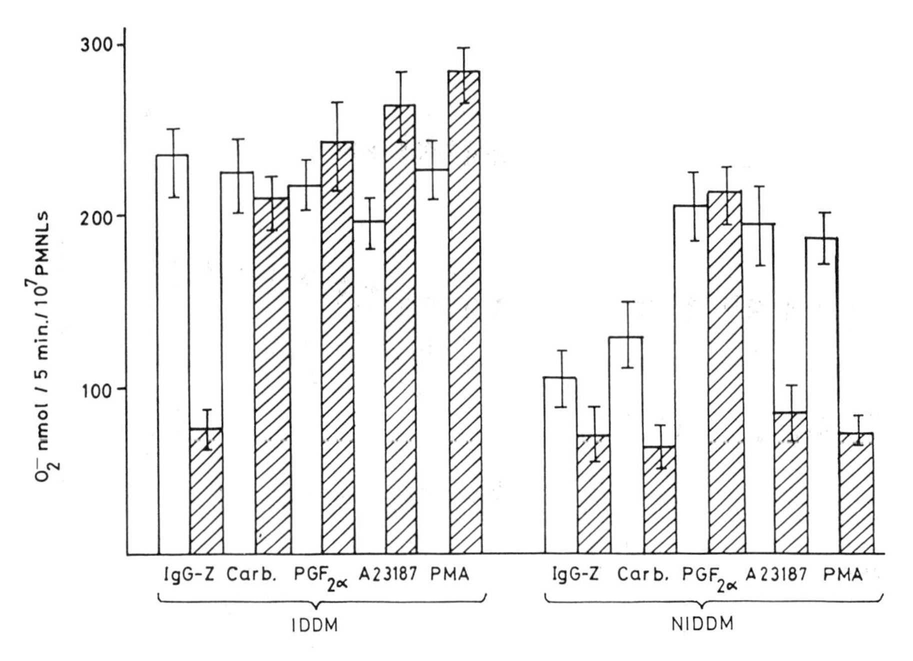

FIGURE 1. Stimulus-induced O_2^- generation of PMNL obtained from patients with IDDM and NIDDM as compared with data from age-matched control subjects. (□) 10 age-matched control subjects in each case; (▨) 5 patients with IDDM and 8 patients with NIDDM, respectively. IgG-Z: zymosan particles coated with IgG; Carb: 1 μM carbachol; $PGF_{2\alpha}$: 5 μM $PGF_{2\alpha}$; A23187: 1 μM of A23187; PMA: 75 μM phorbol myristate acetate. Each value represents the mean ±SD.

Figure 1 demonstrates that not only the cells of both diabetes mellitus groups, but the granulocytes of healthy aged subjects responded with a weak respiratory burst to the stimulation through Fc–C3b and cholinergic receptors. Direct stimulation of calmodulin (calcium ionophore) and PKC (PMA) resulted in a normal response in the granulocytes of aged donors as well as in the cells of IDDM patients. In the case of NIDDM, the applied induction of the respiratory burst except $PGF_{2\alpha}$ was almost completely ineffective. The normal response to $PGF_{2\alpha}$ makes probable that NADPH oxidase enzyme which is first of all responsible for respiratory burst is not impaired itself in the granulocytes of patients with NIDDM. However, it is probable that this agent also acts through the cleavage of phosphatidyl inositol, but the further means of signal transmission in the cell is short and unknown.[53]

On the basis of our results, several conclusions were obtained:

1. Significant decrease of the respiratory burst in both groups of diabetes mellitus is in good agreement with the clinical experience of reduced resistance against infectious diseases. The result is also comparable to the diminished intracellular killing of granulocytes, which was demonstrated in the first series of our experiments.
2. The result that the weak superoxide production in IDDM patients appears only after stimulation through Fc–C3b receptors indicates that the process

of signal transmission is impaired at the coupling of FeC3b receptors; other systems such as calmodulin and PKC are probably intact.

3. Impaired function of PKC and–or the calcium–calmodulin system seems to be a possible explanation of the altered signal transmission in the granulocytes of NIDDM patients.

Finally, it should be noted that in our earlier studies we found similar unresponsiveness of guanylate cyclase activation in PMNL of patients with NIDDM. In addition, impairment of both protein kinase C-triggered adenylate cyclase activation and the calmodulin-dependent Ca^{2+} efflux from these PMNL was demonstrated (unpublished data).

5. CONCLUSION

Signs of decreased effector functions in PMNL obtained from patients suffering from both forms of diabetes mellitus, independent of the autoimmune or non-autoimmune origin of the disease, are similar or even identical: diminished oxidative burst, intracellular killing activity, and increased susceptibility of patients with diabetes mellitus to various infections. On the basis of our results in patients with IDDM, these failures are a consequence of the impaired Fc–C3b receptor function in phagocytic cells, whereas in the NIDDM group the function of intracellular regulatory systems is disturbed. It may be hypothesized that even in some endocrine disorders, it must taken into account that not only the hormonal level in circulation, receptor density, and affinity at the periphery, but also the postreceptor coupling and signal transmission through the Ca^{2+}–calmodulin–protein kinase C systems play an important role in the pathomechanism of these disorders.

REFERENCES

1. Report of WHO Study Group (1985) World Health Organization. Technical Report Series 727.
2. Esman V: Polymorphonuclear leukocyte in diabetes mellitus. *J Clin Chem Clin Biochem* **21:**561–567, 1983.
3. Wilson RM, Reeves WG: Neutrophil phagocytosis and killing in insulin-dependent diabetes. *Clin Exp Immunol* **63:**478–484, 1986.
4. Davidson NJ, Sowden JM, Fletscher J: Defective phagocytosis in insulin controlled diabetics: Evidence for a reaction between glucose and opsonising proteins. *J Clin Pathol* **37:**783–786, 1984.
5. Lohmann D, Krug J, Lampeter EF, et al: Cell-mediated immune reactions against B cells and defect of suppressor cell activity in Type 1 (insulin-dependent) diabetes mellitus. *Diabetologia* **29:**421–425, 1986.
6. Dryberg T: Humoral autoimmunity in the pathogenesis of insulin dependent diabetes mellitus. *Acta Endocrinol* **280(suppl):**9–29, 1986.
7. Komori K, Nakayama K, Aoki S, et al: Effect of anti-insulin antibody on insulin binding to liver membranes: Evidence against antibody-induced enhancement of insulin binding to the insulin receptor. *Diabetologia* **29:**447–452, 1986.
8. McEvoy RE, Witt ME, Ginsberg-Fellner F, et al: Anti-insulin antibodies in children with Type 1 diabetes mellitus. *Diabetes* **35:**634–641, 1986.

9. Blecher M: Speculations on potential anti-receptor autoimmune disease, in Pitman (ed): Receptors, antibodies and disease. Ciba foundation Symposium 90, London, 1982 pp. 279–300.
10. Tosi R, Vela M, Adorno D, et al: Radioimmunoassay typing gives a more precise definition of the HLA association of Type 1 (insulin-dependent) diabetes. *Diabetologia* **29**:430–433, 1986.
11. Hitman, GA: Progress with the genetics of insulin-dependent diabetes mellitus. *Clin Endocrinol* **25**:463–472, 1986.
12. Oberg G, Hällgren R, Moberg L, et al: Bactericidal protein and neutral proteases in diabetes neutrophils. *Diabetologia* **29**:426–429, 1986.
13. Keily MK, Brown JM, Thong YH: Neutrophil and monocyte adeherence in diabetes mellitus, alcoholic cirrhosis, uremia and elderly patients. *Int Arch Allergy Appl Immunol* **78**:132–138, 1985.
14. Babior BM: The respiratory burst of phagocytes. *J Clin Invest* **73**:599–604, 1984.
15. Halliwell B, Gutteridge JMC: Oxygen toxicity, oxygen radicals, transition, metals and disease. *Biochem J* **219**:1–11, 1984.
16. Rosen H, Klebanoff S: Bactericidal activity of superoxide anion generating system. *J Exp Med* **149**:29–38, 1979.
17. Bagdadc JD: Phagocytic and microbicidal function in diabetes mellitus. *Acta Endrocrinol* **83(suppl)**:27–33, 1976.
18. Nolan CM, Beaty HN, Bagdade JD: Further characterization of the impaired byctericidal function of granulocytes in patients with poorly controlled diabetes. *Diabetes* **27**:889–894, 1978.
19. Notsu K, Oka N, Note S, et al: Islet cell antibodies in the Japanese population and subjects with Type 1 (insulin-dependent) diabetes, *Diabetologia* **28**:660–662, 1985.
20. Takahashi A, Tsujihata M, Yokota A, et al: A new method of detection of islet cell antibodies (ICA) using peroxidase-labeled protein A, and incidence of ICA in Type 1 (insulin-dependent) diabetes. *Diabetologia* **29**:378–382, 1986.
21. Karjalainen J, Knip M, Mustonen A, et al: Relation between insulin antibody and complement-fixing islet cell antibody at clinical diagnosis of IDDM. *Diabetes* **35**:620–622, 1986.
22. Grunfeld C, Jones DS, Shigenaga JK: Autoantibodies against the insulin receptor. *Diabetes* **34**:205–211, 1985.
23. Shoelson SE, Marshall S, Horikoshi H, et al: Antiinsulin receptor antibodies in an autoantiidiotypes. *J Clin Endocrinol Metab* **63**:56–61, 1986.
24. Proietto J, Nankervis A, Aitken P, et al: Insulin resistance in cirrhosis: Evidence for a post-receptor defect. *Clin Endocrinol* **21**:677–688, 1984.
25. Olczak SA, Greenwood RH, Hales CN: Post-receptor insulin resistance after diazoxide in non-insulin dependent diabetes. *Horm Metab Res* **18**:38–41, 1986.
26. Fövényi J, Tótpál K, Thaisz E, et al: Non-specific cellular immunity in Type I and II diabetes *Exp Clin Endocrinol* **83**:203–206, 1984.
27. Negishi K, Waldeck N, Chandy G, et al: Natural killer and islet cell activities in Type 1 (insulin-dependent) diabetes, *Diabetologia* **29**:352–357, 1986.
28. Zier KS, Leo MM, Spielman RS, et al: Decreased synthesis of Interleukin 2 (IL-2) in insulin-dependent diabetes mellitus, *Diabetes* **33**:552–555, 1984.
29. Ilonen J, Surcel HM, Mustonen A, et al: Lymphocyta subpopulations at the onset of Type 1 (insulin-dependent) diabetes. *Diabetologia* **27**:106–108, 1984.
30. Rich S, O'Neill G, Daimasso AP, et al: Complement and HLA. *Diabetes* **34**:504–509, 1985.
31. Pozzili P, Sensi M, Al-Sakkaf L, et al: Prospective study of lymphocyte subsets in subjects genetically susceptible to Type 1 (insulin-dependent) diabetes. *Diabetologia* **27**:132–135, 1984.
32. Marcelli-Barge A, Poinier JC, Schmid M, et al: Genetic polymorphism of the fourth component of complement and type 1 (Insulin-dependent) diabetes. *Diabetologia* **27**:116–117, 1984.
33. Topliss D, How J, Lewis M, et al: Evidence for cell-mediated immunity and specific suppressor T lymphocyta dysfunction in Graves' disease and diabetes mellitus. *J Clin Endocrinol Metab* **57**:700–705, 1983.
34. Rodier M, Andary M, Richard JL, et al: Peripheral blood T-cell subsets studied by monoclonal antibodies in Type 1 (insulin-dependent diabetes: Effect of blood glucose control. *Diabetologia* **27**:136–138, 1984.

35. Prud'homme GJ, Colle E, Fuks A, et al: Cellular immune abnormalities and autoreactive T lymphocytes in insulin-dependent diabetes mellitus in rats. *Immunol Today* **6:**160–162, 1985.
36. Sundsmo JP, Papin RA, Wood L, et al: Complement activation in Type 1 human diabetes. *Clin Immunol Immunopathol* **35:**211–225, 1985.
37. Shah SV, Wallin JD, and Eilen SD: Chemiluminescence and superoxide anion production by leukocytes from diabetic patients. *J Clin Endrocrinol Metab* **57:**402–409, 1983.
38. Matkovics B, Varga SzI, Szabó L, et al: The effect of diabetes on the activities of the peroxide metabolism enzymes. *Horm Metab Res* **14:**77–79, 1982.
39. Markert M, Cech P, Frei J: Oxygen metabolism of phagocytosing human polymorphonuclear leukocytes in diabetes mellitus. *Blut* **49:**447–455, 1984.
40. Nath N, Chari SN, Rathi AB: Superoxide dismutase in diabetic polymorphonuclear leukocytes. *Diabetes* **33:**586–589, 1984.
41. Theete LG, Couch RK, Buse MG, et al: The protective role of copper-zinc superoxide dismutase against alloxan-induced diabetes: morphological aspects. *Diabetologia* **28:**677–682, 1985.
42. Thomas G, Skrinska V, Lucas FV, et al: Platelet glutathione and thromboxane synthesis in diabetes. *Diabetes* **34:**951–954, 1985.
43. Loven D, Schedl H, Wilson H, et al: Effect of insulin and oral glutathione on glutathione levels and superoxide dismutase activities in organs of rats with streptozocin-induced diabetes. *Diabetes* **35:**503–507, 1986.
44. Yamamura M, Boler J, Valdimarsson H: A ^{51}chromium release assay for phagocytic killing of Candida albicans. *J Immunol Methods* **13:**227–234, 1979.
45. Tanabe T, Kabayashi Y, Usui T: Enhancement of human neutrophil oxygen consumption by chemotactic factors. *Experientia* **39:**604–611, 1983.
46. Cohen HJ, Chovaniek ME: Superoxide generation by digitonin-stimulated guinea pig granulocytes. *J Clin Invest* **40:**1081–1086, 1978.
47. Pick E, Keisari Y: Superoxide anion and hydrogen peroxide production by chemically elicited peritoneal macrophages - Induction by multiple non-phagocytic stimuli. *Cell Immunol* **59:**301–312, 1981.
48. Paglia DE, Valentine WN: Studies on the quantitative and qualitative characterization of erythrocytes glutathione peroxidase. *J Lab Clin Med* **70:**150–158, 1967.
49. Weiss SJ, Regiani S: Neutrophils degrade subendothelial matrices in the presence of alpha-1-proteinase inhibitor: Cooperative use of lysosomal proteinases and oxygen metabolites. *J Clin Invest* **73:**1297–1303, 1984.
50. Weiss SJ, Curnutte JT, Regiani S: Neutrophil-mediated solubilization of the subendothelial matrix: Oxidative and nonoxidative mechanisms of proteolysis used by normal and chronic granulomatous disease of phagocytes. *J Immunol* **136:**636–641, 1986.
51. Fülöp T, Fóris G, Wórum I, et al: Some variations of PMNL functions with aging. *Mech Aging Dev* **29:**1–7, 1985.
52. Shah SV, Wallin JD, Cruz FC: Impaired oxidative metabolism by leukocytes from renal transplant recipients: A potential mechanism for the increased susceptibility to infections. *Clin Nephrol* **21:**89–92, 1984.
53. Kurose H, Ui M: Dual pathways of receptor-mediated cyclic GMP generation in NG 108-15 cells as differentiated by susceptibility to islet-activating protein, partussis toxin. *Arch Biochem Biophys* **238:**424–434, 1985.

18

The Respiratory Burst of Fertilization

ERIC E. TURNER and BENNETT M. SHAPIRO

1. INTRODUCTION AND HISTORICAL PERSPECTIVE

The activation of a transient burst of cellular respiration was first studied using the system of fertilization as developed in the early part of this century by Warburg,[1,2] Shearer,[3] and others. Biochemists were attracted to the fertilized egg as a key to cellular mechanisms, for the response of an egg to the fertilizing sperm is one of the most dramatic events in biology. Marine invertebrates have been the classic system for the study of these phenomena, because of the large quantity of gametes that can be obtained, and since fertilization occurs in the open sea under conditions that can be readily reproduced in the laboratory. If the fusion of the plasma membranes of egg and sperm is considered the moment of fertilization, the events that follow can be divided into two broad categories: early events, which occur within a few minutes of fertilization and consist primarily of mechanisms to prevent the entry of additional sperm; and later events, which follow the establishment of a block to polyspermy and initiate the developmental program.

Prevention of polyspermy is a critical consideration for the embryo; fertilization by multiple sperm results in abnormal cleavage and death.[4] The initial block to polyspermy appears to be a transient depolarization of the normal negative egg resting potential.[5] Subsequently, a permanent structural block to polyspermy is provided by the elevation of the fertilization envelope, an extracellular protein coat that surrounds the embryo until it hatches at the blastula stage.

The fertilization envelope is assembled within minutes of fertilization from the extracellular protein matrix of the unfertilized egg (vitelline layer) and the contents of the cortical granules, which are released in a wave of exocytosis[6] (Fig. 1). The proteins are initially assembled by noncovalent interactions mediated by divalent cations; the structure is then covalently crosslinked via bonds between adjacent tyrosine residues.[7,8] The crosslinking reaction is carried out by ovoperoxidase, a

ERIC E. TURNER and BENNETT M. SHAPIRO • Department of Biochemistry, University of Washington, Seattle, Washington 98195.

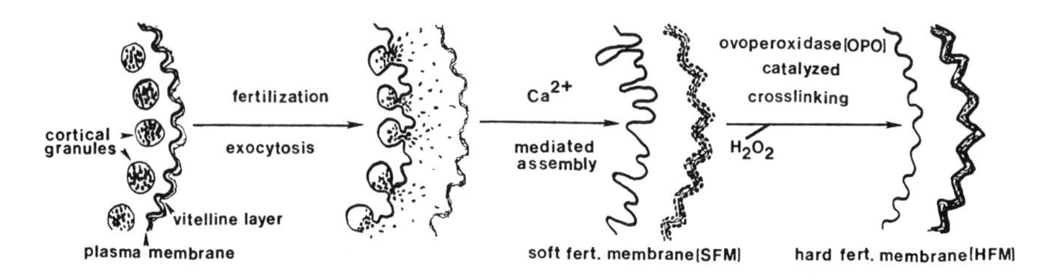

FIGURE 1. Cortical granule exocytosis and fertilization membrane assembly.

70,000-M_r heme protein that is also a cortical granule component,[9] using H_2O_2 produced by the egg.

Crosslinking occurs in the first 15 min following fertilization, during a period of extremely rapid oxygen uptake by the embryo called the respiratory burst of fertilization (Fig. 2). The burst respiration is relatively cyanide insensitive and is superimposed on the cyanide sensitive mitochondrial respiration of the embryo, which is also activated and persists at high levels far into development. The magnitude of the cyanide insensitive burst of oxygen consumption appears to be stoichiometrically equivalent to the H_2O_2 produced by the egg as substrate for the crosslinking reaction.[10] In addition to its role in the formation of a structural barrier to

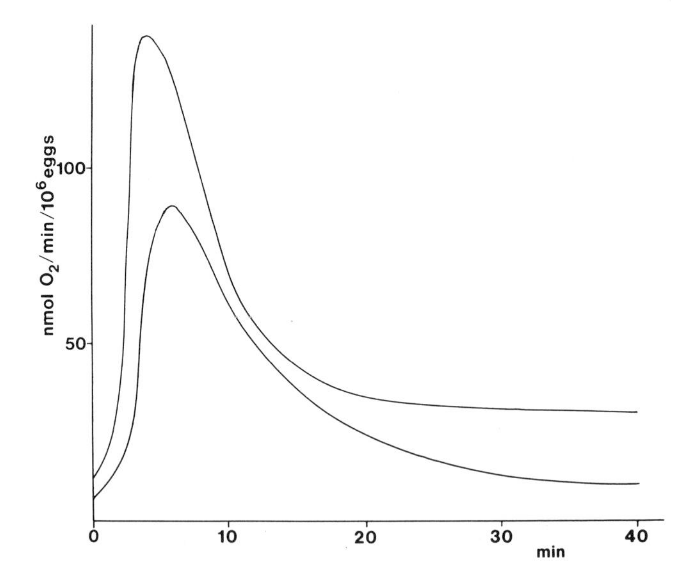

FIGURE 2. The respiratory burst of fertilization. 2.8×10^5 eggs of *Strongylocentrotus purpuratus* were fertilized ($t = 0$) with 1 μl sperm in a Clark O_2 electrode containing 5 ml Millipore-filtered seawater with 20 mM Tris, pH 8.0, at 10°C. The lower trace represents respiration in the presence of 2 mM NaCN added 1 min after insemination. The initial O_2 concentration was 286 μM.

polyspermy, the hydrogen peroxide produced may have a direct cytotoxic effect on supernumary sperm.[11]

The increase in oxygen uptake at fertilization was first observed by Warburg,[1] who used titrimetric methods to measure oxygen consumption in eggs of the sea urchin *Arbacia punctulata* and observed a six- to sevenfold increase in the rate of oxygen uptake in the first hour following fertilization when compared with the rate observed in unfertilized eggs. Later, using the manometric device that bears his name, Warburg[2] also demonstrated a four- to sixfold increase in respiration in *Strongylocentrotus purpuratus;* in subsequent work, others used this device to repeat Warburg's observations in *Arbacia.*[12,13] These experiments did not attempt to resolve events during the first few minutes after fertilization; monitoring began 15 min or more after insemination. Warburg[14] and Loeb and Wasteneys[15] also showed that parthenogenic activation of fertilization envelope formation with butyric acid causes a similar increase in respiration. Using the eggs of the sea urchin *Echinus microtuberculatus,* Shearer[3] was able to fertilize eggs within the Warburg manometer and was first to demonstrate a transient burst of respiration which lasted approximately 15 min, followed by a plateau rate that was still six times that of the unfertilized egg.

Several other early investigators examined respiration before and after fertilization in eggs of invertebrates and fish. These experiments showed severalfold activations of respiration after fertilization in the echinoids *Echinus miliaris*[16] and *Paracentrotus lividus*[17] and a twofold activation in two species of ascidians.[18,19] By contrast, no change or even a slight decrease in respiration has been observed at fertilization in many other species, including annelid worms of the genera *Nereis*[20,21] and *Chaetopterus;*[12,13] the starfish *Marthasterias glacialis;*[22] the marine bivalves *Cumingia tellinoides,*[23] *Ostrea commercialis,*[24] and *Ostrea virginica*[12]; and the teleost fishes *Fundulus heteroclitus*[25] and *Oryzias latipes.*[26] These and many other early respiration studies have been summarized by Rothschild,[27] Monroy,[28] and Czihak.[29]

Except as noted, respirometer measurements in these studies were conducted with the initial data collected from 15 min to 2 hr after insemination, which does not offer adequate time resolution to discern a transient elevation of respiration associated with fertilization envelope elevation. A few attempts to resolve respiratory changes immediately after insemination include the work of Shearer,[3] experiments by Laser and Rothschild[30] that demonstrated a transient burst of respiration following fertilization in eggs of *Echinus miliaris,* and work by Boyd[31] indicating that in the fish *Fundulus* respiration may be transiently activated as it is in the echinoids. Most studies conducted with adequate time resolution and sensitivity to detect the transient respiratory burst that accompanies the other early events of fertilization have been conducted with the oxygen electrode, first applied about 25 years ago, and are discussed below.

The best overall generalization regarding increased respiration at fertilization appears to be that eggs shed in the mature state, primarily echinoids, have low levels of respiration prior to fertilization, with a severalfold activation when fertilized. Eggs shed and fertilized in earlier stages, including the germinal vesicle stage or metaphase of the first or second meiotic division, have higher baseline levels of respiration but no activation at fertilization.[24,29] Lindahl and Holter[32] confirmed this with the time course of oxygen consumption during egg maturation in a single species, *P. lividus*. They observed that the respiration of primary oocytes is clearly higher than unfertilized eggs and even somewhat exceeds that of fertilized eggs. However, considering a transient burst of respiration at fertilization rather than a long-term rise, too few studies have been done with sufficient time resolution to make reliable generalizations about the species distribution of this phenomenon. These early studies document a respiratory burst as well as prolonged activation of respiration in sea urchin eggs but generally do not confirm or exclude a transient burst of respiration in other species.

2. THE RESPIRATORY BURST AND RELATED EVENTS AT FERTILIZATION

2.1. Initiation and Kinetics of the Respiratory Burst

Detailed examination of the time course of the respiratory burst became possible with polarographic recording of oxygen consumption. Ohnishi and Sugiyama[33] used a rotating platinum electrode to measure oxygen uptake at fertilization in *Pseudocentrotus depressus* eggs, and demonstrated a transient burst of respiration beginning 30 sec postinsemination and lasting about 4 min, temporally associated with the elevation of the fertilization envelope. The maximal rate of respiration at 90 sec was about three times the plateau rate at 10 min postinsemination and 15 times the unfertilized rate. The maximum respiratory rate was 1.1 pmoles/min per egg. Similar results were obtained with the urchin *Hemicentrotus pulcherrimus*. They also demonstrated that the characteristic acid efflux seen on fertilization slightly precedes initiation of the respiratory burst and that the respiratory burst is also initiated by parthenogenic activation with sodium cholate or butyric acid, in both cases associated with fertilization envelope elevation.

Epel[34] used a Clark-type oxygen electrode to demonstrate a transient burst of respiration in eggs of *S. purpuratus,* which began about 60 sec postinsemination and lasted about three minutes. The maximal rate of respiration was 6.2 times that of the unfertilized egg, or 60 fmoles/min per egg.

Rates for peak oxygen consumption per egg vary widely among respiration studies, due in part to differences in temperature and in the size of eggs of various species. Epel also showed that fertilization is accompanied by a net conversion of NAD^+ to $NADP^+$ by an NAD-kinase[35] and that a net conversion of pyridine nucleotides to the reduced form occurs beginning about 40 sec postinsemination, slightly before the increase in respiration.[34] These changes in egg metabolism are discussed in more detail below. In addition, Epel[36] demonstrated that the transient increase in respiration that follows fertilization in *S. purpuratus* is not controlled by

ADP availability, as relative ADP levels do not increase at the time of maximum respiratory activity.

Horowitz[37] used the rotating platinum O_2 electrode to measure respiration in three species of marine invertebrate: the sea urcin *Arbacia punctulata,* the asteroid *Asterias forbesii,* and the surf clam *Spisula solidissima.* As demonstrated by previous investigators, *Arbacia* displayed both a long-term activation of respiration and a transient burst. In *Asterias,* respiratory rates also increased within a minute of insemination, but returned to unfertilized levels by 6 min, explaining why earlier experiments with asteroids[22] failed to demonstrate burst respiration. In *Spisula* this study failed to demonstrate prolonged activation of respiration at fertilization; because data collection began 6 min after insemination, a transient burst of respiration was neither demonstrated nor ruled out.

Foerder et al.[10] advanced the understanding of several aspects of the respiratory burst of fertilization by showing its relationship to hydrogen peroxide production and fertilization envelope crosslinking, to chemiluminescence, and its relationship to cortical granule exocytosis. Each of these aspects is considered in turn in the sections that follow.

2.2. Role of Exocytosis and Intracellular Ca^{2+} in the Respiratory Burst

Like many other processes of exocytosis, the cortical granule release appears to be mediated by Ca^{2+}.[38] One of the best pieces of evidence for this is the activation of exocytosis, respiration, and early development by the calcium ionophore A23187.[39] Extracellular Ca^{2+} is not needed for activation, suggesting that it is released from intracellular stores.[40] At least two calcium-dependent early events in egg activation appear to be calmodulin mediated, the exocytosis itself[41] and activation of NAD kinase.[42] Calcium-mediated exocytosis has been demonstrated in isolated cortical granule-plasma membrane preparations.[43,44] Exocytosis proceeds as a wave from the point of sperm entry; this has been best demonstrated in eggs of the medaka fish with injection of the calcium photoprotein aequorin, which emits light upon Ca^{2+} binding and makes visible a wave of Ca^{2-} release originating at the point of sperm entry.[45] Thus, release of Ca^{2+} in one region of the egg may result in a release in adjacent areas in a self-propagating fashion.

Steinhardt and Epel[39] showed that activation of both a transient respiratory burst and a prolonged increase in respiration occurred with parthenogenic activation using A23187; although the respiratory burst was somewhat delayed, it was of normal magnitude. Foerder et al.[10] repeated this result and demonstrated that A23187 also initiated H_2O_2 production, as measured by formate oxidation. Like the respiration which follows insemination, the respiratory burst stimulated by A23187 activation is also largely cyanide insensitive (E. Turner, unpublished results). Steinhardt[46] has also shown in the sea urchin *L. pictus* that in addition to calcium release, intracellular alkalinization is required for activation of burst respiration. Alkalinization following fertilization is dependent on acid efflux via an amiloride-sensitive Na^+/H^+ exchange mechanism.[47,48] Activation of eggs by A23187 in sodium-free seawater does not activate respiration[39] until Na^+ is added back to the medium.[46] Treatment of eggs in Na^+-free seawater with 10 mM NH_4Cl, which artificially alkalinizes the cytoplasm, has little effect alone but permits activation of a normal amount of respiration by ionophore. The cyanide sensitivity of respiration and extent of H_2O_2 production have not been tested under these conditions.

Exocytosis of cortical granules can be inhibited by agents such as urethane and the tertiary amine local anesthetics procaine and tetracaine, which disrupt the attachment of cortical granules to the plasma membrane[49,50] and result in the retention of cortical granules within the cytoplasm of the fertilized egg; these eggs undergo early cleavage on schedule with cortical granules distributed between the blastomeres.[43,51] Inhibition of exocytosis in this manner completely blocks the transient respiratory burst of fertilization.[10] This finding suggests that if the local anesthetics act specifically and inhibit only exocytosis, a rise in intracellular Ca^{2+} per se may not be sufficient to trigger the respiratory burst and that the burst may be intimately linked to exocytosis itself; alternatively, procaine may act via alteration of internal pH. The question of Ca^{2+} release in eggs treated with local anesthetics has not been addressed directly, nor is it known whether calcium ionophore could activate the respiratory burst without exocytosis in procaine-treated eggs.

Another possible mechanism for activation of the respiratory burst involves the recently discovered role of inositol 1,4,5-triphosphate (IP_3) metabolism in egg activation. Triphosphoinositide and diphosphoinositide levels increase at fertilization in sea urchin eggs[52]; microinjection of IP_3 into unfertilized sea urchin eggs initiates cortical granule exocytosis, apparently by stimulating release of Ca^{2+} from an intracellular store[53] and mimics the changes in electrical potential of the egg plasma membrane (fertilization potential) that follow sperm penetration.[54] Application of micromolar Ca^{2+} to the isolated cortex of the egg (a preparation of cortical granules adherent to plasma membrane) also stimulates breakdown of phosphatidylinositol 4,5-biphosphate.[53] Because either appears to be sufficient to initiate the self-propagating wave of cortical granule exocytosis, it is not clear which is the primary triggering event that follows fusion of sperm and egg. It has not been shown whether IP_3 injection activates the respiratory burst of fertilization or peroxide production. Thus, further investigation will be required to dissect the intimate relationship between IP_3 and Ca^{2+} release, exocytosis, and activation of the respiratory burst of fertilization.

2.3. Activation of NAD Kinase and the Pentose Phosphate Pathway

Other early events in the fertilization of sea urchin eggs that may relate to the mechanism of the respiratory burst are coordinated changes in pyridine nucleotide and carbohydrate metabolism. A number of investigators explored the pathways of carbohydrate metabolism in sea urchin eggs, particularly the relative importance of the glycolytic pathway versus the hexose monophosphate (HMP) shunt or pentose phosphate pathway. Lindberg and Ernster[55] demonstrated the preferential metabolism of HMP shunt metabolites in homogenates of the unfertilized eggs of *Strongylocentrotus drobachiensis.* Iodoacetic acid, an inhibitor of glycolysis, did not inhibit oxygen uptake by these preparations. By contrast, Cleland and Rothschild[56] demonstrated active glycolysis in egg homogenates from another urchin, *Echinus esculentus.* Sea urchin eggs contain enzymes for both glycolysis and the HMP shunt, but the latter are in excess.[57] Krahl[58] used isotopically labeled glucose to demonstrate that the HMP shunt is the predominant pathway of glucose metabolism in unfertilized and newly fertilized eggs of *A. punctulata* but that glycolysis becomes increasingly significant later in development. Isono and Yasumasu[59] showed that the fraction of glucose metabolized by the HMP shunt increased from 0.61 in unfertilized eggs of the sea urchin *P. depressus* to 0.77 immediately after fertilization, to a maximum of 0.83 from the 16-cell stage until hatching, with glycolysis then becoming increasingly important later in development. Total flux through the pentose phosphate pathway increases abruptly at fertilization in proportion to the large increase in respiration. On the basis of measurements of respiratory quotient, lipid metabolism does not become significant until after hatching.[60]

Control of glucose entry into the HMP shunt is at the glucose 6-phosphate dehydrogenase (G6PD) step. At fertilization, this enzyme shows a marked increase in activity[61] associated with a shift from particulate to soluble form.[62] More recent studies have shown that much of the solubilization of G6PD occurs within 30 sec of insemination,[63] a time course that suggests that the rise in intracellular free Ca^{2+} that follows fertilization may play a role in activating the enzyme; egg activation with ionophore A23187 also promotes solubilization. Alkalinization of the intracellular space with ammonia does not promote solubilization or activation of G6PD, and the solubilization is not inhibited by transfer of fertilized eggs to Na^+-free medium, indicating that the Na^+–H^+ exchange following fertilization probably does not directly control G6PD. Alkalinization of the intracellular space of eggs with the weak bases ammonia or procaine has been shown, like fertilization, to increase reduced pyridine nucleotide levels in eggs of *Lytechinus pictus,*[64] and a rise in pyridine nucleotides appears to be necessary for the initiation of DNA synthesis. However, activation by weak bases gives a relatively large increase in NADH and a small increase in NADPH compared to fertilization.

A related event at fertilization is the activation of the enzyme NAD-kinase, which carries out the reaction

$$NAD^+ + ATP \rightarrow NADP^+ + ADP$$

Krane and Crane[65] noted an increase in NADPH at fertilization in the eggs of *A. punctulata* and the surf clam *Spisula solidissima.* Epel[35] showed that in eggs of *S. purpuratus* within 60 sec of insemination, the sum of NAD^+ and NADH levels declined from 72.4 nmoles/10^6 eggs to 52.2 nmoles/10^6 eggs, accompanied by a stoichiometrically equivalent increase in the sum of $NADP^+$ and NADPH from 13.2 nmoles/10^6 eggs to 34.4 nmoles/10^6 eggs. Total reduced pyridine nucleotides increased from 12.2 nmoles/10^6 eggs to 31.1 nmoles/10^6 eggs. Both the NAD-kinase reaction and the net reduction of total pyridine nucleotides are initiated at the time of, or slightly preceding, the respiratory burst. More recently, the NAD-kinase has been shown to be under the control of Ca^{2+} and calmodulin.[42]

The overall effect of the activation of NAD-kinase activity at fertilization, the solubilization and activation of G6PD, and the increased flux through the pentose phosphate pathway is to greatly increase the available amount of NADPH in the cell. To the extent that studies have been conducted with adequate time resolution, events that promote the formation of NADPH precede and continue throughout the period of the transient respiratory burst of fertilization. Without direct evidence, this strongly suggests that the H_2O_2-producing oxidase of the respiratory burst uses NADPH as a reductant for molecular oxygen, in a system analogous to that of the superoxide-producing NADPH oxidase of the neutrophil.

3. PEROXIDE PRODUCTION AND FERTILIZATION ENVELOPE HARDENING

3.1. Assembly and Crosslinking of the Fertilization Envelope

The fertilization envelope of the sea urchin egg is an extracellular protein matrix assembled from proteins exocytosed from the cortical granules of the egg and the extracellular vitelline layer.[66] It provides a permanent block to polyspermy, and protects the embryo until hatching at the blastula stage. The initial assembly of cortical granule proteins into the fertilization envelope requires millimolar levels of divalent cations found in seawater.[67] One of the structural proteins included in the assembly process, proteoliasin, has been isolated and characterized.[68] Proteoliasin mediates the assembly of the crosslinking enzyme ovoperoxidase into the fertilization envelope. Following the initial Ca^{2+}-mediated assembly, the fertilization envelope undergoes a series of morphological and biochemical transitions (Fig. 1). The early rearrangement of the fertilization envelope results in a change in the appearance of egg microvillar casts on the membrane surface from rounded igloo forms to sharp tent shapes.[69] The ability of glycine ethyl ester to block this transition suggests that transglutaminase may play a role in the transition. A subsequent transition is the "hardening" or crosslinking of the fertilization envelope by ortho-ortho dityrosyl bonds[7,8] using ovoperoxidase and hydrogen peroxide as shown below:

OH OH
27,500 daltons
Ovoperoxidase
H_2O_2 $2H_2O$
OH OH

This reaction is blocked by peroxidase inhibitors such as cyanide and aminotriazole, leaving a "soft" fertilization envelope that, unlike the final structure, is soluble in sulfhydryl reagents.[7] This crosslinking process has been demonstrated in vitro using soft fertilization envelopes and hydrogen peroxide.[67]

An ovoperoxidase has also been identified in mammals by histochemical methods in the cortical granules of unfertilized mouse eggs and in cortical granule exudate and the zona pellucida (the mammalian analogue of the fertilization envelope) of eggs activated by A23187.[70] Mammalian ovoperoxidase appears to have a role in hardening the zona pellucida of the mouse egg[71] analogous to its role in the sea urchin. Respiration changes following fertilization and H_2O_2 production by mammalian eggs have not been reported.

3.2. Characteristics of Ovoperoxidase

The role of ovoperoxidase as the crosslinking enzyme of the fertilization envelope is well established. The inhibitors of the crosslinking or hardening reaction in vivo[7] and of the in vitro peroxidase activity are virtually identical. Deits et al.[9] detail the purification and basic characterization of ovoperoxidase, a glycoprotein of 70,000 M_r, with one Fe-heme prosthetic group per molecule. Among the previously characterized peroxidases, the spectral and kinetic parameters of ovoperoxidase are most similar to lactoperoxidase found in mammals. In addition to the crosslinking reaction and activity in standard peroxidase assays such as guaiacol oxidation, ovoperoxidase is able to catalyze oxidation of iodide using hydrogen peroxide. This ability has suggested that ovoperoxidase may also act in a halogen-mediated mechanism for killing of supernumary sperm, analogous to the antimicrobial activity of the polymorphonuclear leukocyte.[72]

Further characterization of ovoperoxidase has demonstrated several interesting properties. Deits and Shapiro[73,74] showed pH-induced hysteretic transitions that affect the catalytic and spectral properties of ovoperoxidase. The activity of ovoperoxidase in the iodide oxidation assay is considerably lower at pH 5.5 than at 8.0. However, on dilution into buffer at acidic pH, ovoperoxidase equilibrated at pH 8.0 will initially exhibit catalytic activity and a Soret spectrum characteristic of the higher pH with a slow (30-sec) decay to the lower equilibrium activity level. Similar hysteretic effects were seen on pH shifts from acidic to neutral; a lag was noted before attaining maximal activity, although spectral changes were immediate. Activity changes with pH shifts in both directions are fully reversible.

Ovoperoxidase is stored in the unfertilized egg in an acidic compartment, the cortical vesicle, and exocytosed into seawater at approximately pH 8. The lag in attaining maximal activity may permit assembly of the ovoperoxidase into the fertilization envelope before the enzyme is fully activated. Deits and Shapiro[74] also showed that these hysteretic transitions are absent in the mature fertilization envelope, and that in the membrane the enzyme is apparently locked into a conformational state that is observed only transiently during pH shifts with the isolated enzyme. Proteoliasin,[68] which mediates assembly of ovoperoxidase into the fertilization envelope, thus may affect the catalytic activity of ovoperoxidase as well as its localization.

In addition to its well-characterized peroxidase activity, ovoperoxidase possesses an NAD(P)H–O_2 oxidoreductase activity dependent on Mn^{2+} and ovothiol, a thiolhistidine cofactor from within the egg.[75] We do not believe that this activity is the physiological oxidase of the respiratory burst (see Section 4.2).

3.3. H_2O_2 Production and Chemiluminescence

Foerder et al.[10] demonstrated the relationship between the respiratory burst of fertilization and production of hydrogen peroxide by the egg. In these studies, oxygen consumption was measured by the Clark O_2 electrode. H_2O_2 production was assayed by the oxidation of formate to CO_2 in the presence of catalase and with scopoletin, which loses its intrinsic fluorescence when oxidized by H_2O_2 and horseradish peroxidase (HRP). Formate oxidation by an egg suspension is clearly greatly increased at fertilization and is inhibited to unfertilized levels by inclusion of the catalase/peroxidase inhibitors cyanide (1 mM) and azide (1 mM). Quantitation of the amount of H_2O_2 produced at fertilization was done with the scopoletin assay; because H_2O_2 produced by the egg has several possible fates, including use by ovoperoxidase in the crosslinking reaction and destruction by egg catalase and other H_2O_2 metabolizing systems, it is difficult to capture all the H_2O_2 produced with the scopoletin–HRPO system. A fertilized egg suspension, with 0.25 mM aminotriazole to inhibit utilization of H_2O_2 by ovoperoxidase, scopoletin, and HRPO gave approximately 104 fmoles H_2O_2/embryo over the 15 min following insemination. However, only 20% of an H_2O_2 standard added to the incubation mixture was recovered; if the H_2O_2 produced during the respiratory burst and that exogenously added are presumed to be metabolized as a common pool, H_2O_2 production during the respiratory burst is estimated as 520 fmoles. Measurements by Foerder et al.[10] with the oxygen electrode also showed oxygen consumption over the first 15 min following insemination of 770 fmoles O_2/embryo. Thus approximately two thirds of the oxygen consumed during this period is reduced to H_2O_2. This is approximately equivalent to the CN^--insensitive portion of oxygen uptake during this time period (Fig. 2).

Chemiluminescence was measured[10] in a liquid scintillation counter in whole eggs at fertilization and in fertilization product, the material exocytosed from the cortical granules at fertilization, which includes soluble ovoperoxidase.[9] These proteins are normally assembled into the fertilization envelope but are released in soluble form if the eggs are treated with trypsin or dithiothreitol prior to fertilization or activation. Chemiluminescence during the respiratory burst appears to be due primarily to reaction of H_2O_2 with the peroxidatic system. Intact eggs showed a burst of chemiluminescence at fertilization which coincided with the bursts of oxygen uptake and H_2O_2 production. Chemiluminescence in eggs is almost entirely abolished by inhibitors of the ovoperoxidase such as aminotriazole, phenylhydrazine and sodium azide. Chemiluminescence was also seen on addition of H_2O_2 to fertilization product, and again inhibitors of the peroxidatic activity of this preparation eliminated the chemiluminescence at similar concentrations. These results are similar to the phagocytosing neutrophil, in which much of the chemiluminescence seen is produced by myeloperoxidase[76,77]; the remainder is produced by the NADPH oxidase of the respiratory burst.

4. POSSIBLE MECHANISMS OF THE RESPIRATORY BURST

The mechanism of the respiratory burst of fertilization and hydrogen peroxide production is unknown, although several enzymes have been proposed, including glucose oxidase, NADPH oxidase, an ovoperoxidase NAD(P)H oxidase activity dependent on the thiolhistidine cofactor ovothiol,[75] lipoxygenase,[78–80] and oxidation of the sea urchin pigment echinochrome.[81–83] Evidence in favor of each of these mechanisms is examined in turn.

4.1. Lipoxygenase

Perry and Epel[83] demonstrated Ca^{2+}-dependent O_2 consumption in *Arbacia punctulata* egg homogenates, which is dependent on the egg pigment echinochrome, and have shown H_2O_2 production from isolated echinochrome and O_2. Because this pigment is absent from the eggs of *S. purpuratus* and many other species, this activity cannot be a universal mechanism for the respiratory burst. Perry and Epel[79,80] have also demonstrated Ca^{2+}-stimulated cyanide-insensitive oxygen consumption in *S. purpuratus* egg homogenates. This activity is stimulated by the addition of arachidonate or phospholipase A_2 and is associated with conversion of exogenous arachidonic acid primarily to HETE, but not prostaglandins, suggesting that it is due to a lipoxygenase. This work is inadequate as an explanation of the respiratory burst in intact eggs, as no estimate is made of the extent of lipid peroxidation in vivo. Furthermore, no mechanism is suggested for the production of H_2O_2 by the egg, and the fact that cyanide-insensitive O_2 uptake and H_2O_2 production appear to be roughly stoichiometrically equivalent is not explained. It is interesting to calculate what fraction of the total egg phospholipid would need to be oxidized by lipoxygenase to account for the respiration observed in whole eggs. *S. purpuratus* eggs contain 8.8 mg phospholipid/10^6 eggs,[29] equivalent to 3.2 mg or 10 μmoles unsaturated fatty acid (based on arachidonyl, palmitoyl PC as a model phospholipid). Foerder et al.[10] estimated burst O_2 uptake as 0.77 μmoles $O_2/10^6$ embryos, about two thirds of which is CN^- insensitive. Assuming one O_2 consumed per unsaturated lipid molecule, peroxidation of more than 5% of the unsaturated phospholipid would be necessary to account for the respiratory burst, making lipid peroxidation an unlikely mechanism for the burst H_2O_2 production. These calculations make evident the necessity of having a very abundant or rapidly mobilized source or reducing equivalents for the respiratory burst. The oxygen consumption observed, 770 fmoles/egg, is far in excess of the total pyridine nucleotide content of the cell (85.6 fmoles[35]) and approximates its ATP content (0.8 pmoles[36]).

Lipid peroxidation or phospholipase A_2 activation may play an important role in triggering cortical granule release. Shimada et al.[84] showed that mellitin, a phospholipase A_2 activator from bee venom, activates fertilization membrane elevation and burst respiration in sea urchin eggs, without stimulating protein or DNA synthesis or subsequent cleavage. Ca^{2+}-dependent phospholipase activity has been demonstrated in sea urchin egg homogenates,[85] and the phospholipase inhibitor quinacrine blocks the release of cortical granules from whole eggs.

4.2. Oxidase Activity of Ovoperoxidase

In searching for possible oxidase substrates for eggs and egg homogenates of *S. purpuratus,* Turner et al.[75] used O_2 consumption in the Clark electrode and spectrophotometric measurement of NAD(P)H consumption to demonstrate NAD(P)H oxidoredeuctase activity in egg homogenates. This was localized to the cortex of the egg (plasma membrane plus cortical granules) but was dependent on a heat-stable egg cofactor subsequently identified as the 4-thiolhistidine ovothiol.[86] Purification of the oxidase from solubilized egg cortices demonstrated that it was identical with the ovoperoxidase. The NAD(P)H–O_2 oxidoreductase activity of ovoperoxidase is clearly distinguished from its peroxidase activity in several ways. The overall reaction is the four-electron reduction of oxygen to water:

$$O_2 + 2NAD(P)H + 2H^+ \rightarrow 2NAD(P)^+ + H_2O$$

As NAD(P)H is a one or two electron donor, the reaction must proceed in multiple steps. Evidence for the presence and importance of intermediate states of oxygen reduction is provided by the inhibition of the reaction by superoxide dismutase, which converts two equivalents of superoxide ($O_2^- \cdot$) to H_2O_2 plus O_2, and inhibition by catalase. Micromolar levels of H_2O_2 also eliminate a lag that occurs before the maximal rate of NADH consumption by ovoperoxidase is achieved. Optimal H_2O_2 concentrations are 10–20 μM, with significant inhibition at higher levels. Another important feature of the activity is its relative insensitivity to cyanide inhibition compared with the peroxidatic activity of the enzyme, while azide and aminotriazole had a similar effect on the two activities. The oxidase activity was dependent on reduced ovothiol and Mn^{2+} for maximal activity. Mn^{2+} was maximally effective at 0.1–1.0 mM, but significant stimulation was detectable at micromolar Mn^{2+}. With homogeneous ovothiol, the kinetics of NADH consumption were saturable in ovothiol with a $K_{0.5}$ value of ~ 200 μM, maximal activity at 1 mM, and slight inhibition at higher concentrations.[86]

Several related compounds with a 4-thiolhistidine nucleus have been reported from marine invertebrates.[86–91] The form present in *S. purpuratus* is 1-methyl-α*N*,α*N*-dimethyl-4-thiolhistidine or ovothiol-C[86,87]; forms in other marine invertebrates differ only in the extent of α*N*-methylation. The ability to confer oxidoreductase activity on ovothiol appears to be a unique property of these compounds; aliphatic thiols such as GSH, L-cysteine, and reduced coenzyme A, as well as the 2-thiolhistidine, ergothioneine, failed to stimulate oxidase activity.

The cyanide insensitivity of the oxidase activity of ovothiol initially suggested that it might be responsible for the respiratory burst of fertilization, but evidence presented here indicates that this is unlikely. Both ovothiol and reduced pyridine nucleotides are intracellular metabolites, while ovoperoxidase is extracellular during the respiratory burst. Total egg pyridine nucleotide[35] and ovothiol[87] do not change significantly at fertilization, although reducing equivalents equal to several times the pyridine nucleotide content or approximately one half the ovothiol content of the egg would be required to account for the oxygen consumption seen in the burst. A second problem is that H_2O_2 is not produced stoichiometrically by the NADPH–oxidase reaction. Although the possibility exists that ovoperoxidase might generate sufficient H_2O_2 to use it in turn for dityrosyl crosslinking without significant amounts leaving the active site of the enzyme, this does not account for the significant levels of H_2O_2 detected following fertilization by Foerder et al.[10] Finally, aminotriazole, which inhibits both NAD(P)H–oxidase and peroxidase activities of ovoperoxidase, has no effect on the respiratory burst O_2 consumption.[87] These results effectively exclude the ovothiol-dependent NAD(P)H–oxidase activity of ovoperoxidase as the oxidase of the respiratory burst. However, they do not necessarily eliminate ovothiol as an electron donor in H_2O_2 generation; this possibility is discussed with the redox properties of ovothiol below.

4.3. Other Oxidases

Several other H_2O_2-producing oxidases are found in nature.[92] The release of a β-1–3-glucanase[93] at fertilization suggested that glucose might be mobilized from a macromolecular store as substrate for a glucose oxidase. However, glucose does not stimulate oxygen consumption by whole fertilized eggs or in an egg cortical preparation.[75]

Other widely distributed H_2O_2-producing oxidases are xanthine oxidase[92] and the fatty acyl-CoA oxidase of the peroxisome.[94] Xanthine and the CoA derivatives of palmitate and arachidonate applied to whole eggs do not augment the respiratory burst, and these substances did not stimulate respiration when added to a preparation of egg cortices.[75] Although these are far from all the oxidase activities described, the magnitude of the respiratory burst limits the possibilities to oxidases that use major metabolites as electron donors.

4.4. Parallels between Fertilization and Phagocytosis

There are striking similarities between the respiratory burst and associated events in eggs and the respiratory burst of the activated polymorphonuclear neutrophil. In response to the ingestion of bacteria or synthetic particles, phagocytic white blood cells (WBC) activate a system of H_2O_2 production involved in bacterial killing[95] and use a cyanide insensitive NADPH-dependent H_2O_2-generating system that forms superoxide ($O_2^- \cdot$) as the initial product of oxygen reduction.[96] Several other parallels between the neutrophil system and fertilization include conserved chemiluminescence, exocytosis of a peroxidase, increased hexose monophosphate shunt activity, and acid efflux.[72] The use of NADPH as a reductant by the strikingly similar neutrophil system, the likelihood based on the magnitude of oxygen consumption that the substrate of the H_2O_2-producing oxidase of the egg uses a major redox metabolite, and the massive activation of NADPH synthesis via the NAD–kinase and hexose monophosphate shunt that occurs following fertilization together suggest that an NADPH oxidase is the most likely candidate for the enzyme of the respiratory burst. However, no direct evidence yet exists for such a mechanism.

5. PRODUCTION OF ACTIVATED OXYGEN AND CONTROL OF TOXICITY

The production of H_2O_2 by cells for a metabolic purpose, such as bacterial killing in phagocytic neutrophils or fertilization envelope crosslinking in the sea urchin egg, presents a special problem because of the toxicity of hydrogen peroxide. Hydrogen peroxide is also produced as a product of peroxisomal oxidation of fatty acids[94] and by mitochondria. All eukaryotic cells have evolved extensive mechanisms to control the toxicity of H_2O_2 and other activated oxygen species such as superoxide ($O_2^- \cdot$) and hydroxyl radical. Catalase, which converts two equivalents of H_2O_2 to O_2 plus H_2O, is ubiquitous, as is superoxide dismutase, which converts two equivalents of $O_2^- \cdot$ to O_2 and H_2O_2. Glutathione and glutathione reductase are found in virtually all eukaryotes[97] and occur in some prokaryotes.[98] Glutathione is involved in a number of reactions that detoxify activated oxygen.[97] Mammals and birds possess an additional selenium-dependent enzyme, glutathione peroxidase,[99] which facilitates the reduction of hydrogen peroxide and organic hydroperoxides to water and the corresponding alcohols, respectively, using GSH as a reductant.[100]

It might be expected that cells that produce H_2O_2 for physiological reasons would have special or amplified systems for activated oxygen detoxification or that such systems would be activated in coordination with H_2O_2 production. The ability of reduced ovothiol to react readily with H_2O_2,[86] forming ovothiol disulfide and H_2O, suggests that the principal role of ovothiol may be in detoxification of hydrogen peroxide rather than its production. As sea urchin eggs contain substantial amounts of glutathione as well as ovothiol,[87,98] the relationship between the metabolism of these two compounds will be of special importance. Ovothiol persists in high levels at least 2 weeks into development of the embryo, long after H_2O_2 production has ceased,[87] but this is consistent with a role of ovothiol in activated oxygen detoxification, as surface seawater in which early development of sea urchins and other marine invertebrates takes place has been demonstrated to contain significant concentrations of H_2O_2 and superoxide.[101]

6. CONCLUSION

Although a respiratory burst clearly occurs at fertilization, prior work sheds little light on the mechanism of oxygen consumption and H_2O_2 production. Nonetheless, a set of guidelines has been developed from these studies that should aid in the search for such an oxidase: (1) the enzyme system must be quickly and transiently activated, beginning within 1 min of fertilization; (2) an increase in free intracellular Ca^{2+}, implicated in several aspects of egg activation, is a good candidate for the mechanism of control; (3) activation of the burst is closely linked to exocytosis and is perhaps dependent on it; (4) the activity is relatively cyanide insensitive; and (5) the magnitude of oxygen consumption is such that the system is the dominant metabolic event in the first 20 min following fertilization, requiring as substrate an abundant or rapidly renewable pool of redox metabolite. There is no direct evidence concerning the nature of the oxidase of the respiratory burst, but an H_2O_2 or superoxide-producing NADPH oxidase is an attractive possibility. Oxidase activities that have been detected in eggs or egg homogenates such as lipoxygenase activity,[79,80] and the ovothiol-dependent NADPH oxidase activity of ovoperoxidase[75] do not appear to be good candidates for the respiratory burst oxidase. The novel amino acid ovothiol does not appear to function in the production of H_2O_2 by the sea urchin egg but may function in the control of activated oxygen toxicity.

ACKNOWLEDGMENTS. We thank Mary Patella for preparing the manuscript. This work was supported by grant GM 07266 from the National Institutes of Health.

REFERENCES

1. Warburg O: Beobachtungen uber die oxydationsprozesse im seeigelei. *Z Physiol Chem* **57:**1–16, 1908.
2. Warburg O: Notizen zur entwicklungsphysiologie des seeigeleies. *Arch Gesamte Physiol* **109:**324–335, 1915.
3. Shearer C: On the oxidation processes of the echinoderm egg during fertilisation. *Proc R Soc Lond B* **93:**213–229, 1922.
4. Lillie F: *Problems of Fertilization*. Chicago, University of Chicago Press, 1919.
5. Jaffe LA: Fast block to polyspermy in sea urchin eggs is electrically mediated. *Nature (Lond)* **261:**68–71, 1976.
6. Kay ES, Shapiro BM: The formation of the fertilization membrane of the sea urchin egg. *Biol Fertil* **3:**45–80, 1985.
7. Foerder CA, Shapiro BM: Release of ovoperoxidase from sea urchin eggs hardens the fertilization membrane with tyrosine crosslinks. *Proc Natl Acad Sci USA* **74:**4214–4218, 1977.

8. Hall HG: Hardening of the sea urchin fertilization envelope by peroxidase-catalyzed phenolic coupling of tyrosines. *Cell* **15:**343–355, 1978.
9. Deits T, Farrance M, Kay ES, et al: Purification and properties of ovoperoxidase, the enzyme responsible for hardening the fertilization membrane of the sea urchin egg. *J Biol Chem* **259:**13525–13533, 1984.
10. Foerder CA, Klebanoff SJ, Shapiro, BM: Hydrogen peroxide production, chemiluminescence, and the respiratory burst of fertilization: Interrelated events in early sea urchin development. *Proc Natl Acad Sci USA* **75:**3183–3187, 1978.
11. Boldt J, Schuel H, Schuel R, et al: Reaction of sperm with egg-derived hydrogen peroxide helps prevent polyspermy during fertilization in the sea urchin. *Gamete Res* **4:**365–377, 1981.
12. Ballentine R: Analysis of the changes in respiratory activity accompanying the fertilization of marine eggs. *J Cell Comp Physiol* **15:**217–232, 1940.
13. Whitaker DM: On the rate of oxygen consumption by fertilized and unfertilized eggs. IV. *Chaetopterus* and *Arbacia punctulata. J Gen Physiol* **16:**475–495, 1933.
14. Warburg O: Uber die Oxydationen in lebenden Zellen nach Versuchen am Seeigelei. *Z Physiol Chem* **66:**305–340, 1910.
15. Loeb J, Wasteneys, H: The influence of hypertonic solution upon the rate of oxidations in fertilized and unfertilized eggs. *J Biol Chem* **14:**469–480, 1913.
16. Borei H: Respiration of oocytes, unfertilized eggs and fertilized eggs from *Psammechinus* and *Asterias. Biol Bull* **95:**124–149, 1948.
17. Brock N, Druckrey H, Herken H: Der stoffwechsel des geschadigten gewebes. III. *Arch Exp Pathol Pharmarkol* **188:**451–464, 1938.
18. Minganti A: Experiments on the respiration of *Phallusia* eggs and embryos (Ascidians). *Acta Embryol Morphol Exp* **1:**150–163, 1957.
19. Lentini R: The oxygen uptake of *Ciona intestinalis* eggs during development in normal and in experimental conditions. *Acta Embryol Morphol Exp* **4:**209–218, 1961.
20. Barron ESG: Studies on cell metabolism. I. The oxygen consumption of *Nereis* eggs before and after fertilization. *Biol Bull* **62:**42–53, 1932.
21. Whitaker DM: On the rate of oxygen consumption by fertilized and unfertilized eggs. III. *Nereis limbata. J Gen Physiol* **15:**191–199, 1931.
22. Tang P: The rate of oxygen consumption of *Asterias* eggs before and after fertilization. *Biol Bull* **61:**468–471, 1931.
23. Whitaker DM: On the rate of oxygen consumption by fertilized and unfertilized eggs. II. *Cumingia tellinoides. J Gen Physiol* **15:**183–190, 1931.
24. Cleland KW: Respiration and cell division in developing oyster eggs. *Proc Linn Soc NSW* **75:**282–295, 1950.
25. Philips FS: Oxygen consumption and its inhibition in the development of *Fundulus* and various pelagic fish eggs. *Biol Bull* **78:**256–274, 1940.
26. Nakano E: Respiration during maturation and at fertilization of fish eggs. *Embryologia* **2:**21–31, 1953.
27. Rothschild L: *Fertilization.* New York, John Wiley & Sons, 1956.
28. Monroy A: *Chemistry and Physiology of Fertilization.* New York, Holt, Rinehart and Winston, 1965.
29. Czihak G: *The Sea Urchin Embryo, Biochemistry and Morphogenesis.* New York, Springer-Verlag, 1975.
30. Laser H, Rothschild L: The metabolism of the eggs of *Psammechinus miliaris* during the fertilization reaction. *Proc R Soc Lond* **126:**539–557, 1939.
31. Boyd M: A comparison of the oxygen consumption of unfertilized and fertilized eggs of *Fundulus heteroclitus. Biol Bull* **55:**92–100, 1928.
32. Lindahl PE, Holter H: Uber die atmung der ovozyten erster ordnung von paracentrotus lividus und ihre veranderung wahrend der reifung. *C R Trav Lab Carlsberg Ser Chim* **24:**49–57, 1941.

33. Ohnishi T, Sugiyama M: Polarographic studies of oxygen uptake of sea urchin eggs. *Embryologia* **8:**79–88, 1963.
34. Epel D: Simultaneous measurement of TPNH formation and respiration following fertilization of the sea urchin egg. *Biochem Biophys Res Comm* **17:**69–73, 1964.
35. Epel D: A primary metabolic change of fertilization: interconversion of pyridine nucleotides. *Biochem Biophys Res Comm* **17:**62–68, 1964.
36. Epel D: Does ADP regulate respiration following fertilization of sea urchin eggs? *Exp Cell Res* **58:**312–319, 1969.
37. Horwitz BA: Rates of oxygen consumption of fertilized and unfertilized *Asterias, Arbacia,* and *Spisula* eggs. *Exp Cell Res* **38:**620–625, 1965.
38. Epel D: Experimental analysis of the role of intracellular calcium in the activation of the sea urchin egg at fertilization, In Subtelny S, Wessels NK (eds): The Cell Surface, New York, Academic 1980, p. 169.
39. Steinhardt RA, Epel, D: Activation of sea-urchin eggs by a calcium ionophore. *Proc Nat Acad Sci USA* **71:**1915–1919, 1974.
40. Steinhardt R, Zucker R, Schatten G: Intracellular calcium release at fertilization in the sea urchin egg. *Dev Biol* **58:**185–196, 1977.
41. Steinhardt RA, Alderton JM: Calmodulin confers calcium sensitivity on secretory exocytois. *Nature* **295:**154–155, 1982.
42. Epel D, Patton C, Wallace RW, Cheung WY: Calmodulin activates NAD kinase of sea urchin eggs: An early event of fertilization. *Cell* **23:**543–549, 1981.
43. Vacquier VD: The isolation of intact cortical granules from sea urchin eggs. Calcium ions trigger granule discharge. *Dev Biol* **43:**62–74, 1975.
44. Haggerty JG, Jackson RC: Release of granule contents from sea urchin egg cortices. *J Biol Chem* **258:**1819–1825, 1983.
45. Ridgway EB, Gilkey JC, Jaffe LF: Free calcium increases explosively in activating medaka eggs. *Proc Natl Acad Sci USA* **74:**623–627, 1977.
46. Whitaker MJ, Steinhardt RA: Ionic regulation of egg activation. *Q Rev Biophys* **15:**593–666, 1982.
47. Johnson JD, Epel D, Paul M: Intracellular pH and activation of sea urchin eggs after fertilization. *Nature* **262:**661–64, 1976.
48. Shen SS, Steinhardt RA: Intracellular pH and the sodium requirement at fertilisation. *Nature* **282:**87–89, 1979.
49. Longo FJ, Anderson E: A cytological study of the relation of the cortical reaction to subsequent events of fertilization in urethane-treated eggs of sea urchin, *Arbacia punctulata. J Cell Biol* **47:**646–665, 1970.
50. Hylander BL, Summers RG: The effect of local anesthetics and ammonia on cortical granule-plasma membrane attachment in the sea urchin egg. *Dev Biol* **86:**1–11, 1981.
51. Sugiyama M: Physiological analysis of the cortical response of the sea urchin egg. *Exp Cell Res* **10:**364–376, 1956.
52. Turner PR, Sheetz MP, Jaffe LA: Fertilization increases the polyphosphoinositide content of sea urchin eggs. *Nature* 310:414–415, 1984.
53. Whitaker M, Irvine RF: Inositol 1,4,5-trisphosphate microinjection activates sea urchin eggs. *Nature* **312:**636–639, 1984.
54. Slack BE, Bell JE, Benos DJ: Inositol-1,4,5-trisphosphate injection mimics fertilization potentials in sea urchin eggs. *Am J Physiol* **250:**C340–C343, 1986.
55. Lindberg O, Ernster L: On carbohydrate metabolism in homogenised sea urchin eggs. *Biochim Biophys Acta* **2:**471–477, 1948.
56. Cleland KW, Rothschild L: The metabolism of the sea urchin egg: Anaerobic breakdown of carbohydrate. *J Exp Biol* **29:**285–294, 1952.
57. Krahl ME, Keltch AK, Walters C, Clowes GHA: Glucose-6-phosphate and 6-phosphogluconate

dehydrogenases from eggs of the sea urchin, *Arbacia punctulata*. *J Gen Physiol* **38:**431–439, 1955.
58. Krahl ME: Oxidative pathways for glucose in eggs of the sea urchin. *Biochim Biophys Acta* **20:**27–32, 1956.
59. Isono N, Yasumasu I: Pathways of carbohydrate breakdown in sea urchin eggs. *Exp Cell Res* **50:**616–626, 1968.
60. Isono N: Carbohydrate metabolism in sea urchin eggs. III. Changes in respiratory quotient during early embryonic development. *Annot Zool Jpn* **36:**126–133, 1963.
61. Backstrom S: Activity of glucose-6-phosphate dehydrogenase in sea urchin embryos of different developmental trends. *Exp Cell Res* **18:**347–356, 1959.
62. Isono N: Studies on glucose-6-phosphate dehydrogenase in sea urchin eggs II. *J Fac Sci Univ Tokyo Sect IV Zool* **10:**67–74, 1963.
63. Swezey RR, Epel D: Regulation of glucose-6-phosphate dehydrogenase activity in sea urchin eggs by reversible association with cell structural elements. *J Cell Biol* **103:**1509–1515, 1986.
64. Whitaker MJ, Steinhardt RA: The relation between the increase in reduced nicotinamide nucleotides and the initiation of DNA synthesis in sea urchin eggs. *Cell* **25:**95–103, 1981.
65. Krane SM, Crane RK: Changes in levels of triphosphopyridine nucleotide in marine eggs subsequent to fertilization. *Biochim Biophys Acta* **43:**369–373, 1960.
66. Chandler DE, Heuser J: The vitelline layer of the sea urchin egg and its modification during fertilization. *J Cell Biol* **84:**618–632, 1980.
67. Kay E, Eddy EM, Shapiro BM: Assembly of the fertilization membrane of the sea urchin: Isolation of a divalent cation-dependent intermediate and its crosslinking in vitro. *Cell* **29:**867–875, 1982.
68. Weidman PJ, Kay ES, Shapiro BM: Assembly of the sea urchin fertilization membrane. Isolation of proteoliaisin, a calcium-dependent ovoperoxidase binding protein. *J Cell Biol* **100:**938–946, 1985.
69. Veron M, Foerder C, Eddy EM, Shapiro BM: Sequential biochemical and morphological events during assembly of the fertilization membrane of the sea urchin. *Cell* **10:**321–328, 1977.
70. Gulyas BJ, Schmell ED: Ovoperoxidase activity in ionophore treated mouse eggs. I. Electron microscopic localization. *Gamete Res* **3:**267–277, 1980.
71. Schmell ED, Gulyas GJ: Ovoperoxidase activity in ionophore treated mouse eggs. II. Evidence for the enzyme's role in hardening the *Zona pellucida*. *Gamete Res* **3:**279–290, 1980.
72. Klebanoff SJ, Foerder CA, Eddy EM, Shapiro BM: Metabolic similarities between fertilization and phagocytosis, *J Exp Med* **149:**938–953, 1979.
73. Deits TL, Shapiro BM: pH-induced hysteretic transitions of ovoperoxidase. *J Biol Chem* **260:**7882–7888, 1985.
74. Deits TL, Shapiro BM: Conformational control of ovoperoxidase catalysis in the sea urchin fertilization membrane. *J Biol Chem* **261:**12159–12165, 1986.
75. Turner EE, Somers CE, Shapiro, BM: The relationship between a novel NAD(P)H oxidase activity of ovoperoxidase and the CN^--resistant respiratory burst that follows fertilization of sea urchin eggs. *J Biol Chem* **260:**13163–13171, 1985.
76. Rosen H, Klebanoff SJ: Chemiluminescence and superoxide production by myeloperoxidase-deficient leukocytes. *J Clin Invest* **58:**50–60, 1976.
77. Rosen H, Klebanoff SJ: Formation of singlet oxygen by the myeloperoxidase-mediated antimicrobial system. *J Biol Chem* **252:**4803–4810, 1977.
78. Perry G, Epel D: Calcium stimulation of a lipoxygenase activity accounts for the respiratory burst at fertilization of the sea urchin egg. *J Biol Chem* **75:**40a, 1977.
79. Perry G, Epel D: Characterization of a Ca^{2+}-stimulated lipid peroxidizing system in the sea urchin egg. *Dev Biol* **107:**47–57, 1985.
80. Perry G, Epel D: Fertilization stimulates lipid peroxidation in the sea urchin egg. *Dev Biol* **107:**58–65, 1985.
81. Perry G: Echinochrome oxidation in *Arabacia punctulata* embryos. *Biol Bull* **151:**423, 1976.

82. Perry G, Epel D: Calcium control of a cyanide insensitive respiration in *Arbacia punctulata* eggs and its relation to echinochrome. *Biol Bull* **149:**441, 1975.
83. Perry G, Epel D: Ca^{2+}-stimulated production of H_2O_2 from naphthoquinone oxidation in *Arbacia* eggs. *Exp Cell Res* **134:**65–72, 1981.
84. Shimada H, Terayama AF, Yasumasu, I: Melittin, a component of bee venom, activates unfertilized sea urchin eggs. *Dev Growth Differ* **24:**7–16, 1982.
85. Ferguson JE, Shen SS: Evidence of phospholipase A_2 in the sea urchin egg: Its possible involvement in the cortical granule reaction. *Gamete Res* **9:**329–338, 1984.
86. Turner EE, Klevit R, Hopkins P, Shapiro BM: Ovothiol: A novel thiohistidine compound from sea urchin eggs that confers NAD(P)H-O_2 oxidoreductase activity on ovoperoxidase. *J Biol Chem* **261:**13056–13063, 1986.
87. Turner EE, Klevit R, Hager LJ, Shapiro BM: The Ovothiols: A family of novel redox-active thiolhistidine compounds from marine invertebrate eggs. *Biochemistry,* **26:**4028–4036, 1987.
88. Palumbo A, d'Ishia M, Misuraca G, and Prota G: Isolation and structure of a new sulphur-containing aminoacid from sea urchin eggs. *Tetrahedron Lett* **23:**3207–3208, 1982.
89. Palumbo A, Misuraca G, d'Ischia M, et al: Isolation and distribution of 1-methyl-5-thiol-L-histidine disulphide and a related metabolite in eggs from echinoderms. *Comp Biochem Physiol* **78B:**81–83, 1984.
90. Pathirana C, Andersen RJ; Imbricatine, an unusual benzyltetrahydroisoquinoline alkaloid isolated from the starfish *Dermasterias imbricata. J Am Chem Soc* **108:**8288–8289, 1986.
91. Rossi F, Nardi G, Palumbo A, Prota G: 5-thiolhistidine, a new amino acid from eggs of *Octopus vulgaris. Comp Biochem Physiol* **80B:**843–845, 1985.
92. Keevil T, Mason HS: Molecular oxygen in biological oxidations—An overview. *Methods Enzymol* **52:**3–40, 1978.
93. Epel D, Weaver AM, Muchmore AV, Schimke RT: β-1,3-glucanase of sea urchin eggs: Release from particles at fertilization. *Science* **117:**294–296, 1969.
94. Lazarow PB, DeDuve C: A fatty acyl-CoA oxidizing system in rat liver peroxisomes; enhancement by clofibrate, a hypolipidemic drug. *Proc Nat Acad Sci USA* **73:**2043–2046, 1976.
95. Klebanoff SJ: Myeloperoxidase-Halide-Hydrogen peroxide antibacterial system. *J Bacteriol* **95:**2131–2138, 1968.
96. Babior BM, Peters WA: The $O_2\cdot^-$-producing enzyme of human neutrophils: Further properties. *J Biol Chem* **256:**2321–2323, 1981.
97. Meister A, Anderson ME: Glutathione. *Ann Rev Biochem* **52:**711–760, 1983.
98. Fahey RC, Newton GL, Arrick B, et al: *Entamoeba histolytica:* A eukaryote without glutathione metabolism. *Science* **224:**70–72, 1984.
99. Stadtman TC: Occurrence and characterization of selenocysteine in proteins. *Methods Enzymol* **107:**576–581.
100. Mannervik B: Glutathione peroxidase. *Methods Enzymol* **113:**490–495, 1985.
101. Petasne RG, Zika RG: Fate of superoxide in coastal seawater. *Nature* **325:**516–518, 1987.

19

The Respiratory Burst and Atherosclerosis

GEORGE PARAGH, ÉVA M. KOVÁCS, JÓZSEF T. NAGY, GABRIELLA FÓRIS, and TAMÁS FÜLÖP, Jr.

1. INTRODUCTION

The tissue-damaging effect of free radicals generated by the respiratory burst is frequently discussed in the literature.[1–5] It is also mentioned in review articles that the reactive oxygen species (ROS) produced by mononuclear and polymorphonuclear phagocytes (PMN) may contribute to injury to the vessel wall, especially in the endothelial cell (EC), in different types of vasculitis, and in the vessels of inflamed regions.[3] But up to the present, the role of ROS generation by monocytes and granulocytes in the development of generalized atherosclerosis has not been taken into serious consideration.

According to a recent hypothesis, the initiation of atherosclerotic plaque formation in the arterial walls must be the result of EC injury.[6–8] It is followed by platelet aggregation on the surface of the damaged EC. Platelet-derived growth factor (PDGF) is released from the aggregated platelets, the most important stimulus of smooth muscle cell (SMC) proliferation.[9] Thus, the most important question in the discussion of atherosclerotic plaque formation is the initial factor that can induce EC damage. The presence of various factors may be supposed to exert their effects in a well-determined chronological order to trigger EC injury, leading to plaque formation. The basis of our concept is the question of whether ROS generation of circulating monocytes and granulocytes with its tissue-damaging effect can be one of the factors inducing atherosclerotic plaque formation.

It must be noted that this possibility has long been considered improbable because in atherosclerosis (in which an inflammatory or autoimmune origin can be excluded) the concentration of free radicals produced by resting phagocytes is too

GEORGE PARAGH, ÉVA M. KOVÁCS, JÓZSEF T. NAGY, GABRIELLA FÓRIS, and TAMÁS FÜLÖP, Jr. • First Department of Medicine, University Medical School, 4012 Debrecen, Hungary.

low to induce EC injury. Therefore, to reach the tissue-damaging concentration of ROS from the beginning, we must concede to leukocyte aggregation and adhesion and a stimulus-producing respiratory burst of adhered cells at the location of plaque formation.

Recently, results were published that throw new light on the development of atherosclerosis. It was discovered that not only can the immune system play an important role in lipid metabolism[10–12] but that EC can be transformed into macrophage-like cells under the effects of immunomediators and that they are themselves able to fulfill immune functions.[13–15] By contrast, it has also been stated that ROS generation of circulating phagocytes and the respiratory burst of ECs and SMCs are accompanied by the altered arachidonic acid (AA) cascade[16,17] as proved by the synthesis of products with toxic and vasoconstrictor effects during the respiratory burst.

2. DUAL ROLE OF THE MONOCYTE–MACROPHAGE SYSTEM IN THE IMMUNE RESPONSE AND IN LIPID METABOLISM

It is well known that macrophages play a key role in the immune response.[18,19] HLA-DR expression as class II major histocompatibility antigen promotes antigen presentation, initiating antigen recognition for T lymphocytes.

On the effect of immunomediators, e.g., lymphokine, γ-interferon (IFN_{γ} macrophages undergo a maturation process resulting in inflammatory or provoked macrophages. These cells can produce immunomediators such as interleukin-1 (IL-1), which increase the proliferation of T lymphocytes.[20] Increased effector functions of provoked macrophages can be manifested in increased Fc and C3b receptor-mediated phagocytosis, elevated lysosomal enzyme activity, and increased ROS generation.[18]

In about a 5 days, a culture of monocytes cells turns to macrophages, displaying the characteristic changes of their surface markers.[21] By contrast, the fact that the monocyte–macrophage system also plays a significant role in lipid metabolism is not well appreciated in immunology. Similar to other eukaryotic cells, circulating monocytes have high-affinity-specific low-density lipoprotein (LDL) receptors on their surfaces.[22] These receptors can bind LDL particles, LDL is incorporated, and its apoprotein component is degraded intracellularly; however, the most significant function of LDL receptors is the regulation of lipid metabolism in the following way. The de novo LDL receptor synthesis as well as the enzymatic activity of

List of abbreviations: AA, arachidonic acid; apoE, apoprotein-E; cGMP, cyclic guanosine monophosphate; DAG, diacyglycerol; EC, endothelial cell; FMLP, *N*-formyl-methionyl-leucyl-phenylalanine; Hch, hypercholesterolemia; high-density lipoprotein; HMG-CoA, hydroxymethyl-glutaryl coenzyme-A; IFN γ, γ-interferon; IL-1, interleukin-1; LDL, low-density lipoprotein; LT leukotriene; NDGA, nordihydro-guaiaretic acid; PDGF, platelet-derived growth factor; PG, prostaglandin; PI, phosphatidylinositol; PLA_2, phospholipase A_2; PMN, polymorphonuclear leukocytes; PT, pertussis toxin; ROS, reactive oxygen species; SMC, smooth muscle cell

hydroxymethylgluthary1 coenzyme-A (HMG-CoA) reductase is decreased, whereas cholesterol esterification is increased when LDL is bound to its specific receptor. During the maturation process of in vitro cultured monocytes, the density of specific LDL receptors is decreased and the so-called scavenger receptors appear on the cell surface.[23,24] These scavenger receptors can bind the LDL only in cationized form.[25] After incorporation of cationized LDL and degradation of the apoprotein component, the intracellularly released cholesterol is deposed in the cell membrane. Excretion of the cholesterol depends on the apoprotein-E (apo-E)-secreting ability of macrophages. The extracellular high-density lipoprotein (HDL) fusing with apo-E can withdraw the integrated cholesterol from the membrane. If apo-E cannot be synthesized in the cell, or the extracellular HDL level is very low, the cholesterol is deposed in the cytoplasm. These cholesterol inclusions lead to foam cell formation, the most important component of atherosclerotic lesions.[26] It must be emphasized that neither the importance nor the regulation of extrahepatic apo-E secretion has yet been elucidated.

It has also been shown experimentally that the scavenger receptor activity of macrophages is decreased by lymphokine.[27] By contrast, the apo-E secreting capacity in macrophages of BCG-sensitized rats is decreased.[28] Both data show that if macrophages fulfill immune functions, the decreased apo-E secretion promotes foam cell formation.

On the basis of these results, it can be supposed that the most important role of the scavenger receptors located on the surface of mononuclear phagocytes is not only the direct reduction of the high serum LDL level, but also the regulation of the extrahepatic apo-E secretion, i.e., preservation of the HDL–LDL ratio. The regulation of apo-E synthesis and cholesterol excretion in mononuclear phagocytes is unknown, but its role in the oxidative processes, especially in the AA cascade, seems probable; foam cell formation induced by modified LDL can be prevented by the leukotriene (LT) synthesis inhibitor nordihydroguaiaretic acid.[29]

3. DUAL ROLE OF EC AND SMC IN THE IMMUNE RESPONSE AND IN LIPID METABOLISM

The concept of the key role of macrophages in lipid metabolism is not widely known in the field of immunology. Even more incomplete is our knowledge of the possible immune functions of EC. Although the initiating functions of EC in atherosclerotic plaque formation have been studied intensively, their role in immune function has not been investigated until recently.[13–15,30]

The first step in the pathogenesis of atherosclerosis is injury to the EC. Thrombus formation at the location of the lesion is considered the second step, leading to PDGF synthesis and SMC proliferation.[7,8] The "benign tumor" formed in arterial walls is responsible for altered matrix production and intensive glycosaminoglycan synthesis.[31]

The high serum cholesterol and LDL level can induce foam cell formation in the region of the lesion. According to the current hypothesis, these cells originate from monocyte–macrophages or macrophage-like cells.[32] Subsequent evidence on the role of the immune system in plaque formation demonstrates that SMC proliferation can be induced not only by PDGF,[33] LDL,[34] serum factors of patients with diabetes[35] or uremia,[36] or insulin,[37] but it can also be induced by factors produced by monocytes.[38] This PDGF-like factor is synthesized by monocytes on the effect of different immunostimulants (IFN_γ, IL-1). Additional surprising reports were published concerning the immunological activation of EC by different immunostimulants, such as IFN_γ, lymphokines, chemotactic factors, and endotoxin.[14,15,39] Like EC, these macrophages possess among the class II major histocompatability (MHC) antigens the HLA-DR, -SB, and -DC, on their surfaces, which can promote antigen presentation and binding of the T helper and cytolytic lymphocytes.[13,14] Similar to macrophages, Fc receptor expression and IL-1 production are also increased in these immunologically activated EC.[40] Owing to intensive oxidative activities, prostaglandin (PG) synthesis[41] and superoxide generation[42] are also increased in these cells. The respiratory burst of these macrophage-like EC can be induced by different chemotactic peptides. Among these peptides are such fibrin peptides as the products of fibrinolysis[43] and the fragments generated from the degradation of elastin of the vessel wall.[44,45] Another essential macrophage-like characteristic of activated EC and SMC is the increase of scavenger LDL receptor expression,[46] which can be explained by the following facts: (1) glycosaminoglycans produced by proliferating SMCs form cationized complexes with LDL particles[47]; and (2) the sialic acid content of the membrane in activated EC and SMC is decreased, thereby altering the polarity of the cell membrane and promoting the binding of modified LDL.[48]

Summarizing these results it is concluded that bacterial endotoxins and IFN_γ production during viral infection as nonspecific agents can indirectly trigger atherosclerotic plaque formation. According to this hypothesis, the binding of cytolytic T lymphocytes can damage EC, ROS is generated by EC itself, and the AA cascade is activated pathologically, which leads to PG and LT production. PDGF can be synthesized in monocytes and macrophages independent of platelet aggregation. In addition, the intensive scavenger receptor activity causes foam cell formation, which is supported by the activation of AA cascade as well.

The fact that HLA-DR expression on the surface of EC promotes antigen presentation and that IL-1 production of these cells enhances lymphocyte proliferation and lymphokine production presumably will inspire further experiments in this field. Up to this point, such a central role of the immune system in the pathogenesis of atherosclerosis can neither be refuted nor confirmed.

4. ROLE OF ROS GENERATION IN ATHEROSCLEROTIC PLAQUE FORMATION

According to the results mentioned so far, EC and SMC as well as circulating monocytes and granulocytes are involved in ROS generation. The different natural chemotactic peptides and immune complexes, as well as incorporation of macromolecules may be the direct triggers of the respiratory burst. Among the chemotactic peptides, the C3a and C5a complement fragments,[49] various oligopeptides originated from lymphokines,[50] fibrin peptides,[43] and fragments of elastin degradation should be mentioned.[45] All these ligands stimulate cells through receptors positively coupling to Ni GTP-binding protein and phosphatidyl inositol (PI) turnover.[51] Diacylglycerol the product of PI cleavage, is a stimulator of the oxidative burst; a substrate of phospholipase A^2 (PLA_2) is also a trigger of AA cascade. The activation of NADPH oxidase is a consecutive process caused by transient AA intermediates or by the AA itself; thus, lipid peroxidation and superoxide production are inseparable processes in the cells.[52] The tissue-damaging mechanism of ROS generated from immunocompetent cells is not detailed here, as it is discussed in several review articles.[2,5] It must be noted that only hydrogen peroxide can produce an extracellular effect, as it is able to penetrate the membrane of living cells.[53]

The most important question is: How do free radicals—released from immunocompetent cells, presumably in low concentrations—influence the process of generalized alteration of arterial walls in atherosclerosis? The in vitro conventional cytotoxic measurements showed the lysis of ^{51}Cr-labeled EC only by high concentrations of hydrogen peroxide.[54] By contrast, low concentration of hydrogen peroxide could inhibit the cyclo-oxygenase activity without any effects on PLA_2 enzyme.[55] This finding is in good agreement with studies in which hydrogen peroxide could inhibit the vasodilatant and antithrombogenic prostacyclin production and could increase LT synthesis simultaneously. Hydrogen peroxide has an effect on the subendothelial matrices inhibiting collagen gelation and hyaluronate polymerization, thereby promoting destruction of the basement membrane.[56] By contrast, hydrogen peroxide inhibits the α-proteinase inhibitor and increases subendothelial matrix degradation in the presence of granulocyte elastase.[57] Increased K^+ and purine efflux from EC can be induced by 30-fold lower concentrations of hydrogen peroxide[54] than those of the concentrations necessary for inducing ^{51}Cr or lactate dehydrogenase release.[58] One explanation for the altered cation transport could be the decrease of EC membrane fluidity after hydrogen peroxide exposition as measured by paramagnetic spectrometry.[59] The EC injury was more serious in glutathione-depleted cells and could be protected by antioxidants.[60]

In spite of its harmful effects, hydrogen peroxide activates guanylate cyclase enzyme in EC which leads to the synthesis of an endothelium-derived vascular relaxant factor.[61]

We must not forget that ROS are able to oxidize other compounds either in the extracellular space or in the ROS-generating cells. The close relationship between the AA cascade and the oxidative burst has already been mentioned.[62–64] Therefore, we would like to emphasize that the activation of the AA cascade itself can not be considered a pathological event.[65] Prostacyclin production in the AA cascade especially seems favorable against the development of atherosclerosis, with its vasodilatory, cytoprotective, and platelet aggregation inhibitory effect. By contrast, prostaglandins, and especially LTC_4 and LTD_4 are able to promote atherosclerotic plaque formation and have an effect opposite that of prostacyclins, particularly to thromboxane I-2. The oxidative burst, i.e., the increased ROS production, upsets the balance between the cyclo-oxygenase and lipoxygenase products of the cells. The increased production of LTC_4 also decreases the free glutathione content of cells by activating *S*-glutathione transferase.[62–64] This latter event explains the impaired natural detoxification capacity of the cells through the glutathione peroxidase system.[66] By means of another amplifier mechanism, the tissue-damaging effect of ROS is increased as a consequence of the interaction of hydrogen peroxide and LDL particles.[67] The lipid components of LDL particles can be converted to toxic lipid peroxides exerting their tissue-damaging effect on every cell possessing scavenger LDL receptors on its surface.[68–70]

The inhibition of NADPH-dependent HMG-CoA reductase enzymatic activity by hydrogen peroxide[71,72] is considered an indirect beneficial influence of the respiratory burst in lipid metabolism, as this enzyme is responsible for intracellular cholesterol synthesis.

Even if the ROS produced by PMN, monocytes, or EC do not play a central regulating role in atherosclerotic alterations, their direct or indirect influences cannot be left out of account in the development of this disorder.

5. LDL PARTICLES AS TRIGGERS OF THE RESPIRATORY BURST

Intracellular metabolism of LDL incorporated through its specific or scavenger receptors is influenced by ROS generation at three points: (1) hydrogen peroxide modified LDL particle, provoking a disorder in its metabolism; (2) the negative coupling of the specific LDL receptor to HMG-CoA reductase is signaled by the ROS produced; and (3) apo-E secretion triggered by scavenger receptors during LDL incorporation is inhibited by lipoxygenation.

To clarify the effect of LDL receptor activation on the respiratory burst, the postreceptorial signal transmission of LDL receptors has been investigated in freshly isolated monocytes of healthy subjects in our laboratory.[73] Initially, the time-course effect of 50 μg/ml native LDL was studied on the level of cyclic nucleotides. This experiment was also performed in the presence of 500 ng/ml pertussis toxin (PT) to clarify the role of the GTP-binding Ni protein[74] in the signal transmission of native LDL receptors (Fig. 1). During LDL binding and incorporation, an early adenylate cyclase activity was found followed by a later increase of the cyclic guanosine monophosphate (cGMP) level. The changes in cyclic nucleotide levels could not be blocked by PT incubation, which indicates the negligible role of Ni protein in the postreceptorial signal transmission of specific LDL receptors.

The increase of cytoplasmic free Ca^{2+} involved in intracellular signal transmission[76] was measured by the Quin-2-fluorescence technique. The intracellular free Ca^{2+} was found to be elevated during the early phase of LDL binding, and the incorporation was accompanied by an intensive Ca^{2+} influx measured by $^{45}Ca^{2+}$ (Table I).

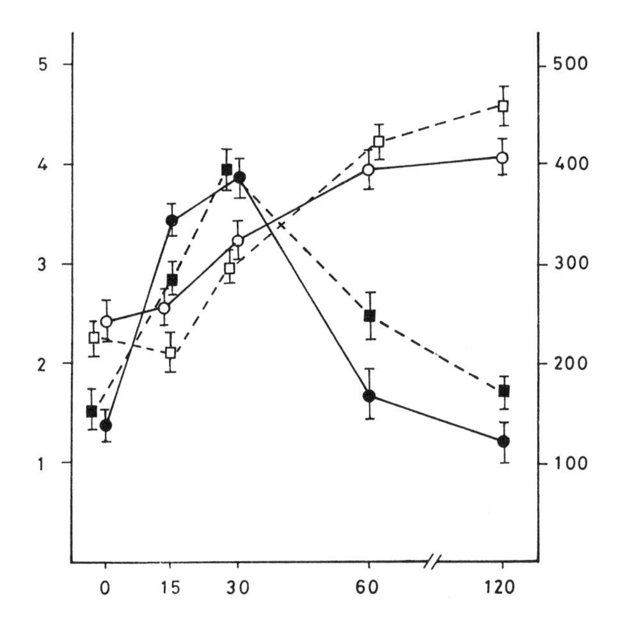

FIGURE 1. Alterations in cyclic nucleotide levels in monocytes stimulated with LDL were independent on pertussis toxin pretreatment. Monocytes were isolated according to the method of Kumagai et al.[75] Incubation was performed in RPMI 1640 medium containing 50 μg protein/ml LDL cAMP (– ● –); cGMP (– ○ –); preincubation for 120 min with 500 ng/ml pertussis toxin cAMP (—■—); cGMP (—□—). Each point represents the mean ±SD of four experiments.

To characterize the effector functions of monocytes, the effect of LDL on the ROS production was studied (Table II). LDL increased the oxygen consumption and superoxide and hydrogen peroxide production in 25–100 μg/ml concentrations but decreased the total glutathion content of monocytes. It should be noted that the signals regulating the incorporation and intracellular degradation of native LDL through its specific receptor are similar to those signals that can be demonstrated during phagocytosis mediated by Fc receptors.[82] The question therefore arises: What are the possible pathways of signal transmission through specific LDL receptors?

The Ca^{2+} mobilization as one of the triggers of the respiratory burst possibly cannot be the consequence of PI cleavage. The high free cytoplasmic Ca^{2+} level is rather due to the Ca channel-opening effect of activated LDL receptors. The activation of $NADPH^{+}$ oxidase and guanylate cyclase can also be stimulated by the products of AA cascade, together with the elevated Ca^{2+} level.[83] By contrast, the inhibition of HMG-CoA reductase enzymatic activity can be explained by the respiratory burst induced by LDL.[72,73]

Hartung et al.[84] reported the induction of the respiratory burst by acetylated and malondialdehyde-LDL in monocyte-derived macrophages. According to the above-mentioned results[73] either specific or scavenger LDL receptor stimulation

TABLE I. LDL-Induced $^{45}Ca^{2+}$ Uptake and Elevation of Cytosolic Free Ca^{2+} Level in Monocytes

LDL (μg protein/m)	$^{45}Ca^{2+}$ uptake (nmoles/10^7 cells/5 min)[a]	$\Delta[Ca^{2+}]_i$[b] (nM)
Medium alone, control	15.6±2.1	Nondetectable
25	27.3±2.6	156±18
50	86.7±5.3	537±41
100	79.3±6.4	598±48

[a]The determination was performed after 5-min incubation with LDL.
[b]The measurement was carried out in 2 min of incubation. ^{45}Ca uptake was determined as described earlier [73], and the alterations in $[Ca^{2+}]_i$ level were measured according to the method of Tsian et al.[77]
[c]Values differ significantly ($p < 0.01$) from those of controls.

can activate the $NADPH^+$ oxidase enzyme. Furthermore, it has been shown that the amount of PG and LTC_4 produced by monocyte-derived macrophages was also increased after incubation with cationized LDL. These results suggest that both specific and scavenger receptor activities can stimulate the respiratory burst and the AA cascade of mononuclear phagocytes. The presence of LDL receptors on the surface of immunologically activated EC and SMC sheds new light on the role of respiratory burst in the development of atherosclerosis.

The fact that GTP-binding Ni protein is not involved in the signal transmission of LDL receptors is crucial. Furthermore, another experimental series (G. Paragh, M. Hauck, G. Fóris) demonstrated that superoxide generation is increased and glutathione content decreased in resting granulocytes obtained from patients with high serum LDL cholesterol level. Presumably that is why patients with hypercholesterolemia (HCh) show reduced superoxide generation after the addition of synthetic chemotactic peptide *n*-formyl-Met-Leu-Phe (FMLP). PT markedly inhibited FMLP-induced superoxide generation[85] in the age-matched control groups, whereas it showed no effects on the FMLP-induced respiratory burst in PMNL of patients with HCh (Fig. 2). Decreased superoxide production during FMLP stimulation of FMLP in the monocytes of patients with high LDL cholesterol level cannot be explained by the diminished $NADPH^+$ oxidase activity because LDL in 50 μg/ml concentration was able to stimulate the superoxide production.

We can conclude that monocytes of patients with high serum cholesterol levels react to stimulation through chemotactic peptide receptors with a diminished respiratory burst. By contrast, increased ROS production was detected after the LDL receptors of the cells were triggered.

A possible explanation for the decreased response to stimulation in this patient group is the damage of the PT-sensitive Ni protein, which could be blocked by oxidants such as *N*-ethylmaleimide.[86] In this manner, the high basal production of ROS as well as the decreased intracellular glutathione level may lead to functional blockade of the Ni protein. Our hypothesis on the damaged function of Ni protein in

TABLE II. Oxidative Burst of Monocytes Triggered by Different Concentrations of LDL[a]

LDL (μg protein/m)	O_2 consumption (nmoles/30 min/10^6 cells)	Superoxide (nmoles/30 min/10^6 cells)	Hydrogen peroxide (nmoles/30 min/10^6 cells)	Glutathione content (ng/10^6 cells/30 min)
Medium, control	15.65 ± 1.16	3.41 ± 0.27	1.45 ± 0.16	305 ± 21
25	81.95 ± 8.48	74.70 ± 3.82[b]	8.08 ± 0.97[b]	218 ± 16
50	78.05 ± 2.34	80.50 ± 6.11[b]	7.61 ± 0.63[b]	142 ± 10
100	97.86 ± 3.25	81.38 ± 7.34[b]	4.98 ± 0.36[b]	136 ± 11

[a]The O_2 consumption was determined according to the method of Tanabe et al.,[78] the superoxide anion generation by the method of Cohen and Chovaniek,[79] the H_2O_2 release by the method of Pick and Keisari,[80] and the glutathione content in cells was detected by the method of Hissin and Hilf.[81]
[b]Values differ significantly ($p < 0.01$) from those of controls.

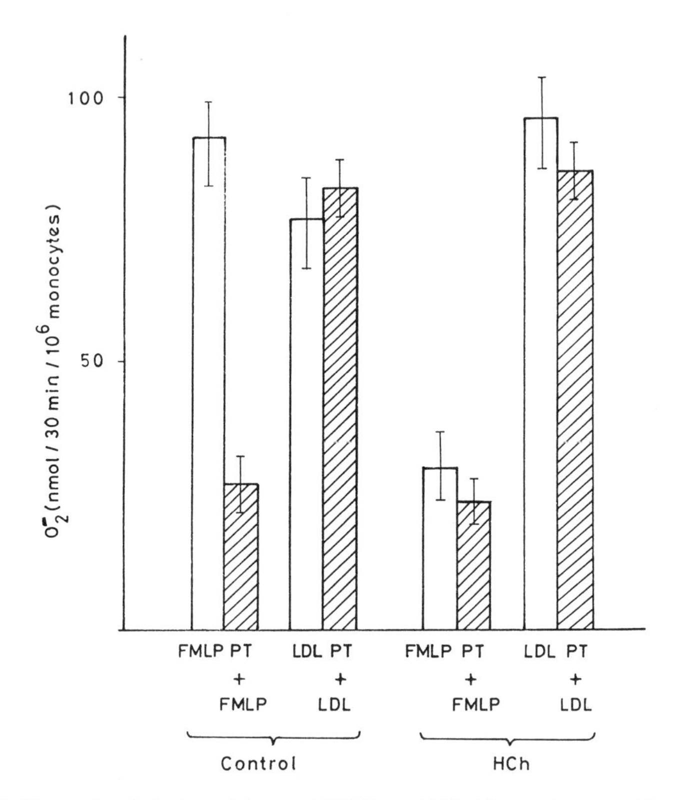

FIGURE 2. The pertussis toxin resistance of FMLP- and LDL-triggered superoxide anion generation by monocytes of patients with hypercholesterolemia (HCh). The produced O_2^- was determined according to the Cohen[79] method of after 30 min incubation. FMLP = 10^{-8} M formyl-Met-Leu-Phe; PT = 500 ng/ml pertussis toxin; LDL = 50 μg protein/ml LDL. Each value represents the mean ±SD. Patients: 18 males (38.6 ± 5.6 years) with >9.0 mmoles/liter serum cholesterol and 11 age-matched male control subjects (41.6 ± 3.6 years) with <5.7 mmoles/serum cholesterol level.

hypercholesterolemia was strengthened by the fact that the respiratory burst triggered through PT-insensitive specific LDL receptors was even higher in monocytes of the patient group than in controls.

6. CONCLUSION

The potential role of the respiratory burst in the development of atherosclerosis is supported by the following facts:

1. ROS even in low concentrations, can induce essential alterations and injuries in EC functions.

2. The toxic effect of ROS is amplified through their increasing effect on lipoxygenation and through their interaction with LDL particles, resulting toxic LDL products.
3. Besides a number of immunostimulants (immune complexes and immunomediators), the peptides (fibrin and elastin fragments) released from the atherosclerotic lesions can also induce ROS generation. Native and cationized LDL-stimulating cells through specific and scavenger LDL receptors are also known to increase the respiratory burst.
4. Circulating immunostimulants and immunomediators (endotoxin, IFN_{γ}, IL-1) promote the macrophage-like alterations of EC. These altered EC can bind cytolytic T lymphocytes, PMNL, and monocytes on their surfaces. By contrast, EC and SMC can themselves produce ROS and LT under immune complex stimulation, chemotactic peptides, and so forth.

REFERENCES

1. Babior BM: The respiratory burst of phagocytes. *J Clin Invest* **73:**598–601, 1984.
2. Slater TF: Free-radical mechanisms in tissue injury. *Biochem J* **222:**1–15, 1984.
3. Brigham KL, and Meyrick B: Granulocyte-dependent injury of pulmonary endothelium: A case of miscommunication. *Tissue Cell* **16:**137–154, 1984.
4. Weitberg AB, Weitzman SA, Clark EP, et al: Effects of antioxidants on oxidant-induced sister chromatid exchange formation. *J Clin Invest* **75:**1835–1841, 1985.
5. Parnham MJ, Lunec J: Meeting report: Free radicals, cell damage and disease. *Agents Actions* **18:**560–562, 1986.
6. Sevitt S: Platelets and foam cells in the evolution of atherosclerosis. *Atherosclerosis* **61:**107–115, 1986.
7. Hammersen F, Hammersen E: The structural reaction pattern of endothelial cells to injurious stimuli. *Agents Actions* **13:**442–450, 1983.
8. Pietila K, Nikkari T: Role of the arterial smooth muscle cell in the pathogenesis of atherosclerosis. *Med Biol* **61:**31–44, 1983.
9. Ross R, Raines EW, Bowen-Pope DF: The biology of platelet-derived growth factor. *Cell* **46:**155–169, 1986.
10. Cuthbert JA, Lipsky PE: Immunoregulation by low density lipoproteins in man: Low density lipoprotein inhibits mitogen-stimulated human lymphocyte proliferation after initial activation. *J Lipid Res* **24:**1512–1524, 1983.
11. Curtis LK, Edgington TS: Regulatory serum lipoproteins: Regulation of lymphocyte stimulation by a species of low density lipoprotein. *J Immunol* **116:**1452–1458, 1976.
12. Antonaci S, Jivillo E, Ventura MT, et al: In vitro effects of human lipoproteins on the immune system in healthy donors: Inhibition of plaque forming cell generation and decreased frequency of NK cells. *Clin Exp Immunol* **56:**677–682, 1984.
13. Pober JS, Gimbrone MA, Collins T, et al: Interactions of T lymphocytes with human vascular endothelial cells: Role of endothelial cells surface antigens. *Immunobiology* **168:**483–494, 1984.
14. Wagner CR, Vetto RM, Burger DR: The mechanism of antigen presentation by endothelial cells. *Immunobiology* **168:**453–469, 1984.
15. Wallis J, Harlan JM: Effector functions of endothelium in inflammatory and immunologic reactions. *Pathol Immunopathol Res.* **5:**73–103, 1986.
16. Pawlowski NA, Kaplan G, Hamill AL: Arachidonic acid metabolism by human monocytes. *J Exp Med* **158:**393–412, 1983.

17. Goetzl EJ: Leukocyte recognition and metabolism of leukotrienes. *Fed Proc* **42:**3128–3131, 1983.
18. Ögmundsdottir HM, Weir DM: Mechanisms of macrophage activation. *Clin Exp Immunol* **40:**223–234, 1980.
19. Cohn ZA: The macrophage—Versatile element of inflammation. *Harvey Lect* **77:**63–80, 1982.
20. Mayernik DG, Hag A, Rinehart JJ: Interleukin 1 secretion by human monocytes and macrophages. *J Leukocyte Biol* **36:**551–557, 1984.
21. Hokland ME: Immunomodulatory effects of interferons on human mononuclear cells with special reference to the expression of cell surface antigens. *Acta Pathol Microbiol Immunol Scand* **286(suppl):**1–35, 1985.
22. Brown MS, Goldstein JL: Receptor-mediated endocytosis: Insights from the lipoprotein receptor system. *Proc Natl Acad Sci USA* **76:**3330–3337, 1979.
23. van Der Schroeff JG, Havekes L, Emeis JJ, et al: Morphological studies on the binding of low density lipoproteins and acetylated low density lipoproteins to the plasma membrane of cultured monocytes. *Exp Cell Res* **145:**95–103, 1983.
24. Traber MG, Kallman B, Kayder HJ: Localization of the binding sites of native and acetylated low density lipoprotein (LDL) in human monocyte-derived macrophages. *Exp Cell Res* **148:**281–292, 1983.
25. Brown MS, Basu SK, Falck JR, et al: The scavenger cell pathway for lipoprotein degradation: Specificity of the binding site that mediates the uptake of negatively-charged LDL by macrophages. *J Supramol Struct* **13:**67–81, 1980.
26. Brown MS, Goldstein JL: Lipoprotein metabolism in the macrophage: Implications for cholesterol deposition in atherosclerosis. *Annu Rev Biochem* **52:**223–261, 1983.
27. Fogelman AN, Seager J, Groopman JE, et al: Lymphokines secreted by an established lymphocyte line modulate receptor-mediated endocytosis in macrophages derived from human monocytes. *J Immunol* **131:**2368–2373, 1983.
28. Takemura R, Werb Z: Modulation of apoprotein E secretion in response to receptor-mediated endocytosis in resident and inflammatory macrophages. *J Exp Med* **159:**167–178, 1984.
29. van Der Schroeff JG, Ravekes D, Weerheim AM, et al: Suppression of cholesteryl ester accumulation in cultured human monocyte-derived macrophages by lipoxygenase inhibitors, *Biochem Biophys Res Commun* **127:**366–372, 1985.
30. Roska AK, Johnson SR, Lipsky PE: Immunologic function of endothelial cells: Guinea pig aortic endothelial cells support mitogen-induced T lymphocyte activation, but not function as antigen-presenting cells. *J Immunol* **132:**136–145, 1984.
31. Wight TN, Ross R: Proteoglycans in primate arteries. II. Synthesis and secretion of glycosaminoglycans by arterial smooth muscle cells in culture. *J Cell Biol* **67:**675–686, 1975.
32. Aqel NM, Ball RY, Waldmann H, et al: Monocyte origin of foam cells in human atherosclerotic plaques. *Atherosclerosis* **53:**265–271, 1984.
33. Ross R, Glomset JA, Kariya B, et al: A platelet-dependent serum factor that stimulates the proliferation of arterial smooth muscle cells in vitro. *Proc Natl Acad Sci USA* **71:**1207–1210, 1974.
34. Fischer-Dzoga K: Cellular proliferation, cellular death and atherosclerosis. *Artery* **5:**222–236, 1979.
35. Ledet T, Fischer-Dzoga K, Wissler RW: Growth of aortic smooth-muscle cells cultured in media containing diabetic and hyperlipidemic serum. *Diabetes* **25:**207–215, 1976.
36. Bagdade JD: Chronic renal failure and atherogenesis. Serum factors stimulate the proliferation of human arterial smooth muscle cells. *Atherosclerosis* **34:**243–248, 1979.
37. Pfeifle B, Ditschuneit H: Effect of insulin on growth of human arterial smooth muscle cells. *Diabetologia* **20:**155–158, 1981.
38. Lopes-Virella MF, Virella G: Immunological and microbiological factors in the pathogenesis of atherosclerosis. *Clin Immunol Immunopathol* **37:**377–386, 1985.
39. Groenewegen G, Buurman WA, van Der Linden CJ: Lymphokines induce changes in morphology and enhance motility of endothelial cells. *Clin Immunol Immunopathol* **36:**378–385, 1985.
40. Libby P, Ordovas JM, Anger KR, et al: Endotoxin and tumor necrosis factor induce interleukin-1 gene expression in adult human vascular endothelial cells. *Am J Pathol* **124:**179–185, 1986.

41. Dejana E, Brevario F, Balconi G, et al: Stimulation of prostacyclin synthesis in vascular cells by mononuclear cell products. *Blood* **64:**1280–1283, 1984.
42. Matsubara T, Ziff M: Increased superoxide anion release from human endothelial cells in response to cytokines. *J Immunol* **137:**3295–3298, 1986.
43. Rowland FN, Donovan MJ, Picciano PT, et al: Fibrin-mediated vascular injury. Identification of fibrin peptides that mediate endothelial cell reaction. *Am J Pathol* **117:**418–428, 1984.
44. Senior RM, Griffin GL, Mecham RP, et al: Val-Gly-Val-Ala-Pro-Gly, a repeating peptide in elastin, is chemotactic for fibroblasts and monocytes. *J Cell Biol* **99:**870–874, 1984.
45. Fülöp T, Jacob MP, Varga ZS, et al: Effect of elastin peptides on human monocytes: $Ca^{2}{}_{+}$ mobilization, stimulation of respiratory burst and enzyme secretion. *Biochem Biophys Res Commun* **141:**92–98, 1986.
46. Van Hinshberg VWM, Emeis JJ, Havekes L: Interaction of lipoproteins with cultured endothelial cells. The endothelial cell, in: *First International Endothelial Cell Symposium of the ETCS, Paris, 1982*. Basel, Karger, 1982, pp 99–112.
47. Salisbury BGJ, Falcone DJ, Minick CR: Insoluble low density lipoprotein-proteoglycan complexes enhance cholesteryl ester accumulation in macrophages. *Am J Pathol* **120:**6–11, 1985.
48. Görög P, Pearson JD: Surface determinants of low density lipoprotein uptake by endothelial cells. *Atherosclerosis* **53:**21–29, 1984.
49. Tonnesen MG, Smedly LA, Henson PM: Neutrophil–endothelial cell interaction. Modulation of neutrophil adhesiveness induced by complement fragments C5a and C5a des arg and FMLP in vitro. *J Clin Invest* **74:**1581–1592, 1984.
50. Fóris G, Hauck M, Dezsö B, et al: Effect of low-molecular weight lymphokine components on the Fc and C3b receptor-mediated macrophage functions. *Cell Immunol* **78:**276–284, 1983.
51. Berridge MJ: Inositol triphosphate and diacylglycerol as second messengers. *Biochem J* **220:**345–360, 1984.
52. Karnovsky ML, Badwey JA: Determinants of the production of active oxygen species by granulocytes and macrophages. *J Clin Chem Clin Biochem* **21:**545–553, 1983.
53. Root RK, Metcalf J, Oshino N, et al: H_2O_2 release from human granulocytes during phagocytosis. *J Clin Invest* **55:**945–955, 1975.
54. Weiss JM, Levine JD, Callahan KS, et al: Role of hydrogen peroxide in neutrophil-mediated destruction of cultured endothelial cells. *J Clin Invest* **68:**714–721, 1981.
55. Whorton AR, Montgomery ME, Kent RS: Effect of hydrogen peroxide on prostaglandin production and cellular integrity in cultured porcine aortic endothelial cells. *J Clin Invest* **76:**295–302, 1985.
56. Greenwald R, Moy WW: Inhibition of collagen gelation by action of superoxide radical. *Arthritis Rheum* **22:**251–257, 1979.
57. Weiss SJ, Regiani S: Neutrophils degrade subendothelial matrices in the presence of alpha-1-proteinase inhibitor. *J Clin Invest* **73:**1297–1303, 1984.
58. Ager A, Gordon JL: Differential effects of hydrogen peroxide on indices of endothelial cell function. *J Exp Med* **159:**592–603, 1984.
59. Freeman BA, Rosen GM, Barber MJ: Superoxide perturbation of the organization of vascular endothelial cell membranes. *J Biol Chem* **261:**6590–6593, 1986.
60. Tsan MM, Danis EH: Enhancement of intracellular glutathione protects endothelial cells against oxidant damage. *Biochem Biophys Res Commun* **127:**270–276, 1985.
61. Thomas G, Ramwell P: Induction of vascular relaxation by hydroperoxides. *Biochem Biophys Res Commun* **139:**102–107, 1986.
62. Granström E: The arachidonic acid cascade. *Inflammation* **8:**515–525, 1984.
63. Piper PJ: Formation and actions of leukotrienes. *Physiol Rev* **64:**744–761, 1984.
64. Lewis RA, Austen KF: The biologically active leukotrienes. *J Clin Invest* **73:**889–897, 1984.
65. Maridonneau-Parini I, Tauber AI: Activation of NADPH-oxidase by arachidonic acid involves phospholipase A_2 in intact human neutrophils but not in the cell-free system. *Biochem Biophys Res Commun* **138:**1099–1105, 1986.
66. Harlan JM, Levine JD, Callahan KS, et al: Glutathione redox cycle protects cultured endothelial cells against lysis by extracellularly generated hydrogen peroxide. *J Clin Invest* **73:**706–713, 1984.

67. Cathcart MK, Morel DW, Chisholm GM III: Monocytes and neutrophils oxidize low density lipoprotein making it cytotoxic. *J Leukocyte Biol* **38:**341–350, 1985.
68. Hinshberg VWM: LDL cytotoxicity. The state of the art. *Atherosclerosis* **53:**113–118, 1984.
69. Morel DW, Hessler JR, Chisholm GM III: Low density lipoprotein cytotoxicity induced by free radical peroxidation of lipid. *J Lipid Res* **24:**1070–1076, 1983.
70. Evensen SA, Galdal KS, Nilsen E: LDL-induced cytotoxicity and its inhibition by anti-oxidant treatment in cultured human endothelial cells and fibroblasts. *Atherosclerosis* **49:**23–30, 1983.
71. Ness GC, McCreery MJ, Samplet CE: Sulfhydryl/disulfide forms of rat liver 3-hydroxy-3-methylglutaryl coenzyme A reductase. *J Biol Chem* **260:**16395–16399, 1985.
72. Harwood HJ, Greene YJ, Stacpoole PW: Inhibition of human 3-hydroxy-3 methylglutaryl coenzyme A reductase activity by ascorbic acid. *J Biol Chem* **261:**7127–7135, 1986.
73. Paragh G, Nagy JT, Szondy É, et al: Immunomodulating effect of low density lipoprotein on human monocytes. *Clin Exp Immunol* **64:**665–672, 1986.
74. Gilman AG: G Proteins and dual control of adenylate cyclase. *Cell* **36:**577–579, 1984.
75. Kumagai I, Itoh K, Hinuma S, et al: Pretreatment of plastic petri dishes with fetal calf serum. A simple method for macrophage isolation. *J Immunol Methods* **29:**17–25, 1979.
76. Streb H, Irvine RF, Berridge MJ, et al: Release of Ca^{2+} from a nonmitochondrial intracellular store in pancreatic acinar cells by inositol-1,4,5-triphosphate. *Nature (Lond)* **306:**67–68, 1983.
77. Tsien RY, Pozzan T, Rink TJ: Calcium homeostasis in intact lymphocytes:cytoplasmic free calcium monitored with a new intracellularly trapped fluorescent indicator. *J Cell Biol* **94:**325–342, 1982.
78. Tanabe T, Kabayashi Y, Usui T: Enhancement of human neutrophil oxygen consumption by chemotactic factors. *Experientia* **39:**604–611, 1983.
79. Cohen HJ, Chovaniek ME: Superoxide generation by digitonin-stimulated guinea pig granulocytes. *J Clin Invest* **40:**1081–1086, 1978.
80. Pick E, Keisari Y: Superoxide anion and hydrogen peroxide production by chemically elicited peritoneal macrophages—Induction by multiple non-phagocytic stimuli. *Cell Immunol* **59:**301–312, 1981.
81. Hissin PJ, Hilf R: Fluorometric method for determination of oxidized and reduced glutathione in tissues. *Anal Biochem* **74:**214–226, 1976.
82. Fülöp T, Fóris G, Leövey A: Age-related changes in cAMP and cGMP levels during phagocytosis in human polymorphonuclear leukocytes. *Mech Ageing Dev* **27:**233–237, 1984.
83. Lad PM, Glovsky MM, Richards JH, et al: Regulation of human neutrophil guanylate cyclase by metal ions, free radicals and the muscarinic cholinergic receptor. *Mol Immunol* **22:**731–739, 1985.
84. Hartung HP, Kladetzky RG, Melnik B, et al: Stimulation of the scavenger receptor on monocytes–macrophages evokes release of arachidonic acid metabolites and reduced oxygen species. *Lab Invest* **55:**209–216, 1986.
85. Okajima F, Katada T, Ui M: Coupling of the guanine regulatory protein to chemotactic peptide receptors in neutrophil membranes and its uncoupling by islet-activating protein, pertussis toxin. *J Biol Chem* **260:**6761–6768, 1985.
86. Wong SKF, Martin BR, Tolkowsky AM: Pertussis toxin substrate is a guanosine 5'[β-thio] diphosphate-, *N*-ethylmaleimide, Mg^{2-} and temperature-sensitive GTP binding protein. *Biochem J* **232:**191–197, 1985.

20

The Respiratory Burst and Aging

TAMÁS FÜLÖP, Jr., GABRIELLA FÓRIS, JÓZSEF T. NAGY, ZSUZSA VARGA, and ANDRÁS LEÖVEY

1. INTRODUCTION

When exposed to certain stimuli, phagocytic cells (neutrophils and mononuclear phagocytes) undergo marked changes in the way they handle oxygen.[1] Their rates of oxygen uptake increase greatly, they begin to produce large amounts of superoxide (O_2^-) and hydrogen peroxide (H_2O_2), and they begin to metabolize large quantities of glucose by way of the hexose monophosphate shunt. Because of the sharp increase in oxygen uptake, this series of changes has come to be known as the respiratory burst. Its purpose is to generate powerful microbicidal agents by the partial reduction of oxygen.

It is well known that elderly persons have a greater susceptibility to morbidity and mortality, because of many bacterial infectious agents.[2] The primary functions of phagocytic cells are generally considered to be the ingestion and killing of pathogenic micro-organisms by the delivery of toxic agents to the phagocytic vacuoles. It is tempting to think that these two phenomena might be tightly linked, because the alterations found with aging[3,4] could be the consequence of changes in respiratory burst.

The reactive oxygen species (ROS) are also produced when phagocytic cells are treated with chemotactic peptides, such as FMLP, aggregated immunoglobulins, arachidonic acid (AA), and soluble stimuli, such as lectins, phorbol myristate acetate (PMA), immune complexes, or Ca ionophore A23187.[5–14] These ROS are also released at the extracellular level, where they may carry out various biological functions.[15]

The production of ROS is not only beneficial to the organism. Mutations, sister chromatid exchanges, chromosomal aberrations, cytotoxicity, carcinogenesis, and

TAMÁS FÜLÖP, Jr., GABRIELLA FÓRIS, JÓZSEF T. NAGY, ZSUZSA VARGA, and ANDRÁS LEÖVEY • First Department of Medicine, University Medical School, 4012 Debrecen, Hungary.

possibly cellular degeneration related to aging may result from the relative overproduction of ROS.[16] This idea is the basis of the free radical theory of aging.[17–19]

However, aerobic organisms have evolved complex mechanisms to prevent uncontrolled oxidative damage, while, at the same time, permitting the presence of sufficient active oxygen for the induction of numerous physiological signals, including those of the immune system.

The enzyme responsible for the production of microbicidal oxidants during the respiratory burst is a membrane-bound oxidase that catalyzes the reduction of O_2 to O_2^- at the expense of NADPH.[20–22] This enzyme is dormant in resting cells but is rapidly activated when cells are exposed to appropriate stimuli.[10–12] The purpose of this chapter is to review our current understanding of the respiratory burst as it relates to the aging process.

2. GENERAL CONCEPTS

2.1. Production of Free Radicals

The respiratory burst results from the activation of an NADPH oxidase that catalyzes the one-electron reduction of O_2:

$$O_2 + NADPH \rightarrow O_2^- + NADP^+ + H^+$$

Most of these O_2^- radicals interact to produce O_2 and H_2O_2, this reaction being accelerated by the cytoplasmic enzyme, superoxide dimutase (SOD):

$$2O_2^- + 2H^+ \rightarrow H_2O_2 + O_2$$

At the same time, glucose is metabolized through the hexose monophosphate shunt in order to regenerate the NADPH consumed by the O_2^--forming enzyme and by the glutathione-dependent H_2O_2 detoxifying system found in the cytoplasm of phagocytic cells.[23]

O_2^- and H_2O_2, immediate products of the respiratory burst, are used primarily as starting materials for the production of other microbicidal oxidants of phagocytes: halogens (myeloperoxidase–H_2O_2–halide system) and oxidizing radicals (singlet oxygen, hydroxyl radical).[24,25]

2.2 Free Radical Theory of Aging

Immune responses are mediated largely through the interactions of phagocytic cells, mainly macrophages (Mϕ) and B and T lymphocytes. The immune functions were extensively studied in aged humans and experimental animals, and the results suggest that the B and T lymphocyte functions are altered in the elderly, particularly the T cell mediated immune response.[26] It is thought that this decline contributes largely to the increased incidence of infections, cancer, and autoimmune disorders.[27]

As knowledge of the free radical reactions in biological systems accumulates, evidence mounts that random free radical reactions, even those beneficial to the organism as a whole, are a major factor in the degradation of multiple systems in the organism, suggesting that the aging process may simply be the sum of these deleterious reactions going on continuously throughout the cells and tissues, as proposed in the free radical theory of aging.[17–19,28]

2.3. The Glutathione Redox Cycle

Naturally, the aerobic cells are well prepared enzymatically for the detoxification of ROS. One of the most important is the superoxide dismutase (SOD), which catalyzes the dismutation of two O_2^- into H_2O_2. The excess of H_2O_2 may be either degraded by catalase, leaking out of the cells but essentially detoxified in the glutathione peroxidase (GSHPx) and glutathione reductase (GSSGR) cycle. GSHPx, together with GSSGR to regenerate reduced glutathione (GSH) from oxidized glutathione (GSSG), is probably a much more significant means of removing excess H_2O_2 in animal cells than is catalase.[29] The GSHPx not only detoxifies that H_2O_2 but also protects the cells from lipid peroxidation and from the harmful effects of lipid peroxides by its reduction to hydroxy acids, thereby terminating the chain reactions of lipid peroxidation.[30]

GSH not only plays a role in protecting cells against the effects of free radicals and of reactive oxygen intermediates (e.g., peroxides) formed in metabolism but functions in reducing disulfide linkages of proteins and other molecules, in the synthesis of the deoxyribonucleotide precursors of DNA. GSH is also transported out of the cells. GSH has a role in the metabolic processing of certain endogenous compounds, such as estrogens, prostaglandins, and leukotrienes (LT). LTA_4, an epoxide derived from arachidonic acid (AA) reacts with GSH to form conjugate LTC_4.[31]

3. THE RESTING OXIDATIVE METABOLISM IN THE PMNL OF THE ELDERLY

The current explanation for aging at the biochemical level is based on the production of free radicals with the subsequent free radical chain reactions ultimately leading to membrane damage. The net result is a decline in cellular integrity and function.[32]

An age-dependent increase was found in the activities of mitochondrial catalase and GSHPx. These findings support earlier observations that aging stimulates the steady-state formation of H_2O_2 and lipid peroxides.[33]

In our laboratory, we measured the O_2 consumption, the production of O_2^- and H_2O_2, as well as the glutathione redox cycle in PMNL of healthy aged subjects (Table I). We demonstrated that the production of O_2^- and H_2O_2 is increased with aging compared with young controls. This corroborates earlier findings that the production of ROS is enhanced with aging.[28] The result is an increase in lipid

TABLE I. Characteristics of Oxidative Processes in Resting Neutrophils Obtained from Healthy Young and Aged Male Subjects[a–c]

Parameters	Subjects	
	Young	Aged
O_2 consumption (nmoles/min/10^7 cells)	6.34±0.52	11.80±0.86
Luminol-dependent chemiluminescence (total count/40 min)	$4.2\times10^6\pm1.6\times10^5$	$11.8\times10^6\pm1.1\times10^5$
Produced H_2O_2 (nmoles/60 min/10^6 cell)	1.08±0.02	1.42±0.03
Glutathione peroxidase (U/10^6 cell)	0.683±0.02	1.108±0.04
Glutathione reductase (U/10^6 cell)	13.30±0.72	10.74±0.65
GS-SG/GSH	0.080	0.348

[a]Each value represents the mean of 10 determinations ±SEM.
[b]Each determination was carried out in triplicate.
[c]All parameters presented obtained from aged subjects differ significantly from control values (young group) at a level of $p < 0.01$.

peroxidation, with the rigidification of cell membranes, damage to the arterial wall, and inactivation of biologically important enzymes via their SH groups. SH groups have been implicated in a variety of physiologically important membrane phenomena.[34,35] A decrease in SH groups was demonstrated with aging,[36] during which the ratio of disulfides/sulfhydryls increases.

It is known that the extracellular activity of H_2O_2 increases with aging in association with the myeloperoxidase halide system.[23,25,37] The induction of this enhanced free radical production is unknown. The clue to this induction could be the increased activity of NADPH-oxidase,[21] normally dormant in resting cells, partly being controlled by the cyclic nucleotide system. The increased cyclic adenosine monophosphate (cAMP) level was shown to inhibit the formation of O_2^-.[38] In our previous study, we showed that in resting PMNL of elderly subjects, the cAMP level was decreased comparing with that in young subjects,[39] which may partially explain the increased production of O_2^- by the activation of NADPH oxidase.

Another explanation for this enhanced free radical production with aging may be the disturbance of the enzymatic detoxification system. The GSHPx activity was found to be increased in PMNL (Table I), while other workers[41,42] demonstrated decreased activity in erythrocytes of elderly subjects. Glutathione reductase activity, which is necessary for the reduction of GSSG in GSH, is significantly decreased in PMNL (Table I), as well as in the erythrocytes of elderly patients.

As a result of these enzymatic variations in resting PMNL with aging, the ratio of GSH/GSSG is decreased by the fact that GSSG accumulates in the cells. The increased reactivity of the H_2O_2 is also related to the decreased GSH/GSSG ratio in the cells with aging.[43,44] This decrease contradicts the basic tendency of the cells, which by the equilibrium position of the GSSGR greatly favors GSSG reduction; the GSH/GSSG ratio in the cells is kept very high.[45,46] Moreover, GSSG itself can be damaging.[47]

Furthermore, the total glutathione concentration is also decreased with aging.

This decrease may be explained by: (1) the reduced glutathione synthesis, (2) the efflux of both forms of glutathione from the cells, and (3) the GSH incorporation in leukotriene (C_4).[23,37,48,49]

In PMNL of the elderly, the GSH, the main substrate for the detoxification of free radicals, is significantly decreased,[50] as well as in other tissues. This loss may explain various alterations found with aging.[44,51] It was shown that PMNL from patients deficient in GSH, however, accumulate H_2O_2 after particle ingestion, have impaired bacterial killing, and demonstrate phagocytosis.[52]

It is also known that an increased cAMP level is necessary for the enhanced production of GSH; since cAMP levels are low in PMNL of the elderly, this could also contribute to the equally low level of GSH as compared with that in young subjects.[51] The enhanced activity of GSHPx with aging can also contribute to the decreased level of GSH.

Moreover, if we consider the results of Reiss and Gershon,[53] who found that the rat cytoplasmic SOD activity was diminished with aging, we can state that the ROS seems to accumulate in the resting aging cells, due to the fact that the detoxifying enzyme systems are defective concommitantly to alterations of the cyclic nucleotide system.

4. THE RESPIRATORY BURST DURING VARIOUS STIMULATIONS

Phagocytosis of bacteria by phagocytic cells is accompanied by a burst of oxidative metabolic changes, including increased O_2 consumption, hexose monophosphate shunt activity, and O_2^- generation.[54,55] Similar metabolic process occurs when phagocytic cells are treated with soluble or particulate stimuli.[5–14]

NADPH-oxidase in the plasma membrane of the cells is responsible for the induction of the respiratory burst. Concepts concerning the NADPH-oxidase have evolved and it is known that different ligands use different pathways for activating NADPH-oxidase, ultimately resulting in the induction of the respiratory burst. These differences in the activation of the respiratory burst were also found with aging.

4.1. Receptor Stimulation of the Respiratory Burst

4.1.1. Stimulation through the Fc_γ Receptor

Interaction of the surface-bound IgG (e.g., opsonized bacteria, fungi, or zymozan) with Fc_γ receptor is sufficient to trigger both phagocytosis and release of AA.[56] Both the Fc and C3b receptors are transmembrane glycoproteins that bind appropriately opsonized particles and that are capable of initiating changes in cellular behavior.[57] Ligation of the Fc receptor by IgG-coated particles results in the concomitant release of H_2O_2 and AA[57,58] as well as other highly bactericidal products.

The age-related alterations in the effector functions of phagocytic cells are controversial.[59] Traditional methods failed to demonstrate altered functions,[60,61] but, using receptor-mediated functions in the study of their functions, either for monocytes or for PMNL, a net decline in effector functions could be demonstrated. The phagocytosis was normal, but the intra- and extracellular killing were altered.[62,63] The intracellular killing of bacteria was found unaltered with aging by other investigators.[3,61] It is known that the phagocytic and effector functions can be decoupled, as in the case of C3b receptors.[65] It appears that the signals for phagocytosis and effector functions with their biochemical and metabolical processes are functionally distinct. Indeed, Fc receptors can generate at least three separate signals (phagocytosis, H_2O_2, and AA generation).

The study of the respiratory burst, crucial to the microbicidal activity of phagocytic cells, gave conflicting results in elderly subjects. Generally, most investigators found, in agreement with our findings, a diminished capacity of PMNL to produce O_2^- and H_2O_2 under phagocytic stimulation.[4,65] Nagel et al.[3] reported decreased O_2^- production, and van Epps et al.[66] found that persons older than 70 years of age have a lower chemiluminescence in response to phagocytic stimulation than do young subjects. Some workers reported an unchanged NBT reduction test with the PMNL of elderly under phagocytic stimulation.[67] Nevertheless, experimental evidence points to diminished production of O_2^- and H_2O_2 under phagocytic stimulation, which might partly explain the decrease of effector functions of phagocytic cells with aging. The question arises as to why the production of free radicals could not be stimulated as efficiently as in young subjects. It is questionable as to whether the alterations observed in the oxidative metabolism of resting PMNL could totally explain the changes found during phagocytic stimulation of PMNL of elderly subjects. We think that the main point for the decreased responsiveness of PMNL with aging is the alteration of the transmembrane signaling following stimulation of Fc_γ and C3b receptors. Thus, the activation of the NADPH-oxidase, the first step in the phagocytic stimulation of the cells through the Fc_γ and C3b receptors may not be effective, explained, as we reported earlier,[63] by the fact that the guanylate cyclase is refractory to any stimulation in PMNL of elderly subjects. It is known that NADPH-oxidase has various triggers; among them the most important are the AA and its metabolic products, proteases, protein kinase C (PK-C), and cGMP.[21] Stimulation of the respiratory burst should be considered via other receptors as well.

4.1.2. Stimulation of Chemotactic and Other Receptors

Binding of chemotactic peptides such as FMLP to their receptors on phagocytic cells stimulates chemotaxis as well as O_2^- production and lysosomal enzyme secretion.[68] Chemotactic peptides rapidly increase intracellular $(Ca^{2+})_i$ in PMNL. Mobilization of Ca^{2+} is likely to play an important role in the activation of phagocytic cells, because the elevation of intracellular Ca^{2+} is necessary for optimal activation of PMNL in response to most stimulating agents.[69–71] Observations in numerous cell types have demonstrated a strong correlation between agonist-induced increases in intracellular $(Ca^{2+})_i$ and production of inositol 1,4,5-triphosphate (IP_3) as a consequence of enhanced polyphosphoinositides hydrolysis.[72] Likewise, chemoattractants have been shown to stimulate a polyphosphoinositide-specific phospholipase C in PMNL, resulting in the formation of IP_3.[73,74] It was therefore hypothesized that guanine nucleotide regulatory (N or G) proteins might mediate the coupling of chemoattractant receptor to phospholipase C.[74–76] Thus, chemoattractants receptors appear to use a pertussis toxin (PT)-sensitive transduction mechanism that is not shared by nonspecific stimuli, such as lectins, PMA, or Ca ionophore.[20,76,77] Data obtained demonstrate that the action of PT on the chemotactic peptide receptor is proximal to the rise of IP_3 and that the regulation of the rise in Ca^{2+}, superoxide production, degranulation, and membrane depolarization is distal to the site of inhibition of the toxin.[78] Moreover, it is known that FMLP receptor occupancy generates another potent stimulatory signal,[79] presumably diacylglycerol (DG), which is also produced by phospholipase C.

FMLP also stimulates lipid turnover directly or indirectly in a Ca^{2+}-dependent manner by activating phospholipase C and A_2.[80,81] Arachidonate is oxygenated by lipoxygenase (LPO) to active products that influence PMNL function.[82] Thus, AA metabolites not only mediate inflammatory responses but also enhance the stimulus–response coupling of human PMNL. Two pathways have been proposed for metabolization of AA from membrane phospholipids either directly via phospholipase A_2 to yield a lysophospholipid and AA or indirectly via phospholipase C followed by lipase action on the resulting DG.[83] Alternatively, DG is phosphorylated by a kinase to generate phosphatidic acid (PA) as part of the PI/PA cycle.

FMLP and Ca ionophores stimulate a transient increase in cAMP, but not of PMA in PMNL,[84] suggesting that cAMP-dependent kinase regulates signaling for the former two stimuli. Interestingly, treatment of PMNL with agents that elevate cAMP levels depresses stimulation of O_2^- generation[38,86] as was also demonstrated in Mφ.[14]

In PMNL of elderly subjects, FMLP, like Fc-receptor stimulation, induced a diminished O_2^- production compared with PMNL of young subjects (Fig. 1). The application of nordihydroguaiaric acid (NDGA), an inhibitor of LPO, further diminished the O_2^- production in PMNL of aged subjects, whereas it showed no effects in PMNL of young subjects (Fig. 2). All these findings clearly suggest that one of the described receptor coupling phenomena is altered with aging and that some products, such as AA and its metabolites (via the LPO pathway) become important in the NADPH-oxidase activation in PMNL of the elderly. We found that the production of leukotienes under receptor stimulation was increased in PMNL of elderly.[85]

The stimulation of the respiratory burst in PMNL through lymphokine[87] and opioid[88] receptors gave the same diminished oxidative metabolism response in elderly as that produced by Fc_γ and FMLP receptors.

In order to understand better the decreased efficacy of O_2^- generation, with aging via receptor stimulation, several transduction mechanism changes may be

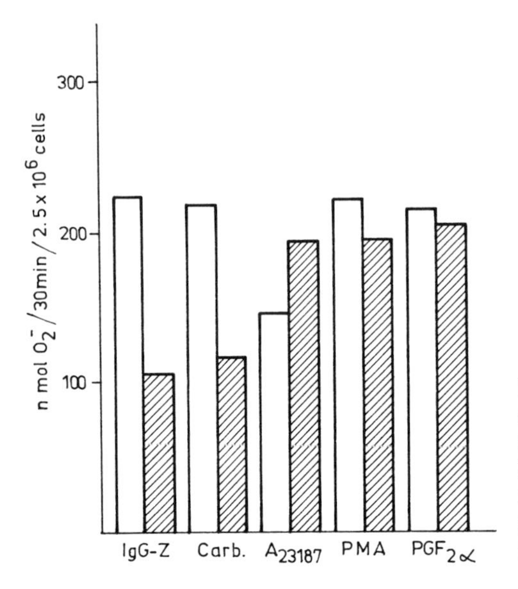

FIGURE 1. O_2^- generation by various stimuli in PMNL of healthy young and aged subjects. (□) young; (▨) aged. Ig-Z, opsonized zymosan; Carb, carbachol; A23187, Ca ionophore A23187; PMA, phorbol myristate acetate; and $PGF_{2\alpha}$, prostaglandine $F_{2\alpha}$.

considered: (1) the guanine nucleotide regulatory protein (N_i) function could be altered, but experimental evidence is still lacking; (2) the breakdown of polyphosphoinositides seems to be altered, because the IP_3 generation was slightly decreased in PMNL of elderly (unpublished data); (3) the Ca^{2+} metabolism is changed, as supported by the fact that the rise of intracellular free $(Ca^{2+})_i$ was less

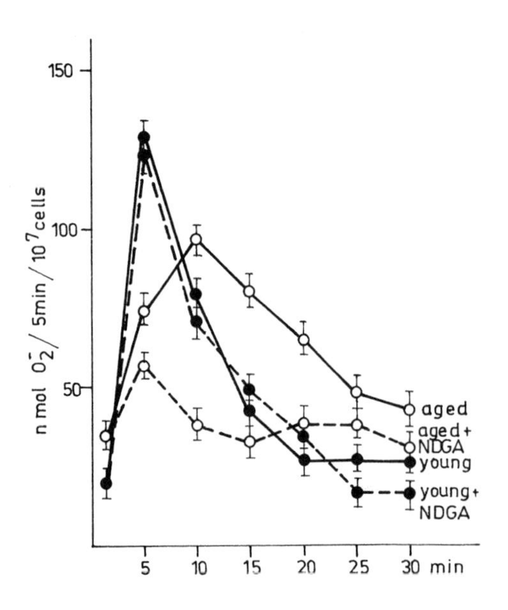

FIGURE 2. Effect of nordihydorguaiaric acid (NDGA) on FMLP-stimulated O_2^- generation in PMNL of young and elderly subjects. (——●——) young, stimulated by FMLP; (- - -●- - -) young, stimulated by FMLP + NDGA; (——○——) aged, stimulated by FMLP; and (- - -○- - -) aged, stimulated by FMLP + NDGA.

in PMNL and T lymphocytes of elderly than in young subjects; and (4) the cyclic nucleotide system could be altered, as is the case in PMNL of elderly.[39,63] However, the data reported in the literature are conflicting, with some investigators describing the irresponsiveness of adenylate cyclase to stimulation,[90,91] while our group found an altered turn-off of the adenylate cyclase.[39] However, it seems evident that the guanylate cyclase is refractory to any stimulation in the PMNL of elderly.[87,88]

All these divergent alterations seem to point to a possible primary change of the guanine nucleotide binding regulatory protein (N_i) at the level of which the transduction of the signal is impaired. It can be hypothesized that the alteration is situated at the SH site of the N_i which is oxidized, but this possibility needs further investigations (Fig. 3).

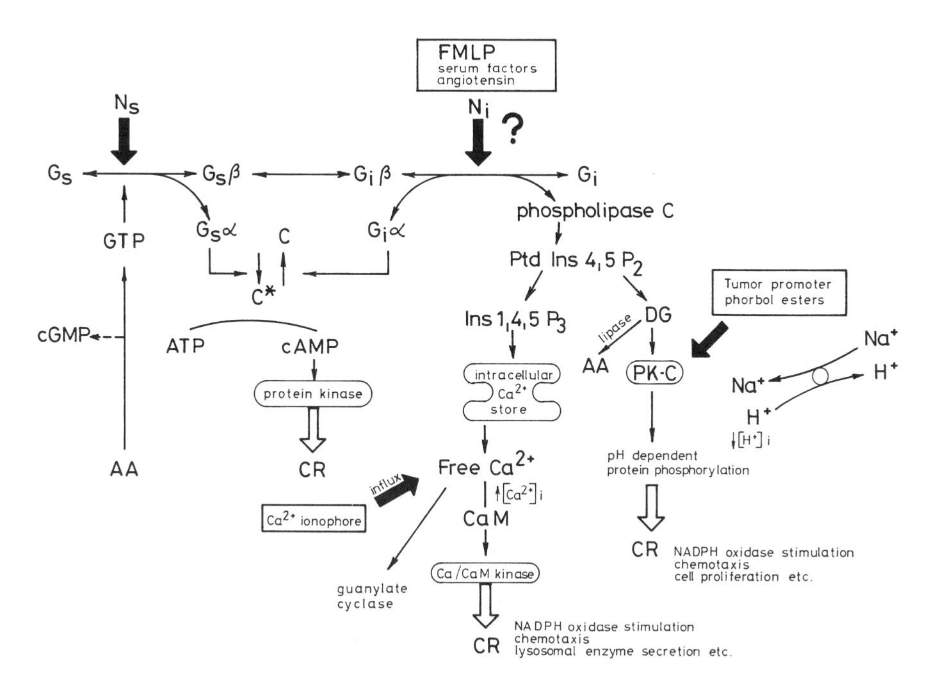

FIGURE 3. Model for receptor and nonreceptor stimulation of various kinase systems leading to different cellular responses, including O_2^- generation. The receptor and the ligand (e.g., FMLP) produces a receptor–ligand complex. This complex induces the exchange of bound GDP with GTP on an N protein, yielding an activated N protein species. The activated N_i protein interacts with polyphosphoinositide-specific phospholipase C. The activated phospholipase cleaves PIP_2 into diacylglycerol (DG) and inositol triphosphate (IP_3). These second messengers (through the protein kinase C and cytosolic free calcium) increase, respectively, mediating the cellular response. The nonreceptor stimulation acts directly on PK-C activity (e.g., PMA) or on calcium metabolism (e.g., calcium ionophore A23187), ultimately leading to the same cellular response as receptor stimulation, concerning the O_2^- generation. The question mark would signify that we do not know the primary alteration in respiratory burst stimulation, in the case of elderly subjects.

4.2. Nonspecific Stimulation of the Respiratory Burst

Soluble agents such as A23187 and PMA do not share the same transduction mechanism as chemotactic peptides and other receptors linked ligands (e.g., opsonized bacteria, lymphokines), even though they stimulate the respiratory burst of phagocytic cells[78] (Fig. 3).

It is well-known that PMA, a direct activator of PK-C,[92] elicits O_2^- production even in intact Mϕ by a process not involving phospholipid breakdown and fatty acid liberation.[14] This suggests a key role in the activation of respiratory burst to the PK-C, activated by the DG.[93–95] Thus, it seems that the stimulation of O_2^- production in alveolar Mϕ by nonphysiological phorbol diesters appears to be independent of any requirement for Ca^{2+} mobilization. The activity of phorbol ester has been attributed to their partial structural analogy to DG, which can activate a Ca^{2+}-dependent phospholipid-dependent protein kinase (PK-C) even at physiological intracellular Ca^{2+} concentration.[70,92]

The Ca ionophore A23187 also stimulates the production of O_2^- by activating NADPH-oxidase via the elevation of cytosolic Ca^{2+}.[96] This is a direct increase of Ca^{2+} not involving the phosphoinositide hydrolysis, acting on a cellular mediator, leading to NADPH-oxidase stimulation.[97] In the case of Ca ionophore A23187, it was suggested that phospholipid-derived AA[14] as well as LPO pathway products,[98] play an active role in the stimulus–response coupling leading to O_2^- production. It was interesting to investigate the modulation of Ca ionophore A23187-induced O_2^- production by NDGA. In the case of the PMNL of young subjects, the NDGA slightly diminished the $O_2{}^-$ production, however the general form of the curve did not change, while in the case of elderly the production was significantly decreased indicating, that in this case the LPO pathway products play a second messenger role. This seems to be a general phenomenon, because the same role of the LPO products was observed in the case of FMLP stimulation of O_2^- production, modulated by NDGA, in PMNL of elderly subjects (Fig. 4).

The experimental evidence then, seems to suggest that PK-C may play a central role in signal transduction of phagocytic cells, particularly for activation of the respiratory burst.[93,99,100] In PMNL of elderly the stimulation of the O_2^- production by such agents as Ca ionophore A23187 or PMA gave the same result as in the case of PMNL of young subjects (Fig. 1). This demonstrates that nonspecific stimulation of the respiratory burst of PMNL of elderly was able to induce a similar production of O_2^- to that in young. These results contradict those found during receptor stimulation. The results of the nonspecific stimulation further suggest that the primary defect in phagocytic cells of the elderly subjects, concerning the respiratory burst stimulation, occurs proximal to the intracellular Ca^{2+} elevation and PK-C activation.

Indeed, our hypothesis based on the experimental results accumulated in recent years is as follows: the defect is situated at the guanine nucleotide regulatory protein (N_i) level. At this level, alterations in the SH groups can be supposed, as these SH groups have been implicated in a variety of physiologically important membrane phenomena.[34,35]

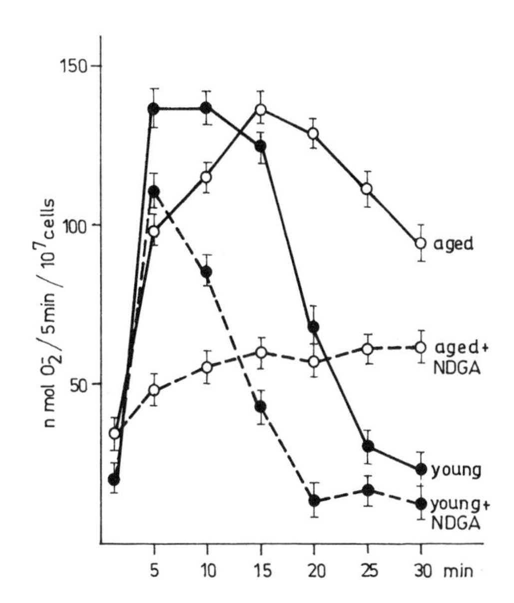

FIGURE 4. Effect of nordihydroguaiaric acid (NDGA) on Ca ionophore A23187 stimulated O_2^- generation in PMNL of young and elderly subjects. (——●——) young, stimulated by Ca ionophore A23187; (- - -●- - -) young, stimulated by Ca ionophore A23187 + NDGA; (——○——) aged, stimulated by Ca ionophore A23187; and (- - -○- - -) aged, stimulated by Ca ionophore A23187 + NDGA.

4.3. Detoxification of Reactive Oxygen Species

More recent evidence also implicates phagocyte-derived O_2 reduction products in the causation of damage to cells and materials in the vicinity of the phagocytosis, such as tumor cells, platelets, chemotactic factors or collagen.[101,102] An important issue in the study of the respiratory burst involves the mechanisms used by the cells to prevent auto-oxidative damage. The principal detoxification systems used to combat this effect are the SOD (for removal of O_2^-) and GSHPx and catalase (which destroy H_2O_2).[12,28] Oxygen free radicals have been implicated as causal factors in many biological processes among others aging.[19,28,103]

It has been shown that the total intracellular ionic strength displays an age-dependent increase mainly attributable to increased intracellular K^+ and Na^+ concentrations in postmitotic cells[104] concomitant with reduced activity of SOD in these cells.[105] Some age-related changes of SOD activity in brain and liver have been reported,[53,106] and studies have demonstrated an association between tissue SOD activity as related to the specific metabolic rate and the life span potential.[107,108]

GSHPx converts H_2O_2 and lipid peroxides to their corresponding alcohols, and at the same time GSH is oxidized to GSSG. In erythrocytes, marked decreases in GSH content, free thiol groups, and GSSG reductase activity were observed as a function of cell age, as well as mouse age. Recently, aging-specific decreases in GSH and total glutathione concentration were found in tissues[44] as well as in PMNL.[50] Enhanced oxidation via the glutathione redox cycle was proposed as to explain the decline of GSH with aging. However, GSHPx and GSSGR activity was found to be decreased with aging. The findings of lower enzyme levels demonstrated a decreased capacity for the GSH-linked detoxification of peroxides in senescent tissues. Collectively, experimental evidence suggests that aging tissues become more susceptible to oxidative damage produced by endogenous peroxides. These decreases indicate that detoxification via GSHPx and GSSGR is impaired with aging.

During specific stimulation, i.e., via receptors, the GSHPx activity that was originally elevated in the PMNL of elderly subjects decreased markedly, while in young persons its activity increased concomitantly to the elevation of H_2O_2. GSSGR demonstrated a delayed but constant increase during stimulation. Thus, the GSSG remained constant, while the GSH further decreased under stimulation. For this reason, other mechanisms, such as GSH synthesis, degradation, or utilization, must account for the GSH decrease.

The decreased responsiveness of GSHPx with aging, under phagocytosis, is probably one of the basic alterations underlying the changes observed in oxidative metabolism during stimulation. In evaluating this decreased responsiveness, the difficulty resides in the fact that besides the decrease in the intracellular Se level or changes in its incorporation to the GSHPx, we do not know any other physiological or pathological circumstances that might play a role in regulating GSHPx activity.[23]

Changes in the glutathione redox cycle are controversial and still poorly studied. However, it is evident from the experimental results that this important detoxifying system does not work properly with aging, and as a consequence the detoxification of free radicals is impaired.

5. CONCLUSION

Nonspecific stimulation of the respiratory burst in the cells of elderly subjects appears to be as effective as in the cells of young subjects. However, its stimulation through receptors is significantly altered, partly explaining the diminished host-defense functions with aging that lead to infections, tumors, and autoimmune disorders. Not only is the production of ROS disturbed, but the detoxifying mechanisms are impaired with aging as well. Experimental evidence suggests that ROS have primarily a harmful role in the aging process: (1) the damage caused by ROS is increased, and (2) their role in host defense is decreased. Unfortunately, our knowledge of these events does not permit us to decide upon the altered mechanism that plays a key role in the immunosenescence of phagocytic cells. Once this key mechanism has been established, we will be able to work toward restoring their immunocompetence. Efforts should be made for continuing our investigations in this direction.

REFERENCES

1. Babior BM, Crowley CA: Chronic granulomatous disease and other disorders of oxidative killing by phagocytes, in Stanburg JB, Wyngaarden JB, Fredrickson DS, et al (eds): *Metabolic Basis of Inherited Diseases*. New York, McGraw-Hill, 1983, pp 1956–1985.

2. Berk SL, Smith JK: Infectious diseases in the elderly. *Med Clin North Am* **67:**273–293, 1983.
3. Nagel JE, Pyle RS, Chrest FJ, et al: Oxidative metabolism and bactericidal capacity of polymorphonuclear leukocytes from normal young and aged adults. *J Gerontol* **37:**529–534, 1982.
4. Corberand J, Ngyen F, Laharrague P, et al: Polymorphonuclear functions and aging in humans. *J Am Geriatr Soc* **29:**391–397, 1981.
5. Romeo D, Jug M, Zabucchi G, et al: Perturbation of leukocyte metabolism by non phagocytosable concanavalin A coupled beads. *FEBS Lett* **42:**90–95, 1974.
6. Johnston RB, Lehmeyer JE, Guthrie LA: Generation of superoxide anion and chemiluminescence by human monocytes during phagocytosis and on contact with surface-bound immunoglobulin G. *J Exp Med* **143:**1551–1563, 1976.
7. Baxter MA, Leslie RGQ, Reeves WG: The stimulation of superoxide anion production in guinea pig peritoneal macrophages and neutrophils by phorbol myristate acetate, opsonized zymosan and IgG2 containing soluble immune complexes. *Immunology* **48:**657–658, 1983.
8. Yagawa K, Kaku M, Ichinose Y, et al: Fc receptor mediated desensitization of superoxide (O_2^-) generation response of guinea pig macrophages and polymorphonuclear leucocytes. *Immunology* **55:**629–638, 1985.
9. McPhail LC, Snyderman R: Activation of the respiratory burst enzyme in human polymorphonuclear leukocytes by chemoattractants and other soluble stimuli. Evidence that the same oxidase is activated by different transductional mechanisms. *J Clin Invest* **72:**192–200, 1983.
10. Cohen HJ, Chovaniec ME: Superoxide generation by digitonin-stimulated guinea pig granulocytes: A basis for continuous assay monitoring superoxide production and for the study of the activation of the generating system. *J Clin Invest* **61:**1088–1096, 1978.
11. Bender JG, McPhail LC, Van Epps DE: Exposure of human neutrophils to chemotactic factors potentiates activation of the respiratory burst enzyme. *J Immunol* **130:**2316–2323, 1983.
12. Pick E, Keisari Y: Superoxide anion and hydrogen peroxide production by chemically elicited peritoneal macrophages. Induction by multiple nonphagocytic stimuli. *Cell Immunol* **59:**301–318, 1981.
13. Palmblad J, Gyllenhammar H, Lindgren JA, et al: Effects of leukotrienes and f-Met-Leu-Phe on oxidative metabolism of neutrophils and eosinophils. *J Immunol* **132:**3041–3045, 1984.
14. Bromberg Y, Pick E: Unsaturated fatty acids as second messengers of superoxide generation by macrophages. *Cell Immunol* **79:**240–252, 1983.
15. Clark RA: Extracellular effects of myeloperoxidase–hydrogen peroxide–halide system, in Weissmann G (ed): *Advances in Inflammation Research.* New York, Raven, 1983, vol 5, p 107.
16. Cerutti PA: Prooxidant states and tumor promotion. *Science* **227:**375–381, 1985.
17. Harman D: Aging: A theory based on free radical and radiation chemistry. *J Gerontol* **11:**298–299, 1956.
18. Harman D, Heidrick ML, Eddy DE: Free radical theory of aging. Effect of free-radical-reaction inhibitors on the immune response. *J Am Geriatr Soc* **25:**400–407, 1977.
19. Harman D: Free radical theory of aging: Nutritional implications. *Age* **1:**143–150, 1978.
20. McPhail LC, Clayton CC, Snyderman R: The NADPH oxidase of human polymorphonuclear leukocytes. *J Biol Chem* **259:**5768–5775, 1984.
21. Babior BM: The respiratory burst of phagocytes. *J Clin Invest* **73:**599–601, 1984.
22. Tsunawaki S, Nathan CF: Enzymatic basis of macrophage activation. *J Biol Chem* **259:**4305–4312, 1984.
23. Baker SS, Cohen HJ: Altered oxidative metabolism in selenium deficient rat granulocytes. *J Immunol* **130:**2856–2860, 1983.
24. Karnofsky JR, Wright J, Miles-Richardson GE: Biochemical requirements for singlet oxygen production by purified human myeloperoxidase. *J Clin Invest* **74:**1489–1495, 1984.
25. Ito M, Karmali R, Krim M: Effect of interferon on chemiluminescence and hydroxyl radical production in murine macrophages stimulated by PMA. *Immunology* **56:**533–541, 1985.
26. Makinodan T, Kay MMB: Age influences on the immune system, in Kungel E, Dixon BC (eds): *Advances of Immunology.* New York, Academic, 1980, p 287.
27. Weksler ME: Senescence of the immune system. *Med Clin North Am* **67:**263–272, 1983.

28. Leibovitz BE, Siegel BV: Aspects of free radical reactions in biological systems: aging. *J Gerontol* **35:**45–56, 1980.
29. Nathan CF, Arrich BA, Murray MW et al: Tumor cell anti oxidant defenses. Inhibition of the glutathione redox cycle enhances macrophage mediated cytolysis. *J Exp Med* **153:**766–779, 1981.
30. Halliwell B: Superoxide dismutase, catalase and glutathione peroxidase: Solutions to the problems of living with O_2. *New Phytol* **73:**1057–1068, 1974.
31. Piper PJ (ed): *SRS-A and Leukotrienes.* New York, Wiley, 1981.
32. Tappel AL: Lipid peroxidation damage to cell components. *Fed Proc* **32:**1870–1875, 1973.
33. Nohl H, Megner D: Do mitochondria produce oxygen radicals in vivo? *Eur J Biochem* **82:**563–567, 1978.
34. Sutherland RM, Rothstein A, Weed RI: Erythrocyte membrane sulfhydryl groups and cation permeability. *J Cell Physiol* **69:**185–196, 1967.
35. Goldman JM, Hadley ME: Sulfhydryl requirement for α-adrenergic receptor activity and MSH action in melanophores. *J Pharm Exp Ther* **182:**93–105, 1972.
36. East EJ, Chang RCC, Yu NT, et al: Human spectroscopic measurement of total sulfydryl in intact lens as affected by aging and ultraviolet radiation. *J Biol Chem* **253:**1436–1441, 1978.
37. Klebanoff SJ: Cytocidal mechanism of phagocytic cells. in Fougereau N, Dausset J (eds): *Immunology 1980.* London, Academic, 1980, vol 2: *Progress Immunology,* p 720.
38. Lehmeyer JE, Johnston RB: Effect of anti-inflammatory drugs and agents that elevate intracellular cyclic AMP on the release of toxic oxygen metabolites by phagocytes: Studies in a model of tissue bound IgG. *Clin Immunol Immunopathol* **9:**482–490, 1978.
39. Fülöp T Jr, Fóris G, Leövey A: Age related changes in cAMP and cGMP levels during phagocytosis in human polymorphonuclear leukocytes. *Mech Ageing Dev* **27:**233–237, 1984.
40. Abraham EC, Taylor JF, Lang CA: Influence of mouse age and erythrocyte age on glutathione metabolism. *Biochem J* **174:**819–825, 1978.
41. Thompson CD, Rea HM, Doesburg VM: Selenium concentrations and glutathione peroxidase activities in whole blood of New Zealand residents. *Br J Nutr* **37:**457–460, 1977.
42. Thompson CD, Rea HM, Robinson MF, et al: Low blood selenium concentrations and glutathione peroxidase activity in elderly people. *Proc Univ Otego Med School* **55:**18–26, 1977.
43. Halliwell B: Biochemical mechanisms accounting for the toxic action of oxygen in living organisms: The key role of superoxide dismutase. *Cell Biol Int Rep* **2:**113–128, 1978.
44. Hazelton GA, Lang CA: Glutathione contents of tissues in the aging mouse. *Biochem J* **188:**25–30, 1980.
45. Joselyn P: *Biochemistry of the Thiol Group.* London, Academic, 1972.
46. Tietze F: Enzymic method for quantitative determination of nanogram amounts of total and oxidised glutathione. *Anal Biochem* **27:**502–522, 1969.
47. Rupriak ATR, Quincey RV: Mechanism of action of a microsomal inhibitor of protein synthesis potentiated by GSSG. *Biochem J* **136:**335–342, 1973.
48. Stjernschantz J: The leukotrienes. *J Med Biol* **62:**215–230, 1984.
49. Ovrenius S, Ormstad K, Thor H, et al: Turnover and functions of glutathione studied with isolated hepatic and renal cells. *Fed Proc* **42:**3177–3188, 1983.
50. Fülöp T Jr, Fóris G, Wórum I, et al: Age related variations of some PMNL functions. *Mech Ageing Dev* **29:**1–7, 1985.
51. Noelle JR, Lawrence DA: Determination of glutathione in lymphocyte and possible association of redox state and proliferative capacity of lymphocytes. *Biochem J* **198:**571–579, 1981.
52. Spielberg SP, Boxer LA, Oliver JM, et al: Oxidative damage to neutrophils in glutathione synthetase deficiency. *Br J Haematol* **42:**215–220, 1979.
53. Reiss U, Gershon D: Rat-liver superoxide dismutase: Purification and age-related modifications. *Eur J Biochem* **63:**617–623, 1976.
54. Sbarra AJ, Karnovsky ML: The biochemical basis of phagocytosis. I. Metabolic changes during the ingestion of particles by polymorphonuclear leukocytes. *J Biol Chem* **234:**1355–1361, 1959.
55. Curnutte JT, Babior BM: Biological defense mechanisms. The effect of bacteria and serum on superoxide production by granulocytes. *J Clin Invest* **53:**1662–1673, 1974.

56. Rouzer CA, Scott WA, Kempe J, et al: Prostaglandin synthesis by macrophages requires a specific receptor-ligand interaction. *Proc Natl Acad Sci USA* **77:**4279–4284, 1980.
57. Unkeles JC, Wright SD: Structure and modulation of Fc and complement receptors. *Contemp Topics Immunobiol* **14:**171–196, 1984.
58. Bonney RJ, Naruns P, Davies P, et al: Antigen–antibody complexes stimulate the synthesis and release of prostaglandins by mouse peritoneal macrophages. *Prostaglandins* **18:**605–618, 1979.
59. Corberand JX, Laharrague PF, Fillola G: Neutrophils are normal in normal aged humans. (Letter to the editor) *J Leukocyte Biol* **40:**333–335, 1986.
60. Perkins EH: Phagocytic activity of aged mice. *J Reticuloendothel Soc* **9:**642–643, 1971.
61. Palmblad J, Haak A: Aging does not change blood granulocyte bactericidal capacity and levels of complement factors 3 and 4. *Gerontology* **24:**381–387, 1978.
62. Fülöp T Jr, Fóris G, Wórum I, et al: Age dependent changes of the Fc_{γ}-receptor-mediated functions of human monocytes. *Int Arch Allergy Appl Immunol* **74:**76–79, 1984.
63. Fülöp T Jr, Fóris G, Wórum I, et al: Age dependent alterations of Fc_{γ}-receptor-mediated effector functions of human polymorphonuclear leukocytes. *Clin Exp Immunol* **61:**425–432, 1985.
64. Aderem AA, Wright SD, Silverstein SC, et al: Ligated complement receptors do not activate the arachidonic acid cascade in resident peritoneal macrophages. *J Exp Med* **161:**617–622, 1985.
65. Fülöp T Jr, Hauck M, Kékessy D, et al: The physiologic significance of the glutathione redox cycle in resting and stimulated human PMNLs. Studies on PMNL obtained from healthy young and aged subjects. (in preparation).
66. Van Epps DE, Goodwin JS, Murphy S: Age dependent variations in polymorphonuclear leukocyte chemiluminescence. *Infect Immun* **22:**57–61, 1978.
67. Eschenbach C, Sebach G, Müller-Lissner St: Nitroblau-tetrazolium-reduktionskapazität von neutrophilen granulocyten. *Klin Woch* **53:**1049–1056, 1975.
68. Snyderman R, Pike MC: Chemoattractant receptors on phagocytic cells, in Paul WE (ed): *Annual Review of Immunology*. Palo Alto, CA, Annual Reviews, 1984, vol 2, p 257.
69. Smolen JE, Korchak HM, Weissmann G: The roles of extracellular and intracellular calcium in lysosomal enzyme release and superoxide anion generation by human neutrophils. *Biochem Biophys Acta* **677:**512–519, 1981.
70. Scully SP, Segal GB, Lichtman MA: Relationship of superoxide production to cytoplasmic free calcium in human monocytes. *J Clin Invest* **77:**1349–1356, 1986.
71. Torres M, Coates TD: Neutrophil cytoplasts: Relationships of superoxide release and calcium pools. *Blood* **64:**891–895, 1984.
72. Berridge MJ, Irvine RF: Inositol triphosphate a novel second messenger in cellular signal transduction. *Nature (Lond)* **312:**315–320, 1984.
73. Dougherty RW, Godfrey PP, Hoyle PC, et al: Secretagogue-induced phosphoinositide metabolism in human leukocytes. *Biochem J* **222:**307–315, 1984.
74. Snyderman R, Smith CD, Verghese MW: Model for leukocyte regulation by chemoattractant receptors: Roles of a guanine nucleotide regulatory protein and polyphosphinositide metabolism. *J Leukocyte Biol* **40:**785–800, 1986.
75. Snyderman R: Regulatory mechanisms of a chemoattractant receptor on leukocytes. *Fed Proc* **43:**2743–2748, 1984.
76. Vergehese MW, Smith CD, Charles LA, et al: A guanine nucleotide regulatory protein controls polyphosphoinositide metabolism Ca^{2+} mobilization and cellular responses to chemoattractants in human monocytes. *J Immunol* **137:**271–275, 1986.
77. Goldman DW, Chang FH, Gifford LA: Pertussis toxin inhibition of chemotactic factor-induced calcium mobilization and function in human polymorphonuclear leucocytes. *J Exp Med* **162:**145–156, 1985.
78. Krause KM, Schlegel W, Wollheim CB, et al: Chemotactic peptide activation of human neutrophils and HL-60 cells. Pertussis toxin reveals correlation between inositol triphosphate generation calcium ion transients and cellular activation. *J Clin Invest* **76:**1348–1354, 1985.
79. Pozzan TP, Lew D, Wollheim CB, et al: Is cytosolic free calcium regulating neutrophil activation? *Science* **221:**1413–1415, 1983.

80. Cockroft S, Bennett JP, Gomperts BD: F-met-leu-phe induced phosphatidylinositol turnover in rabbit neutrophils is dependent on extracellular calcium. *FEBS Lett* **110:**115–120, 1980.
81. Takenawa T, Hanma Y, Nagai Y: Role of Ca^{2+} in phosphatidylinositol response and arachidonic acid release in formylated tripeptide or Ca^{2+} ionophore A23187-stimulated guinea pig neutrophils. *J Immunol* **130:**2849–2857, 1983.
82. Wynkoop EM, Broekman J, Korchak HM, et al: Phospholipid metabolism in human neutrophils activated by N-formyl-methionyl-leucyl-phenylalanine. *Biochem J* **236:**829–837, 1986.
83. Hamma Y, Onozaki K, Hashimoto T, et al: Differential activation of phospholipids metabolism by formylated peptide and ionophore A23187 in guinea pig peritoneal macrophages. *J Immunol* **129:**1619–1624, 1982.
84. Smolen JE, Weissman G: Stimuli which provoke secretion of azurophil enzymes from human neutrophils induce increments in adenosine cyclic 3′-5′-monophosphate. *Biochim Biophys Acta* **672:**197–206, 1981.
85. Fülöp T Jr, Varga Z, Fóris G, Kékessyl D: unpublished data.
86. Simchowitz L, Fischbein LC, Spilberg I, et al: Induction of transient elevation in intracellular levels of cAMP by chemotactic factors: An early event in human neutrophil activation. *J Immunol* **124:**1482–1491, 1980.
87. Hauck M, Fülöp T Jr, Fóris G, et al: Divergent effects of human lymphokine-derived oligopeptides on PMNL function of young and aged healthy subjects. *Int J Immunopharmacol* **9:**3–8, 1987.
88. Fülöp T Jr, Kékessy D, Fóris G: Impaired coupling of naloxone sensitive opiate receptors to adenylate cyclase in PMNL of aged male subjects. *Int J Immunopharmacol* **9:**651–657, 1987.
89. Miller RA: Immunodeficiency of ageing: Restorative effects of phorbol ester combined with calcium ionophore. *J Immunol* **137:**805–808, 1986.
90. McLauglin B, O'Malley K, Cotter TG: Age related differences in granulocyte chemotaxis and degranulation. *Clin Sci* **70:**59–62, 1986.
91. Tam CF, Walford RL: Cyclic nucleotide levels in resting and mitogen-stimulated spleen cell suspensions from young and aged mice. *Mech Ageing Dev* **7:**309–320, 1978.
92. Castagne M, Takai Y, Kaibuchi, et al: Direct activation of calcium activated, phospholipid dependent protein kinase by tumor promoting phorbol esters. *J Biol Chem* **257:**7847–7853, 1982.
93. Wilson E, Olcott MC, Bell RM, et al: Inhibition of the oxidative burst in human neutrophils by sphingoid long chain basis. *J Biol Chem* **261:**12616–12623, 1986.
94. Gerard C, McPhail LC, Marfat A, et al: Role of protein kinases in stimulation of human polymorphonuclear leukocyte oxidative metabolism by various agonists. Differential effects of a novel protein kinase inhibitor. *J Clin Invest* **77:**61–65, 1986.
95. Cox CC, Dougherty RW, Ganong BR, et al: Differential stimulation of the respiratory burst and lysosomal enzyme secretion in human polymorphonuclear leukocytes by synthetic diacylglycerols. *J Immunol* **136:**4611–4616, 1986.
96. McPhail L, Henson P, Johnston R: Respiratory burst enzyme in human neutrophils. *J Clin Invest* **67:**710–716, 1981.
97. McPhail LC, Henson PM, Johnston RB: Respiratory burst enzyme in human neutrophils. Evidence for multiple mechanisms of action. *J Clin Invest* **67:**710–716, 1981.
98. Williams JD, Lee TH, Lewis RA, et al: Intracellular retention of the 5-lipoxygenase pathway product leukotriene B_4, by human neutrophils activated with unopsonized zymosan. *J Immunol* **134:**2624–2630, 1985.
99. Fujita I, Irita K, Takeshige K, et al: Diacylglycerol, 1-oleyl 2 acetyl-glycerol, stimulates superoxide-generation from human neutrophils. *Biochem Biophys Res Commun* **120:**318–324, 1984.
100. Wolfson M, McPhail LC, Nasrallah VN, et al: Phorbol myristate acetate mediates redistribution of protein kinase C in human neutrophils: Potential role in the activation of the respiratory burst enzyme. *J Immunol* **135:**2057–2062, 1985.
101. Nathan CF, Bruckner LM, Silverstein SC, et al: Extracellular cytolysis by activated macrophages and granulocytes. II. Hydrogen peroxide as a mediator of cytotoxicity. *J Exp Med* **149:**84–100, 1979.

102. Clark RA, Klebanoff SJ: Chemotactic factor inactivation by the myeloperoxidase–hydrogen peroxide–halide system: An inflammatory control mechanism. *J Clin Invest* **64**:913–920, 1979.
103. Pryor WA: Free radical pathology. *Chem Eng News* **49**:34–51, 1971.
104. Zs.Nagy I, Pieri C, Givli C, et al: Effects of centrophenoxine on the monovalent electrolyte contents of the large brain cortical cells of old rats. *Gerontology* **25**:94–102, 1979.
105. Semsei I, Zs.Nagy I: Effects of ionic strength on the activity of superoxide dismutase in vitro. *Arch Gerontol Geriatr* **3**:287–295, 1984.
106. Vanella A, Geremia E, D'Urso, et al: Superoxide dismutase activities in aging rat brain. *Gerontology* **28**:108–113, 1982.
107. Tolmasoff JM, Ono T, Cutler RG: Superoxide dismutase correlation with life-span and specific metabolic rate in primate species. *Proc Natl Acad Sci USA* **77**:2777–2781, 1980.
108. Cufler RG: Superoxide dismutase, longevity and opecific metabolic rate. *Gerontology* **29**:113–120, 1983.

21

The Respiratory Burst and the Onset of Human Labor, Preterm Labor, and Premature Rupture of the Membranes

CURTIS L. CETRULO, ANTHONY J. SBARRA, ANJAN CHAUDHURY, MARK T. PETERS, CHARLES J. LOCKWOOD, GAIL THOMAS, JOSEPH L. KENNEDY, Jr., FARID LOUIS, CHRIS J. SHAKR, and HORN-WEN WANG

1. INTRODUCTION

The respiratory burst is implicated in many different and diverse areas of biology. This should not be surprising. If phagocytosis and/or pinocytosis are truly the mechanisms that cells use for eating and drinking, one should expect no boundaries. In the field of obstetrics, the respiratory burst may be implicated in the onset of human labor, preterm labor, and premature rupture of the fetal membranes. The common denominator in all these events could well be brought about by phagocytosis or a phagocytosis-like phenomenon and the accompanying respiratory burst.

For the past several years, our laboratory has attempted to develop the concept that the interaction of particulate and/or nonparticulate material with fetal membrane cells (i.e., amnion, decidua) can result in the release of their lysosomal enzymes, in particular, phospholipase A_2 and peroxidase. Either of these lysosomal

CURTIS L. CETRULO, ANJAN CHAUDHURY, MARK T. PETERS, CHARLES J. LOCKWOOD, CHRIS J. SHAKER, and HORN-WEN WANG • Department of Maternal Fetal Medicine, St. Margaret's Hospital for Women, Boston, Massachusetts 02125 and Department of Obstetrics and Gynecology, Tufts University School of Medicine, Boston, Massachusetts 02125 ANTHONY J. SBARRA and GAIL THOMAS • Department of Medical Research and Laboratories, St. Margaret's Hospital for Women, Boston, Massachusetts 02125 and Department of Obstetrics and Gynecology, Tufts University School of Medicine, Boston, Massachusetts 02125 JOSEPH L. KENNEDY, Jr. • Department of Pediatrics, St. Margaret's Hospital for Women, Boston, Massachusetts 02125 FARID LOUIS • Department of Pathology, St. Margaret's Hospital for Women, Boston, Massachusetts 02125.

enzymes may contribute to the initiation of normal and preterm labor. Peroxidase may also promote rupturing of the fetal membranes prematurely. This chapter reviews our findings in each of these areas.

2. PHAGOCYTOSIS AND THE ONSET OF HUMAN LABOR

2.1. Experimental Model

Lysosomal enzymes have been shown to play critical roles in two apparently unrelated physiological events. First, it has been demonstrated that the lysosomal enzyme myeloperoxidase (MPO), present in polymorphonuclear leukocytes (PMN), is activated as a result of phagocytosis. It, along with H_2O_2 and a halide, principally chloride, form a potent antimicrobial system.[1,2] This enzyme appears to alter fetal membrane integrity and predispose to rupture. Second, the lysosomal enzyme, phospholipase A_2, present in fetal membranes,[3] bacteria,[4] and PMN,[5] catalyzes the hydrolysis of glycerophospholipids, releasing the polyunsaturated fatty acid, arachidonic acid, from the *sn*-2 position of the glycerol moiety.[6] This liberated arachidonic acid is the obligate percursor of prostaglandins[7] and the limiting factor for biosynthesis of prostaglandin E_2 (PGE_2) and $PGF_{2\alpha}$.[7] Production of these prostaglandins will, in turn, initiate myometrial contractions and cervical changes, leading to labor.

The activation of MPO in the PMN is attributed to the interaction of particulate and/or certain nonparticulate materials with the cell. For example, surfactants and fatty acids have been shown to stimulate PMN metabolically and to release lysosomal enzymes.[8] It was recently shown that release of lysosomal enzymes as monitored by *N*-acetylglucosaminidase release is approximately threefold greater from amnion tissue collected from women in labor than that of women not in labor.[9] This finding suggests that alteration of fetal membranes, especially lysis of amnion lysosomes, is an early event in the biochemical reactions that lead to myometrial contractions and onset of labor.

The precise mechanisms that lead to the activation of fetal membrane lysosomes, especially in the amnion, are unknown but are thought to be the critical step in the initiation of labor.[10] In this section, we develop the concept that amnion lysosomes are activated as a result of the interaction of amniotic fluid surfactant(s) with amnion cells. In this regard, we have developed an in vitro experimental model that will permit us to study amnion lysosomal enzyme release under controlled conditions. A layer of human amnion membrane mounted on a specially designed reaction vessel serves as the reaction surface. The rate of release of the lysosomal marker enzyme, *N*-acetylglucosaminidase, was measured both in the absence and in the presence of particulate material, amniotic fluid centrifuged amniotic fluid, lecithin, and lysolecithin.

2.2. Interaction of Particulate Material and Surfactants with Amnion Cells

This experimental model has permitted the study of amnion lysosomal enzyme activity under controlled conditions. A monolayer of amnion cells was prepared by mounting the membrane on a specimen container that was open at both ends. The amnion cell surface served as the bottom of the vessel and the reaction site. Using this system, we have demonstrated that amnion lysosomal enzyme release can be induced by a phagocytosis-like phenomenon: The addition of polystyrene latex spherules to amniotic membrane incubating in pseudoamniotic fluid[11] results in activation of lysosomal enzymes, as reflected in the release of the lysosomal marker enzyme, *N*-acetylglucosaminidase.[12] This release is significantly greater than that noted from amnion tissue incubated in the absence of latex particles. Interestingly, this increase is noted only with amnions of placentas derived from patients who have not experienced labor. Amnions collected from the placentas of patients who have experienced labor and delivered vaginally do not show a stimulated release of lysosomal enzyme in the presence of particles. However, the release of the enzyme from amnions collected from placentas of subjects who have experienced labor is significally greater than what is noted in amnions collected from placentas of subjects not having experienced labor and delivered by cesarean section.

Using a somewhat different in vitro model, Schwartz and co-workers[9] have also demonstrated a significantly higher release of *N*-acetylglucosaminidase from amnions obtained from the placentas of women in labor as compared with amnions collected from placentas of women not in labor. It would appear that once amnion cells have been subjected to labor, the release of enzyme has been maximally activated and no further stimulation could be achieved by the addition of particles. These observations indicate that amnion cells can be stimulated or activated by particles to release lysosomal enzyme to a comparable level of amnion cells that have been under the influence of labor. Similar results were obtained when particles were added at 30 min, wherein each reaction vessel was used as its own control. The triggering agent(s) in this in vitro model appears to be the interaction of particles with amnion cells. This observation is compatible with what has been noted when particles interact with PMN.[1,2]

Furthermore, it has been shown that the PMN glycolytic pathway inhibitor, iodoactate, inhibits particle engulfment and the stimulating effects of particle addition, whereas the respiratory inhibitor, sodium azide, does not.[1,2] In a similar fashion, iodoacetate completely abolishes the stimulating effect of particles on amnion cells, indicating that glycolytic energy is necessary for the observed increased activity. The small inhibitory effect observed with azide suggests a possible, although minor, role for oxidative energy.

The surfactants are also able to interact with PMN, resulting in increased oxidative metabolic activity and enzyme release. Since amnion cells appear to interact with particulate material and release lysosomal enzymes as PMN do, the ability of amnion cells to interact with surfactant-rich amniotic fluid collected from term pregnancies was tested. A considerable release of *N*-acetylglucosaminidase, a typical lysosomal marker enzyme, from amnion cells was noted when term amniotic fluid was incubated with chorioamniotic membranes. However, when amniotic fluid was previously centrifuged at high speed to remove colloidal surfactants and other particulate materials and then tested, less release was noted. Interestingly, the addition of latex particles to centrifuged amniotic fluid resulted in an increased release of the enzyme. However, particulate material did not affect enzyme release when added to uncentrifuged amniotic fluid. This finding suggests that maximal release had already occurred. These observations would suggest that surfactants and/or other materials removed by centrifugation were able to interact with the cells and activate the release of lysosomal enzymes. Attempts were made to determine whether different surfactants possessed varying potencies for the release of lysosomal enzymes from amnion. Lecithin and lysolecithin were added to amnions in equal concentrations, but only lysolecithin was found to be active. This is not surprising, as lysolecithin is known to be a membrane perturber.

2.3. Release and Function of Amnion Phospholipase A_2: Initiation of Labor at Term

The hypothesis that lysosomal enzymes of the chorioamniotic membranes play a major role in initiating human labor has been extensively advanced by the work of MacDonald et al.[10] These investigators proposed that the release of the lysosomal enzyme, phopholipase A_2, from the amniocyte into the extracellular space of the membrane results in the liberation of fatty acids from the *sn*-2 position of glycerophospholipids. The enzyme appears to have the highest specificity for phosphatidylethonalanine (PE), the primary phospholipid in fetal membranes, enriched with arachidonic acid. This high degree of enzyme specificity is not exhibited by liver phospholipase A_2.[13] One might postulate that amnion lysosomal phospholipase A_2 may have the unique function of catalyzing the release of membrane-bound arachidonic acid. Interestingly, PE has been detected in higher concentrations in term amniotic fluids than from amniotic fluids collected from early pregnancies.[14]

The increased release of phospholipase A_2 from lysosomes not only leads to direct initiation of labor via prostaglandin synthesis but to a number of secondary reactions promoting further lysis of the lysosomes as well. All the products of phospholipase A_2 activity (lysolecithin, lysoethanolomine, arachidonic acid, and prostaglandins) are potent lytic agents resulting in a positive-feedback chain reaction. Thus, even a small initial increase in phospholipase A_2 activity could result in an exponential burst in lysosome release and prostaglandin synthesis, leading to labor and spontaneous delivery.

From the above information, detailed by Sbarra et al.,[15] it may be postulated that the onset of normal labor may be initiated by substance(s) in term amniotic fluids. While the identity of these substance(s) remains unknown, the possibility that they are surfactants is suggested by the fact that amnion cells, analogous to PMN, released increased amounts of the lysosomal enzymes when stimulated by surfactants. Teleologically, this would be an attractive hypothesis allowing the fetus to initiate labor in the mother by way of surfactant production. When pulmonary maturity is achieved, the fetal alveolar cells liberate surfactants in the amniotic fluid, which in turn activate the amnion lysosomal enzyme, phospholipase A_2. This enzyme is then able to catalyze the hydrolysis of glycerophospholipids releasing arachidonic acid. Arachidonic acid, the obligate precursor of prostaglandins would then participate in the biosynthesis of PGE_2 and $PGF_{2\alpha}$. Myometrial contractions, cervical change, and the onset of labor would result. Since normal labor generally occurs after the surge of lecithin in amniotic fluid (i.e., 35–36 weeks), it would appear that a critical level of substances (i.e., surfactants) would need to be reached before enough phospholipase A_2 is released so that sufficient net synthesis of prostaglandins could occur in order for labor to be initiated.

3. PHAGOCYTOSIS AND PREMATURE RUPTURE OF THE MEMBRANES

3.1. Etiology of PROM

Premature rupture of the membranes (PROM)—the rupture of the fetal membranes prior to the onset of labor—is a major problem in obstetrics and neonatology. While its etiology is not entirely clear, an infectious process has long been suspected. Recent data appear to support this concept. It has been reported that acute inflammation of the placental membranes is twice as common when membranes rupture within 4 hr before labor than when they rupture after the onset of labor.[16] Furthermore, in an investigation of cord immunoglobulins in neonates delivered after preterm PROM, two patterns of response to presumed infection can be identified, suggesting an infectious cause for PROM. If the interval between rupture and the onset of labor is considered, there was both clinical and immunological evidence of an elevated IgA peak within 12 hr of rupture in 142 of 227 (63%) cord bloods. A second peak, after 72 hr of rupture, was also identified. The first peak suggests that the infection was present before the rupture and might have been the direct cause of the rupture.[17] In another study, it was shown that patients with PROM were more likely to have anaerobic bacteria isolated in endocervical cultures than were women without PROM.[18] These data, although suggestive of an infectious process in association with PROM, do not suggest the mechanism(s) by which membranes are ruptured.

The lysosomal enzyme peroxidase, first described in phagocytes and later in macrophages, constitutes a potent antimicrobial system when combined with H_2O_2 and a halide.[2,19] This system is also cytotoxic and tumorcidal.[2] Furthermore, peroxidase, in the presence of H_2O_2 and a halide, can oxidatively cleave peptide bonds and thus could lead to the breakdown of membranes.[20] This cytotoxic system could conceivably be activated as a result of phagocytosis and could function to attack fetal membranes and lead to PROM. Some evidence to support this hypothesis exists. It is known that cervical mucus and endometrial tissue display peroxidase activity. Lysosomes and lysosomal enzymes found in amnion epithelial cells[3] should suggest that these cells should have peroxidase activity. It is also known that many different bacteria produce H_2O_2. From the above, it appears that the reactants of the peroxidase–H_2O_2–halide system are present either in the proximity of or in the fetal membranes, or both. They potentially could have a role in PROM, especially when activated by phagocytosis and infection.[21]

3.2. Role of Peroxidase in PROM

Peroxidase activity has been demonstrated in fetal membranes and placental macrophages. The total increased enzymatic activity was noted only in patients following labor and could be due either to increased synthesis, to increased release, or to neutralization of an inhibitor. The highest increase is found in the decidua, followed by chorion and amnion. Labor is associated with increased prostaglandin concentrations, which may be a consequence of the release of lysosomal enzymes, in particular, phospholipase A_2. Arachadonic acid, the rate-limiting substrate for prostaglandin synthesis, is presumably made available by the action of phospholipase A_2 on phospholipids. It is therefore reasonable to assume that if this lysosomal enzyme is released, other lysosomal enzymes such as peroxidase would also be released.

The placental macrophages from the patients without labor have significantly more peroxidase activity than do the other three tissues combined. This is not so in the tissues from patients following labor. In the latter, the decidua appears to have more activity than the amnion, chorion, and placental macrophages combined. Furthermore, the macrophages in this instance show less activity than do the macrophages of patients without labor. These observations suggest that along with an increase in enzymatic activity, a mobilization of enzyme has also occurred, possibly due to labor, resulting in depletion in the amnion and chorion. It may be postulated that labor has in some way exposed the membranes to peroxidase action in vivo, so that additional activity of the peroxidase–H_2O_2–halide system can no longer be demonstrated in vitro. It is tempting to speculate that the increase movement of activity in the direction of the amnion from the placental macrophages has physiological implications. For instance, the peroxidase could act in concert with H_2O_2 and a halide as a cytotoxic agent, thereby weakening the membranes and leading to actual rupture. A significant increase in protein release from chorioamnions collected from deliveries without labor and incubated with peroxidase–H_2O_2–halide system is noted as early as 6 hr of incubation and increases as incubation time increased up to 120 hr as compared with controls. No significant increase between experimental and controls is noted in membranes collected from deliveries without labor with either spontaneous or surgically ruptured membranes. This would again suggest that labor has an effect on fetal membranes, increasing the susceptibility of the membrane collagen to peroxidase attack. The physiological significance of this peroxidase gradient phenomenon is unclear.

In an effort to determine the nature of the peroxidase noted in fetal membranes and placental macrophages, the decarboxylase activity of the enzyme was compared with the peroxidase obtained from human PMN. We noted that the enzyme has minimal decarboxylase activity compared with the PMN peroxidase. This finding would suggest that halides, other than chloride, would be more active in the cytotoxic system.

3.3. Surfactant and Fetal Membrane Interactions

Surfactants can be effective in initiating the release of lysosomal enzymes. Some of these released enzymes (i.e., phospholipases) can cleave arachadonic acid from phospholipids. Since the resultant lysocompounds formed, in particular lysolecithins, are known to disrupt membranes, we examined the effect of lysolecithin on fetal membranes. We did not notice any increase in protein release with lysolecithin, compared with controls. However, we noted a significant decrease in bursting pressure required for the lysolecithin treated membranes when compared with controls (i.e., incubation in the absence of added lysolecithin). This finding is of interest and may be of physiological significance. In addition, since amniotic fluid collected from term pregnancies contains abundant amounts of lecithin, we examined the effect of phospholipase A_2 and amniotic fluid on the bursting pressure of fetal membranes so treated. As would be expected, the decreased bursting pressures noted suggest that the phospholipase A_2 reacted with the amniotic fluid lecithin and converted it into lysolecithin. From these data, it is tempting to speculate that both the peroxidase–H_2O_2–halide system and lysolecithin and/or lysolecithin-like compounds have physiological significance. Perhaps both systems working in concert are involved with the weakening and bursting of fetal membranes.

Premature rupture could also be associated with an infectious process, with the infectious agent acting as the trigger.[21] Interestingly, Bejar et al.[4] showed that microorganisms frequently isolated from the vaginal–cervical area also possess phospholipase A_2 activity. Alternatively, maternal mediators of infection, such as interleukin-1 (IL-1) and tumor necrosis factor, may facilitate lysis of lysosomes with the release of both phospholipase A_2 and peroxidases.

4. EFFECT OF BACTERIAL GROWTH IN FETAL MEMBRANES

4.1. Experimental Model

In an effort to study the role of infection in PROM more directly, we examined the effect of bacterial growth on the bursting pressure of chorioamniotic membranes. We have shown that organisms suspended in tissue culture medium and allowed to grow overnight in reaction vessels will weaken the membranes, as evidenced by a significantly decreased bursting pressure. This was true with either a gram-positive–catalase-negative (group B streptococcus) (GBS) or a gram-negative–catalase-positive organism (*Escherichia coli*). When the organisms were allowed to grow in the vessels containing bacteriologic medium (brain heart infusion), however, the membranes are not weakened. The bursting pressure of the chorioamniotic membranes in control and in experimental vessels is similar.

4.2. Nature of Effect

At least two possible explanations for these findings may be offered. First, for weakening to occur, it would appear that organisms and metabolically active membranes must both be present. It is known that membranes in tissue culture medium remain viable for 2–3 weeks.[22] In bacteriologic medium, metabolic activity and viability of membranes would not be expected to be optimally preserved. Second, bacteria are able to grow in brain heart infusion (BHI) but are unable to weaken the membranes. This would suggest that the medium is in some way inhibiting the mechanism(s) responsible for rupture. Possibly the engulfment and/or interaction of bacteria with decidual cells will activate the peroxidase and, in conjunction with H_2O_2 and a halide, this cytotoxic system would become operative. The H_2O_2 source would depend on the organism involved. *E. coli* is catalase positive, and any H_2O_2 produced would be expected to be destroyed. The organism, however, does weaken membranes when allowed to grow in vessels containing tissue culture medium. The interaction of the organisms with fetal membrane cells could result in H_2O_2 production by membrane cells and activation of the cytotoxic system. With GBS (catalase-negative), H_2O_2 would be expected to be produced and not destroyed. In this case, the H_2O_2 could be postulated to be produced by either the organisms or the fetal membrane cells, or both. It is known that the peroxidase system is inhibited by a multitude of substances. Azide, cyanide, 3-amino-1,2,4-triazole, thiocyanate, methemazole, thiorea, and sulfonamindes are some specific inhibitors. Nonspecifically, the system has been reported to be inhibited by protein at high concentrations.[23]

The addition of 20–60% BHI to tissue culture medium inhibited the reduced bursting pressures noted when either *E. coli* or GBS was allowed to grow in the tissue culture medium without added BHI. It should be noted that, as expected, the added BHI does not inhibit bacterial growth. The inhibition noted could be at the peroxidase level, due to the relatively high protein concentration and/or the thio compounds present in BHI. If the peroxidase–H_2O_2–halide system is inhibited by BHI in the reaction vessels, the purified system should and is found to be inhibited in vitro. This would suggest that although microorganisms can grow in reaction vessels containing either medium, the growth of organisms in BHI alone is not sufficient to cause a weakening of the membranes; it also suggests that it is the induction of lysosomal peroxidase that damages the membrane.[24]

Further studies showed that no GBS extracellular stable product(s) in sufficient concentration could affect bursting pressures. Also, incubation of heat-killed bacteria with membranes does not result in increased weakening. On the other hand, it has been suggested that the frailty of fetal membranes may be related to the enzymatic depolymerization of collagen.[25] A recent report supports this concept. It has been shown that collagenase-producing bacteria can reduce the tensile strength of chorioamniotic membranes. However, GBS, which has been shown to weaken membranes, has been reported to produce elastase.[26] Galask et al.[27] described the attachment of *E. coli* and GBS to chorioamniotic membranes. They noted that GBS had a greater capacity to attach and invade than *E. coli*.[27]

5. SUMMARY AND CONCLUDING REMARKS

The concept that phagocytosis and/or a phagocytosis-like phenomenon is the common denominator between the onset of human term labor and preterm labor is proposed. In addition, it is reasonable to consider that an infectious process is also involved in the genesis of PROM. It is apparent that organisms can interact with fetal membranes, weaken them, and with minimum applied pressure, cause them to rupture. Furthermore, it appears that at least two different mechanisms may be operative. Certainly the report that some collagenase producing bacteria are able to reduce the tensile strength of chorioamniotic membrane is important and must be

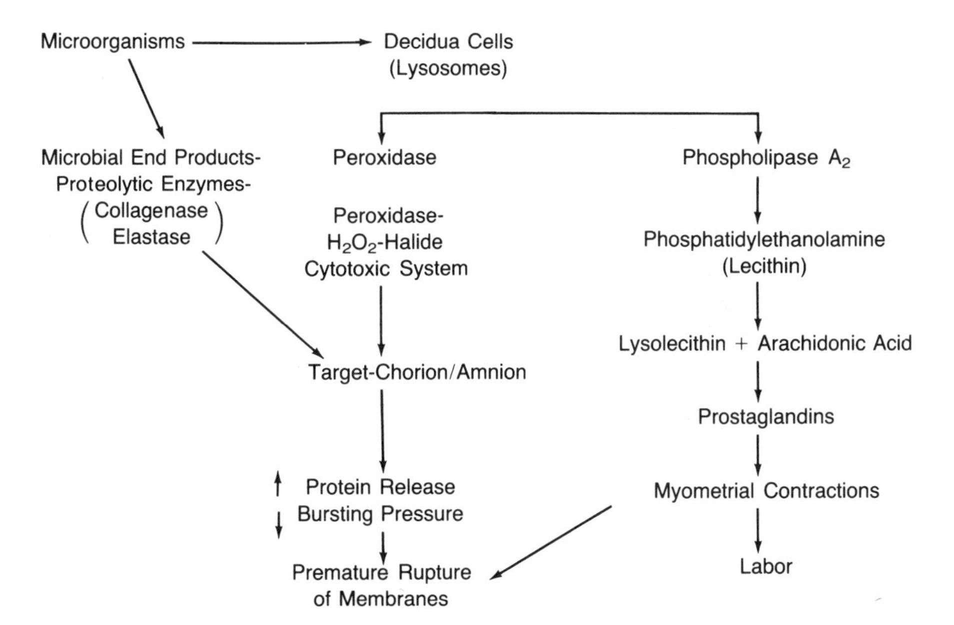

FIGURE 1. Postulated sequence of events leading to premature rupture of the membranes (PROM) and labor.

fully explored. However, our data presently suggest that the membranes themselves are participating in the breakdown. A summary of these events are diagrammatically presented in the figure.

A number of questions still remain unanswered at the present time, especially in regard to PROM. For example, since amniotic fluids have antibacterial activity, does PROM result only from an ascending route? What is the role of cervical mucus, known to contain peroxidase in PROM? Is the mechanism of rupture the same in all cases and is a common denominator in all cases any microorganism(s) or is it a specific one? Is there a dose effect? Can we hypothesize that patients with PROM are already either colonized, or sub-clinically or clinically infected? Can large inocula of organisms overcome the antibacterial effect of amniotic fluids and cause rupture? Can inhibitors of the peroxidase-H_2O_2-halide system and/or proteolytic enzymes (i.e., collagenese) provide a vantage point from which PROM prevention strategies may be developed?

ACKNOWLEDGMENTS. Material for these studies is reprinted with permission from the American College of Obstetricians and Gynecologists (Obstetrics and Gynecology 70:107–110, 1987) and The C. V. Mosby Company (Am J. Obs. & Gyn 146:622, 1983 and 153:38, 1985).

REFERENCES

1. Klebanoff, SJ: Myeloperoxidase-mediated cytotoxic systems, in Sbarra AJ, Strauss R (eds): *The Reticuloendothetial System.* New York, Plenum, 1980, vol 2: *Biochemistry and Metabolism,* pp 279–308.
2. Sbarra AJ, Selvaraj RJ, Paul B, et al: in Altura BM, Saba TM (eds): *Pathophysiology of the Reticuloendothelial System.* New York, Raven, 1981, pp 9–30.
3. Schwartz BE, Schultz FM, MacDonald PC, Johnson JM: Initiation of human parturition. IV. Demonstration of phospholipase A_2 in the lysosomes of human fetal membranes. *Am J Obstet Gynecol* **125:**1089–1092, 1976.
4. Bejar R, Curbelo V, Davis C, Gluck L: Premature labor. II. Bacterial sources of phospholipase. *Obstet Gynecol* **57:**479–482, 1981.
5. Elsbach P, Weiss J: Lipid metabolism by phagocytic cells, in Sbarra AJ, Strauss R (eds): *The Reticuloendothelial System.* New York, Plenum, 1980, vol 2: *Biochemistry and Metabolism,* pp 91–119.
6. MacDonald PC, Schultz FM, Duenhoelter JH, et al: Initiation of human parturition. I. Mechanism of action of arachidonic acid. *Obstet Gynecol* **44:**629–636, 1974.
7. Bergstrom S, Danielsson H, Samuelsson B: The enzymatic formation of prostaglandin E_2 from arachidonic acid prostaglandins and related factors. *Biochem Biophys Acta* **90:**207–210, 1964.
8. Kakinuma K: Effects of fatty acids on the oxidative metabolism of leukocytes. *Biochem Biophys Acta* **348:**76–85, 1974.
9. Schwartz BE, MacDonald PC, Johnson JM: Initiation of Human Parturition. XI. Lysosomal enzyme release *in vitro* from amnions obtained from laboring and nonlaboring women. *Am J Obstet Gynecol* **137:**21–24, 1984.
10. MacDonald PC, Porter JC, Schwartz BE, Johnson JM: Initiation of parturition in the human female. *Semin Perinatol* **2:**273–286, 1978.

11. Schwartz AL, Forster CS, Smith PA, Liggins GC: Human amnion metabolism. I. *In vitro* maintenance. *Am J Obstet Gynecol* **127:**470–474, 1977.
12. Levvy GA, Conchie J: Mammalian glycosidases and their inhibition by aldonolactones. *Methods Enzymol* **8:**580–585, 1966.
13. Okasaki T, Okita JR, MacDonald PC, Johnson JM: Initiation of human parturition. X. Substrate specificity of phosphalipase A_2 in human fetal membranes. *Am J Obstet Gynecol* **130:**432–438, 1978.
14. Kulovich MV, Hallman M, Gluck L: The lung profile. I. Normal pregnancy. *Am J Obstet Gynecol* **135:**57–63, 1979.
15. Sbarra AJ, Selvaraj RJ, Cetrulo CL, et al: Phagocytosis and onset of human labor. *Am J Obstet Gynecol* **146:**622–629, 1983.
16. Naeye RL: Factors that predispose to premature rupture of fetal membranes. *Obstet Gynecol* **60:**93–98, 1982.
17. Cederquist LL, Zerroudakis IA, Ewoll LC, Litwin SD: The relationship between prematurely ruptured membranes and fetal immunoglobin production. *Am J Obstet Gynecol* **134:**784–788.
18. Creatsas G, Pavlatos M, Lolis D, et al: Bacterial contamination of the cervix and premature rupture of membranes. *Am J Obstet Gynecol* **139:**522–525, 1981.
19. Klebanoff SJ, Clark RA, Rosen H: Myeloperoxidase-mediated cytotoxicity. in Schult J, Ahmad J (eds): *Cancer Enzymology*. Academic Press, NY, pp. 267–288, 1976.
20. Selvaraj RJ, Paul BB, Strauss RR, Sbarra AJ: Peptide cleavage and oxidative decarboxylation by the MPO–H_2O_2–halide system. *Infect Immun* **9:**255–260, 1974.
21. Sbarra AJ, Selvaraj RJ, Cetrulo CL, et al: Infection and Phagocytosis as possible mechanisms of rupture in premature rupture of the membrane. *Am J Obstet Gynecol* **153:**38–43, 1985.
22. Burgos H, Page FW: The maintenance of human amniotic membranes in culture. *Br J Obstet Gynecol* **88:**284–300.
23. Klebanoff SJ, Clark RA: *The Neutrophil: Function and Clinical Disorders*. Amsterdam, North-Holland, 1978.
24. Sbarra AJ, Thomas GT, Cetrulo CL, et al: Effect of bacterial growth in the bursting pressure of fetal membranes in vitro. *Obstet Gynecol* **70:**107–110, 1987.
25. Artal R, Burgeson RE, Hobel CJ, et al: An *in vitro* model for the study of enzymatically mediated biochemical changes in the chorioamniotic membranes. *Am J Obstet Gynecol* **133:**656–659, 1979.
26. McGregor JA, Lawellin D, Franco-Buff A, et al: Protease production by microorganisms associated with reproductive tract infection. *Am J Obstet Gynecol* **154:**109–114, 1986.
27. Galask RP, Varner MW, Petzold CR, et al: Bacterial attachment to the chorioamniotic membranes. *Am J Obstet Gynecol* **148:**915–928, 1984.

22

Tuftsin

Biochemical and Biological Aspects

VICTOR A. NAJJAR

1. INTRODUCTION

Tuftsin is a tetrapeptide (Thr-Lys-Pro-Arg) that is a part of specific leukophilic γ-globulin termed leukokinin.[1] It represents residues 289–292 of the heavy chain of γ-globulin.

Two enzymes are responsible for releasing tuftsin as the free and active tetrapeptide. A splenic enzyme cleaves at the arginyl residue. In the absence of a functional spleen, a deficient state is produced.[2–10] The other, a membrane enzyme, cleaves at the α-amino group of threonine.[11,12]

Tuftsin in the underivatized state stimulates all recognized functions of phagocytic cells. This results in augmentation of motility, phagocytosis and pinocytosis.[1,8,13] This activation leads to increased bactericidal activity over and above the direct effect of tuftsin on the bacterial organism[14] both in vitro and in vivo.[15–17] Despite the fact that tuftsin at a high concentration of 69 μg/ml exerts a strong direct bactericidal effect on many organisms, including *Pneumococcus,*[14] it is unlikely that it plays a role in its mechanism of action, given the low serum levels of tuftsin of about 250–500 μg/ml. However, since its toxicity is almost nonexistent, one cannot rule out the possibility that at high doses it might well have therapeutic application in infection.

Tuftsin also exerts strong immunogenic stimulation on T cells that are triggered by more favorable processing and presentation of antigen by adherent macrophages.[18] There is also considerable augmentation of antibody-forming cells, amounting to more than threefold of the control.[15]

Tuftsin has also been shown to rejuvenate age-depressed macrophage and T-

VICTOR A. NAJJAR • Department of Molecular Biology and Microbiology, Tufts University School of Medicine, Boston, Massachusetts 02111.

cell cytolysis in mice, to a remarkable degree. This results in a greater cytolytic effect fully equal to that of similar cells obtained from young mice.[15]

The most interesting effect of tuftsin is its antineoplastic activity in mice[19–25] and in humans.[19–21] In addition, a preliminary report on a patient with acquired immune-deficiency syndrome (AIDS) noted a remarkably favorable response to tuftsin IV injection.[26]

Of practical interest is the definite increase in the production of superoxide anion, hydrogen peroxide, and presumably hydroxyl radical in tuftsin-treated macrophages.[27–29] This along with the augmentation in vivo and in vitro of tumor necrosis factor by tuftsin[30] may play a role in the increased killing activity of macrophage toward bacteria as well as toward neoplastic cells.[30] The thrust of this chapter is to give the reader a general view of the properties and biological activities of tuftsin.

2. GENERAL CHARACTERISTICS OF TUFTSIN

Tuftsin forms a part of a carrier protein leukokinin, a leukophilic γ-globulin. It is present in close proximity to the attachment of the carbohydrate portion. The only active form of tuftsin is the free underivatized tetrapeptide. Any modification, such as removing the threonine residue to yield Lys-Pro-Arg or replacing a residue by another, such as Thr-Glu-Pro-Arg, to name a few, results in an inactive peptide that actually inhibits the stimulation effect of tuftsin.[2] However, it does not inhibit basal levels of phagocytosis or motility.[10,11]

One of the newly discovered characteristics of tuftsin is its participation in the binding of the first component of the complement (C1) to antigen-antibody aggregates. It also inhibits C1-mediated hemolysis.[31]

The two enzymes that release tuftsin from leukokinin have not yet been fully characterized. The enzyme present in the spleen, tuftsin–endocarboxypeptidase, cleaves tuftsin at its carboxy-terminal arginyl bond to form leukokinin-S (S for spleen). The enzyme is a trypsinlike enzyme, but with an apparent high specificity. It acts only on a specific fraction of γ-globulin that stimulates leukophilic granulocytes, monocytes, and macrophages.[32,33] It does not cleave tuftsin from any other IgG fraction, even though it also contains the tetrapeptide.[11,12,34]

The second, a membrane enzyme leukokininase, has its active site exposed to the external environment. It acts specifically on leukokinin-S to cleave tuftsin at its amino-terminal threonine. It has a sharp pH optimum at 6.7 in 0.1 M phosphate buffer.[11]

The two enzymatic steps that release the free and active tetrapeptide are illustrated as follows:

```
H—Ala-Lys-Thr-Lys-Pro-Arg-Glu-Gln—OH
          ↑                  ↑
(Second cleavage      (First cleavage
leukokininase)        tuftsin-endocarboxypeptidase)
```

Tuftsin is broken down by a series of enzymes that are present in the cytosol of rabbit peritoneal polymorphonuclear leukocytes (PMN). The most active cytosol enzyme is an aminopeptide. The cluster of such enzymes yields Lys-Pro-Arg, Lys-Pro, Pro-Arg, Thr, Lys, Pro, and Arg.[35] An active membrane aminopeptidase has been described in guinea pig PMN neutrophils.[36]

Tuftsin is a specific biological entity endowed with the unique function of stimulating the phagocytic cell in all its known functions.[24,37] This is supported by several observations:

1. There are enzymes that yield free tuftsin from the carrier molecule leukokinin.
2. Tuftsin is active only in the free underivatized state.
3. Any change in its structure yields an inactive, poorly active or often an inhibitory analogue.
4. A human congenital tuftsin deficiency syndrome has been found to exhibit severe and recurrent infections in which the mutant peptide proved inhibitory in all patients with the deficiency that have so far been studied in the United States as well as in those reported from Japan.[38] In one such case, we isolated and identified the structure of the mutant peptide, Thr-Glu-Pro-Arg.[37] This was indeed strongly inhibitory to tuftsin activity, as was determined before its structure was known.[2–9] Here, the triplets coding for lysine AAA or AAG through a base change coded for glutamate, GAA or GAG.
5. Finally, patients who have undergone elective splenectomy for one reason or another, as can be expected, also have tuftsin deficiency with its attendant symptoms of frequent severe and at times fatal infections.[10,39–41,87]

3. PHAGOCYTOSIS

It was the property of stimulation of phagocytosis that made the isolation of tuftsin possible.[1] At first, we used the whole cytophilic leukokinin-S to assess the stimulation of phagocytosis only to discover, by kinetic studies, that the whole carrier molecule was not necessary. Upon binding to the phagocytic cell, leukokinin-S is acted on by the enzyme leukokininase, a trypsinlike peptidase with a pH optimum of 6.9 that cleaves the H—lysylthreonyl—OH bond between residues 288 and 289. This releases tuftsin, which stimulates phagocytosis to the full extent of leukokinin-S.

The stimulation of phagocytosis has been reported by numerous investigators.[1,10,11,32–34,38,42] The first analogue-synthesized Thr-Lys-Pro-Pro-Arg was a strong inhibitor of tuftsin stimulation for both phagocytosis and motility.[2,12] In a similar manner, tuftsin stimulated the reduction of nitroblue tetrazolium.[10]

Pinocytosis was assessed by tuftsin stimulation of the uptake of radiolabeled bovine serum albumin (BSA) by rabbit peritoneal granulocytes. No other cell was susceptible to this stimulation, such as L1210, $3T_6$ and $3T_{12}$ cells.[43]

4. MOTILITY

Motility was first assessed by the vertical movement of granulocytes from blood buffy coat using glass capillary tubes.[44] It was possible to show that motility was strongly stimulated by tuftsin. The pentapeptide analogue Thr-Lys-Pro-Pro-Arg is a complete tuftsin inhibitor of motility. With tuftsin at a concentration of 5 μM, granulocytes moved 430 nm in 6 hr. At 25 μM, motility attained 500 μm in 3 hr and 700 μm in 6 hr. In a mixture of 8 μM pentapeptide in 25 μM tuftsin, motility was reduced to control levels of 300 μm in 3 hr, a complete inhibitory effect.[12] In addition to these effects tuftsin increases the longevity of human PMN cells.[9]

It is of interest to point out that the pentapeptide did not inhibit basal phatocytic activity or basal motility. It only inhibited the stimulation above basal values. This indicates that following activation by tuftsin, the stimulatory effect follows a pathway that is not identical to that of the basal activity pathway. This is now under investigation.

Motility and chemotaxis were also assessed by kinetic measurements in Boyden chambers, demonstrating defective motility and chemotaxis in Hodgkin disease and systemic lupus erythematosus (SLE). Both motility and chemotaxis were restored by tuftsin treatment of the cells.[45–47] Stimulation of migration and chemotaxis was also reported using a different methodology.[48,48a]

5. TOXICITY OF TUFTSIN

Early studies of tuftsin showed that 2 mg/kg IV in the dog and 22 mg/kg IV in the rat exhibited no alteration in blood pressure, respiratory rate, heart rate, and electrocardiogram (ECG).[9,12] In mice the LD_{50} was found to be 2.4 μg/kg body weight, and no toxicity was found at 2 μg/kg.[19,20]

6. IMMUNOGENIC ACTIVITY

Immunogenic stimulation was investigated by two approaches. The first is to allow mouse macrophages to process a T-dependent antigen in the presence of and absence of tuftsin. The washed macrophages are then allowed to interact with spleen cells. The latter seem to acquire a specific message such that when killed by irradiation and injected into the footpad of mice, the macrophages transmit their message to the lymph cells of the draining lymph nodes. The latter respond to the addition of the original T-dependent antigen by considerable uptake of [^{3}H]thymidine. When tuftsin is present during antigen processing, the uptake amounts to more than sevenfold that of the control, in which tuftsin is omitted.[18]

The second approach is a direct assay of the number of antibody-forming spleen cells after the injection of antigen to mice injected previously with tuftsin. Here T-dependent antigen, TNP-KLH, injection was most effective when injected

7–10 days following administration of tuftsin. The tuftsin-injected mice yielded more than three times the number of antibody-forming spleen cells as compared with control mice not receiving tuftsin. T-independent antigen responded in a more or less similar manner. Administration of the tetrapeptide also augmented cytotoxic activity of spleen cells and enhanced antibody-dependent cell-mediated immunity.[15,42]

Modulation of these various facets of immunity was further emphasized by the demonstration that IV or IP injections of tuftsin in mice, dog, and human subjects consistently resulted in leukocytosis, with no change in the differential white blood cell count (WBC).[20] It was further shown that tuftsin stimulated cytolysis of in vivo-activated macrophages toward P185 and Cloudman S-91 cells.[25]

7. ANTIBACTERIAL EFFECTS OF TUFTSIN

Through activation, tuftsin renders macrophages highly bactericidal.[16,17] This was shown for a number of bacteria: *Escherichia coli, Staphylococcus aureus, Salmonella typhimurium,* and *Listeria monocytogenes.*

A surgical approach was used to assess the antibacterial activity of tuftsin through phagocytic stimulation using what is termed *the mouse septic model.* Here ligation of the mouse cecum was followed by puncture with a standard-gauge needle. A definite favorable outcome following tuftsin injection was obtained. It was concluded that tuftsin can ensure protection against sepsis.[49]

Unexpectedly, it was find that at ~60 μg/ml, tuftsin exerted a potent bactericidal effect in the absence of phagocytic cells.[14] In view of the low (if any) toxicity of tuftsin, it might well be feasible to use it in high doses to treat localized susceptible infection in man.

8. ANTINEOPLASTIC ACTIVITY IN MICE

Several types of murine cancer were used in this study. In every case, there was a positive effect with tuftsin treatment. It was found that the results parallel the immunological status of the animal at the time of cancer cell implantation. In general, a better effect is noted when tuftsin is given a few days (3–5) before the injection of tuftsin (day 0). Further correlation with immunity was obtained with Abelson's murine leukemia virus. This virus must be injected a few hours after birth in order to obtain viral tumors within about 1 week. The immune system of such young animals is considerably underdeveloped. Consequently, no favorable effect of tuftsin was anticipated and none was obtained (V.A. Najjar and A. Bizinkauskas, unpublished data). In other murine cancers, a good effect in syngeneic mice was noted in all the cell lines studied.[25,48]

Murine leukemia L1210 cells were more responsive to tuftsinyltuftsin than to tuftsin.[24] Similarly, murine virus-induced sarcoma was responsive to tuftsin but even more so to Lys[4] tuftsinyltuftsin.[50] Murine melanoma B16/tB also responded well to tuftsin and tuftsinyltuftsin.[24] Rous murine sarcoma and fibrosarcoma cells showed good survival with tuftsin,[25] and the development of Rauscher virus leukemia was also inhibited by the tetrapeptide.[51]

Lymphoma cells CH1 B cell gave erratic results. When tuftsin was given IV, most animals showed no tumor growth for 150 days. However, when the tetrapeptide was given IP, little therapeutic effects was noted.[25]

The antineoplastic activity of tuftsin in humans has been reported in terminal patients who did not respond to conventional therapy. Some good effects were observed in three patients, one with a pharyngeal tumor, another with a non-small cell carcinoma of the lungs, and a third patient with myelogenous leukemia.[19–21]

An interesting result on the effect of tuftsin on virus expression was obtained. Endogenous xenotropic murine retrovirus expression was enhanced in a dose-dependent fashion.[52] *Actinomycin D* inhibited the process. Tuftsin stimulated the synthesis of RNA, DNA, and protein in Kirsten sarcoma virus-transformed BALB/c cells.

9. RESTORATION OF DISEASE-DEPRESSED AND AGE-DEPRESSED FUNCTIONS BY TUFTSIN

Monocyte motility and chemotaxis as well as phagocytosis are depressed in Hodgkin disease and SLE.[45–47] These problems were alleviated when the cells were incubated with tuftsin before testing.

Measured by the leading front, chemotactic activity yielded a migration with an average of 39 μm without tuftsin, and 56 μm with tuftsin. The total distance of migration averaged 586 μm for the control group and 1808 μm for the tuftsin-treated group.

T-cell and macrophage cytolysis are depressed in old age but are regenerated by tuftsin injection.[15,53] In one strain of mice, 22% developed tumors at age 12–16 months, whereas in those given tuftsin starting at the age of 6 month none developed tumors. Similarly, another strain C3H/OuJ $W^x/+$ developed breast cancer—about 100% at age 10 months. However, weekly tuftsin injections, 5 μg/mouse starting at age 3 months, resulted in the complete absence of breast cancer.[54]

10. EFFECT OF TUFTSIN ON CYCLIC NUCLEOTIDE LEVELS

Tuftsin treatment of human PMN leukocytes and mouse macrophages raises the level of cyclic guanosine monophosphate (cGMP) 80–90% and lowers the level of cyclic adenosine monophosphate (cAMP) bx 20–25%. The kinetics of phagocytic stimulation corresponded directly to the cyclic nucleotide levels. Calcium release from preloaded cells is also augmented by tuftsin, which may have some bearing on cyclic nucleotide levels.[55]

11. INDUCTION OF TUMOR NECROSIS FACTOR BY TUFTSIN IN VIVO AND IN VITRO

Induction in vivo of tumor necrosis factor (TNF) by tuftsin was studied in mice C3H/HeJ. Each mouse received only one 25-μg IP injection of tuftsin. The mice were then sacrificed at various times. Samples of serum, unelicited peritoneal cells, and splenic cells were obtained. All cells were cultivated for 5 days in RPMI-1640. Cytolysis by the supernatants of these cells as well as serum were tested with [^{3}H]thymidine-labeled L929 cells.[30]

Induction in vitro of HL60 cells was performed by incubation with various concentrations of tuftsin in the culture medium. Supernatants were assayed for cytolytic activity. Cytolysis stimulation by all supernatants was evident at a tuftsin concentration of 0.1 μg/ml and attained a maximum at 10 μg/ml. Cytolysis in serum was augmented 40% by tuftsin injection in mice. Similarly, the supernatants of peritoneal macrophages and of splenic adherent cells showed an augmentation of 70% and 33%, respectively.[30]

12. REACTIVE OXYGEN COMPOUNDS

Highly reactive oxygen compounds were generated at high levels following tuftsin stimulation of macrophages. There was a rapid response to various concentrations of tuftsin. Optimum concentration was 375 nM. However, at higher concentrations, 500 to 600 nM, there was complete suppression of superoxide formation. This is not surprising, since high concentrations of tuftsin suppress all activities so far studied.

The superoxide formed during stimulation of the macrophage accounts for 90% of the total.[27–29] The secretion of active oxygen compounds by macrophages and granulocytes has been well documented. Tuftsin has been shown to augment considerably the production of superoxide ($O_2^- \cdot$) anion and hydrogen peroxide (H_2O_2).[27–29] The latter is presumably formed by dismutation of $O_2^- \cdot$ by superoxide dismutase. H_2O_2 together with $O_2^- \cdot$ can interact through trace metal catalysis and give rise to hydroxyl radical $OH \cdot$. All three oxygen compounds are highly active. In addition to these are hypochlorites produced by the oxidation of chloride ion by H_2O_2 through a peroxidase-catalyzed reaction.[56,57]

Other sources of leukocyte oxidants are the chloramines. These contribute to the cytotoxic activity of phagocytic cells. The chloramines are formed by the interaction between HOCl and amines.[58] Together these reactive compounds are quite formidable and may well take part in the demise of bacterial and neoplastic cells. For the latter, TNF very likely plays a significant role.[30]

The respiratory burst, which is insensitive to cyanide, takes place during phagocytosis by all types of phagocytes; all these compounds are augmented as a result of the ingestion of various types of particles. By contrast, tuftsin stimulates

$O_2^-\cdot$ and H_2O_2 directly and in the absence of decernible particles in the medium. Tuftsin therefore acts as a different biochemical trigger.

It appears that $O_2^-\cdot$ is not only formed by NADPH oxidation,[59] but also directly through the pathway of purine catabolism. Adenosine is deaminated to inosine, which leads to hypoxanthine. The oxidation of hypoxanthene to xanthene yields $O_2^-\cdot$. Further oxidation of xanthene also yields $O_2^-\cdot$ with the formation of uric acid. It is significant that tuftsin-stimulated macrophages results in activation of this purine pathway to produce augmentation of $O_2^-\cdot$ secretion. The catabolic pathway appears to be the source of at least 90% of the superoxide anion.[27,28]

13. ANALOGUES AND THEIR EFFECTS

The first analogue to be synthesized was the result of the insertion of an extra proline residue to form Thr-Lys-Pro-Pro-Arg.[12] This pentapeptide was found to possess very strong binding affinity to tuftsin receptors—approximately four times that of tuftsin.[37,60] It strongly inhibited tuftsin stimulation of phagocytosis and stimulation of motility. In either case, the basal activity was not inhibited. These results demonstrate that mere binding, although necessary, is insufficient to initiate tuftsin effects on both motility and phagocytosis. It may be that proper binding initiates subsequent steps that eventually lead to the final biological effect. However, improper binding by the pentapeptide seems to stop at that stage and goes no further. All modifications of the structure of tuftsin that lead to stronger binding than tuftsin also show strong inhibition of its activity.[37,60] It is possible that such tight binding may result in conformational alteration in the receptor not conducive to further propagation of the signal. This inhibitory characteristic is fully shared, particularly by two biologically interesting analogues. Thr-Glu-Pro-Arg is one such analogue. This mutant tetrapeptide was isolated from one patient with congenital tuftsin deficiency syndrome. In all IgG samples from that patient, this tetrapeptide had replaced tuftsin completely in the Fc portion of the heavy chain. In these patients, there was a profound defect in immunity resulting in frequent severe infections.[3–8,38] The structural alteration represents a point mutation with a replacement of lysine by glutamate residue.[37] The other oligopeptide of interest is the octapeptide tuftsinyltuftsin, Thr-Lys-Pro-Arg-Thr-Lys-Pro-Arg. It represents merely two tuftsin tetrapeptides linked head to tail. This oligopeptide, synthesized in the author's laboratory, proved the strongest inhibitor of tuftsin activity with the strongest affinity to the receptor amounting to 10 times that of tuftsin.[60] Its antineoplastic activity against certain tumors is greater than that of tuftsin, presumably because little if any tripeptide (Lys-Pro-Arg) inhibitor is formed. This tripeptide is formed by splitting off threonine from tuftsin by a deaminase enzyme. This tripeptide is a strong inhibitor of tuftsin activity.

14. MOLECULAR MIMICRY BETWEEN RECEPTOR AND ANTIBODY

The binding affinity of several tuftsin analogues to specific tuftsin receptor on rabbit peritoneal granulocytes was compared with that of rabbit antituftsin antibody binding site. The oligopeptides were (1) Thr-Lys-Pro, (2) Thr-Lys-Pro-Arg, (3) Thr-Glu-Pro-Arg, (4) Thr-Lys-Pro-Pro-Arg, (5) Ala-Lys-Thr-Lsy-Pro-Arg, (6) Ala-Lys-Thr-Lys-Pro-Arg-Glu-Ala_3, and (7) Thr-Lys-Pro-Arg-Thr-Lys-Pro-Arg tuftsinyltuftsin.[60] Peptides (1) and (6) did not show any binding affinity to either receptor or antibody. Peptide (5) bound less than tuftsin to both receptor and antibody. Tetrapeptide (3) and pentapeptide (4) bound receptor and antibody equally strongly and approximately four times that of tuftsin itself. Finally, octapeptide (7) had the highest affinity of all peptide analogues, including tuftsin, for both antibody and receptor.[60] In every case, the binding affinity to the receptor and to the antibody was almost the same for each analogue.

Scatchard analyses of receptor binding to tuftsin and antibody binding to tuftsin showed a K_d value of 6.1×10^{-8} M for the receptor and 3.9×10^{-8} M for the antibody. The difference is well within experimental error.[60,61]

For the production of antibody to tuftsin, the tetrapeptide was attached to the carrier antigen at its amino-terminal, permitting the free carboxy-terminal arginine to be the first residue to interact with the antibody site. As we have seen, there is a great deal of similarity between the affinity of the antibody site and receptor site. It follows that the orientation of tuftsin as it enters the receptor site is similar as well, with the arginine residue leading the entry into the receptor-binding site.[61]

15. TUFTSIN CONFORMATION

There has not been firm agreement on the conformation of tuftsin in solution. A quasicyclic molecule, β-turn[62] and a hairpin with split ends have been proposed.[63] We have used [^{1}H]- and [^{13}C]-NMR to study the structure of both the tetrapeptide and the pentapeptide ($Pro^{3,4}$). Our results indicate that tuftsin has no preferred conformation in aqueous solution.[64] However, in dimethyl sulfoxide, tuftsin conformation is more ordered than in water, but not its pentapeptide analogue, which has no preferred conformation. In any event, it is unlikely that a β-turn is the preferred conformation in an aqueous solution.

16. SYNTHESIS OF TUFTSIN

Tuftsin has been synthesized by a variety of methods. The first we used was with the solid-phase method of Merrifield.[65] We have introduced a helpful modification of this method[66] in the synthesis of tuftsin based on deprotection by trifluoromethane sulfonic acid.[66a] The guanido group of arginine is protected by tosyl and the α-amino by tertiary butyloxycarboxyl (BOC); such a protected arginine is esterfied to a chloromethyl solid support consisting of polystyrene[1,9] divinylbenzene beads. The second step is the formation of a peptide bond between the deprotected α-amino group of arginine and the carboxyl group of proline, with dicyclohexylcarbodiimide (DCC). The BOC of proline is then removed and the third amino acid lysine, duly protected at its N^{ϵ} amino by benzyloxycarbonyl (Z) and at α-amino by BOC. A peptide bond is then formed again through the action of DCC. Finally, threonine, protected at its hydroxyl group by a benzyl ether link (Bzl) and by BOC on its α-amino group, is bonded by DCC to the deprotected α-amino group of lysine after the usual removal of BOC. All protecting groups and the ester linkage to the resin are cleaved off by one single reagent, trifluoromethanesulfonic acid in anisole. The product is then purified on a Dowex AG 50W-X4 column.[11,12] The yield of purified tuftsin was 60–75%. Its purity is then tested by high-performance liquid chromatography (HPLC).[32] Almost all analogues were synthesized in our laboratory using this methodology. Synthesis using polymeric reagents was pioneered at the Weizmann Institute.[67] The soluble peptide chain is elongated at its amino-terminal through interaction with an excess of insoluble active esters. With this method, tuftsin synthesis was most efficient, with an overall yield of 92% before deprotection and 80% for free tuftsin.[67]

A recent method of conventional liquid-phase synthesis has been reported. It is successfully accomplished through the use of activated *N*-hydroxysuccinimide ester (OSu) of the constituent amino acids. The nitroarginine carboxy-terminal residue is protected at the carboxyl end as the OBzl ester, lysine as the N^{ϵ}-Z and, as usual with the classic method of synthesis, the hydroxyl group of threonine needs no protection. All α-amino groups are BOC protected. This BOC was removed with 50% trifluoroacetic acid (TFA) in CH_2Cl_2. The rest of the protecting groups were removed by hydrogenolysis. With this method,[17] it was possible to prepare as much as 50 g of pure tuftsin.

17. METHODS FOR SYNTHESIS OF RADIOACTIVE TUFTSIN

Synthesis of radioactive tuftsin was conditioned by the availability of [^{3}H]-Arg as the only radioactive residue available displaying sufficient specific activity. Two approaches to synthesis of radioactive suftsin were developed. One is to synthesize the protected tripeptide H-N^{α}-BOC-*O*-Bzl-Thr-N^{ϵ}-Z-Lys-Pro-OH. The carboxyl group of the proline residue is then esterified to *N*-hydroxysuccinimide by DCC. Once purified, it is then reacted with [2,3,4,5-^{3}H]arginine.[68] We have confirmed this finding but encounter difficulties that were not easy to rectify. This arises when the amino group of arginine attacks the succinimidyl moiety at the carboxyl nitrogen bond instead of the carboxyl ester of proline and results in the incorporation of the linear succinimide into the tetrapeptide. On thin layer chromatography, a blue ninhydrin color results.[69]

A more profitable approach is to synthesize tuftsin in the usual manner by solid phase[65,66] or by the classic route[68,70] and to use 3,4-dehydroproline instead of proline.[71] This is followed by catalytic reduction with tritium gas.[72,73]

Radioactive tuftsin is a necessary reagent for evaluating the level of tuftsin in various situations such as receptor–tuftsin interaction and assay, as well as the use of radioimmunoassay (RIA). For the latter, a tuftsin-specific antibody should be generated by linking a derivatized α-amino of tuftsin to a carrier protein. Antibodies in a suitable subject such as rabbit can then be raised successfully.[67,70] Some samples of antihuman IgG contain antibody to tuftsin, since leukokinin carries tuftsin cleaved at the carboxy-terminal arginine. In essential features, it thus resembles tuftsin hapten on a carrier protein.

18. TUFTSIN RECEPTORS

Early studies indicated that neuraminidase forms a part of tuftsin receptor of the granulocytic leukocyte, since treatment of these cells with neuraminidase abolished tuftsin stimulation.[74] Similarly, thiol treatment of membrane receptors diminishes binding to tuftsin, but not completely.[75] These findings are consistent with the fact that any alteration in the structure of the receptor or the structure of tuftsin itself adversely affects biological activity. Tuftsin receptors gave a dissociation constant of 13×16^{-8} M and are present in about 50,000 receptors per PMN.[55,68,76] Human monocytes yielded a similar K_d value with 100,000 receptors per cell. These results were duly confirmed with a K_d value of 6.1×10^{-8} M and 40,000 receptors per PMN and 80,000 number of receptors on mouse peritoneal macrophages.[37,61] The K_M of receptor binding and that of phagocytosis rate is 1.17×10^{-7} M and 1.11×10^{-7} M, respectively, essentially identical to the K_d of binding.[61,66] In simple terms, this means that the amount of receptor occupancy by tuftsin is directly related to the rate of phagocytosis.

Work with lymphocyte preparations showed minimal binding. However, totally different values for lymphocyte binding were obtained using [^{14}C]methyl ester and [^{125}I]tuftsinyltyrosine.[77] These results refer only to the derivatized analogues and not to underivatized tuftsin. Such modifications may give false values and yield an abnormally high binding capacity, since many analogues bind tuftsin more strongly than tuftsin.[60,78]

Successful purification of tuftsin receptor from rabbit peritoneal granulocytes on affinity columns was recently reported.[79] Affigel solid support, activated by ester linkage to *N*-hydroxysuccinimide, was then coupled to the α-amino group of the pentapeptide, Thr-Lys-Pro-Pro-Arg. This pentapeptide binds the receptor about four times more avidly than tuftsin. Receptors were solubilized with 3-[(3-cholamidopropyl)dimethylammonio]propane sulfonate (CHAPS), passed through the column and washed with buffers at pH 5.0 and 8.0. The receptor bound to the coupled pentapeptide was then eluted with free pentapeptide or tuftsin. The latter is then completely removed by a few milligrams of Dowex AG 50W-X4. The receptor was then subjected to concentration dialysis. After complete reduction with mercaptoethanol and alkylation with iodoacetamide, it was analyzed by sodium dodecylsulfate polyacrylamide gel electrophoresis (SDS–PAGE). Two bands were obtained with a relative molecular weight of 52,000 and 62,000 M_r. Reduced and alkylated receptor preparations lose the ability to bind [^{3}H-Arg^4]tuftsin. However, if the underivatized receptor preparation is analyzed directly on SDS–PAGE, two closely adjacent bands appear with an mobility equivalent to ~60,000 M_r that, upon transfer to nitrocellulose paper, bound [^{3}H]tuftsin, which was readily recorded on photographic film.[79]

Gel filtration on Sephacryl S-300 demonstrated two peaks of [^{3}H]tuftsin binding, corresponding to 500,000 and 250,000 M_r, with the former predominant. These two oligomeric peaks are in a state of equilibrium, since rechromatography of either results in the same two peaks.

Electron microscopic examination indicated an apparently pure and uniformly spherical molecule. This molecule has a diamter of 104 Å. On the basis of a sphere, a rough molecular weight of 480,000 has been calculated from the formula

$$M_r/\text{pN} = 4/3\pi r^3$$

The value obtained is a crude estimate but agrees well with the value obtained with gel electrophoresis.

Receptors on macrophages have been shown to bind tuftsin rhodamine complex and to undergo internalization within a period of 5 min.[67] A similar extensive study was reported with fluorescein isothiocynate-labeled human PMN and monocytes.[80]

19. OTHER STUDIES

Other cogent and interesting studies include the following:

1. Interleukin-2 (Il-2) formation by spleen cells after IV injection of tuftsin was demonstrated. Spleen cells from mice 3 days after tuftsin injection spontaneously produced a similar level of Il-2 as Concanavalin A (Con A)-stimulated cells.[42]
2. Central effects were observed in the rat after intracerebroventricular injection of tuftsin.[81–83] Significant antinociceptive activity was also observed with tuftsin, [DArg^4], and [D-Arg^3]tuftsin. Naloxone failed to modify the reaction at a dose of 2.5 mg/kg IP 30 min before injection of tuftsin or its analogues. This indicates that this effect is not mediated by opiate receptors. Furthermore, tuftsin affects exploratory and locomotor activity of the rat.

3. The effect of tuftsin on differentiation of undifferentiated monocyte cell line $P388_{D1}$ has been reported. This cell line as such exhibits no cytotoxicity against tumor cells. However, upon differentiation, induction of cytotoxicity can be obtained. Tuftsin at 5, 10, and 20 μg/ml stimulated several observable differentiated properties along with cytotoxicity.[48] These are very interesting results indeed.

The effect of tuftsin on bone marrow colony formation as assayed by colony-forming units (CFU-C) in soft agar was reported by the same investigators. At 0.5 μg tuftsin per culture, colony-stimulating activity was equivalent to that obtained with colony-stimulating factor. Colonies showed differentiated mononuclear cells or granulocytes.[48]

20. TUFTSIN DEFICIENCIES

The existence of a congenital deficiency tuftsin syndrome with evidence of a mutated tetrapeptide is good proof that tuftsin is a true biological entity with a naturally assigned function.

20.1. Congenital Tuftsin Deficiency

Congenital tuftsin deficiency and the acquired splenic type were actually discovered before tuftsin was identified as an oligopeptide.[2–6,8,9] Twenty-two patients of the congenital type have now been identified in the United States and Japan.[3–6,38,84,85] The symptoms exhibited are those of recurrent severe infections in early childhood but milder signs and symptoms in the adult. The symptoms in childhood include upper respiratory tract infections, pneumonia, severe and generalized eczematous skin infection, lymphadenities, and draining lymphnode abscesses. The mutant peptide Thr-Glu-Pro-Arg was isolated from a patient; its sequence was determined, and the peptide was synthesized.[37] In this patient, a young woman whose mother was also deficient, there was complete absence of the normal tetrapeptide.

In all cases reported, a mutant tetrapeptide, like the tetrapeptide Thr-Glu-Pro-Arg, is inhibitory to tuftsin[8] and binds receptors about four times as avidly. Because of this strong binding, radioreceptor assay or RIA yield high false values that are expected and diagnostic.[78]

In our laboratory, we were able to confirm these results with radioactive tuftsin [^{3}H-Pro^{3}]tuftsin. Our values for normal adults were 250–500 ng/ml.[78] Four patients with clinically proven congenital tuftsin deficiency syndrome were studied with RIA and radioreceptor assay. In both systems, the values obtained were very high: 1500–2,500 ng/ml.[78] This false value was indeed expected. The mutant peptides, which were present in all patients and which may well be [Glu^{2}]tuftsin, bind tuftsin antibody and tuftsin receptor with an affinity of about four times as strong as tuftsin itself. Consequently, such high values, although indeed misleading, are strictly diagnostic.

In all such cases, only one of the parents had tuftsin deficiency. This may indicate that the allele is dominant and that the patient is heterozygous with dominance. However, the presence of only the mutant peptide with no trace of tuftsin would be attributable to allelic exclusion.

20.2. Acquired Tuftsin Deficiency

The defect of acquired tuftsin deficiency was first shown in splenectomized dogs.[86] It has since been shown in splenectomized human subjects and splenectomized rats.[3,5–7,87,88] It was noted that after splenectomy, the carrier leukokinin lost its phagocytosis stimulatory activity. The spleen, whether that of a dog or man, has an enzyme that cleaves tuftsin at the carboxy-terminal arginyl (Arg^4) bond. This cleavage does not occur in the absence of the spleen. The same observation was made in humans who had undergone elective splenectomy.[40,41,86–89] Defective activity was also found in patients who suffer from infiltrative disease of the spleen, such as extensive thrombosis (as in sickle cell disease) and leukemic infiltration. An extensive study of acquired tuftsin deficiency was recently reported. Serum tuftsin levels in patients with elective splenectomy, splenectomy after traumatic rupture of the spleen, and sickle cell disease have been reported.[8,10,41,88] These are lowest in elective splenectomy and low in sickle cell disease. However, patients with traumatic splenectomy usually have normal levels. It is assumed that splenic tissue becomes implanted in the omentum in sufficient quantity to yield normal levels of tuftsin. Normal levels averaged 229 ng/ml after treatment of serum with trypsin. Similar values were obtained in adults and children.[40,70] In sickle cell disease, a value of 154 ng/ml was obtained. In elective splenectomy for spherocytosis or for staging Hodgkin disease, a value of 87–112 ng/ml was obtained. We have previously shown through phagocytosis assay that tuftsin deficiency, whether acquired or congenital, results in extremely defective phagocytosis.[2–8,84]

21. MECHANISM OF ACTION

Simple binding to the receptor is necessary but is insufficient to augment the biological activities of tuftsin. We therefore sought out other possible mechanisms that follow receptor binding. We found that following tuftsin treatment, cyclic adenylate levels in rabbit peritoneal granulocytes were not sufficiently altered.[89] However, other workers have shown significant though slight lowering of intracellular cAMP along with a definite elevation of cGMP. With the change in cyclic nucleotide levels, tuftsin caused an increased release of intracellular Ca^{2+} of preloaded PMN leukocytes.[55] It is possible that inappropriate binding as might occur with 4–10 times avidity of the ligand, fails to trigger any response in the levels of cAMP or cGMP. This remains to be determined.

ACKNOWLEDGMENTS. This work was supported by grant AI09116 from the U.S. Public Health Service, grant 1-556 from the March of Dimes Birth Defects Foundation, and grant RDP-32E from the American Cancer Society.

REFERENCES

1. Najjar VA, Nishioka, K: Tuftsin—A physiological phagocytosis stimulating peptide. *Nature (Lond)* **228:**672–673, 1970.
2. Najjar, VA: The physiological role of γ-globulin, in Meister DA (ed): *Advances in Enzymology*. New York, Wiley, 1974, vol 41, pp 129–174.
3. Najjar VA: Defective phagocytosis due to deficiencies involving the tetrapeptide tuftsin. *J Pediatr* **87:**1121–1124, 1975.
4. Najjar, VA: The physiological role of membrane γ-globulin interaction, in: Chapman D, Wallach DFH (eds): *Biological Membranes*. New York, Academic, 1976, vol 3, pp 191–240.
5. Najjar, VA: Molecular basis of familial and acquired phagocytosis deficiency involving the tetrapeptide, Thr-Lys-Pro-Arg, tuftsin, *Exp. Cell Biol* **46:**114–126, 1978.
6. Najjar, VA: The clinical and physiological aspects of tuftsin deficiency syndromes exhibiting defective phagocytosis, *Klin. Wochenschr* **57:**751–756, 1979.
7. Najjar VA: Biochemical aspects of tuftsin deficiency syndrome. *Med Biol* **59:**134–138, 1981.
8. Najjar VA, Constantopoulos A: A new phagocytosis stimulating tetrapeptide hormone, tuftsin and its role in disease. *J Reticuloendothel Soc* **12:**197–215, 1972.
9. Nishioka K, Constantopoulos A, Satoh PS, Najjar VA: The characteristics, isolation and synthesis of the phagocytosis stimulating peptide tuftsin. *Biochem Biophys Res Commun* **47:**172–179, 1972.
10. Spirer Z, Zakuth V, Golander A, et al: The effect of tuftsin on the nitrous blue tetrazolium reduction of normal human polymorphonuclear leukocytes. *J Clin Invest* **55:**198–200, 1975.
11. Nishioka K, Constantopoulos, A, Satoh P, Mitchell W, Najjar VA: Characteristics and isolation of the phagocytosis-stimulating peptide, tuftsin. *Biochim Biophys Acta* **310:**217–229, 1973.
12. Nishioka K, Satoh P, Constantopoulos A, Najjar VA: The chemical synthesis of the phagocytosis stimulating tetrapeptide tuftsin (Thr-Lys-Pro-Arg) and its biological properties, *Biochim Biophys Acta* **310:**230–237, 1973.
13. Najjar VA, Fridkin M, (eds): Antineoplastic Immunogenic and Other Effects of the Tetrapeptide Tuftsin: A Natural Macrophage Activator, *Ann NY Acad Sci* **419:**273, 1983.
14. Blok-Perkowska D, Muzalewski F, Konopinska D: Antibacterial properties of tuftsin and its analogs. *Antimicrob Agents Chemother* **25:**134–136, 1984.
15. Florentin I, Bruley-Rosset M, Kiger N, et al: In vivo immunostimulation by tuftsin. *Cancer Immunol Immunother* **5:**211–216, 1978.
16. Martinez J, Winternitz F, Vindel J: Nouvelles synthèses et propriétés de la tuftsine. *Eur J Med Chem Chim Ther* **12:**511–516, 1977.
17. Martinez J, Winternitz F: New synthetic and natural tuftsin-related compounds and evaluation of their phagocytosis-stimulating activity. *Ann NY Acad Sci* **419:**23–34, 1983.
18. Tzehoval E, Segal S, Stabinsky Y, et al: Tuftsin (an Ig-associated tetrapeptide) triggers the immunogenic function of macrophages: General implications to activation of programmed cells. *Proc Natl Acad Sci USA* **75:**3400–3404, 1978.
19. Catane R, Schlanger S, Gottlieb P, et al: Toxicology and antitumor activity of tuftsin in mice. *Proc Am Assoc Cancer Res Am Soc Clin Oncol* **22:**371–373, 1981.
20. Catane R, Schlanger S, Weiss L, et al: Toxicology and antitumor activity of tuftsin. *Ann NY Acad Sci* **419:**251–260, 1983.
21. Catane R, Sulkes A, Buziely B, et al: Initial clinical studies with tuftsin. *Int J Immunotherapy* **11:**81–85, 1986.
22. Najjar VA: Tuftsin (Thr-Lys-Pro-Arg) a natural activator of phagocytic cells with antibacterial and

antineoplastic activity, in Torrence PF (ed): *Biological Response Modifiers*. New York, Academic, 1985, pp 141–169.
23. Najjar VA, Konopińska D, Chaudhuri MK, et al: Tuftsin a natural activator of phagocytic functions including tumoricidal activity. *Mol Cell Biochem* **41**:3–12, 1981.
24. Najjar VA, Linehan L, Konopińska D: The antineoplastic effects of tuftsin and tuftsinyltuftsin on B16/5B melanoma and L1210 cells. *Ann NY Acad Sci* **419**:261–267, 1983.
25. Nishioka K, Babcock GF, Phillips JH, et al: In vivo and in vitro antitumor activities of tuftsin. *Ann NY Acad Sci* **419**:234–241, 1983.
26. Marilus R, Spirer Z, Michaeli D, et al: First case of aids in a homosexual in Israel. Results of different therapeutic regimens. *Israel J Med Sci* **20**:249–251, 1984.
27. Tritsch GL, Niswander PW: Positive correlation between superoxide release and intracellular adenosine deaminase activity during macrophage membrane perturbation regardless of nature or magnitude of stimulus. *Mol Cell Biochem* **49**:49–52, 1982.
28. Tritsch GL, Niswander PW: Modulation of macrophage superoxide release by purine metabolism. *Life Sci* **32**:1359–1370, 1983.
29. Hartung HP, Toyka KV: Augmentation of oxidative and arachidonate metabolism in macrophages by tuftsin (Thr-Lys-Pro-Arg). *Agents Actions* **15**:38–39, 1984.
30. Wleklik MS, Luczak M, Najjar VA: Tuftsin induced tumor necrosis activity. *Mol Cell Biochem* **75**:169–174, 1987.
31. Lukas TJ, Munoz H, Erickson BW: Inhibition of C1-mediated immune hemolysis by monomeric and dimeric peptides from the second constant domain of human immunoglubulin G. *J Immunol* **127**:2555–2560, 1981.
32. Constantopoulos A, Najjar VA: Tuftsin, a natural and general phagocytosis stimulating peptide affecting macrophages and polymorphonuclear granulocytes. *Cytobios* **6**:97–100, 1972.
33. Hisatsune K, Nozaki S, Ishikawa T, et al: A biochemical study of the phagocytic activities of tuftsin and its analogues. *Ann NY Acad Sci* **419**:205–213, 1983.
34. Satoh PS, Constantopoulos A, Nishioka K, Najjar VA: Tuftsin, threonyl-lysyl-prolyl-arginine, in Meinhofer J (ed): *Chemistry and Biology of Peptides*. Ann Arbor, Michigan, Ann Arbor Science, 1971, pp 403–408.
35. Rauner RA, Schmidt JJ, Najjar VA: Proline endopeptidase and exopeptidase activity in polymorphonuclear granulocytes. *Mol Cell Biochem* **10**:77–80, 1976.
36. Nagoaka I, Yamashita T: Inactivation of phatocytosis-stimulating activity of tuftsin by polymorphonuclear neutrophils. A possible role of leucine aminopeptidase as an ecto-enzyme. *Biochim Biophys Acta* **675**:85–93, 1981.
37. Najjar VA, Chaudhuri MK, Konopińska D, et al: Tuftsin (Thr-Lys-Pro-Arg), a physiological activator of phagocytic cells: A possible role in cancer suppression and therapy, in Hersh MA, Chirigos MA, Mastrangelo MJ (eds): *Augmenting Agents in Cancer Therapy*. New York, Raven, 1981, pp 459–478.
38. Inada K, Nemoto N, Nishijima A, Wada S, Hirata M, Yoshida M: in Kokobun Y, Kobayashi N (eds): *Phagocytosis: Its Physiology and Pathology*. Baltimore, University Park Press, 1977, pp 101–108.
39. Constantopoulos A, Najjar VA, Wish JB, et al: Defective phagocytosis due to tuftsin deficiency in splenectomized subjects. *Am J Dis Child* **125**:663–665, 1973.
40. Spirer Z, Zakuth V, Bogair N, Frikin M: Radioimmunoassay of the phagocytosis stimulating peptide tuftin in normal and splenectomized subjects. *Eur J Immunol* **7**:69–74, 1977.
41. Spirer Z, Weisman Y, Zakuth V, et al: Decreased serum tuftsin concentrations in sickle cell disease. *Arch Dis Child* **55**:566–567, 1980.
42. Florentin I, Martinez J, Maral J, et al: Immunopharmacological properties of tuftsin and of some analogues. *Ann NY Acad* **419**:177–191, 1983.
43. Najjar VA, Schmidt J: The chemistry and biology of tuftsin, in Pick E (ed): *Lymphokine Reports*. New York, Academic, 1980, vol 1, pp 157–159.
44. Ketchel MM, Favour CB: Acceleration and inhibition of migration of human leucocytes in vitro by plasma protein fractions. *J Exp Med* **101**:647–663, 1955.

45. Kavai M, Lukacs K, Szegedi G, et al: Chemotactic and stimulating effect of tuftsin and its analogues on human monocytes. *Immunol Lett* **2:**219–224, 1981.
46. Lukacs K, Berenyi E, Kavai M, et al: Potentiation of the defective monocyte chemotaxis in Hodgkin's disease by in vitro tuftsin treatment. *Cancer Immunol Immunother* **15:**162–163, 1983.
47. Lukacs K, Szabo G, Sonkoly I, et al: Stimulating effect of tuftsin and its analogues on the defective monocyte chemotaxis in systemic lupus erythematosus. *Immunopharmacology* **7:**171–178, 1984.
48. Babcock GF, Amoscato AA, Nishioka K: Effect of tuftsin on the migration chemotaxis, and differentiation of macrophages and granulocytes. *Ann NY Acad Sci* **419:**64–74, 1983.
48a. Beretz A, Hiller Y, Gottlieb P, et al: The effect of tuftsin, its analogs and its conjugates with formyl chemotactic peptide on chemotaxis of human monocytes, in Sakakibara S (ed): *Peptide Chemistry.* Osaka, Protein Research Foundation, 1982, pp 207–212.
49. Baker CC, Chaudry IH, Gaines HO, Baue AD: Evaluation of factors affecting mortality rate after sepsis in a murine cecal ligation and puncture model. *Surgery* **94:**331–335, 1983.
50. Konopinska D, Luczak M, Wleklik M, et al: Elongated tuftsin analogues-synthesis and biological investigation. *Ann NY Acad Sci* **419:**35–43, 1983.
51. Knyszynski K, Gottlieb P, Fridkin M: Inhibition by tuftsin of Rauscher virus leukemia development in mice. *J. Natl Cancer Inst* **71:**87–90, 1983.
52. Suk WA, Long CW: Enhancement of endogenous xenotropic murine retrovirus expression by tuftsin. *Ann NY Acad Sci* **419:**75–86, 1983.
53. Bruley-Rosset M, Hercend T, Rappaport H, Mathé G: Immunorestorative capacity of tuftsin after long-term administration to aging mice. *Ann NY Acad Sci* **419:**242–250, 1983.
54. Fridkin M, Najjar VA: Tuftsin: Its chemistry, biology and clinical potential, in Sela M (ed): *Critical Reviews of Biochemistry.* Boca Raton, Florida, CRC Press, in press.
55. Goldman R, Bar-Shavit Z: On the mechanism of the augmentation of the phagocytic capability of phagocytic cells by tuftsin, substance P, neurotensin, and kentsin and the interrelationship between their receptors. *Ann NY Acad Sci* **419:**143–155, 1983.
56. Klebanoff S: Myeloperoxidase-mediated cytotoxic systems, in Sbarra AJ, Strauss RR (eds): *Reticuloendothelial System.* New York, Plenum, 1980, vol 2, pp 279–308.
57. Sbarra AJ, Selvary RJ, Paul BB, et al: Chlorination, decarboxylation and bactericidal activity mediated by the MPO–H_2O_2-Cl^- system. *Adv Exp Med* **73:**191–203, 1976.
58. Thomas EL, Jefferson MM, Grisham MB: Myeloperoxidase-catalyzed incorporation of amines into protein: Role of hypochorous acid and dichloroamines. *Biochemistry* **24:**6299–6308, 1982.
59. Badwey JA, Karnovsky ML: Active oxygen species and the functions of phagocytic leukocytes. *Annu Rev Biochem* **49:**695–726, 1980.
60. Chaudhuri MK, Konopińska D, Bump NJ, Najjar VA: The similarity between tuftsin (Thr-Lys-Pro-Arg) receptors and tuftsin antibody: A case of induced molecular mimicry. *Ann NY Acad Sci* **419:**135–142, 1983.
61. Najjar VA, Bump NJ, Lee J: Isolation and characterization of tuftsin receptor, in Martinez J, Castro B (eds): *Forum Peptides.* Le Cap d'Agde, France, 1985, pp 17–21.
62. Siemion IZ, Lisowski M, Konopinska D, Nawrocka E: ^{13}C nuclear magnetic resonance and circular dichroism studies of the tuftsin conformation in water. *Eur J Biochem* **112:**339–343, 1980.
63. Fitzwater S, Hodes ZI, Scheraga HA: Conformational energy study of tuftsin. *Macromolecules* **11:**805–811, 1978.
64. Blumenstein M, Layne PP, Najjar VA: Nuclear magnetic resonance studies on the structure of the tetrapeptide tuftsin, L-threonyl-L-lysyl-L-prolyl-L-arginine, and its pentapeptide analogue L-threonyl-L-lysyl-L-prolyl-L-prolyl-L-arginine. *Biochemistry* **18:**5247–5253, 1979.
65. Merrifield RB: Solid-phase peptide synthesis. III. An improved synthesis of bradykinin, *Biochemistry* **3:**1385–1390, 1964.
66. Chaudhuri MK, Najjar VA: The solid phase synthesis of tuftsin and its analogs. *Anal Biochem* **95:**305–310, 1979.
66a. Yajima H, Ogawa H, Watanabe H, et al: Studies on peptides. XLVIII. Application of the trifluoromethane-sulphonic acid procedure to the synthesis of tuftsin. *Chem Pharm Bull* **23:**371–374, 1975.

67. Gottlieb P, Stabinsky Y, Zakuth V, et al: Synthetic pathways to tuftsin and radioimmunoassay. *Ann NY Acad Sci* **419:**12–22, 1983.
68. Fridkin M, Gottlieb P: Tuftsin, Thr-Lys-Pro-Arg. *Mol Cell Biochem* **41:**73–98, 1981.
69. Savrda J: An unusual side reaction of succinimidyl esters during peptide synthesis. *J Org Chem* **42:**3199–3200, 1977.
70. Fridkin M, Stabinsky Y, Zakuth V, Spirer Z: Tuftsin and some analogs. Synthesis and interaction with human polymorphonuclear leukocytes. *Biochim Biophys Acta* **496:**203–211, 1977.
71. Amoscato AA, Babcock GF, Nishioka K: Synthesis and biological activity of [L-3,4-Dehydroproline[3]]-tuftsin, *Peptides* **5:**489–494, 1984.
72. Najjar VA, Bump NJ, Lee J, Wleklik M: Tuftsin (Thr-Lys-Pro-Arg) a natural drug for immunostimulation; structure and function relationships, in Makriyannis A (ed): *New Methods in Drug Research.* Barcelona, Spain, J. R. Prous Science Publishers, vol 2, in press.
73. Najjar VA, Konopińska D, Lee L: Tuftsin, in Disabato G, Everse J (eds): *Phagocytosis and Cell-Mediated Cytotoxicity.* New York, Academic, 1986, vol 132, pp 318–325.
74. Constantopoulos A, Najjar VA: The requirement for membrane sialic acid in the stimulation of phagocytosis by the natural tetrapeptide, tuftsin. *J Biol Chem* **248:**3819–3822, 1973.
75. Bump NJ, Chaudhuri MK, Munson D, et al: Further studies on tuftsin, a natural activator of phagocytic cells. *J Immunol Immunopharmacol* **5:**8–13, 1985.
76. Gottlieb P, Stabinsky Y, Hiller Y, et al: Tuftsin receptors. *Ann NY Acad Sci* **419:**93–106, 1983.
77. Nair RMG, Ponce B, Fudenberg HH: Interactions of radiolabeled tuftsin with human neutrophils. *Immunochemistry* **15:**901–907, 1978.
78. Najjar VA, Bump NJ: A stimulator of all known functions of macrophage, in Fenichel RL, Chirigos MA (eds): *Immunomodulating Agents.* New York, Dekker, 1984, pp 229–242.
79. Bump NJ, Lee J, Wleklik M, et al: Isolation and subunit composition of tuftsin receptor. *Proc. Natl Acad Sci USA* **83:**7187–7191, 1986.
80. Amoscato AA, Davies PJA, Babcock GF, Nishioka K: Receptor-mediated internalization of tuftsin. *Ann NY Acad Sci* **419:**114–134, 1983.
81. Herman ZS, Stachura Z, Krezeminski T, et al: Central effects of tuftsin. *Ann NY Acad Sci* **419:**156–163, 1983.
82. Valdman AV, Kozlovskaia MM, Ashmarin IP, et al: [Central effects of the tetrapeptide tuftsin]. *Biull Eksp Biol Med (Moscow)* **92:**31–33, 1981.
83. Valdman AV, Bondarenko NA, Kozlovskaia MM, et al: [Comparative study of the psychotropic activity of tuftsin and its analogs]. *Biull Eksp Biol Med (Moscow)* **93:**49–52, 1982.
84. Constantopullos A, Najjar VA: Tuftsin deficiency syndrome, a report of two new cases. *Acta Paediatr Scand* **62:**645–648, 1973.
85. Najjar VA: Tuftsin, in *Drugs of the Future.* Barcelona, Spain, J. R. Prous Science Publishers, 1987.
86. Najjar VA, Fidalgo BV, Stitt E: The physiological role of the lymphoid system. VII. The disappearance of leucokinin activity following splenectomy. *Biochemistry* **7:**2376–2379, 1968.
87. Spirer Z: The role of the spleen in immunity and infection. *Adv Pediatr* **27:**55–88, 1980.
88. Spirer Z, Zakuth V, Orda R, et al: Acquired tuftsin deficiency. *Ann NY Acad Sci* **419:**220–226, 1983.
89. Constantopoulos A, Najjar VA: Adenyl cyclase of polymorphonuclear leucocytes. *Nature New Biol* **243:**268–269, 1973.

Index